THERAPEUTIC MODULATION of CYTOKINES

Pharmacology and Toxicology: Basic and Clinical Aspects

Mannfred A. Hollinger, Series Editor
University of California, Davis

Published Titles

Inflammatory Cells and Mediators in Bronchial Asthma, 1990, Devendra K. Agrawal and Robert G. Townley
Pharmacology of the Skin, 1991, Hasan Mukhtar
In Vitro *Methods of Toxicology*, 1992, Ronald R. Watson
Basis of Toxicity Testing, 1992, Donald J. Ecobichon
Human Drug Metabolism from Molecular Biology to Man, 1992, Elizabeth Jeffreys
Platelet Activating Factor Receptor: Signal Mechanisms and Molecular Biology, 1992, Shivendra D. Shukla
Biopharmaceutics of Ocular Drug Delivery, 1992, Peter Edman
Beneficial and Toxic Effects of Aspirin, 1993, Susan E. Feinman
Preclinical and Clinical Modulation of Anticancer Drugs, 1993, Kenneth D. Tew, Peter Houghton, and Janet Houghton
Peroxisome Proliferators: Unique Inducers of Drug-Metabolizing Enzymes, 1994, David E. Moody
Angiotensin II Receptors, Volume I: Molecular Biology, Biochemistry, Pharmacology, and Clinical Perspectives, 1994, Robert R. Ruffolo, Jr.
Angiotensin II Receptors, Volume II: Medicinal Chemistry, 1994, Robert R. Ruffolo, Jr.
Chemical and Structural Approaches to Rational Drug Design, 1994, David B. Weiner and William V. Williams
Biological Approaches to Rational Drug Design, 1994, David B. Weiner and William V. Williams
Direct Allosteric Control of Glutamate Receptors, 1994, M. Palfreyman, I. Reynolds, and P. Skolnick
Genomic and Non-Genomic Effects of Aldosterone, 1994, Martin Wehling
Human Growth Hormone Pharmacology: Basic and Clinical Aspects, 1995, Kathleen T. Shiverick and Arlan Rosenbloom
Placental Toxicology, 1995, B. V. Rama Sastry
Stealth Liposomes, 1995, Danilo Lasic and Frank Martin
TAXOL®: Science and Applications, 1995, Matthew Suffness
Endothelin Receptors: From the Gene to the Human, 1995, Robert R. Ruffolo, Jr.
Serotonin and Gastrointestinal Function, 1995, Timothy S. Gaginella and James J. Galligan
Drug Delivery Systems, 1995, Vasant V. Ranade and Mannfred A. Hollinger
Alternative Methodologies for the Safety Evaluation of Chemicals in the Cosmetic Industry, 1995, Nicola Loprieno
Phospholipase A_2 in Clinical Inflammation: Molecular Approaches to Pathophysiology, 1995, Keith B. Glaser and Peter Vadas
Ryanodine Receptors, 1995, Vincenzo Sorrentino
Experimental Models of Mucosal Inflammation, 1996, Timothy S. Gaginella
ImmunoPharmaceuticals, 1996, Edward S. Kimball

Pharmacology and Toxicology: Basic and Clinical Aspects

Mannfred A. Hollinger, Series Editor
University of California, Davis

Forthcoming Titles

Alcohol Consumption, Cancer and Birth Defects: Mechanisms Involved in Increased Risk Associated with Drinking, Anthony J. Garro
Antibody Therapeutics, William J. Harris and John R. Adair
Brain Mechanisms and Psychotropic Drugs, A. Baskys and G. Remington
Chemoattractant Ligands and Their Receptors, Richard Horuk
CNS Injuries: Cellular Responses and Pharmacological Strategies, Martin Berry and Ann Logan
Muscarinic Receptor Subtypes in Smooth Muscle, Richard M. Eglen
Neural Control of Airways, Peter J. Barnes
Pharmacological Regulation of Gene Expression in the CNS, Kalpana Merchant
Pharmacology in Exercise and Sports, Satu M. Somani
Pharmacology of Intestinal Secretion, Timothy S. Gaginella
Placental Pharmacology, B. V. Rama Sastry
Receptor Characterization and Regulation, Devendra K. Agrawal
Receptor Dynamics in Neural Development, Christopher A. Shaw

THERAPEUTIC MODULATION of CYTOKINES

Edited by

Brian Henderson, Ph.D.
Boissard Professor of Biochemistry
Eastman Dental Institute for Oral Health Care Sciences
and University College of London
London, United Kingdom

Mark W. Bodmer, Ph.D.
Director of Biology
Celltech Therapeutics Ltd.
Slough, United Kingdom

CRC Press
Boca Raton New York London Tokyo

Library of Congress Cataloging-in-Publication Data

Therapeutic modulation of cytokines / editors, Brian Henderson, Mark Bodmer.
p. cm. -- (Pharmacology and toxicology)
Includes bibliographical references and index.
ISBN 0–8493–8381–1
1. Cytokines--Inhibitors--Therapeutic use--Testing. 2. Cytokines---Pathophysiology. 3. Cytokines--Physiological effect.
I. Henderson, Brian, Ph.D. II. Bodmer, Mark. III. Series: Pharmacology & toxicology (Boca Raton, Fla.)
RM282.C95T48 1996
616.07′9--dc20 95–21372
CIP

International Standard Book Number 0-8493-8381-1
Library of Congress Card Number 95-21372
Printed in the United States of America 1 2 3 4 5 6 7 8 9 0
Printed on acid-free paper

SERIES PREFACE

The series, *Pharmacology and Toxicology: Basic and Clinical Aspects*, has been created in recognition of the fact that, from time to time, a new area of interest within a discipline matures to a critical mass that merits organization and integration of the respective observations into a free-standing monograph. In order for such an undertaking to be successful, each editor and/or author must be qualified to identify and select sources best suited to communicate essential aspects of that subject. Such is the case with *Therapeutic Modulation of Cytokines.* Brian Henderson and Mark W. Bodmer have assembled a list of international contributors whose expertise in their respective areas is well known.

Mannfred A. Hollinger, Ph.D.
Series Editor
Professor
Department of Medical Pharmacology and Toxicology
University of California, Davis
Davis, California

PREFACE

In 1888, William Welch, who later became dean of John Hopkins Medical School, suggested that fever associated with infection may be a result of the generation of 'ferments' (cytokines) by host cells exposed to bacteria or their products. However, it was not until the late 1940s and early 1950s that Welch's prescience received experimental verification with the description of a heat-labile pyrogen produced by leukocytes. This endogenous pyrogen had to be rediscovered by the immunologists before the explosion of research into the biology of the cytokines could begin. In 1979, endogenous pyrogen, by now in multiple guises, was renamed interleukin-1.

Since the mid-1970s evidence has accumulated that cytokines have profound actions on cells, consistent with these proteins having a role in disease pathology. During the 1980s and 1990s it was conclusively demonstrated that cytokines were being produced locally or systemically in most major diseases. It was also established during this period that local or systemic injection of cytokines could induce the symptoms of human diseases, and that antibody-mediated blockade of cytokines could ameliorate a range of experimental models of tissue pathology. Thus, cytokines now meet Koch's postulates (or in the pharmacological context — Dale's criteria) as mediators of local and systemic pathology in diseases ranging from cancer to arthritis and from psoriasis to septic shock.

By the early 1980s there was sufficient evidence that cytokines were key mediators of tissue pathology and the pharmaceutical industry took its first tentative steps toward the development of cytokine therapeutics. Since then there has been an explosion of interest in the therapeutic modulation of cytokines and this has occurred hand-in-hand with the development of the biopharmaceutical industry. The use of cytokines as therapeutics also developed during this period and is explored in this volume in the context of the potential treatment of immune and inflammatory diseases by the administration of 'antagonist' cytokines.

Initial efforts to develop cytokine-modulating therapeutics were devoted to the generation of antagonists of IL-1 receptor binding, but little advance was, or has been, made in this respect, and most work on antagonists is now devoted to the biology of the natural antagonist, the interleukin-1 receptor antagonist. In retrospect, the failure to develop low molecular mass antagonists is not surprising given the high affinity of cytokine receptors and the likelihood that cytokines form multiple-point attachments to their receptors. Since these early attempts by the drug industry, a more Catholic approach has been taken to the development of cytokine therapeutics and, as described in this volume, a range of potential therapeutic modalities ranging from the blockade of cytokine-induced intracellular signaling with low molecular mass compounds to the use of cytokine-neutralizing monoclonal antibodies and soluble receptors is now under investigation. Most of these potential therapeutic innovations rely heavily on the current advances in molecular biology and biopharmaceutical technology. It is heartening to realize that while the first efforts to develop cytokine-modulating therapeutics began only 10 years ago there are already a number of distinct therapeutic modalities in clinical trial, and the editors predict that cytokine-blocking pharmaceutical agents will be in clinical use within the next 5 to 10 years.

This volume begins with chapters which review our current understanding of the physiological and pathophysiological roles of cytokines and the potential for therapeutic manipulation of either cytokine synthesis or cytokine activity. The second section describes the development of low molecular mass inhibitors of cytokine synthesis. This includes discussions of the mechanisms of action of the potent immunosuppressants cyclosporin and FK506, the development of inhibitors of cytokine-converting proteases, and the application of oligonucleotide-based transcription and translation inhibitors. The capacity to neutralize cytokines once they have been released from activated cells is detailed in the third section, which deals with naturally occurring cytokine inhibitors and cytokine-neutralizing antibodies. This area is currently nearest application in the clinic although the requirement for non-oral administration of these agents could limit their usefulness. The fourth section explores the mechanisms of cytokine-induced intracellular signaling. The final section in this book reviews the biology of cytokines such as transforming growth factors-beta and interleukin-10 which have the capacity to downregulate cellular activation induced by other proinflammatory cytokines such as interleukin-1. The potential therapeutic potential of using cytokines to block cytokines brings the discussion in this volume to an end.

This monograph is not a *vade mecum* of cytokine biology and cytokine modulation. Rather, it is a series of interlacing monographs which, as well as introducing the physiology and pathophysiology of cytokines, provide a glimpse of the enormous therapeutic potential of modulating the synthesis or action of cytokines. In this respect this monograph covers the variety of technical approaches currently available in this field and those in the foreseeable future.

THE EDITORS

Brian Henderson, Ph.D., is Professor of Biochemistry and Head of the Maxillofacial Surgery Research Laboratory at the Eastman Dental Institute, University College London.

Professor Henderson graduated in Biochemistry in 1972 from Strathclyde University, Glasgow, and obtained his Ph.D. in Biochemistry in 1977 from Brunel University, London.

Professor Henderson has worked both in academia (McGill University, Montreal; Kennedy Institute of Rheumatology, London; and King's College, London) and industry — Wellcome Research Laboratories, Kent, U.K.; and Celltech Therapeutics, Slough, U.K. He is a member of the Biochemical Society, Royal Microscopical Society, British Society for Rheumatology, and British Society for Immunology. He is a recipient of grants from many research foundations including the Medical Research Council, Wellcome Trust, Nuffield Trust, Wolfson Foundation, and the Arthritis and Rheumatism Council for Research.

Professor Henderson has published 4 books and 120 papers and reviews. His current research interests are in the therapeutic modulation of cytokines and in defining the role that cytokines play in the interaction between commensal bacteria and host cells.

Mark W. Bodmer, Ph.D., is Director of Biology, Celltech Therapeutics Ltd., a U.K.-based biotechnology/pharmaceutical company.

Dr. Bodmer graduated in 1979 from Cambridge University obtaining a B.A. in Natural Sciences. He completed his Ph.D. at the Department of Genetics, Cambridge University in 1982, on molecular evolution of *Drosophila.* He joined Celltech in 1982 initially as a gene cloner and sequencer but became interested in improving the pharmaceutical properties of monoclonal antibodies by engineering at the DNA level. He was rapidly promoted through Team Leader and Head of Therapeutic Area to his current position.

Dr. Bodmer has published over 50 scientific papers and patents in the fields of antibody engineering and cytokine suppression. He is a member of the Biochemical Society (U.K.), British Society for Immunology, the AAAS, and the New York Academy of Sciences. His current areas of interest include the role of cytokines in immune and inflammatory diseases and the application of recombinant DNA technology as potential therapeutics.

CONTRIBUTORS

Jacques Banchereau, Ph.D.
Schering Plough, Laboratory for Immunological Research
Dardilly, France

C. Frank Bennett, Ph.D.
ISIS Pharmaceuticals
Carlsbad, California

Mark W. Bodmer, Ph.D.
Celltech Therapeutics Ltd.
Slough, United Kingdom

Francine Brière, Ph.D.
Schering Plough Laboratory for Immunological Research
Dardilly, France

Robert V. Bundick, B.Sc., M.Sc.
Department of Biochemistry
Fisons Pharmaceuticals PLC
Loughborough, United Kingdom

Christophe Caux, Ph.D.
Schering Plough Laboratory for Immunological Research
Dardilly, France

Richard W. Chapman, Ph.D.
Schering Plough Research Institute
Kenilworth, New Jersey

Stanley T. Crooke, M.D., Ph.D.
ISIS Pharmaceuticals
Carlsbad, California

Francis M. Cuss, M.D.
Schering Plough Research Institute
Kenilworth, New Jersey

Francesco S. Di Giovine, M.D., Ph.D.
Section of Molecular Medicine
Department of Medicine and Pharmacology
University of Sheffield
Royal Hallamshire Hospital
Sheffield, United Kingdom

Gordon W. Duff, Ph.D., FRCP
Section of Molecular Medicine
Department of Medicine and Pharmacology
University of Sheffield
Royal Hallamshire Hospital
Sheffield, United Kingdom

Robert W. Egan, Ph.D.
Schering Plough Research Institute
Kenilworth, New Jersey

Anne-Catherine Fluckiger, Ph.D.
Schering Plough Laboratory for Immunological Research
Dardilly, France

Roland Foulkes, Ph.D.
Celltech Therapeutics Ltd.
Slough, United Kingdom

Brian Henderson, B.Sc., P.G.Dipl., Ph.D.
Cytokine Biology Group
Maxillofacial Surgery Research Unit
Eastman Dental Institute for Oral Health Care Sciences
University College London
London, United Kingdom

Gerald M. Hochwald, M.D.
Departments of Neurology, Physiology, and Neuroscience
New York University School of Medicine
New York, New York

Andrew Roland MacKenzie, B.Sc., Ph.D.
Sandoz Research Institute Berne Ltd.
Berne, Switzerland

Anthony Meager, Ph.D.
Division of Immunobiology
National Institute for Biological Standards and Control
Herts, United Kingdom

John B. Mee, B.Sc.
Section of Molecular Medicine
Department of Medicine and Pharmacology
University of Sheffield
Royal Hallamshire Hospital
Sheffield, United Kingdom

Pierre Merville, Ph.D.
Schering Plough Laboratory for Immunological Research
Dardilly, France

Douglas K. Miller, Ph.D.
Inflammation Research
Merck Research Laboratories
Rahway, New Jersey

Steven M. Opal, M.D.
Brown University School of Medicine
Infectious Disease Division
Memorial Hospital of Rhode Island
Pawtucket, Rhode Island

Michael A. Palladino, Ph.D.
Ixsys Inc.
San Diego, California

E. Roy Pettipher, B.Sc., Ph.D.
Central Research Division
Pfizer Inc.
Groton, Connecticut

Françoise Rousset, Ph.D.
Schering Plough Laboratory for Immunological Research
Dardilly, France

Jeremy Saklatvala, B.Sc., MBBS, Ph.D.
Cytokine Laboratory
The Babraham Institute
Cambridge, United Kingdom

Jerzy A. Schmidt, B.Sc., Ph.D.
Department of Biochemistry
Fisons Pharmaceuticals PLC
Loughborough, United Kingdom

G. Jeanette Thorbecke, M.D., Ph.D.
Department of Pathology and the Comprehensive Kaplan Cancer Center
New York University School of Medicine
New York, New York

Shelby P. Umland, Ph.D.
Schering Plough Research Institute
Kenilworth, New Jersey

CONTENTS

Cytokine Physiology and Pathophysiology

Chapter 1

INTRODUCTION TO CYTOKINES AND THEIR PHYSIOLOGY

Anthony Meager

CONTENTS

I. INTRODUCTION TO CYTOKINES

As multicellular organisms became morphologically and physiologically more complex, a range of secreted 'effector' molecules evolved to mediate cell-to-cell communication, both locally and at a distance. Such molecules are known or believed to be involved in (1) cell proliferation leading to the development of tissues and organs, (2) the integrated control of physiological systems, and (3) the maintenance and defense of the whole organism. Some of these molecules, known as hormones, are produced by specialized cells in the endocrine system and act specifically, often at long range, on their target cells. Others are produced by one or more cell types outside of the endocrine system and usually induce local cellular responses, although they can also spread to distant cells via the circulation. The word 'cytokine' was coined in the mid-1970s by Dr. S. Cohen and has since become the generic term for nonendocrine effector molecules, particularly those of the protein variety. Originally, mediators involved in immunity and inflammation were designated as cytokines, whereas other mediators associated with proliferation and chemotaxis were called polypeptide growth factors. This distinction has faded and in essence is fairly meaningless, and so both categories are now joined to give a large superfamily of cytokines with wide-ranging activities.

0-8493-8381-1/96/$0.00+$.50

A. NOMENCLATURE AND CHARACTERISTICS OF CYTOKINES AND THEIR RECEPTORS

Historically, cytokines have been the 'fruit' of four areas of research: virology for interferons, immunology for interleukins, hematology for colony-stimulating factors, and cell biology for polypeptide growth factors. Although a few cytokines, e.g., interferons, were identified and characterized before 1980, it has only really been since the recent upsurge in recombinant DNA technology that a large and still growing number of diverse cytokines have been cloned. Their strange-sounding and often inappropriate names have arisen from the particular biological activities first ascribed or associated with them before the cloning era (Table 1). Such names, however, have stuck, along with a terrible mass of acronyms, despite the fact that the majority of cytokines can effect more than one biological response, at least as revealed by *in vitro* testing.

Cytokines may be reasonably described as secreted small- to medium-sized proteins or glycoproteins that act as intercellular mediators of cellular changes in metabolism and function. Another characteristic of cytokines is that they exercise their effects in cells via specific cell-surface receptors. Many of these 'cytokine receptors' have recently been cloned, characterized as transmembrane glycoproteins, and shown to have both the capacity to specifically bind their cognate ligands to their extracellular domain(s) and to transmit the 'informational' content of the bound cytokine into the cytoplasm via their intracellular component(s) (see Chapter 12). Although there is now a plethora of data on the structure of cytokines, often at the tertiary (3-D) level, such information has yielded little in the way of elucidating how they can act as specific intercellular messengers. However, increasing knowledge of the structures of cytokine receptors has allowed a number of groupings to be made. First, on the basis of the types of extracellular domains of receptors, which appear to be limited in number, cytokines may be grouped into four or five 'families': those that bind to receptors contained in the immunoglobulin (Ig) superfamily, those that bind to the so-called 'hematopoietin' or cytokine receptor superfamily, those that bind to a 'cysteine-rich' (nerve-growth factor) receptor superfamily, those that bind to the so-called G-protein binding receptor superfamily, and those that bind to receptor tyrosine- or serine-kinases or as yet unclassified receptors (Table 2). Several other cytokine receptors, particularly in the immunoglobulin superfamily, e.g., FGF-R and M-CSF-R, contain tyrosine kinases or are linked to nonreceptor tyrosine kinases, such as IL-2βR, IL-3R, and GM-CSF.[88–90] Most, if not all, of the tyrosine kinase receptors are associated with mitogenic responses, and many are also identified with protooncogene products (Table 2).

Thus, while cytokines appear to have divergent structures, the limited common features of their receptor-binding domains suggest that they should have conserved structural elements to allow high-affinity receptor binding. With the elucidation of the tertiary (3-D) structure of individual cytokines, it has begun to emerge that the cytokines binding to members of the hematopoietin/cytokine superfamily do have a shared four-α-helical bundle structure.[93,102] Interestingly, the simple equation of one cytokine ligand per receptor-binding domain does not always hold, and several examples exist where two (or more) molecularly related cytokines (which may be only distantly related by amino acid sequence) bind to the same receptor or receptors. For example, the many IFN-α subtypes, together with IFN-β and IFN-ω (Table 1), probably bind to the same IFN-α receptor.[92,93] Similarly, IL-1α, IL-1β, and the IL-1 receptor antagonist (IL-1ra) all bind to two distinct IL-1 receptors, IL-1RI and IL-1RII[39,86] (reviewed in detail in Chapter 9). This situation also occurs for TNF-α and TNF-β. In these cases the two receptors are distinct, remain spatially separated, and are present in differing proportions depending upon cell type. In

TABLE 1
Cytokines and Biological Activities

Cytokine Name, Abbreviation, and Alternative Names	Biological Activities	Ref.
Epidermal growth factor (EGF) [urogastrone]	Mitogenic and differentiating activity in a wide variety of cell types, both *in vitro* and *in vivo*	Reviewed in 1
Transforming growth factor alpha (TGF-α)	Same as EGF (may be fetal analogue of EGF); bone resorption and wound healing	Reviewed in 2
Amphiregulin (AR)	Tumor cell growth/differentiation	3
Heregulin (HRG-α) [Neu differentiation factor]	Tumor cell growth/differentiation	4,5
Betacellulin	Tumor cell growth/differentiation	6
Insulin-like growth factor-I (IGF-I) [somatomedin c]	Mediator of growth hormone activity. Tumor cell growth/differentiation	Reviewed in 7,8
Insulin-like growth factor-II (IGF-II) [multiplication stimulating activity]	Similar to IGF-I (may be fetal analogue of IGF-I)	7,8
Platelet-derived growth factor (PDGF), 3 isoforms, PDGF-AA, PDGF-AB, and PDGF-BB	Mitogenic for smooth muscle cells and variety of tumor cells; stimulator of chemotaxis of connective tissue cells; wound healing	Reviewed in 9
Vascular endothelial cell growth factor (VEGF) [vasculotropin]	Mitogenic for vascular endothelial cells (vasculogenesis, 'angiogenesis'), wound healing	Reviewed in 10
Platelet-derived endothelial cell growth factor (PD-ECGF) [gliostatin]	Mitogenic and differentiating activity in endothelial cells and neurons	11
Acidic fibroblast growth factor (aFGF) [prostatropin, endothelial cell growth factor, eye-derived growth stimulating factor-1]	Strong mitogenic activity for several cell types, including fibroblasts, endothelial cells, neuroectodermal cells; wound healing	Reviewed in 12
Basic fibroblast growth factor (bFGF) [as aFGF]	Same as for aFGF	Reviewed in 12
Keratinocyte growth factor (KGF)	Mitogenic activity for keratinocytes; wound healing	13
FGF-related protooncogene products, INT-2, hst/KS3, FGF-5, FGF-6	Mitogenic activity in embryonic and adult tissues; growth factors for various tumor cells	14
Hepatocyte growth factor (HGF) [scatter factor]	Mitogenic activity for hepatocytes *in vitro;* promoter of motility and matrix invasion of epithelial cells	15a,b
Mammary-derived growth factor (MDGF-1)	Stimulator of growth of mammary epithelial cells	16
Transforming growth factor beta-1 (TGF-β_1) [cartilage inducing factor-1]	Mitogenic or antiproliferative activity dependent upon cell type; regulatory activity in extracellular matrix formation; antiinflammatory, wound healing, etc.	Reviewed in 17,18

TABLE 1
Cytokines and Biological Activities (continued)

Cytokine Name, Abbreviation, and Alternative Names	Biological Activities	Ref.
Transforming growth factor beta-2 (TGF-β_2) [cartilage inducing factor-2]	As TGF-β_1	17,18
Transforming growth factor beta-3 (TGF-β_3)	Similar to TGF-β_1, but expression confined to mesenchymal cells and therefore *in vivo* role may differ	17,18
Also TGF-β_4 (avian) and TGF-β_5 (amphibian) plus other members of the TGF-β superfamily which include:- Müllerian inhibiting substance (MIS), *Drosophila* decaplentaplegic gene complex (DPP-C), *Xenopus* Vg-1 product, inhibins activins, bone morphogenetic proteins (BMP2–7), osteogenin, osteogenic protein (OP-1) and growth/differentiation factors (GDF-1, 3, and 9)	Wide range of growth and differentiating activities	See 17,20,21
Nerve growth factor (NGF)	Differentiation/survival of sensory, sympathetic, and cholinergic neurons	Reviewed in 22
Brain-derived neurotrophic factor (BDNF)	Differentiation/survival of sensory, cholinergic, dopaminergic neurons, and retinal ganglion cells	22
Neurotrophins (NT-3, NT-4, and NT-5)	Differentiation/survival of sensory neurons	22
Ciliary neurotrophic factor (CNTF)	Differentiation/survival of sensory, motor, cholinergic neurons; differentiation of glial cell progenitors into type 2 astrocytes	See 23,116
Oncostatin M (OSM)	Normal cell growth/differentiation Growth inhibitory for some tumor cells	24,25 (For comparison with LIF, G-CSF, and IL-6)
Leukemia inhibitory factor (LIF) [human interleukin with differentiating activity, HILDA; cholinergic neuronal differentiation factor]	Stimulates proliferation of murine embryonal stem (ES) cells; inducer of myeloid cell differentiation Stimulator of cholinergic function of neurons; very pleiotropic	26
Granulocyte-colony-stimulating factor (G-CSF) [pluripoietin, differentiation factor]	Stimulates mainly proliferation/-differentiation of neutrophils (granulocytes, precursors, and mature cells); differentiation of leukemia cell lines	Reviewed in 27,28
Granulocyte-macrophage-colony stimulating factor (GM-CSF)	Stimulates proliferation of granulocytic, monocytic, and other hematopoietic progenitors; activates mature granulocytes and monocytes Very pleiotropic	27,29,30

TABLE 1
Cytokines and Biological Activities (continued)

Cytokine Name, Abbreviation, and Alternative Names	Biological Activities	Ref.
Macrophage-colony-stimulating factor (M-CSF) [colony stimulating factor-1, CSF-1]	Stimulates monocytic colony growth, supports survival of monocytes *in vitro* Activates mature monocytes	27,31
Stem cell factor (SCF) [mast cell growth factor, MCGF; kit ligand, KL]	Synergizes with other cytokines to stimulate growth of myeloid, lymphoid, and erythroid progenitors; stimulates mast cell proliferation	32,33, Reviewed in 34. See also 169
Erythropoietin (EPO)	Stimulates proliferation of committed erythroid progenitors; induces hemoglobin formation	35,36
Interleukin-1 alpha (IL-1α) [endogenous pyrogen, EP; lymphocyte activating factor, LAF; hematopoietin-1, HP-1]	Activates T-cells; induces growth factors and inflammatory mediators Synergizes with certain hematopoietic growth factors. Very pleiotropic — multiple effects on lymphoid and other nonhematopoietic cells. Induces fever	37,38
Interleukin-1 beta (IL-1β) [same as IL-1α; osteoclast-activating factor, OAF]	Same or similar to IL-1 α	37,38
Interleukin-1 receptor antagonist (IL-1ra)	Inhibits IL-1α/β actions by binding to IL-1 receptors	39
Interleukin-2 (IL-2) [T-cell growth factor-I, TCGF-I; killer helper factor, KHF]	Stimulates proliferation of Th-cells proliferation/differentiation of CTL, and also acts on NK cells, B-cells and macrophages; induces other cytokines	40
Interleukin-3 (IL-3) [multiple colony-stimulating factor, multi-CSF, hematopoietic cell growth factor, HCGF]	Stimulates early growth of granulocyte, monocyte, erythroid, and megakaryocyte precursor cells; also supports mast cell growth. Growth factor for leukemia-derived cell lines	41
Interleukin-4 (IL-4) [B-cell stimulating factor-1, BSF-1; T-cell growth factor-II, TCGF-II; macrophage activation factor, MAF]	Stimulates or aids proliferation of B-cells, and mast cells. Promotes B-cell Ig class switching and induces IgE secretion. Induces class II MHC on B-cells and macrophages	42,43
Interleukin-5 (IL-5) [T-cell replacing factor, TRF1; eosinophil differentiation factor, EO-DF; IgA-enhancing factor]	Induces proliferation and differentiation of eosinophil progenitors. Enhances IgA production. Differentiation factor for B-cells	42,44
Interleukin-6 (IL-6) [B-cell stimulating factor-2, BSF-2; hybridoma-plastocytoma growth factor, HPGF; hepatocyte stimulating factor, HSF]	Stimulates differentiation of B-cells. Enhances Ig secretion by B-cells. Co-factor for hematopoietic colony growth. Stimulates hepatocytes to produce acute phase proteins; growth factor for B-cell lines	45
Interleukin-7 (IL-7) [lymphopoietin-1, LP-1; thymocyte growth factor, THGF; pre-B cell growth factor, PBGF]	Supports growth of immature T and B lymphocytes; co-stimulatory activity on mature T-cells	46, See 47

TABLE 1
Cytokines and Biological Activities (continued)

Cytokine Name, Abbreviation, and Alternative Names	Biological Activities	Ref.
Interleukin-8 (IL-8) [macrophage-derived neutrophil chemotactic factor, MDNCF; neutrophil activating factor-1, NAF-1]	Chemotactic for neutrophils and T-cells. Activates neutrophils, but not lymphocytes and monocytes	48,49
IL-8 is a member of the small cytokine or chemokine superfamily. There are two subgroups based on the positions of cysteine residues. IL-8 is a member of the c-x-c subgroup which also contains platelet factor 4, (PF4), β-thromboglobulin, (βTG), GRO-α, GRO-β, IP-10, etc. The c-c subgroup contains macrophage chemoattractant protein-1 (MCP-1), RANTES, macrophage inflammatory proteins (MIP-1α, MIP-1β), I-309, etc.	Proinflammatory chemotactic and activating actions for neutrophils, lymphocytes, and monocytes/macrophages	
Interleukin-9 (IL-9) [T-cell growth factor-III, TCGF-III; P40]	Sustains antigen-independent growth of certain Th-cells; enhances mast cell proliferative response to IL-2 Stimulates erythroid burst-forming activity	50, See 51
Interleukin-10 (IL-10) [cytokine synthesis inhibitory factor, CSIF]	Inhibits cytokine synthesis (e.g., IFN-γ) by Th1 cells and CTLs; suppresses antigen-dependent proliferation of Th1 cells	52,53,252
Interleukin-11 (IL-11)	Enhances megakaryocytopoiesis *in vitro*. Synergizes with IL-3 to stimulate proliferation of early hematopoietic progenitors; enhances B-cell Ig production	54,55
Interleukin-12 (IL-12) [natural killer stimulatory factor, NKSF; cytotoxic lymphocyte maturation factor, CLMF]	Stimulates cytokine-synthesis (e.g., IFN-γ) by T-cells and natural killer (NK) cells. Enhances proliferation of Th1 cells *in vitro*	56,57
Interleukin-13 (IL-13)	IL-4-like regulator of inflammatory and immune responses; anti-inflammatory effects on monocytes, induction of IgG4, IgE, and proliferation of B-lymphocytes	58,172,173
Tumor necrosis factor alpha (TNF-α) [cachectin] Tumor necrosis factor beta (TNF-β) [lymphotoxin-α, LT-α]	TNF-α and TNF-β are similar multifunctional and pleiotropic stimulators of cellular responses. Cytolytic to certain tumour cells. Inducers of cytokines and cell adhesion molecules. Upregulate class I MHC antigens. Regulators of proliferation/differentiation in lymphocytes and hematopoietic progenitors	59–62

TABLE 1
Cytokines and Biological Activities (continued)

Cytokine Name, Abbreviation, and Alternative Names	Biological Activities	Ref.
Also lymphotoxin beta (LT-β), a 33-kDa membrane glycoprotein with homology to TNF-α, TNF-β,	Complexes with LT-α in cell membranes	63
and T-cell antigen gp39 the CD40 ligand	Interaction between gp39 and CD40 stimulates B-cell proliferation	64
Interferon alpha (IFN-α) Subclass I, 15 or more molecularly related subtypes [leukocyte interferon; type I interferon; acid-stable interferon] Subclass II, one or more molecularly related subtypes depending on species [interferon omega, IFN-ω] Subclass III, several molecularly related subtypes so far only found in ruminants,	All interferons induce antiviral activity in a wide variety of cell types; they are multifunctional and pleiotropic regulators of immune functions. Stimulators of differentiation in many cells. Antiproliferative for certain tumor cells *in vitro* Augment class I MHC antigen expression	65–68
e.g., bovines [interferon *tau,* IFN-τ; ovine trophoblastic protein-1, OTP-1]	Interferon *tau* acts as antiluteolytic agent in domestic ruminant species	68
Interferon beta (IFN-β) [fibroblast interferon; type I interferon; acid stable interferon]	Similar to IFN-α subclass I	See IFN-α
Interferon gamma (IFN-γ) [immune interferon; type II interferon; macrophage activating factor, MAF]	Antiviral activity. Factor involved in macrophage activation. Inducer of class II MHC antigens. Synergizes with anticellular effect of TNF-α in certain tumor cell lines. Co-stimulator for proliferation of B-lymphocytes. Enhances IgG2a secretion by B-cells	65,67,69,70
Macrophage migration inhibitory factor (MIF)	Inhibition of macrophage migration, involvement in delayed hypersensitivity and cellular immunity	71

other cases, however, different receptor chains, which individually exist as low-affinity receptors for particular cytokines, combine to form high-affinity receptors, e.g., the IL-2 high-affinity receptor is comprised of three components, IL-2αR, IL-2βR, and IL-2γR.[89,90,127] There are also examples where a common, nonspecific receptor chain is shared by different cytokines to form high-affinity receptors by combination with distinct, specific receptor chains for each cytokine. They are IL-3, IL-5, GM-CSF,[85,103,104] CNTF, OSM, LIF and IL-6[84,85] and probably IL-2, IL-4, IL-7 and IL-15. This, therefore, is one reason why different cytokines can mediate the same biological response or are associated with the same biological activity; e.g., IL-3, IL-5, and GM-CSF all exhibit a very similar biological activity in eosinophils;[85] CNTF, OSM, LIF, and IL-6 all induce acute-phase protein synthesis in liver hepatocytes.[85,105] Following on from this, it is not surprising that considerable overlap in the biological activities of many different cytokines has been demonstrated *in vitro,* suggesting an overall redundancy in communicative power among these potent effector molecules.

TABLE 2
Classes of Cytokine Receptors

Immunoglobulin Receptor Superfamily[72,74,75,80]	Hematopoietin/ Cytokine Receptor Superfamily[73–75,80,174]	Nerve-Growth Factor Receptor Superfamily[74–76,80]	G-Protein Coupled Receptor Superfamily[77,78]	Other Receptor Tyrosine or Serine Kinases Plus Unclassified[79]
[a]FGF-R (flg and bek)[81]	Class I:	[a]NGF-R (trk)[22,94,99]	IL-8R and other members of small cytokine/-chemokine family receptors	[a]EGF-R and related receptors (erbB, erbB2, erbB3)[4,97]
[a]α/β PDGF-R[9,82]	EPO-R			
[a]M-CSF-R (fms)[83]	G-CSF-R	TNF-RI and II[95,96]		
[a]SCF-R (kit)[80,169]	[b]GM-CSF-R (fps/fes)[88]	Fas antigen		[a]IGF-1R
CTNF-R[23]		CD40		[a]VEGF-R (flk-1)[98]
OSM-R[85]	[b]IL-2β-R (lck)	etc.		
LIF-R[85]	IL-2γ-R[89,90,127]			[a]HGF-R (met)[99]
IL-6R[84]				TGF-β-RI and TGF β-RII[100,101,165]
IL-12R[87]	[b]IL-3R (fps/fes)[88]			
IL-1RI and II[86]	IL-4R			IL-2αR[89,90,127]
	IL-5R			
	IL-7R			
	IL-9R[91]			
	Class II:			
	IFN-α-R[92,93]			
	IFN-γ-R[92,93]			

[a] These receptors contain intracellular tyrosine kinase domains.
[b] These receptors are coupled to tyrosine kinases.

It has been remarked that cytokines bear analogies to symbols or signs (with directional input)[106] or represent letters or words of an intercellular language.[107] These are interesting comparisons, but it is unclear why so many different cytokines, which would in the above context be represented by dissimilar symbols/signs/words, apparently have the same meaning. For example, many cytokines are directly mitogenic for several cell types. Apart from the receptor 'promiscuity' mentioned earlier, the fact that many receptors contain or are directly coupled to tyrosine kinases suggests another reason for the overlap of cytokine activities. Nevertheless, cytokine activation of receptor tyrosine kinases does not always have the same result. For instance, in the rat pheochromocytoma cell line PC12, EGF which acts through the *erb*B tyrosine kinase (Table 2) is mitogenic, whereas NGF which acts through the *trk* tyrosine kinase leads to cell differentiation.[108] Most evidence to date has confirmed that the intracellular signaling pathways from the EGF and NGF receptors are grossly similar, with the same cellular enzymes and proteins being involved.[108] It is presumed that there are subtle differences that result in bifurcation of the common signaling pathway, and that the discriminatory power to do this resides within the cell (and not, or less so, at the level of the receptor and its ligand).[109] It is therefore all the more remarkable that if cytokines have ambiguous informational content and serve merely as receptor triggers, that so many of them have been conserved and have given rise to what appears to be a hugely redundant communication system.

Many of the cellular responses evoked by cytokines were originally demonstrated *in vitro*. However, such responses are often manifested with a single cytokine applied at doses which might be considered as nonphysiological. Moreover, cells in culture, in particular tumor-derived cell lines, will have different and sometimes abnormal growth and survival requirements to those that may be conceived to exist for cells present in 'structured' tissues and organs or in more complex environments, such as the blood. In the case of the hematopoietic process, i.e., the continual production of mature blood erythrocytes and immunocytes from bone marrow stem cells, virtually every cytokine that has been characterized, including all the colony-stimulating factors and interleukins, has been shown to have regulatory effects *in vitro*.[110] The involvement of such a vast array of cytokines in hematopoiesis *in vivo* has, however, been much harder to demonstrate, and it is unclear whether those cytokines which are produced as a result of physical injury or infection play any significant role in the steady-state regeneration of blood cells. Nevertheless, the pleiotropic nature of cytokines and their apparent redundancy of biological activities shown *in vitro* pose problems when it comes to considering their likely or potential physiological roles.

In some cases, the restricted expression of certain cytokine receptors to a distinct cell lineage or type has served to help define physiological involvement, but has not answered the question as to whether such-or-such a cytokine alone is sufficient for a particular process or response to occur *in vivo*. Since cytokines have been shown to interact in certain combinations *in vitro*, by synergistically enhancing cell responses or proliferation or by inducing the synthesis of other cytokines which can subsequently act in an autocrine manner on the cultured cells,[79] the idea has arisen that cytokines rarely act alone and together form a network of regulatory mediators, now termed the 'cytokine network'.[111] *In vivo*, it is widely believed that cytokines act in complex, but predetermined mixtures in the presence of other noncytokine mediators, e.g., hormones and neurotransmitters, and can induce cells to produce 'new' cytokines that act in a paracrine (or autocrine) manner. It has been hypothesized that some cytokines act as 'competence' factors, i.e., they induce in cells a state of readiness to respond to secondary stimuli, or as 'progression' factors, i.e., they induce cells activated by competence factors to proliferate, differentiate, or complete metabolic steps and structural changes involved in cell movement/migration, apoptosis (cell suicide), oncogenesis (cell transformation to a tumorigenic phenotype), and antimicrobial activity.[79,112] Thus, the apparent huge redundancy in the communicative power of cytokines begins to disappear if in 'real life' (as opposed to *in vitro*) their actions are directed by a cell-lineage-restricted expression of receptors; they act in an ordered fashion of competence and progression factors, being individually only 'players' in a wider program required to orchestrate cellular changes; and they are themselves subject to a regulatory machinery that controls their synthesis, secretion, and actions.

Regarding the synthesis of cytokines, it is clear that most, but not all, cytokines are inducible proteins and often are not present at all (in an organism) until an appropriate stimulus for their synthesis is received. Their gene expression is thus tightly regulated, and even when triggered may last for only a short time. In addition, while many cytokines are true secreted proteins, several, as with hormones, are initially synthesized in the form of inactive precursors (Table 3) and require proteolytic cleavage either intracellularly or extracellularly to become active. However, some cytokines are known to exist as membrane-bound homologues, e.g., M-CSF, SCF, and TNF-α which may also be biologically active. Bioavailability of cytokines is probably also subject to the presence of a number

TABLE 3
Factors Affecting Cytokine Secretion and Action

1. Synthesis in a latent (inactive) form and presence of activating enzymes, e.g.,
   IL-β, converting enzyme[113,197]
   TGF-β, plasmin[114]
   HGF, urokinase[115]
2. Lack of a leader amino acid sequence, e.g., FGF,[12] CNTF[116]
3. Molecules that bind to cytokines:
   Binding proteins, e.g., α_2-macroglobulin[117,123]
   Proteoglycans, e.g., decorin (TGF-β)[158]
   Other, e.g., heparin (FGF, VEGF, etc.)[118,119,120]
   Receptors,[80,104] e.g., circulating soluble, circulating autoantibodies[121]
4. Cytokine receptor antagonists, e.g., IL-1ra[39]
5. Regulation of cytokine receptor expression, i.e., physical numbers of appropriate binding and transducing components in a cell
6. Cell matrix, i.e., components thereof and composition[107,122]
7. Presence and regulation of intracellular signaling pathways linked to cytokine receptors[82,90,109,125,126]

of molecules with which noncovalent interaction can occur (Table 3). Their actions are also likely to be regulated by these binding molecules and also by other mechanisms, as summarized in Table 3. The potential uses of cytokine network interactions for therapy are described in Chapters 14 and 15.

II. CYTOKINE PHYSIOLOGY

Most cytokines have been well conserved throughout evolution, going back as far as the invertebrates,[124] suggesting that they have important physiological roles. However, the physiology of cytokines can only be studied in animals and man. Since cytokines can often be demonstrated to effect a range of biological responses, it is not clear in many cases whether a particular cytokine has a major physiological role or whether it has several subordinate or accessory roles; to date it has been difficult to establish or determine the physiological importance of cytokines. Clearly, many cytokines, especially those produced by the immune system, have a role in host defense and are present during pathophysiological responses. However, it is unknown in many instances whether such cytokines act in a physiological manner during normal development and in the maintenance of homeostatic processes. The main difficulty in obtaining scientific evidence for the involvement of cytokines in physiological processes is that the latter are difficult to model *in vitro* or in a manner that is accessible to investigation. Because of this, probably the best alternative for determining the physiological roles of cytokines is the study of animals and man, where some deficiency of a particular cytokine or its receptor, either inherited or acquired or induced, is manifest. The deficiency could be either in the synthesis or action of the cytokine and there are a number of situations in which this happens (Table 4).

A. WAYS OF STUDYING CYTOKINE PHYSIOLOGY

1. Natural Genetic Deficiencies

Natural genetic defects which result in a loss of specific cytokines and their receptors appear to be relatively uncommon. While there often is variation in the levels of cytokine production among individuals, complete loss is extremely rare. Loss of IL-2 production, for example, has been reported in some patients suffering from severe combined immunodeficiency disease (SCID),[128,137] but this could be for a variety of reasons not involving

TABLE 4
Cytokine Deficiencies for the Study of Cytokine Physiology

Loss of cytokine production:
- Natural genetic defects, e.g., M-CSF — op/op mouse, SCF — sl/sl mouse
- Gene knock-out mice → loss of specific cytokine gene expression, many examples see Table 5
- Antisense oligonucleotides → disruption of cytokine synthesis
- Loss of cytokine-producing cells, e.g., loss of kidney cells and EPO production, loss of T-cells and IL-2,3,4,5 etc., IFN-γ, TNF-β production

Loss of cytokine action:
- Natural genetic defects, e.g., in receptors for SCF, w/w mouse
- Gene knock-out mice → loss of specific cytokine receptor gene expression, growing number of examples, see Table 5
- Antisense oligonucleotides → disruption of cytokine receptor synthesis
- Neutralization of cytokine or receptor by: specific antibodies, soluble receptors, viral homologues of receptors, and receptor antagonists
- Loss of responder cell → e.g., loss of T-cells in SCID animals and AIDS patients

mutation or deletion of the IL-2 gene, such as lack of formation of T-cells[129] or presence of abnormal transcription factors.[132] However, an important new discovery has been the finding that some patients with a particular form of SCID, which is genetically linked to the X-chromosome (XSCID), are characterized by a lack of expression of the IL-2γR.[130] The gene coding for IL-2γR is located on the X-chromosome and, in XSCID, mutations in the gene resulting in stop codons have been found which result in truncated (nonfunctional) IL-2γR molecules. The result of this defect is the marked reduction or absence of mature T-cells, indicating that IL-2γR has a vital physiological role in the development of T-cells in the thymus.[130]

The nerve-growth factor receptor superfamily[74–76,80] contains a number of cell surface glycoproteins, other than NGF-R, TNF-RI, and TNF-RII (Table 2), for which ligands have not yet been found or only recently characterized. One of these glycoproteins, CD40, is present on human B-cells and is involved in T-cell mediated activation which results in the switch from IgM expression to that of other immunoglobulin isotypes, e.g., IgG and IgE.[131,156,157,164] The ligand for CD40 has recently been cloned and characterized as a further membrane protein (gp39), expressed on activated T-cells, which shows significant homology to TNF-α and TNF-β.[64,166] Therefore, gp39 might be considered to be a membrane-bound cytokine. Furthermore, defective expression of gp39 has been found to underlie an X-chromosome-linked immunoglobulin deficiency characterized by undetectable serum levels of IgG, IgA, and IgE, but normal or elevated levels of IgM.[133] It is clear that B-cells in affected individuals, though normal in numbers, are exclusively IgM/IgD positive and do not express any surface IgG, IgE, or IgA. Thus, gp39 plays an important role as a (co)-stimulator of production of particular classes of Ig, e.g., gp39 drives IgE production in interleukin-4 (IL-4)-treated B-cells.[164]

Further natural defects in human cytokine genes have not yet been found, but there are other interesting examples of this kind in mice. The membrane tyrosine kinase receptor, c-kit (Table 2) is known to be coded for by a gene in the dominant white spotting (W)-locus present on mouse chromosome 5.[80,134] Mutations in the W-locus have been found to have pleiotropic effects on embryonic development, regulation of hematopoiesis, and development of melanocytes. Mice which inherit two mutant alleles, resulting in loss of c-kit expression/function, are severely anemic. The defect(s) in the W-locus appears strongly associated with bone marrow stem cells,[80] thus indicating that c-kit expression/function is required for stem cell proliferation and development. Furthermore, a corresponding defect in the so-called Steel (SL) locus, which appears intrinsic to the stromal layer of bone marrow, has a similar effect on stem cell growth. It is now known that the gene product of the SL locus is stem cell growth factor (SCF), a cytokine that is active in both membrane-bound and soluble forms, and which is the cognate ligand for c-kit.[135,136,169] It is clear, therefore, that in the mouse the c-kit/SCF complex plays an essential role in hematopoiesis and begs the question as to whether comparable defects in the human genes for c-kit and SCF underlie inherited hematopoietic disorders. However, no evidence has yet been found to support the involvement of such defects in human congenital anemias such as Blackfan-Diamond anemia, where responsiveness of early hematopoietic progenitors to SCF has been demonstrated.[34]

Platelet-derived growth factor (PDGF), which has potent mitogenic activity in a variety of cell types, also binds to a receptor of the tyrosine kinase group in the Ig receptor superfamily (Table 2). In fact, there are two receptor chains, αPDGF-R and βPDGF-R, which can form both functional homodimers and heterodimers.[9,82] Interestingly, the mouse patch (Ph) mutation (see Table 5) has been shown to result from a complete deletion of genomic sequences coding for αPDGF-R.[134] Since presumptive homozygotes with this deletion display gross anatomical abnormalities and die midway through gestation, αPDGF-R expression would appear to have an important role(s) in embryogenesis.[134]

Another example of a mutant mouse which can be directly linked to a cytokine gene defect is the osteopetrotic, autosomally recessive op/op mouse.[34,138] In this mutant mouse, osteopetrosis, resulting from defective bone resorption, is strongly linked with an absence of bone-resorbing osteoclasts in these animals. (Osteoclasts are macrophage-like cells which derive from the monocyte-macrophage lineage of hematopoietic progenitor cells.) The discovery, then, in 1990 that op/op mice were totally deficient in macrophage colony-stimulating factor (M-CSF), a cytokine known to stimulate monocytic colony growth *in vitro* and to support and activate mature monocytes,[27,31] gave a rational explanation for the absence of osteoclasts.

It has been shown that the loss of M-CSF synthesis results from a frameshift mutation in the M-CSF gene located on mouse chromosome 3.[138,139] Moreover, it has been proved that the lack of M-CSF is the cause of the congenital osteopetrosis in op/op mice by curing the osteoclast deficiency with injections of recombinant M-CSF.[140,141] The presence of M-CSF also reversed the deficiency of monocytes and peritoneal macrophages in op/op mice. However, not all macrophage populations appear to be critically dependent on M-CSF. In the op/op mouse phagocytes in the thymic cortex, lymph node medulla, liver (Kupffer cells), lung (alveolar macrophages), brain (microglia), as well as dendritic cells (specialized antigen-presenting cells for T-cell immunity) are relatively normal in numbers, suggesting that other cytokines, e.g., the granulocyte-macrophage colony-stimulating factor (GM-CSF), are more likely to be physiologically important to their development and survival.[142] A corresponding mutation in the human M-CSF gene, which could potentially give rise to congenital osteopetrosis, e.g., juvenile osteopetrosis, which embraces a range of autosomal recessive disorders characterized by a lack of osteoclast-mediated bone resorption in children, has yet to be found.[34]

One further example of a natural mutation is the loss of Fas antigen expression — Fas antigen is a member of the nerve growth factor receptor superfamily[74–76] (Table 2) which mediates apoptosis, i.e., programed cell death, in lymphoproliferation minus (lpr^-) mice.[143] The Fas antigen, whose ligand has been recently characterized, is a receptor that has an intracellular domain showing strong homology with that of the TNF-RI receptor. Mouse Fas antigen is coded for by a gene on chromosome 19 located within the lpr locus, and mice that are lpr^- either lack Fas mRNA due to a rearrangement in the second intron of the Fas gene or express a nonfunctional Fas due to a point mutation converting *ile* to *asp* in the middle of the intracellular domain.[143] In normal mice, Fas antigen is found in the thymus, heart, lungs, and ovary, but in lpr⁻ mice deficient in Fas there is an accumulation of immature (malignant) $CD4^-CD8^-$T-lymphocytes and autoreactive T-cells that lead to autoimmune disease, indicating that Fas has an important role in the thymic (negative) selection and development of mature T-lymphocytes.

2. Knock-Out Mice

The mutations that have been described are function knock-outs where, because of disruptions in specific cytokine or receptor genes, functional cytokine or receptor proteins are absent. It is now possible to artificially create knock-out mutations in mice by using the strategy of gene targeting in embryonic stem (ES) cells[144,148] which can be genetically manipulated in culture. A nonfunctional gene is targeted by a vector to a particular gene in the mouse ES cell genome resulting in disruption of gene expression, and cell lines selected for the disrupted gene are injected into blastocysts. Programmed breeding of the resulting chimeric animals and selection of offspring carrying the disrupted gene in the germline eventually leads to null or knock-out mice which are homozygous for the disrupted gene, as long as this is not prenatally lethal.

To date, several different knock-out mice with disrupted cytokine or receptor genes have been bred and, rather surprisingly, some of these mice are both healthy and fertile. For instance, knock-out mice for the transforming growth factor alpha (TGF-α) gene are only distinguishable from their littermates by a pronounced waviness of the hair and whiskers.[145,146] TGF-α is known to act as a potent mitogen for many epithelial and mesenchymal cells — its action being mediated by the epidermal growth factor receptor (EGF-R), the product of the c-erbB tyrosine kinase protooncogene — and has been strongly associated with neoplastic transformation.[2,79,147] The TGF-α mRNA has been found in a wide variety of normal embryonic and adult tissues, suggesting that TGF-α must have important physiological roles.[145–147] Nevertheless, the finding that TGF-α gene knock-out mice develop normally in many respects indicates that TGF-α is not essential for embryogenesis or growth of tissues and organs to maturity. However, closer examination of the skin of these mice has shown that there is a gross derangement of hair follicles throughout the dermal layer suggesting that TGF-α is a physiological regulator responsible, in part, for modeling skin architecture.[145,146] Wound healing is apparently normal in these mice, however, suggesting that TGF-α, which has been shown to accelerate wound healing in animal model systems,[147] is not critically involved in this process. It has been speculated[145] that epidermal growth factor (EGF), which binds to the same receptor and has similar activities, may be upregulated in the TGF-α gene knock-out mice to compensate for the lack of TGF-α. Also, the ligands that bind to EGF-R could largely be functionally (physiologically) redundant in most tissues, but presently there is a paucity of data on this.

The reason for the waviness of the hair of TGF-α gene knock-out mice is not understood, but the similarity in appearance to that of the mutant waved-1 (wa-1) mouse suggested that the latter had an abnormal TGF-α gene (see Table 5). However, two groups[145,146] have shown that the wa-1 defect results from reduced expression of TGF-α mRNA and not to a deletion in the coding region.

Transforming growth factor beta (TGF-β), while having some biological properties in common with TGF-α, is molecularly quite different from the latter. TGF-β_1 is the first representative member of a growing superfamily of molecularly related biologically active proteins (Table 1) effecting a wide spectrum of activities, and is itself one of the most pleiotropic cytokines described.[17–21] It has been variously reported to affect cell growth and differentiation in numerous tissues and also to inhibit inflammatory and immune responses (see Chapter 14). Its main physiological role may well be in tissue remodeling,[107] but it could also be involved in wound healing.[18,163] The recent creation of

TABLE 5
Natural and Induced Defects in Cytokines and Cytokine Receptor Genes and Their Consequences

Species	Cytokine or Receptor	Locus/Phenotype	Results	Ref.
Man	IL-2Rγ	XSCID	Absence of mature T-cells	130
	CD-40 ligand (gp39)	X-linked Ig deficiency	Absence of IgG, IgA, and IgE	133,156,157
Mouse	SCF	Steel (SL)	Macrocytic anemia, lack of cutaneous melanocytes, sterility	34,135,136
	SCF-R (c-kit)	White-spotting (W)	Same as for SCF deficiency	
	αPDGF-R	Ph-locus	Anatomical abnormalities, fetal death	134
	M-CSF	op-Locus	Osteopetrosis, lack of osteoclasts	138–140
	Fas antigen	lpr-Locus	Lack of thymic selection of T-lymphocytes	143
Knock-out mouse	TGF-α	Similar to waved-1 (wa-1)	Wavy hair, deranged hair follicles	145,146
	TGF-β_1	—	Multifocal inflammatory syndrome 2–3 weeks from birth, uniformly lethal	149,150
	α-Inhibin (member of the TGF-β superfamily)	—	Normal to birth, develop gonadal tumors	249
	LIF	—	Reduced stem cell survival, impaired thymic functions, female sterility	170
	IL-2	—	Lack of mature T-cell responsiveness, high IgG levels	151
	IL-4	—	Lack of Th2 cell response, no IgE or IL-5	154
	IFN-γ	—	Defective immune responses to pathogens, impaired macrophage functions, no IgG2a	167
	IFN-γR	—	Similar to IFN-γ deficiency	168
	TNF-RI	—	Resistance to LPS Susceptibility to *Listeria*	250

TGF-β_1 gene knock-out mice[149,150] has opened up a way to examine critically the putative physiological roles of TGF-β_1. The first remarkable finding is that TGF-β_1 null phenotype mice develop normally to term, thus suggesting that TGF-β_1 is dispensable or superfluous during embryogenesis.[159] This may again be indicative of cytokine redundancy, where possibly TGF-β_2 or TGF-β_3, which have very similar properties to TGF-β_1,[160,161] may substitute for TGF-β_1. An alternative explanation could be that in some way the maternal decidua contributes TGF-β_1 to the developing fetuses. Although as yet unproven, the latter explanation looks attractive since within a few weeks of birth all of the TGF-β_1 gene knock-out mice develop a diffuse, uniformly lethal, inflammatory syndrome characterized by huge infiltrates of mononuclear cells in the heart, lungs, and gut.[149,150] The precise mechanism of the disruption of control of inflammatory and immune responses requires further investigation, but it appears that the presence of TGF-β_1 is required from birth to suppress the inappropriate activation of granulocytes, lymphocytes, and monocytes in these knock-out mice. It will be of considerable interest to see whether injection of TGF-β_1 can stop the inflammatory syndrome from developing.

Leukemic inhibitory factor (LIF) is another cytokine exerting a multitude of biological activities[26] and for which a physiological role has been hard to define. However, it has the unique property of being able to maintain murine embryonal stem (ES) cells in cycle so that they retain their pluripotentiality indefinitely; without LIF ES cells differentiate into a wide variety of cell types. This activity suggests that LIF could have a role in the maintenance of stem cell pools *in vivo.* LIF gene knock-out mice, like the TGF-α and TGF-β_1 knock-out animals, develop normally, indicating that LIF is nonessential for embryogenesis. It has been found, however, that these LIF-deficient mice have greatly decreased stem cell numbers in their spleens and bone marrow.[170] Stem cells have either differentiated or disappeared, however, those stem cells that remain are still pluripotent suggesting that LIF, while necessary for their survival, is not required for their differentiation. The LIF-deficient mice also show impairment of thymic T-lymphocyte stimulation, due probably to the lack of LIF production by thymic epithelial cells.[170] An earlier report[171] indicated that maternal LIF expression was required for blastocyst implantation, and the additional finding that LIF-deficient female, but not male mice were sterile would appear to confirm this.[170] Thus, the normal *in vivo* roles of LIF are now better understood, the LIF-gene knock-out mice having contributed to an understanding of its functions.

Several gene knock-out mice have been created for cytokines which are intrinsic to cell-mediated immunity. One of these is interleukin-2 (IL-2), a cytokine produced by activated T-lymphocytes and natural killer (NK) cells, which acts mainly as a growth factor for mature T-lymphocytes and an activator of NK and lymphokine activated killer (LAK) cells.[40] The IL-2 null phenotype mice develop normally with no effects being observed on the early development and inactivation of T-lymphocytes.[151] However, a lack of response of mature T-cells to polyclonal activators, e.g., phytohemagglutinin, together with a lack of T-cell-mediated 'help' for B-cell differentiation were manifest in these mice. There were also excessively high levels of IgG_1, IgG_{2a}, and IgG_{2b}.[151] It has not yet been reported whether the IL-2 gene knock-out mice are more susceptible to infections, as has been found in one child with a defect in IL-2 mRNA production.[128]

IL-2, IFN-γ, and TNF-β production, at least in the mouse, is mainly confined to the Th1 subset of T-lymphocytes, whereas IL-4, together with IL-5 and IL-10, is produced by the Th2 subset.[152,153] Both Th1 and Th2 cell-derived cytokines can provide 'help' for B-cell proliferation and differentiation, but Th1 by producing IFN-γ and TNF-β which activate macrophage cytotoxicity, are linked with delayed-type hypersensitivity (DTH) responses, while Th2, by producing IL-4 which mediates Ig class switching in B-lymphocytes to IgE, are associated with the allergic response.[164] There is evidence that the two subsets are mutually inhibitory since IFN-γ inhibits Th2 cell growth and IL-4 and IL-10 inhibit cytokine synthesis by Th1 cells[153] (reviewed in detail in Chapter 15). IL-12, produced by activated macrophages, stimulates Th1 cell proliferation, whereas IL-4 stimulates Th2 cell growth and cytokine production.[57,162] In the case of IL-4 gene knock-out mice, it has been found that they generally are unable to mount a Th2 cell type of immune response.[154,155] For example, IgE was undetectable in these mice. In addition, when the mice were parasitized with *N. brasiliensis,* a nematode, only a weak eosinophilia was observed. Since eosinophil proliferation is dependent on IL-5,[44] this finding indicates that IL-4 is required for inducing/augmenting Th2 cytokine production.[155]

Interferon gamma (IFN-γ), as previously mentioned, is synthesized by Th1 cells but is also produced by cytotoxic T lymphocytes and NK cells. It exerts pleiotropic effects including activation of macrophage and NK functions, induction of MHC class II antigen expression, Ig class-switching of B-lymphocytes to IgG_{2a}, and antiviral activity (Table 1).[65,67,69,70] Two types of knock-out mice, one with a disrupted IFN-γ gene[167] and the other with a disrupted IFN-γ receptor gene,[168] have recently been developed. In both cases the mice developed normally and were healthy. However, both exhibited a defect in natural resistance when challenged with pathogens,[167,168] e.g., increased susceptibility to infection with *Listeria monocytogenes* and vaccinia virus. In the IFN-γ gene knock-out mice there was, as could have been predicted from the biological effects of IFN-γ, impaired production of macrophage microcidal products and reduced expression of macrophage MHC class II antigen expression. NK cell activity, but not CTL activity, was also reduced in these mice, while splenocytes proliferated in an uncontrolled manner in response to mitogen or alloantigen. In the IFN-γ-R gene knock-out mice there was a lack of an antigen-specific IgG_{2a} response. It therefore has been clearly demonstrated that IFN-γ is essential for immune responses to pathogens by several cell types.

Tumor necrosis factor alpha (TNF-α) and TNF-β, which have been widely implicated in host defense and inflammatory reactions,[59–62] have two distinct receptors, TNF-RI and TNF-RII[95,96] (Table 2), whose roles remain rather uncertain. However, recent information from T Mak's group[250] indicates that knock-out mice for TNF-RI (55 kDa receptor) are much more resistant to lipopolysaccharide (LPS) and therefore to endotoxic shock than control mice, but in contrast they become highly susceptible to *Listeria monocytogenes* infection (Table 5). Thus, it would appear that TNF-RI has a crucial role in host defense against microorganisms and their pathogenic factors, as well as in pathological responses to septic challenge.

In conclusion, knock-out mice are providing advances in our knowledge of the physiological roles of some cytokines and their receptors. It should be pointed out, however, that adaptation in the knock-out mice, e.g., by possible compensatory functions of closely related cytokines or the maternal replacement of absent cytokines in the fetus, could have a bearing on normal development and immune responses and perhaps obscure certain physiological roles of cytokines. In addition, it is uncertain how far the physiological roles for cytokines that have been derived from knock-out mice can be transferred to the equivalent human cytokines, since these may have subtle differences in function from murine cytokines. It is clear, however, that knock-out mice guarantee that all traces of a particular cytokine activity are removed, whereas other investigational methods for blocking cytokine actions, e.g., injection or infusion of neutralizing antibodies, may not be 100% effective in removing cytokine activity.

3. Loss of Cytokine-Producing Cells

Loss of a particular cytokine can result from deficiencies other than loss of functional cytokine genes, e.g., loss of cytokine producer cells. A classic case of the loss of cytokine production due to reduced numbers of producer cells is that of anemia occurring in patients with kidney disease, where erythropoietin (EPO) production is severely reduced as a consequence.[35,36] EPO production is also controlled by the amount of oxygen in the blood, such that inspiration of decreased atmospheric oxygen at high altitude leads to decreased oxygen delivery to the tissues and temporary anemia. In response to this low oxygen tension, kidney cells increase their production of EPO, leading to increased numbers of circulating erythrocytes and thus more oxygen delivery. Since the anemic condition of renal failure patients is also overcome by injection of recombinant EPO, it is clear that EPO has a major physiological role in the regulation of circulating erythrocyte numbers. Its action is likely to be at the level of stimulation of proliferation and differentiation of early erythroid progenitor cells. The numbers of colony-forming-unit erythroid (CFU-E) cells have been shown to increase in response to EPO *in vitro,*[35] suggesting that their precursors, the so-called mature burst-forming-unit erythroid (BFU-E) cells, are the earliest EPO-responsive progenitor cells. Possibly other cytokines, e.g., IL-3 and SCF, are involved at earlier stages of erythroid lineage development from bone marrow stem cells.

Loss of cytokine producer cells often occurs in severe-combined immunodeficiency diseases (SCID), e.g., in adenosine deaminase (ADA) deficiency,[129] where T-cells are reduced or absent, or in HIV-induced acquired immunodeficiency disease syndrome (AIDS) where the subset of $CD4^+$ T-cells is markedly reduced in numbers. Since, however, mature T-cells produce several different cytokines when activated and these have pleiotropic activities, it is usually difficult to assign physiological roles to particular cytokines in conditions which have multifactorial pathologies. The selective ablation of interleukin-2 (IL-2)-producing cells can also now be achieved by developing transgenic mice with the herpes simplex virus/1-thymidine kinase (HSV-TK) gene under the control of the murine IL-2 promoter in IL-2-producing cells, which renders them sensitive to the cytotoxic effects of the antiviral drug ganciclovir.[175]

4. Cytokine Neutralization

Other ways for defining the physiological roles of cytokines have mainly relied on the findings of experiments where a particular cytokine is effectively neutralized *in vivo,* essentially by preventing its interaction with its cognate receptor. There are a variety of means for achieving this, including neutralizing anti-cytokine antibodies, soluble cytokine receptors and other cytokine binding molecules, and receptor antagonists (Table 4). Such investigations are carried out in animals, mainly mice, but on most occasions ablation of cytokine actions has only been carried out under experimental conditions in which pathological conditions have been induced, e.g., microbial challenge, tissue injury, graft-vs.-host disease, or autoimmune disease. However, there are some exceptions to this where cytokine activity has been abrogated under normal conditions. For example, application of an anti-c-kit (SCF-R) monoclonal antibody to embryonic and neonatal mice acted as an antagonistic blocker of c-kit function[80] and led to interference of the development of melanocytes at two stages of life, one embryonic and one postnatal.[176] In a similar type of study, an antitumor necrosis factor alpha (TNF-α) antiserum was administered to mice *in utero* and following birth.[177] This led to a severe but transient growth retardation, already manifest at birth, taking at least eight weeks for correction to occur. Growth retardation in this case appeared to correlate with decreased levels of insulin-like growth factor-I (IGF-I). In addition, an unrelated marked atrophy of the thymus, spleen, and lymph nodes was observed, suggesting that TNF-α has a physiological role in the early development of lymphoid tissues.[177] In a further example of this kind, continuous anti-interleukin-10 (IL-10) antibody administration was found to deplete mice of an unusual subset of B-lymphocytes, known as Ly-1 B-cells, which have the distinctive property of continuous self-replenishment.[178] IL-10 has been shown to block the production of Th1 T-cell cytokines[52,53] such as IL-2 and IFN-γ, and it was found that in the anti-IL-10-treated mice endogenous levels of IFN-γ were probably related to the Ly-1 B-cell depletion, since anti-IFN-γ administration largely restored Ly-1 B-cell numbers.[178] Other sequelae of the anti-IL-10 treatment have also been observed, e.g., elevated circulating levels of TNF-α and IL-6, susceptibility to LPS-induced shock, and high $IgG_{2a/b}$ levels.[251] In contrast, preliminary data from IL-10-deficient knock-out mice[252] have not confirmed the involvement of IL-10 in the regulation of Ly-1 B-cells, nor have raised levels of TNF-α, IL-6, and IgG been found, suggesting that either compensatory mechanisms emerge in the knock-out mice or that the anti-IL-10 treatment is complicated by immune responses against the xenogenic anti-IL-10 antibodies. The therapeutic effects of neutralizing cytokines are described in Chapters 9–11.

There are a good number of examples of cytokines having pathophysiological roles or disease correlates, many of which are reviewed elsewhere in this volume, which may help to shed some light on the complex and versatile nature of cytokines in health and disease. For instance, a plethora of reports on interleukin-1 (IL-1) and tumor necrosis factor alpha (TNF-α) has indicated that these cytokines, which have overlapping biological activities,[60,179] are required for normal host defense as they augment antimicrobial, antiviral, and antitumor activity of immunocytes, enhance neutrophil adherence to endothelial cells, and are involved in granuloma formation.[38,184] However, over-production of IL-1 and TNF-α leads to a predominance of proinflammatory effects which are correlated with disease states such as hypotension and sepsis.[180–182] In contrast, IL-1 and TNF-α secreted at low levels have several potential physiological roles, including temperature regulation, bone and cartilage remodeling, regulation of early hematopoietic cell progenitors, and regulation of extracellular matrix production. For example, IL-1 levels have been reported to be elevated following hard exercise, ultraviolet light exposure, and during the normal luteal phase of women, and is likely in all of these cases to be partly responsible for the temperature increases that are observed.[182]

That IL-1 and TNF are critically involved in normal host defense mechanisms and pathological conditions is suggested by numerous studies where the activities of these cytokines have been abrogated in experimentally induced disease states in animals. For example, neutralizing antibodies against IL-1 and TNF or their receptors suppress cell-mediated immunity *in vivo* and increase susceptibility to pathogens such as *Listeria monocytogenes*[185,186] and can attenuate inflammatory responses.[187–189] The IL-1 receptor antagonist (IL-1ra) and soluble forms of the IL-1R can also produce similar effects[39,190] (see Chapter 9). Rather interestingly, further supportive evidence for the roles of cytokines in host defense has been revealed by a set of recent findings that members of certain virus families, the pox viruses and herpes viruses in particular, encode soluble homologues of cytokines or their receptors, suggesting that viruses have evolved means for countering host immune responses that would otherwise inhibit viral growth.[191–193] The pox viruses have in general been found to express soluble homologues of cytokine receptors; e.g., vaccinia and cowpox viruses encode a soluble form of IL-1RII (see Table 6)[194,195] and myxoma virus encodes a soluble form of IFN-γR[196] which act to bind their respective cytokines and prevent them from binding to their respective surface membrane receptors present in target cells. Thus, cytokine-driven immune responses are downregulated, presumably to the advantage of the virus. Besides the IL-1RII homologue, cowpox virus has also been found to contain a gene that codes for an inhibitor, a serpin, of the interleukin-1β converting enzyme,[197,198] and this again would prevent an IL-1β-driven immune response from being mounted (further review of IL-1β convertase is given in Chapter 7). In the case of the herpes virus group the Epstein-Barr virus (EBV) expresses a homologue of IL-10 (see Table 6), a cytokine that has been shown to inhibit cytokine synthesis by activated Th1 lymphocytes.[153,191] Thus EBV, by producing a form of IL-10, turns down the antiviral T-cell response directed against itself and, in addition, by the stimulatory effect of IL-10 on Th2 lymphocytes[153] augments the 'help' for B-cell proliferation, i.e., increases its target cells for infection. Other viruses encode cytokine-like proteins and receptors that have a more direct effect on the proliferation of the cells they infect. In particular, retroviruses, which require cell division in order to replicate, encode cytokines and cytokine receptor homologues, e.g., the v-sis oncogene product of simian sarcoma virus is identical to PDGF-B chain (see Tables 1 and 6) and the v-fms and v-kit oncogene products of feline sarcoma virus are mutated versions of M-CSF-R (c-fms) and SCF-R (c-kit) (see Tables 2 and 6)[193,199] which promote hyperplasia of target cells. These findings suggest that the host counterparts of these viral oncogenes, which are themselves cytokines or cytokine receptors, have critical roles in the normal regulation of cell proliferation.

TABLE 6
Viral Homologues of Cytokines and Their Receptors That Counteract Immunity

Virus Family	Virus	Homologue Designation	Cytokine or Soluble Receptor	Ref.
Pox viruses	Shope fibroma, myxoma	T2	TNF-RII	76,95,191,192
	Myxoma	M-T7	IFN-γR	196
	Vaccinia, cowpox	B15R	IL-1 RII	194,195
		B18R	?	
	Vaccinia	VGF	EGF	1,79
Herpes viruses	Epstein-Barr virus	BCRF1	IL-10	191,192
	Cytomegalovirus	US28	MIP-1αR	193
	Herpes virus, saimiri (monkey)	ECRF3	IL-8R	193
Retroviruses	Simian sarcoma virus	v-sis	PDGF-B (c-sis)	199
	McDonough feline sarcoma virus	v-fms	M-CSF-R (c-fms)	199
	Hardy-Zuckerman-4 feline sarcoma virus	v-kit	SCF-R (c-kit)	199

To date, there is a lot of evidence for the involvement of cytokines in host defense, and the observation that many cytokines are produced in quantity only in infections and invasive diseases suggest that they act mainly to counteract disturbances of homeostatic mechanisms. Besides the proinflammatory cytokines IL-1 and TNF-α, interferons IFN-α and IFN-β fall into such a category of host defense cytokines because of their proven antiviral activity.[65–68] However, like IL-1 and TNF-α, they could also be involved in other physiological processes.[200] Nevertheless, it has been difficult to show that IFN-α and IFN-β are produced under normal physiological conditions, although potential exogenous stimuli, e.g., gut bacteria, foreign antigens, etc., may be presumed to act as inducers. It is probable that IFN-α, IFN-γ, and other cytokines are present at various stages of pregnancy and could have activities, other than antiviral and antimicrobial, which impinge on the development of the fetus.[201] In the case of ruminants, a special form of interferon designated IFN-τ (Table 1) appears to have evolved from IFN-ω and exceptionally acts as the hormone to prevent corpus leuteum regression in the first few weeks of pregnancy.[68] IFN-τ appears to be similar or identical to what was known in older terminology as ovine trophoblastic protein-1 (OTP-1).

5. Transgenic Mice and Constitutive Cytokine Production

Further information on the physiological roles of cytokines may be deduced from the study of transgenic animals[148] or animals engrafted with cytokine-producing cells, although here since the cytokine in question is often produced constitutively at high levels by a particular tissue or cell type, the outcome may be just as likely to suggest pathophysiological roles.[236] Since much of this type of work has been carried out with particular strains of mice, e.g., Balb/c, the genetic background can also affect the result. There are now many examples whereby a cytokine gene has been introduced into the mouse genome to produce transgenic mice and some of these, with details of regulation and site of expression together with the outcome, are summarized in Table 7.

By and large, where there is overexpression due to a strong enhancer/promoter there is target cell hyperplasia and a marked tendency for oncogenic activation and the

TABLE 7
Cytokines and Cytokine Receptor Genes in Transgenic Mice

Cytokine (or Receptor)	Transgene	Site of Expression	Results	Ref.
TGF-α	Mouse mammary tumor virus (MMTV) promoter — TGF-α	Liver, pancreas, mammary gland	Hyperplasia and neoplasia in mammary glands, pancreas, liver	203–206
	Human keratin 14 (k14) — TGF-α	Epidermis of skin	Wrinkled, flaky skin (like psoriasis?), epidermal hyperproliferation and thickening, weaker and small mice with gradual recovery during postnatal development	202
erbB2 (neu) — member of the EGF-R family; receptor for hereregulin[4,5]	MMTV-erbB2	Mammary epithelium	Breast tumors	207
KGF (member of the FGF superfamily)	K14-KGF	Epidermis of skin	Wrinkled skin, thickened epidermis, suppression of hair follicle morphogenesis, suppression of adipogenesis	208
TGF-β_1	Regulatory elements of pregnancy-responsive whey-acidic protein (WAP) gene — TGF-β_1	Mammary gland	Inhibition of alveolar development and lactation	209
EPO	Complete human EPO gene with some 3′ and 5′ flanking regions	Liver	Polycythemia, increased numbers of erythroid precursors, increased serum HuEPO levels, responsive to anemia	210
GM-CSF	Moloney retrovirus promoter — GM-CSF	Macrophages	Fatal cachectic state, multiple inflammatory foci, blindness, gut complications, etc. Elevated levels of IL-1, TNF, in serum	211–214

IL-2	MHC class I (H-2K^d) promoter — human IL-2	mRNA found in many organs	Baldness, interstitial pneumonia due to inflammatory infiltrate, immunodeficiency, but not autoimmunity, expansion of NK cells, but not T cells	215, 216
	Rat insulin promoter — murine IL-2	Pancreatic β-islet cells	Lethal pancreatitis, but no diabetes	217
IL-4	Ig H chain promoter/ enhancer — murine IL-4	(a) Thymus + T cells	Thymic hypoplasia, inflammatory ocular lesion (blepharitis), raised IgE and IgG1	218
		(b) B cells	Raised IgE and IgG1, but no blepharitis	219
	lck Promoter — murine IL-4	Thymocytes	Thymic hypoplasia, but no change in IgE and IgG1 (IL-4 acting locally?)	220
IL-4R (soluble form)	Murine metallothionien (MT) promoter complex murine soluble IL-4R	Unknown	Normal numbers of T and B cells, normal levels of IgE and IgG1, deficiency in cardiac allograft rejection (IL-4 role in development of alloreactivity?)	221
IL-5	Human CD2 promoter — murine IL-5	T-cells	Eosinophilia, massive eosinophil infiltration, but mice otherwise normal	222,223
	MT promoter — murine IL-5	Blood, liver, spleen, kidney and bone marrow	Eosinophilia plus massive tissue and organ infiltration, more CD5$^+$ B cells in spleen, autoantibodies	224
IL-6	Human Ig H chain promoter/enhancer — human IL-6	B-cells	IgG1 plasmacytosis, infiltration of plasma cells into lung, spleen and kidney. Generation of plasmacytomas when transgene introduced into Balb/c genetic background	84,225, 226
	Human keratin (K14) promoter — murine IL-6	Skin, tongue	Growth retardation and poor hair growth, epidermal scaliness (tail, paws) but no changes typical of psoriasis (IL-6 stimulates keratinocyte proliferation *in vitro*)	227
IL-7	Murine IL-7	Lymphoid organs	Increased numbers of both B- and T-cells	228
	Ig heavy (H) chain promoter — murine IL-7	Bone marrow, lymph nodes, spleen, thymus, and skin	Reduction of CD4$^+$ CD8$^+$ (double-positive) thymocytes, progressive cutaneous disorder involving dermal lymphoid infiltrate, lympho-proliferative disorder leading to B- and T-cell lymphomas (physiologic role of IL-7 in T-cell development, but able to act as oncogene?)	229

TABLE 7
Cytokines and Cytokine Receptor Genes in Transgenic Mice (continued)

Cytokine (or Receptor)	Transgene	Site of Expression	Results	Ref.
IFN-α1	MT-promoter — murine IFN-α1	Only in testis	Male sterility	230
IFN-β	MT-promoter — human IFN-β	Unknown, IFN-β present in serum	Enhances viral resistance	231
IFN-β	MT-promoter — murine IFN-β	Constitutive in testis	Male sterility, but IFN-β induced IFN-α in this case	232,233
IFN-γ	MHC class II/insulin promoter — murine IFN-γ	Pancreatic islet	Diabetes, probably by cytotoxic lymphocyte (CTL) actions	234,235
TNF-α	Human TNF-α (deregulated)	Inducible in macrophages	Hair loss	
	Human β-globin gene and human TNF-α	All organs	Chronic inflammatory polyarthritis, prevented by neutralizing antibody to TNF-α	237
	Above and human CD2	T-cell (not inducible in macrophages)	Wasting syndrome and hair loss, widespread vascular thromboses, inducible only via TNF-RI	238
	Murine TNF-α- promoter and CAT reporter	Constitutive in thymus, inducible in macrophages	Physiological role in thymus development of T-cells?	239,240
			Induced by LPS in kidney, lung, pancreas, heart, and spleen, but not liver	
	KT14 promoter — human TNF-α	Keratinocytes (present in serum)	Wasting, no fat, thin skin, blackened gut. mRNA for IL-1α and IL-6 induced in keratinocytes. No arthritis	241

appearance of tumors to occur, e.g., TGF-α,[202–206] erbB2 (neu),[207] KGF,[208] IL-5,[222] IL-6,[225–227] and IL-7.[228,229] However, there are also likely to be more 'global' effects of cytokine expression, often resulting in disturbances to tissue morphogenesis and metabolic mechanisms (Table 7). Similar findings have resulted in mice engrafted with cells producing a particular cytokine.[236] For example, in mice overexpressing GM-CSF either because of transgenesis or repopulation of bone marrow with GM-CSF-producing hematopoietic cells, a rapidly fatal wasting (cachectic) state develops with multiple inflammatory foci in skeletal and cardiac muscle tissue.[211] Such animals also can develop a range of other pathologies including retinal detachment in the eye, gut congestion, hemorrhage, and fibrosis in the peritoneal and pleural cavities.[211–214] It is likely that the excess levels of GM-CSF overstimulate macrophages to produce other cytokines such as IL-1, TNF-α, and IFN-γ, also at high levels, which have proinflammatory and ultimately toxic actions. In the case of mice engrafted with LIF-producing cells, an even more extraordinary series of disease states has been shown to develop including loss of all body fat, hyperirritability, osteosclerosis with excess bone formation, compensatory extramedullary hematopoiesis in the liver and spleen, cirrhosis, elevated serum calcium levels, calcification of the liver, heart, and skeletal muscle, pancreatitis, thymus atrophy, etc.[242,243]

Some of these lesions could be caused indirectly; however, the receptors for LIF are present on many cell types besides macrophages, megakaryocytes, and hematopoietic progenitors, suggesting that some pathological processes are due to direct LIF action. Transgenic and engrafted mice thus provide experimental models for the preclinical testing of cytokines since they can predict whether a particular cytokine acts *in vivo* in a manner similar to that previously noted during *in vitro* studies, and also reveal the potential pathological consequences of excess cytokine levels (Table 7). They are, however, less likely to provide evidence for the physiological roles of cytokines. Nonetheless, controlled expression of a cytokine in a particular tissue may sometimes give an indication of its physiological involvement. For example, when TGF-β_1 cDNA was placed under the control of the regulatory elements of the pregnancy-responsive mouse whey-acidic protein gene, lactation-deficient females of transgenic lines resulted.[209] This suggests that TGF-β_1 has a role in the morphogenesis and function of the mammary gland. There are also possible roles for IL-4,[218,220] IL-7,[229] and TNF-α[239] in thymic regulation of evolving T-lymphocytes. Elevated constitutive expression of TNF-α (and IFN-γ) in the thymuses of Down's syndrome (trisomy 21) children also appears to be associated with abnormal thymic anatomy and thymocyte inactivation, suggesting its involvement at low constitutive levels in normal thymic development and function.[244]

Studies using novel retrovirus cytokine gene transfer systems have yielded similar but occasionally distinct results in the transgenic mice experiments. For example, infection of murine bone marrow cells with a recombinant retrovirus expressing IL-4 led to constitutive long-term expression of IL-4 with consequent changes to T-cell development and Ig synthesis; levels of serum IgG1 and, with antigenic stimulation, IgE, were increased[245] (see Table 7 for transgene experiments and comparison of results). By contrast, the case of infection of bone marrow cells with an IL-6 expressing recombinant retrovirus, a fatal (transplantable) myeloproliferative disease with massive neutrophil infiltration of lungs, liver, and lymph nodes developed within four weeks of engraftment, but unlike the IL-6 transgene experiments (Table 7) no plasmacytosis occurred.[246] Such differences suggest the importance of the tissue and cellular localization of cytokine expression, together with quantitative and temporal elements, on the outcome of the experiment. Furthermore, other reports suggest IL-6 has a role in angiogenesis[247] and in acute phase protein synthesis.[248] So, does IL-6 have a unique physiological role or is it physiologically promiscuous depending upon context?

III. CONCLUSIONS

Cytokine biology has proved itself to be undeniably complex, being underscored by biochemical redundancy and pleiotropic activities. The translation of cytokine *in vitro* effects to the *in vivo* situation has not proved to be straightforward for most cytokines because of the numerous interactions, both among cytokines themselves and cytokines and other biological effector molecules, that are possible. While some biological activities shown *in vitro* can be demonstrated *in vivo,* suggesting potential physiological roles, there remain questions about the type(s) of expressing and responding cells, sites, levels and duration of expression, developmental stages, and genetic background which complicate matters and lead to 'unsafe' interpretations. The advent of advanced molecular genetics, e.g., knock-out mice, has given some insight into whether or not cytokines are critically involved in embryogenesis; so far, in most cases it would appear that they are not. However, many cytokines seem to emerge as important mediators of cellular responses after birth. If the regulation of hematopoiesis and the development of a fully competent, activatable immune system are perceived as physiological processes, then cytokines are crucial mediators of these and perhaps they should stop being thought of as defense mediators outside the normal physiology of the host. Whether particular cytokines act alone or together in one or more physiological processes remains an open question. It is clear that the loss of a cytokine or its receptor will in most cases have some effect that, despite biochemical and biological redundancy among cytokines and their receptors, is frequently not entirely compensated for and is manifested by susceptibility to infection by pathogens or have other pathological consequences which can lead to lethality (Table 5). It is therefore presumed most cytokines are required and have one or more physiological roles in an integrated system of intercellular communication and interaction, the integrity of which is essential for survival. The present fragmentary evidence for defined physiological roles of most cytokines casts a shadow of uncertainty on where the clinical usefulness of these powerful biological molecules will lie.

ACKNOWLEDGMENT

I am indebted to Miss Deborah Kirk for the excellent typing of this chapter.

REFERENCES

1. **Carpenter, G. and Zendegui, J. G.,** Epidermal growth factor, its receptor and related proteins. *Exp. Cell Res.,* 164, 1, 1986.
2. **Derynck, R.,** Transforming growth factor alpha. *Cell,* 54, 593, 1988.
3. **Shoyab, M., Plowman, G. D., McDonald, V. L., et al.,** Structure and function of human amphiregulin: a member of the epidermal growth factor family. *Science,* 243, 1074, 1989.
4. **Holmer, W. E., Sliwkowski, M. X., Akita, R. W., et al.,** Identification of heregulin, a specific activator of $p185^{erbB2}$. *Science,* 256, 1205, 1992.
5. **Peles, E., Bacus, S. S., Koski, R. A., et al.,** Isolation of the Neu/HER-2 stimulatory ligand: a 44 kDa glycoprotein that induces differentiation of mammary tumour cells. *Cell,* 69, 205, 1992.
6. **Sasada, R., Ono, Y., Taniyama, Y., et al.,** Cloning and expression of cDNA encoding human betacellulin, a new member of the EGF family. *Biochem. Biophys. Res. Commun.,* 190, 1173, 1993.

7. **Daughaday, W. H. and Rotwein, P.,** Insulin-like growth factors I and II. Peptide, messenger ribonucleic acid and gene structures, serum and tissue concentrations. *Endocrine Rev.,* 10, 68, 1989.
8. **Gilmour, R. S., Prosser, C. G., Fleet, I. R., et al.,** From animal to molecule: aspects of the biology of insulin-like growth factors. *Br. J. Cancer,* 58, Suppl. IX, 23, 1988.
9. **Heldin, C.-H.,** Structural and functional studies on platelet-derived growth factor. *EMBO J.,* 11, 4251, 1992.
10. **Ferrara, N., Houck, K., Jakeman, L., and Leung, D. W.,** Molecular and biological properties of the vascular and endothelial growth factor family of proteins. *Endocrine Rev.,* 13, 18, 1992.
11. **Asai, K., Nakanishi, K., Isobe, I., et al.,** Neurotrophic action of gliostatin on cortical neurons: identity of gliostatin and platelet-derived endothelial cell growth factor. *J. Biol. Chem.,* 267, 20311, 1992.
12. **Thomas, K. A. and Gimenez-Gallego, G.,** Fibroblast growth factors: broad spectrum mitogens with potent angiogenic activity. *Trends Biochem. Sci.,* 11, 81, 1986.
13. **Aaronson, S. A., Bottaro, D. P., Miki, T., et al.,** Keratinocyte growth factor. A fibroblast growth factor family member with unusual target cell specificity. *Ann. N.Y. Acad. Sci.,* 638, 62, 1991.
14. **Thomas, K. A.,** Transforming potential of fibroblast growth factor genes. *Trends Biochem. Sci.,* 13, 327, 1988.

15a. **Gherardi, E. and Stoker, M.,** Hepatocyte growth factor — scatter factor: mitogen, motogen, and met. *Cancer Cells,* 3, 227, 1991.

15b. **Nakamura, T.,** Structure and function of hepatocyte growth factor. *Prog. Growth Factor Res.,* 3, 67, 1991.

16. **Bano, M., Worland, P., Kidwell, W. R., et al.,** Receptor-induced phosphorylation by mammary-derived growth factor 1 in mammary epithelial cell lines. *J. Biol. Chem.,* 267, 10389, 1992.
17. **Massaguáe, J.,** The transforming growth factor-beta family. *Ann. Rev. Cell. Biol.,* 6, 597, 1990.
18. **Sporn, M. B. and Roberts, A. B.,** Transforming growth factor-β: recent progress and new challenges. *J. Cell Biol.,* 119, 1017, 1992.
19. **Burt, D. W.,** Evolutionary grouping of the transforming growth factor-β superfamily. *Biochem. Biophys. Res. Commun.,* 184, 590, 1992.
20. **Meager, A.,** *Cytokines.* Open University Press, Buckingham, U.K., 1990, chap. 2, pp. 55, 56.
21. **McPherron, A. C. and Lee, S.-J.,** GDF-3 and GDF-9: two new members of the transforming growth factor-β superfamily containing a novel pattern of cysteines. *J. Biol. Chem.,* 268, 3444, 1993.
22. **Bradshaw, R. A., Blundell, T. L., Lapatto, R., et al.,** Nerve growth factor revisited. *Trends Biochem. Sci.,* 18, 48, 1993.
23. **Davis, S., Aldrich, T. H., Valenzuela, D. M., et al.,** The receptor for ciliary neurotrophic factor. *Science,* 253, 59, 1991.
24. **Malik, N., Kallestad, J. C., Gunderson, N. L., et al.,** Molecular cloning, sequence analysis, and functional expression of a novel growth regulator, oncostatin M. *Mol. Cell. Biol.,* 9, 2847, 1989.
25. **Rose, T. M. and Bruce, G.,** Oncostatin M is a member of a cytokine family that includes leukaemia-inhibitory factor, granulocyte colony-stimulating factor, and interleukin-6. *Proc. Natl. Acad. Sci. U.S.A.,* 88, 8641, 1991.
26. **Hilton, D. J.,** LIF: lots of interesting functions. *Trends Biochem. Sci.,* 17, 72, 1992.
27. **Metcalf, D.,** Control of granulocytes and macrophages: molecular, cellular and clinical aspects. *Science,* 254, 529, 1991.
28. **Layton, J. E.,** Granulocyte colony-stimulating factor: structure, function and physiology. *Growth Factors,* 6, 179, 1992.
29. **Gasson, J. C.,** Molecular physiology of granulocyte-macrophage colony-stimulating factor. *Blood,* 77, 1131, 1991.
30. **Ruef, C. and Coleman, D. L.,** Granulocyte-macrophage colony-stimulating factor: pleiotropic cytokine with potential clinical usefulness. *Rev. Infect. Dis.,* 12, 41, 1990.
31. **Rettenmier, C. W., Roussel, M. F., and Sherr, C. J.,** The colony-stimulating factor-1 (CSF-1) receptor (c-fms, proto-oncogene product) and its ligand. *J. Cell Sci.,* Suppl. 9, 27, 1988.
32. **Witte, O. N.,** Steel locus defines new multipotent growth factor. *Cell,* 63, 5, 1990.

33. **Huang, E., Nocka, K., Beyer, D., et al.,** The hematopoietic growth factor KL is encoded by the Steel locus and is the ligand of the c-kit receptor, the gene product of the W locus. *Cell,* 63, 225, 1990.
34. **Moore, M. A. S.,** Clinical applications of positive and negative hematopoietic stem cell regulators. *Blood,* 78, 1, 1991.
35. **Koury, M. J. and Bondurant, M. C.,** The molecular mechanism of erythropoietin action. *Eur. J. Biochem.,* 210, 649, 1992.
36. **Jelkmann, W.,** Erythropoietin: structure, control of production and function. *Physiol. Rev.,* 72, 449, 1992.
37. **Di Giovine, F. S. and Duff, G. W.,** Interleukin-1: the first interleukin. *Immunol. Today,* 11, 13, 1990.
38. **Dinarello, C. A.,** Interleukin-1 and its biologically related cytokines. In *Advances in Immunology 44,* Dixon, F. J., Ed., Academic Press, San Diego, 1989, p. 153.
39. **Arend, W. P.,** Interleukin-1 receptor antagonist. A new member of the interleukin-1 family. *J. Clin. Invest.,* 5, 1445, 1991.
40. **Smith, K. A., Ed.,** *Interleukin-2.* Academic Press, San Diego, 1988, chap. 1–3.
41. **Schrader, J. W., Ed.,** *Interleukin-3: the Panspecific Hemopoietin. Lymphokines 15.* Academic Press, New York, 1988.
42. **Möller, G., Ed.,** *IL-4 and IL-5: Biology and Genetics.* chap. beginning p. 5, 77, 137, 189 for IL-4: chap. beginning p. 29, 107, 137, 189 for IL-5. *Immunol. Rev.,* 102, 1988.
43. **Boulay, J.-L. and Paul, W. E.,** The interleukin-4 family of lymphokines. *Curr. Opinion Immunol.,* 4, 294, 1992.
44. **Sanderson, C. J.,** Interleukin-5, eosinophils and disease. *Blood,* 79, 3101, 1992.
45. **Le, J. and Vilck, J.,** Interleukin-6: A multifunctional cytokine regulating immune reactions and the acute phase protein response. *Lab. Invest.,* 61, 588, 1989.
46. **Goodwin, R. G., Lupton, S., Schmierer, A., et al.,** Human interleukin-7: molecular cloning and growth factor activity on human and murine B-lineage cells. *Proc. Natl. Acad. Sci. U.S.A.,* 86, 302, 1989.
47. **Rich, B. E., Campos-Torres, J., Tepper, R. I., et al.,** Cutaneous lymphoproliferation and lymphomas in interleukin-7 transgenic mice. *J. Exp. Med.,* 177, 305, 1993.
48. **Sherry, B. and Cerami, A.,** Small cytokine superfamily. *Curr. Opinion Immunol.,* 3, 56, 1991.
49. **Schall, T. J.,** Biology of the RANTES/SIS cytokine family. *Cytokine,* 3, 165, 1991.
50. **Yang, Y.-C., Ricciardi, S., Ciarletta, A., et al.,** Expression cloning of a cDNA encoding a novel human hematopoietic growth factor: human homologue of murine T-cell growth factor P40. *Blood,* 74, 1880, 1989.
51. **Houssiau, F. A., Renauld, J.-C., Fibbe, W. E., et al.,** IL-2 dependence of IL-9 expression in human T lymphocytes. *J. Immunol.,* 148, 3147, 1992.
52. **De Waal Malefyt, R., Yssel, H., Roncarolo, M.-G., et al.,** Interleukin-10. *Curr. Opinion Immunol.,* 4, 314, 1992.
53. **Howard, M. and O'Garra, A.,** Biological properties of interleukin-10. *Immunol. Today,* 13, 198, 1992.
54. **Paul, S. R., Bennett, F., Calvetti, J. A., et al.,** Molecular cloning of a cDNA encoding interleukin-11, a stromal cell-derived lymphopoietic and hematopoietic cytokine. *Proc. Natl. Acad. Sci. U.S.A.,* 87, 7512, 1990.
55. **Quesniaux, V. F. J., Mayer, P., Liehl, E., et al.,** Review of a novel hematopoietic cytokine, interleukin-11. *Int. Rev. Exp. Pathol.,* 34, 205, 1993.
56. **Gately, M. K., Wolitzky, A. G., Quinn, P. M., and Chizzonite, R.,** Regulation of human cytolytic lymphocyte responses by interleukin-12. *Cell Immunol.,* 143, 127, 1992.
57. **Scott, P.,** IL-12: initiation cytokine for cell-mediated immunity. *Science,* 260, 496, 1993.
58. **Minty, A., Chalon, P., Derocq, J.-M., et al.,** Interleukin-13 is a new human lymphokine regulating inflammatory and immune responses. *Nature,* 362, 248, 1993.
59. **Camussi, G., Albano, E., Tetta, C., and Bussolino, F.,** The molecular action of tumour necrosis factor-α. *Eur. J. Biochem.,* 202, 3, 1991.
60. **Jäättelä, M.,** Biological activities and mechanisms of action of tumour necrosis factor-α/cachectin. *Lab. Invest.,* 64, 724, 1991.

61. **Beutler, B., Ed.,** *Tumour Necrosis Factors: The Molecules and Their Emerging Role in Medicine.* Raven Press, New York, 1992.
62. **Ruddle, N. H.,** Tumour necrosis factor (TNF-α) and lymphotoxin (TNF-β). *Curr. Opinion Immunol.,* 4, 327, 1992.
63. **Browning, J. L., Ngam-ek, A., Lawton, P., et al.,** Lymphotoxin β, a novel member of the TNF family that forms a heteromeric complex with lymphotoxin on the cell surface. *Cell,* 72, 847, 1993.
64. **Hollenbaugh, D., Grosmaire, L. S., Kullas, C. D., et al.,** The human T cell antigen gp39, a member of the TNF gene family, is a ligand for the CD40 receptor: expression of a soluble form of gp39 with B cell co-stimulatory activity. *EMBO J.,* 11, 4313, 1992.
65. **De Maeyer, E. and De Maeyer-Guignard, J., Eds.,** *Interferons and Other Regulatory Cytokines.* John Wiley & Sons, New York, 1988.
66. **Meager, A.,** Antiviral, antimicrobial and antitumour cytokines. In: *Cytokines.* John Wiley & Sons, Chichester, U.K., 1990, chap. 6.
67. **Sen, G. C. and Lengyel, P.,** The interferon system: a bird's eye view of its biochemistry. *J. Biol. Chem.,* 267, 5017, 1992.
68. **Roberts, R. M., Cross, J. C., and Leaman, D. W.,** Interferons as hormones of pregnancy. *Endocrine Rev.,* 13, 432, 1992.
69. **Meager, A.,** Mediators of immune responses and inflammatory reactions. In: *Cytokines.* John Wiley & Sons, Chichester, U.K., 1990, chap. 5.
70. **De Maeyer, E. and De Maeyer-Guignard, J.,** Interferon-γ. *Curr. Opinion Immunol.,* 4, 321, 1992.
71. **Weiser, W. Y., Temple, P. A., Witek-Giannotti, J. S., et al.,** Molecular cloning of a cDNA encoding a human macrophage migration inhibitory factor. *Proc. Natl. Acad. Sci. U.S.A.,* 86, 7522, 1989.
72. **Hunkapiller, T. and Hood, L.,** Diversity of the immunoglobulin gene superfamily. *Adv. Immunobiol.,* 44, 1, 1989.
73. **Cosman, D., Lyman, S. D., Idzerda, R. L., et al.,** A new cytokine superfamily. *Trends Biochem. Sci.,* 15, 265, 1990.
74. **Gillis, S.,** Cytokine receptors. *Curr. Opinion Immunol.,* 3, 315, 1991.
75. **Foxwell, B. M. J., Barrett, K., and Feldmann, M.,** Cytokine receptors: structure and signal transduction. *Clin. Exp. Immunol.,* 90, 161, 1992.
76. **Mallett, S. and Baclay, A. N.,** A new superfamily of cell surface proteins related to the nerve growth factor receptor. *Immunol. Today,* 12, 220, 1991.
77. **Murphy, P. M., Özçelik, T., Kenney, R. T., et al.,** A structural homologue of the N-formyl peptide receptor: characterisation and chromosome mapping of a peptide chemoattractant receptor family. *J. Biol. Chem.,* 267, 7637, 1992.
78. **Neote, K., DiGregorio, D., Mak, J. Y., et al.,** Molecular cloning, functional expression and signalling characteristics of a C-C chemokine receptor. *Cell,* 72, 415, 1993.
79. **Meager, A.,** Cell proliferation, growth factors, and oncogenes. In: *Cytokines.* John Wiley & Sons, Chichester, U.K., 1990, chap. 2.
80. **Olsson, I., Gullberg, U., Lantz, M., and Richter, J.,** The receptors for regulatory molecules of hematopoiesis. *Eur. J. Haematol.,* 48, 1, 1992.
81. **Robinson, J.,** Multiple receptors found for the growing FGF family. *Trends Pharmacol. Sci.,* 12, 123, 1991.
82. **Williams, L. T.,** Signal transduction by the platelet-derived growth factor receptor. *Science,* 243, 1564, 1989.
83. **Sherr, C. J.,** Colony-stimulating factor-1 receptor. *Blood,* 75, 1, 1990.
84. **Kishimoto, T., Akira, S., and Taga, T.,** Interleukin-6 and its receptor: a paradigm for cytokines. *Science,* 258, 593, 1992.
85. **Miyajima, A., Hara, T., and Kitamura, T.,** Common subunits of cytokine receptors and the functional redundancy of cytokines. *Trends Biochem. Sci.,* 17, 378, 1992.
86. **Chizzonite, R., Truitt, T., Killan, P. L., et al.,** Two high-affinity interleukin-1 receptors represent separate gene products. *Proc. Natl. Acad. Sci. U.S.A.,* 86, 8029, 1989.
87. **Merberg, D. M., Wolf, S. F., and Clark, S. C.,** Sequence similarity between NKSF and the IL-6/G-CSF family. *Immunol. Today,* 13, 17, 1992.

88. **Hanazono, T., Chiba, S., Sasaki, K., et al.,** c-fps/fes Protein-tyrosine kinase is implicated in a signalling pathway triggered by granulocyte-macrophage colony-stimulating factor and interleukin-3. *EMBO J.,* 12, 1641, 1993.
89. **Waldmann, T. A.,** The interleukin-2 receptor. *J. Biol. Chem.,* 266, 2681, 1992.
90. **Minami, T., Kono, T., Yamada, K., and Taniguchi, T.,** The interleukin-2 receptors: insights into a complex signalling mechanism. *Biochim. Biophys. Acta,* 1114, 163, 1992.
91. **Renauld, J.-C., Druez, C., Kermouni, A., et al.,** Expression cloning of the murine and human interleukin-9 receptor cDNAs. *Proc. Natl. Acad. Sci. U.S.A.,* 89, 5690, 1992.
92. **Bazan, J. F.,** Shared architecture of hormone binding domains in Type I and II interferon receptors. *Cell,* 61, 753, 1990.
93. **Bazan, J. F.,** Hemopoietic receptors and helical cytokines. *Immunol. Today,* 11, 350, 1990.
94. **Klein, R., Jing, S., Nanduri, V., et al.,** The trk proto-oncogene encodes a receptor for nerve growth factor. *Cell,* 65, 189, 1991.
95. **Sprang, S. R.,** The divergent receptors for TNF. *Trends Biochem. Sci.,* 15, 366, 1990.
96. **Tartaglia, L. A. and Goeddel, D. V.,** Two TNF receptors. *Immunol. Today,* 13, 151, 1992.
97. **Plowman, G. D., Whitney, G. S., Neubauer, M. G., et al.,** Molecular cloning and expression of an additional epidermal growth factor receptor-related gene. *Proc. Natl. Acad. Sci. U.S.A.,* 87, 4905, 1990.
98. **Millauer, B., Wizigmann-Voor, S., Schuürch, H., et al.,** High affinity VEGF binding and developmental expression suggest Flk-1 as a major regulator of vasculogenesis and angiogenesis. *Cell,* 72, 835, 1993.
99. **Park, M.,** Lonesome receptors find their mates. *Curr. Biol.,* 1, 248, 1991.
100. **Lin, H. Y., Wang, X.-F., Ng-Eaton, E., et al.,** Expression cloning of the TGF-β type II receptor, a functional transmembrane serine/threonine kinase. *Cell,* 68, 775, 1992.
101. **Wrana, J. L., Attisano, L, Cáarcamo, J., et al.,** TGF-β signals through a heteromeric protein kinase receptor complex. *Cell,* 71, 1003, 1992.
102. **Milburn, M. V., Hassell, A. M., Lambert, M. H., et al.,** A novel dimer configuration revealed by the crystal structure at 2.4 Å resolution of human interleukin-5. *Nature,* 363, 172, 1993.
103. **Nicola, N. A. and Metcalf, D.,** Subunit promiscuity among hemopoietic growth factor receptors. *Cell,* 67, 1, 1991.
104. **Lopez, A. F., Elliott, M. J., Woodcock, J., and Vadas, M. A.,** GM-CSF, IL-3, and IL-5: cross-competition on human hemopoietic cells. *Immunol. Today,* 13, 495, 1992.
105. **Schooltink, H., Stoyan, T., Roeb, E., et al.,** Ciliary neutrotrophic factor induces acute-phase protein expression in hepatocytes. *FEBS Lett.,* 314, 280, 1992.
106. **Kawade, Y.,** Cytokine network in analogy to language — a general view of interferon research from a distance. *J. Interferon Res.,* 10, 101, 1990.
107. **Nathan, C. and Sporn, M.,** Cytokines in context. *J. Cell Biol.,* 113, 981, 1991.
108. **Chao, M. V.,** Growth factors signalling: where is the specificity? *Cell,* 68, 995, 1992.
109. **Cantley, L. C., Augur, K. R., Carpenter, C., et al.,** Oncogenes and signal transduction. *Cell,* 64, 281, 1991.
110. **Metcalf, D.,** Hemopoietic regulators. *Trends Biochem. Sci.,* 17, 286, 1992.
111. **Burke, F., Naylor, M. S., Davies, B., and Balkwill, F.,** The cytokine wall chart. *Immunol. Today,* 114, 165, 1993.
112. **O'Keefe, E. J. and Pledger, W. J.,** A model of cell cycle control: sequential events regulated by growth factors. *Mol. Cell. Endocrinol.,* 31, 167, 1983.
113. **Nett, M. A., Cerretti, D. P., Berson, D. R., et al.,** Molecular cloning of the murine IL-1β converting enzyme cDNA. *J. Immunol.,* 149, 3254, 1992.
114. **Lyons, R. M., Gentry, L. E., Purchio, A. F., and Moses, H. L.,** Mechanism of activation of latent recombinant transforming growth factor β1 by plasmin. *J. Cell Biol.,* 110, 1361, 1990.
115. **Naldini, L., Tamagnone, L., Vigna, E., et al.,** Extracellular proteolytic cleavage by urokinase is required for activation of hepatocyte growth factor/scatter factor. *EMBO J.,* 11, 4825, 1992.
116. **Lo, D. C.,** A central role for ciliary neurotrophic factor? *Proc. Natl. Acad. Sci. U.S.A.,* 90, 2557, 1993.
117. **James, K.,** Interactions between cytokines and α_2-macroglobulin. *Immunol. Today,* 11, 163, 1990.
118. **Ruoslahti, E. and Yamaguchi, Y.,** Proteoglycans as modulators of growth factor activities. *Cell,* 64, 867, 1991.

119. **Gitay-Goren, H., Soker, S., Vlodavsky, I., and Neufeld, G.,** The binding of vascular endothelial growth factor to its receptor is dependent on cell surface-associated heparin-like molecules. *J. Biol. Chem.,* 267, 6093, 1992.
120. **Jackson, R. L., Busch, S. J., and Cardin, A. D.,** Glycosaminoglycans: molecular properties, protein interactions, and role in physiological processes. *Physiol. Rev.,* 71, 481, 1991.
121. **Bendtzen, K., Svenson, M., Jønsson, V., and Hippe, E.,** Autoantibodies to cytokines — friends or foes? *Immunol. Today,* 11, 167, 1990.
122. **Tanaka, Y., Adams, D. H., and Shaw, S.,** Proteoglycans on endothelial cells present adhesion-inducing cytokines to leukocytes. *Immunol. Today,* 14, 111, 1993.
123. **James, K., van den Haan, J., and Farmer, K.,** Preliminary studies on the interaction of TNF-α and IFN-γ with α_2-macroglobulin. *Immunol. Lett.,* 32, 49, 1992.
124. **Beck, G. and Habicht, G. S.,** Primitive cytokines: harbingers of vertebrate defence. *Immunol. Today,* 12, 180, 1991.
125. **Pazin, M. J. and Williams, L. T.,** Triggering signalling cascades by receptor tyrosine kinases. *Trends Biochem. Sci.,* 17, 374, 1992.
126. **Vairo, G. and Hamilton, J. A.,** Signalling through CSF receptors. *Immunol. Today,* 12, 362, 1991.
127. **Taniguchi, T. and Minami, Y.,** The IL-2/IL-2 receptor system: a current overview. *Cell,* 73, 5, 1993.
128. **Weinberg, K. and Parkman, R.,** Severe combined immunodeficiency due to a specific defect in the production of interleukin-2. *N. Engl. J. Med.,* 322, 1718, 1990.
129. **Arnaiz-Villena, A., Timón, M., Rodriguez-Gallego, C., et al.,** Human T-cell activation deficiencies. *Immunol. Today,* 13, 259, 1992.
130. **Noguchi, M., Yi, H., Rosenblatt, H. M., et al.,** Interleukin-2 receptor γ chain mutation results in X-linked severe combined immunodeficiency in humans. *Cell,* 73, 147, 1993.
131. **Noelle, R. J., Ledbetter, J. A., and Aruffo, A.,** CD40 and its ligand, an essential ligand-receptor pair for thymus-dependent B-cell activation. *Immunol. Today,* 13, 431, 1992.
132. **Castigli, E., Pahwa, R., Good, R. A., et al.,** Molecular basis of a multiple lymphokine deficiency in a patient with severe combined immunodeficiency. *Proc. Natl. Acad. Sci. U.S.A.,* 90, 4728, 1993.
133. **Fuleihan, R., Ramesh, N., Loh, R., et al.,** Defective expression of the CD40 ligand in X chromosome-linked immunoglobulin deficiency with normal or elevated IgM. *Proc. Natl. Acad. Sci. U.S.A.,* 90, 2170, 1993.
134. **Stephenson, D. A., Mercola, M., Anderson, E., et al.,** Platelet-derived growth factor receptor α-subunit gene (Pdgfra) is deleted in the mouse patch (Ph) mutation. *Proc. Natl. Acad. Sci. U.S.A.,* 88, 6, 1991.
135. **Flanagan, J. G. and Leder, P.,** The kit ligand: a cell surface molecule altered in steel mutant fibroblasts. *Cell,* 63, 185, 1990.
136. **Anderson, D. M., Lyman, S. D., Baird, A., et al.,** Molecular cloning of mast cell growth factor, a haematopoietin that is active in both membrane bound and soluble forms. *Cell,* 63, 235, 1990.
137. **Pahwa, R., Chatila, T., Pahwa, S., et al.,** Recombinant interleukin-2 therapy in severe combined immunodeficiency disease. *Proc. Natl. Acad. Sci. U.S.A.,* 86, 5069, 1989.
138. **Yoshida, H., Hayashi, S.-I., Kunisada, T., et al.,** The murine mutation osteopetrosis is in the coding region of the macrophage colony stimulating factor gene. *Nature,* 345, 442, 1990.
139. **Wiktor-Jedrzejczak, W., Bartocci, A., Ferrante, A. W., et al.,** Total absence of colony-stimulating factor 1 in the macrophage-deficient osteopetrotic (op/op) mouse. *Proc. Natl. Acad. Sci. U.S.A.,* 87, 4828, 1990.
140. **Felix, R., Cecchini, M. G., and Fleisch, H.,** Macrophage colony stimulating factor restores *in vivo* bone resorption in the op/op osteopetrotic mouse. *Endocrinology,* 127, 2592, 1990.
141. **Kodama, H., Yamasaki, A., Nose, M., et al.,** Congenital osteoclast deficiency in osteopetrotic (op/op) mice is cured by injections of macrophage colony-stimulating factor. *J. Exp. Med.,* 173, 269, 1991.
142. **Witmer-Pack, M. D., Hughes, D. A., Schuler, G., et al.,** Identification of macrophages and dendritic cells in the osteopetrotic (op/op) mouse. *J. Cell Sci.,* 104, 1021, 1993.
143. **Watanabe-Fukunaga, R., Brannan, C. I., Copeland, N. G., et al.,** Lymphoproliferation disorder in mice explained by defects in Fas antigen that mediates apoptosis. *Nature,* 356, 314, 1992.
144. **Koller, B. H. and Smithies, O.,** Altering genes in animals by gene targetting. *Annu. Rev. Immunol.,* 10, 705, 1992.

145. **Mann, G. B., Fowler, K. J., Gabriel, A., et al.,** Mice with a null mutation of the TGF-α gene have abnormal skin architecture, wavy hair and curly whiskers and often develop corneal inflammation. *Cell,* 73, 249, 1993.
146. **Luetteka, N. C., Qiu, T. H., Peiffer, R. L., et al.,** TGF-α deficiency results in hair follicle and eye abnormalities in targetted and waved-1 mice. *Cell,* 73, 263, 1993.
147. **Derynck, R.,** The physiology of transforming growth factor-α. *Adv. Cancer Res.,* 58, 27, 1992.
148. **Kollias, G. and Grosveld, F.,** The study of gene regulation in transgenic mice. In: *Transgenic Animals,* Grosveld, F. and Kollias, G., Eds., Academic Press, London, 1992, chap. 4.
149. **Shull, M. M., Ormsby, I., Kier, A. B., et al.,** Targeted disruption of the mouse transforming growth factor-β1 gene results in multifocal inflammatory disease. *Nature,* 359, 693, 1992.
150. **Kulkarni, A. B., Huh, C.-G., Becker, D., et al.,** Transforming growth factor-β1 null mutation in mice causes excessive inflammatory response and early death. *Proc. Natl. Acad. Sci. U.S.A.,* 90, 770, 1993.
151. **Schorle, H., Holtsche, T., Hünig, T., et al.,** Development and function of T cells in mice rendered interleukin-2 deficient by gene targetting. *Nature,* 352, 621, 1991.
152. **Mosmann, T. R. and Coffman, R. L.,** Th1 and Th2 cells: different patterns of lymphokine secretion lead to different functional properties. *Annu. Rev. Immunol.,* 7, 145, 1989.
153. **Mosmann, T. R. and Moore, K. W.,** The role of IL-10 in cross-regulation of Th1 and Th2 responses. *Immunoparasitol. Today,* 12, A49–A53, March 1991.
154. **Kühn, R., Rajewsky, K., and Müller, W.,** Generation and analysis of interleukin-4 deficient mice. *Science,* 254, 707, 1991.
155. **Kopf, M., LeGros, G., Bachmann, M., et al.,** Disruption of the murine IL-4 gene blocks Th2 cytokine responses. *Nature,* 362, 245, 1993.
156. **Korthäuer, U., Graf, D., Mages, H. W., et al.,** Defective expression of T-cell CD40 ligand causes X-linked immunodeficiency with hyper-IgM. *Nature,* 361, 539, 1993.
157. **DiSanto, J. P., Bonnefoy, J. Y., Gauchat, J. F., et al.,** CD40 ligand mutations in X-linked immunodeficiency with hyper-IgM. *Nature,* 361, 541, 1993.
158. **Border, W. A., Noble, N. A., Yamamoto, T., et al.,** Natural inhibitor of transforming growth factor-β protects against scarring in experimental kidney disease. *Nature,* 360, 361, 1992.
159. **Erikson, H. P.,** Gene knockouts of c-src, transforming growth factor-β1, and tenascin suggest superfluous, non-functional expression of proteins. *J. Cell Biol.,* 120, 1079, 1993.
160. **Barnard, J. A., Lyons, R. M., and Moses, H. L.,** The cell biology of transforming growth factor β. *Biochim. Biophys. Acta,* 1032, 79, 1990.
161. **Meager, A.,** Assays for transforming growth factor β. *J. Immunol. Methods,* 141, 1, 1991.
162. **Romagnani, S.,** Induction of Th1 and Th2 responses: a key role for the 'natural' immune response? *Immunol. Today,* 13, 379, 1992.
163. **Border, W. A. and Ruoslahti, E.,** Transforming growth factor-β in disease: the dark side of tissue repair. *J. Clin. Invest.,* 90, 1, 1992.
164. **Geha, R. S.,** Atopic allergy and other hypersensitivities. *Curr. Opinions Immunol.,* 4, 779, 1992.
165. **Ebner, R., Chen, R.-H., Shum, L., et al.,** Cloning of a type I TGF-β receptor and its effect on TGF-β binding to the type II receptor. *Science,* 260, 1344, 1993.
166. **Gauchat, J.-F., Aubry, J.-P., Mazzei, G., et al.,** Human CD40-ligand: molecular cloning, cellular distribution and regulation of expression by factors controlling IgE production. *FEBS Lett.,* 315, 259, 1993.
167. **Dalton, D. K., Pitts-Meek, S., Keshav, S., et al.,** Multiple defects of immune cell function in mice with disrupted interferon-γ gene. *Science,* 259, 1739, 1993.
168. **Huang, S., Hendriks, W., Althage, A., et al.,** Immune response in mice that lack the interferon-γ receptor. *Science,* 259, 1742, 1993.
169. **Galli, S. J., Tsai, M., and Wershil, B. K.,** The c-kit receptor, stem cell factor and mast cells: what each is teaching us about the others. *Am. J. Pathol.,* 142, 965, 1993.
170. **Escary, J.-L., Perreau, J., Duménil, D., et al.,** Leukaemia inhibitory factor is necessary for maintenance of haemopoietic stem cells and thymocyte stimulation. *Nature,* 363, 361, 1993.
171. **Stewart, C. L., Kaspar, P., Brunet, L. J., et al.,** Blastocyst implantation depends on maternal expression of leukaemia inhibitory factor. *Nature,* 359, 76, 1992.

172. **Punnonen, J., Aversa, G., Cocks, B. G., et al.,** Interleukin-13 induces interleukin-4-dependent IgG4 and IgE synthesis and CD23 expression by human B cells. *Proc. Natl. Acad. Sci. U.S.A.,* 90, 3730, 1993.
173. **McKenzie, A. N. J., Culpepper, J. A., deWaal Malefyt, R., et al.,** Interleukin-13, a T-cell-derived cytokine that regulates human monocyte and B-cell function. *Proc. Natl. Acad. Sci. U.S.A.,* 90, 3735, 1993.
174. **Cosman, D.,** The haemopoietin receptor superfamily. *Cytokine* 5, 95, 1993.
175. **Minasi, L. E., Kamogawa, Y., Carding, S., et al.,** The selective ablation of interleukin-2 producing cells isolated from transgenic mice. *J. Exp. Med.,* 177, 1451, 1993.
176. **Nishikawa, S., Kusakabe, M., Yoshinaga, K., et al.,** In utero manipulation of coat colour formation by a monoclonal anti-c-kit antibody: two distinct waves of c-kit-dependency during melanocyte development. *EMBO J.,* 10, 2111, 1991.
177. **De Kossodo, S., Grau, G. E., Daneva, T., et al.,** Tumour necrosis factor α is involved in mouse growth and lymphoid tissue development. *J. Exp. Med.,* 176, 1259, 1992.
178. **Ishida, H., Hastings, R., Kearney, J., and Howard, M.,** Continuous anti-interleukin-10 antibody administration depletes mice of Ly-1 B cells but not conventional B cells. *J. Exp. Med.,* 175, 1213, 1992.
179. **Le, J. and Vilçek, J.,** Tumour necrosis factor and interleukin-1: cytokines with multiple overlapping biological activities. *Lab. Invest.,* 56, 234, 1987.
180. **Standiford, T. J. and Strieter, R. M.,** TNF and IL-1 in sepsis: good cytokines gone bad. *J. Lab. Clin. Med.,* 120, 180, 1992.
181. **Weinberg, J. R., Boyle, P., Meager, A., and Guz, A.,** Lipopolysaccharide, tumour necrosis factor and interleukin-1 interact to cause hypotension. *J. Lab. Clin. Med.,* 120, 205, 1992.
182. **Beutler, B.,** TNF in pathophysiology-biosynthetic regulation. *J. Invest. Dermatol.,* 95, 581, 1990.
183. **Rothwell, N. J.,** The endocrine significance of cytokines. *J. Endocrinol.,* 128, 171, 1991.
184. **Porter, A. G.,** The prospects for therapy with tumour necrosis factors and their antagonists. *Trends Biotechnol.,* 9, 158, 1991.
185. **Bromberg, J. S., Chavin, K. D., and Kunkel, S. L.,** Anti-tumour necrosis factor antibodies suppress cell-mediated immunity *in vivo. J. Immunol.,* 148, 3412, 1992.
186. **Rogers, H. W., Sheehan, K., Brunt, L. M., et al.,** Interleukin-1 participates in the development of anti-Listeria responses in normal and SCID mice. *Proc. Natl. Acad. Sci. U.S.A.,* 89, 1011, 1992.
187. **Gershenwald, J. E., Fong, Y., Fahey, T. J. III, et al.,** Interleukin-1 receptor blockade attenuates the host inflammatory response. *Proc. Natl. Acad. Sci. U.S.A.,* 87, 4966, 1990.
188. **McIntyre, K. W., Stepan, G. J., Kolinsky, K. D., et al.,** Inhibition of interleukin-1 (IL-1) binding and bioactivity *in vitro* and modulation of acute inflammation *in vivo* by IL-1 receptor antagonist and anti-IL-1 receptor monoclonal antibody. *J. Exp. Med.,* 173, 931, 1991.
189. **Evans, T., Carpenter, A., Silva, A., and Cohen, J.,** Differential effects of monoclonal antibodies to tumour necrosis factor alpha and gamma interferon on induction of hepatic nitric oxide synthase in experimental Gram-negative sepsis. *Infect. Immunol.,* 60, 4133, 1992.
190. **Fanslow, W. C., Sims, J. E., Sassenfled, H., et al.,** Regulation of alloreactivity *in vivo* by a soluble form of the interleukin-1 receptor. *Science,* 248, 739, 1990.
191. **Barinaga, M.,** Viruses launch their own 'star wars'. *Science,* 258, 1730, 1992.
192. **Gooding, L. R.,** Virus proteins that counteract host immune defences. *Cell,* 71, 5, 1992.
193. **Murphy, P. M.,** Molecular mimicry and the generation of host defence protein diversity. *Cell,* 72, 823, 1993.
194. **Spriggs, M. K., Hruby, D. E., Maliszewski, C. R., et al.,** Vaccinia and cowpox viruses encode a novel secreted interleukin-1 binding protein. *Cell,* 71, 145, 1992.
195. **Alcami, A. and Smith, G. L.,** A soluble receptor for interleukin-1β encoded by vaccinia virus: a novel mechanism of virus modulation of the host response to infection. *Cell,* 71, 153, 1992.
196. **Upton, C., Mossman, K., and McFadden, G.,** Encoding of a homolog of the IFN-γ receptor by myxoma virus. *Science,* 258, 1369, 1992.
197. **Cerretti, D. P., Kozlosky, C. J., Mosley, B., et al.,** Molecular cloning of the interleukin-1β converting enzyme. *Science,* 256, 97, 1992.
198. **Ray, C. A., Black, R. A., Kronheim, S. R., et al.,** Viral inhibition of inflammation: cowpox virus encodes an inhibitor of the interleukin-1β converting enzyme. *Cell,* 69, 597, 1992.

199. **Cooper, G. M.,** Retroviral oncogenes. In: *Oncogenes,* Jones and Bartlett, Boston, 1990, chap. 3.
200. **Bocci, V.,** Roles of interferon produced in physiological conditions. A speculative review. *Immunology,* 61, 1, 1988.
201. **Hunt, J. S.,** Immunobiology of pregnancy. *Curr. Opinion Immunol.,* 4, 591, 1992.
202. **Vasar, R. and Fuchs, E.,** Transgenic mice provide new insights into the role of TGF-alpha during epidermal development and differentiation. *Genes Dev.,* 5, 714, 1991.
203. **Halter, S. A., Dempsey, P., Matsiu, Y., et al.,** Distinctive patterns of hyperplasia in transgenic mice with mouse mammary tumour virus transforming growth factor-alpha. Characterisation of mammary gland and skin proliferations. *Am. J. Pathol.,* 140, 1131, 1992.
204. **Sandgron, E. P., Luetteke, N. C., Palmiter, R. D., et al.,** Overexpression of TGF-α in transgenic mice: induction of epithelial hyperplasia, pancreatic metaplasia and carcinoma of the breast. *Cell,* 61, 1121, 1990.
205. **Jhappen, C., Stahle, C., Harkins, R. N., et al.,** TGF-α overexpression in transgenic mice induces liver neoplasia and abnormal development of the mammary gland and pancreas. *Cell,* 61, 1137, 1990.
206. **Matsui, Y., Halter, S. A., Holt, J. T., et al.,** Development of mammary hyperplasia and neoplasia in MMTV-TGF-α transgenic mice. *Cell,* 61, 1147, 1990.
207. **Guy, C. T., Webster, M. A., Schaller, M., et al.,** Expression of the neu protooncogene in the mammary epithelium of transgenic mice induces metastatic disease. *Proc. Natl. Acad. Sci. U.S.A.,* 89, 10578, 1992.
208. **Guo, L., Yu, Q.-C., and Fuchs, E.,** Targetting expression of keratinocyte growth factor to keratinocytes elicits striking changes in epithelial differentiation in transgenic mice. *EMBO J.,* 12, 973, 1993.
209. **Jhappen, C., Geiser, A. W., Kordon, E. C., et al.,** Targetting expression of transforming growth factor β1 transgene to the pregnant mammary gland inhibits alveolar development and lactation. *EMBO J.,* 12, 1835, 1993.
210. **Semenza, G. L., Traystman, M. D., Gearhart, J. D., and Antonarakis, S. E.,** Polycythemia in transgenic mice expressing the human erythropoietin gene. *Proc. Natl. Acad. Sci. U.S.A.,* 86, 2301, 1989.
211. **Lang, G. A., Metcalf, D., Cuthbertson, R. A., et al.,** Transgenic mice expressing a haemopoietic growth factor gene (GM-CSF) develop accumulations of macrophages, blindness, and a fatal syndrome of tissue damage. *Cell,* 51, 675, 1987.
212. **Gearing, A. J., Metcalf, D., Moore, J. G., and Nicola, N. A.,** Elevated levels of GM-CSF and IL-1 in the serum, peritoneal and pleural cavities of GM-CSF transgenic mice. *Immunology,* 67, 216, 1989.
213. **Cuthbertson, R. A., Lang, R. A., and Coghlan, J. P.,** Macrophage products IL-1 alpha, TNF alpha and bFGF may mediate multiple cytopathic effects in the developing eyes of GM-CSF transgenic mice. *Exp. Eye Res.,* 51, 335, 1990.
214. **Elliott, M. J., Strasser, A., and Metcalf, D.,** Selective upregulation of macrophage function in granulocyte-macrophage colony-stimulating factor transgenic mice. *J. Immunol.,* 147, 2957, 1991.
215. **Ishida, Y., Nishi, M., Taguchi, O., et al.,** Effects of the deregulated expression of human interleukin-2 in transgenic mice. *Int. Immunol.,* 1, 113, 1989.
216. **Ishida, Y., Nishi, M., Taguchi, O., et al.,** Expansion of natural killer cells but not cells in human interleukin-2/interleukin-2 receptor (Tac) transgenic mice. *J. Exp. Med.,* 170, 1103, 1989.
217. **Allison, J., Malcolm, L., Chosich, N., and Miller, J. F.,** Inflammation but not autoimmunity occurs in transgenic mice expressing constitutive levels of interleukin-2 in islet beta cells. *Eur. J. Immunol.,* 22, 1115, 1992.
218. **Tepper, R. I., Levinson, D. A., Stanger, B. Z., et al.,** IL-4 induces allergic-like inflammatory disease and alters T cell development in transgenic mice. *Cell,* 62, 457, 1990.
219. **Burstein, H. J., Tepper, R. I., Leder, P., and Abbas, K.,** Humoral immune functions in IL-4 transgenic mice. *J. Immunol.,* 147, 2950, 1991.
220. **Lewis, D. B., Yu, C. C., Forbush, K. A., et al.,** Interleukin-4 expressed *in situ* selectively alters thymocyte development. *J. Exp. Med.,* 173, 89, 1991.
221. **Maliszewski, C. R., Morrissey, P. J., Fanslow, W. C., et al.,** Delayed allograft rejection in mice transgenic for a soluble form of the IL-4 receptor. *Cell Immunol.,* 143, 434, 1992.

222. **Dent, L. A., Strath, M., Mellor, A. L., and Sanderson, C. J.,** Eosinophilia in transgenic mice expressing interleukin-5. *J. Exp. Med.,* 172, 1425, 1990.
223. **Hitoshi, Y., Yamaguchi, N., Korenaga, M., et al.,** *In vivo* administration of antibody to murine IL-5 receptor inhibits eosinophilia in IL-5 transgenic mice. *Int. Immunol.,* 3, 135, 1991.
224. **Tominaga, A., Takaki, S., Koyama, N., et al.,** Transgenic mice expressing a B cell growth and differentiation factor gene (interleukin-5) develop eosinophilia and autoantibody production. *J. Exp. Med.,* 173, 429, 1991.
225. **Suematsu, S., Matsuda, T., Aozasa, K., et al.,** IgG1 plasmacytosis in interleukin-6 transgenic mice. *Proc. Natl. Acad. Sci. U.S.A.,* 86, 7547, 1989.
226. **Suematsu, S., Matsusaka, T., Matsuda, T., et al.,** Generation of plasmacytomas with the chromosomal translocation τ (12;15) in interleukin-6 transgenic mice. *Proc. Natl. Acad. Sci. U.S.A.,* 89, 232, 1992.
227. **Turksen, K., Kupper, T., Degenstein, L., et al.,** Interleukin-6: insights to its function in skin by overexpression in transgenic mice. *Proc. Natl. Acad. Sci. U.S.A.,* 89, 5068, 1992.
228. **Samaridis, J., Casorati, G., Traunecker, A., et al.,** Development of lymphocytes in interleukin-7-transgenic mice. *Eur. J. Immunol.,* 21, 453, 1991.
229. **Rich, B. E., Campos-Torres, J., Tepper, R. I., et al.,** Cutaneous lymphoproliferation and lymphomas in interleukin-7 transgenic mice. *J. Exp. Med.,* 177, 305, 1993.
230. **Hekman, A. C., Trapman, J., Mulder, A. H., et al.,** Interferon expression in the testes of transgenic mice leads to sterility. *J. Biol. Chem.,* 263, 12151, 1988.
231. **Chen, X. Z., Yun, J. S., and Wagner, T. E.,** Enhanced viral resistance in transgenic mice expressing the human beta-1 interferon. *J. Virol.,* 62, 3883, 1988.
232. **Iwakura, Y., Asano, M., Nishimune, Y., and Kawade, Y.,** Male sterility of transgenic mice carrying exogenous mouse interferon-beta gene under the control of the metallothionein enhancer-promoter. *EMBO J.,* 7, 3757, 1988.
233. **Asano, M., Hayashi, M., Yoshida, E., et al.,** Induction of interferon-alpha by interferon-beta, but not of interferon-beta by interferon-alpha, in the mouse. *Virology,* 176, 30, 1990.
234. **Sarvetnick, N., Liggitt, D., Pitts, S. L., et al.,** Insulin-dependent diabetes mellitus induced in transgenic mice by ectopic expression of class II MHC and interferon gamma. *Cell,* 52, 773, 1988.
235. **Sarvetnick, N., Shizuru, J., Liggitt, D., et al.,** Loss of pancreatic islet tolerance induced by β-cell expression of interferon-γ. *Nature,* 346, 844, 1990.
236. **Metcalf, D.,** The consequences of excess levels of haemopoietic growth factors. *Br. J. Haematol.,* 75, 1, 1990.
237. **Keffer, J., Orobert, L., Caslaris, H., et al.,** Transgenic mice expressing human tumour necrosis factor α: a predictive genetic model of arthritis. *EMBO J.,* 10, 4025, 1991.
238. **Probert, L., Keffer, J., Corbella, P., et al.,** Wasting, ischaemia and lymphoid abnormalities in mice expressing T-cell targetted human tumour necrosis factor transgenes. *J. Immunol.,* 151, 1894, 1993.
239. **Giroir, B. P., Brown, T., and Beutler, B.,** Constitutive synthesis of tumour necrosis factor in the thymus. *Proc. Natl. Acad. Sci. U.S.A.,* 89, 4864, 1992.
240. **Giroir, B. P., Johnson, J. H., Brown, T., et al.,** The tissue distribution of tumour necrosis factor biosynthesis during endotoxaemia. *J. Clin. Invest.,* 90, 693, 1992.
241. **Cheng, J., Twksen, K., Yu, Q.-C., et al.,** Cachexia and graft-vs.-host-disease-type skin changes in keratin promoter-driven TNF-α transgenic mice. *Genes Dev.,* 6, 1444, 1992.
242. **Metcalf, D. and Gearing, D. P.,** Fatal syndrome in mice engrafted with cells producing high levels of the leukaemia inhibitory factor. *Proc. Natl. Acad. Sci. U.S.A.,* 86, 5948, 1989.
243. **Metcalf, D. and Gearing, D. P.,** A myelosclerotic syndrome in mice engrafted with cells producing high levels of leukaemia inhibitory factor (LIF). *Leukemia,* 3, 847, 1989.
244. **Murphy, M., Friend, D. S., Pike-Nobile, L., and Epstein, L. B.,** Tumour necrosis factor-α and IFN-γ expression in human thymus: location and overexpression in Down Syndrome (TRISOMY 21). *J. Immunol.,* 149, 2506, 1992.
245. **Chambers, C. A., Kang, J., and Hozumi, N.,** Long-term expression of IL-4 *in vivo* using retroviral-mediated gene transfer. *J. Immunol.,* 149, 2899, 1992.
246. **Hawley, R. G., Fong, A. Z. C., Burns, B. I., and Hawley, T. S.,** Transplantable myeloproliferative disease induced in mice by an interleukin-6 retrovirus. *J. Exp. Med.,* 176, 1149, 1992.

247. **Motro, B., Itin, A., Sachs, L., and Keshnet, E.,** Pattern of interleukin-6 gene expression *in vivo* suggests a role for this cytokine. *Proc. Natl. Acad. Sci. U.S.A.,* 87, 3092, 1990.
248. **Heinrich, P. C., Castell, J. V., and Andus, T.,** Interleukin-6 and the acute phase response. *Biochem. J.,* 265, 621, 1990.
249. **Matzuk, M. M., Finegold, M. J., Su, J.-G. J., et al.,** α-Inhibin is a tumour-suppressor gene with gonadal specificity in mice. *Nature,* 360, 313, 1992.
250. **Pfeffer, K., Matsuyama, T., Kündig, T. M., et al.,** Mice deficient for the 55kd tumour necrosis factor receptor are resistant to endotoxic shock, yet succumb to *L. monocytogenes* infection. *Cell,* 73, 457, 1993.
251. **Ishida, H., Hastings, R., Thompson-Snipes, L., and Howard, M.,** Modified immunological status of anti-IL-10 treated mice. *Cell. Immunol.,* 148, 371, 1993.
252. **Moore, K. W., O'Garra, A., de Waal Malefyt, R., et al.,** Interleukin-10. *Annu. Rev. Immunol.,* 11, 165, 1993.

Chapter 2

IMMUNOREGULATORY CYTOKINES

Francesco S. di Giovine, John B. Mee, and Gordon W. Duff

CONTENTS

0-8493-8381-1/96/$0.00+$.50

I. INTRODUCTION

Immune responses are based initially on the recruitment and activation of lymphocytes, followed by selective expansion and differentiation of lymphocyte clones in response to signals from cell surface antigen receptors and receptors for accessory molecules such as the cytokines.[1]

The identification of this role of cytokines started with the recognition of *in vitro* effects on lymphocytes mediated by soluble factors.[2] The "factors" were named descriptively, and terms like BAF (B-cell activating factor), LAF (lymphocyte activating factor), BCGF (B-cell growth factor), BCDF (B-cell differentiating factor), and TRF (T-cell replacing factor) were invented.[3] Often, these represented different actions of the same protein entity or the outcome of complex cytokine effects. Concomitantly, other families of cytokines emerged from other research fields, including the interferons, tumor necrosis factors, colony stimulating factors, and many others.[4]

More than 100 cytokines are now fully characterized. Those that are derived from leukocytes and act in inflammatory or immune responses are classified as "interleukins" when their primary structure is known.

Many cytokines are central to the function of the innate immune system. The induction of interleukin-1 (IL-1) and tumor necrosis factor (TNF) by microbial agents is antigen-independent and MHC unrestricted. Their powerful functions range from induction of adhesion molecules and chemotactic cytokines in the local environment to systemic actions such as fever, the acute-phase response in the liver, and activation of bone marrow.

In the adaptive immune response, humoral and cellular mechanisms are also regulated by cytokines. T-lymphocytes can be divided first into those with primitive antigen receptors composed of γ and δ chains (γ/δ T-cells) and those with more diverse antigen receptors including α and β chains (α/β T-cells). The latter can be subdivided, after thymic selection, into those that are CD8 positive (historically called "cytotoxic" T-cells) or CD4 positive (historically called "helper" T-cells). Helper T-cells recognize antigen incorporated in MHC class II molecules, while CD8 positive T-cells recognize antigen in a complex with MHC class I.

Following activation through the antigen receptor, helper T-cells (T_H) produce a large number of immunoregulatory cytokines. On the basis of the cytokines produced, Mosmann and colleagues at DNAX first reported[5] a subdivision of murine T_H cells into T_H1 and T_H2 types. T_H1 lymphocytes produce IL-2, IFN-γ, and TNF-β, while T_H2 cells produce IL-4, IL-5, IL-10, and IL-13 (see Table 1).

Because of the biological actions of these cytokines, activation of T_H1 cells would be expected to generate cell-mediated immune response with increased macrophage effector

TABLE 1
Cytokine Secretion and T-Cell Functional Subsets

	CTL	T_H1	T_H0	T_H2	T_Hp
IL-2	+/–	++	++	–	++
IFN-γ	++	++	++	–	–
TNF-β	+	++		–	
GM-CSF	+	++	++	+	–
TNF-α	+	++		+	
IL-3	+	++	++	++	–
IL-4	–	–	++	++	–
IL-5	–	–	++	++	–
IL-6	–	–		++	
IL-10	–	–	++	++	
IL-13	–	–		++	–

functions, CTL activation, and T-cell proliferation. This is an effective response to intracellular pathogens, but can also occur in type IV delayed-type hypersensitivity reactions. Activation of T_H2 cells, on the other hand, will potentiate eosinophil, mast cells, and B-cell antibody responses (IgE, IgG2a production and polyclonal stimulation) which are effective in the course of bacterial infection or against soluble toxins, but are also associated with types I, II, and III hypersensitivity reactions.

T_H1 and T_H2 subsets display a certain degree of cross regulation: T_H2-derived IL-4 and IL-10 inhibit T_H1 cytokine synthesis,[6] and IFN-γ from T_H1 cells inhibits IL-4-induced T_H2 proliferation[7] and IL-4-induced antibody class switching in B cells.[8] The mutually inhibitory nature of these T_H subsets results in predominantly cell-mediated or antibody-mediated immune responses *in vivo*.

In the human system, the basic principles of cross-regulation as seen in the mouse seem to be confirmed, although many T_H clones are of mixed phenotype and cytokine regulation appears to be dose and time dependent.[9] Studies in humans have confirmed that the phenotype, as defined in mouse T helper clones, represents an extreme stage of T-cell differentiation, i.e., the effector branch of the system. There is no doubt that earlier T_H phenotypes do exist, and represent different stages of T-cell differentiation.

A scheme of T-lymphocyte differentiation has been recently proposed (Figure 1). The resting, "preactivation" T-cell (T_HP) on first contact with antigen produces only IL-2 and becomes susceptible to further stimulation (T_H0 stage). In the presence of cytokines such as IL-12 from monocytes and IFN-γ from NK cells, the switch to T_H1 is more frequent (as in the case of a viral infection).[10] Cytokines such as IL-4, produced by T_H2 cells, basophils, and mast cells drive differentiation towards the T_H2 type and inhibit the development of T_H1 cells.[11]

This chapter describes the main characteristics of some of the cytokines that play a role in immunoregulation. The roles of other cytokines in the regulation of the immune response are reviewed in Chapters 6, 9, 14, and 15.

II. THE INTERLEUKIN-1 FAMILY

Interleukin-1 is one of the ancestral growth factors of the immune system, having evolved long before the subdivision of lymphocytes into T- and B-cells.[12] Initially described as a macrophage product, it is now evident that most cells can produce IL-1

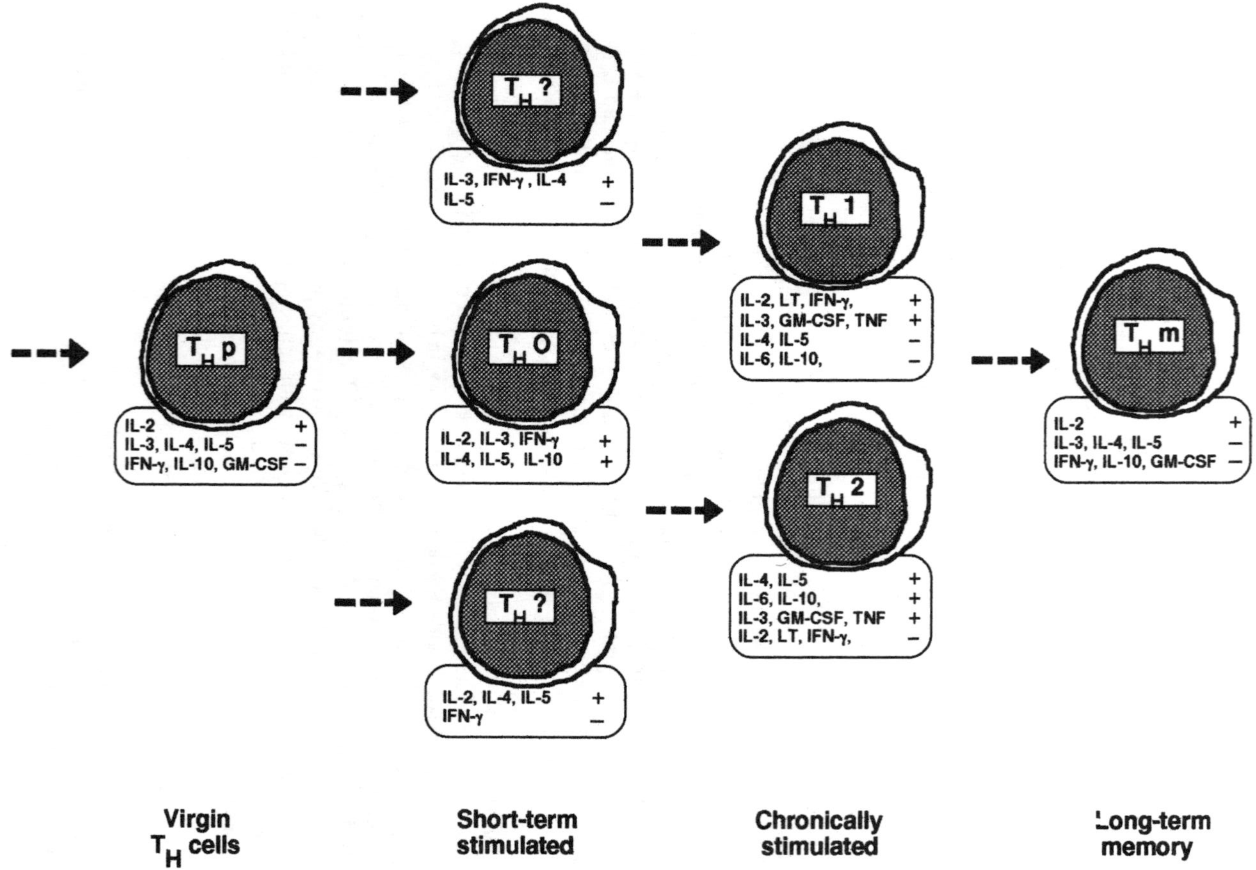

FIGURE 1. Relationship between the various cytokine secretion phenotypes. A scheme of the T-helper-cell functional subdivision is shown. Cytokines produced following activation are indicated in the box below each cell type. Resting cells ("virgin" T-cells and "memory" T-cells) only express IL-2 protein; all intermediate differentiation/activation stages have definite patterns of cytokine production. (Modified from Mosmann, T. R. and Moore, K. W., *Immunoparasit. Today,* A49-A53, 1991.)

TABLE 2
Interleukin-1

	IL-1α	IL-1β	IL-1RA
Protein size	17.5 kDa	17.3 kDa	17.0 kDa
Molecular form	159 aa (mature peptide)	153 aa (mature peptide)	152 aa (extracellular form)
Gene (location, structure)	2q13–21, 7 Exons	2q13–21, 7 Exons	2q14–21, 4 Exons
Source	Monocytes/ macrophages Endothelial cells Glial cells Keratinocytes Epithelial cells	Monocytes/ macrophages Endothelial cells Glial cells and many other cell types	Monocytes/ macrophages Endothelial cells Glial cells Keratinocytes
Stimuli	Microbial products TNF, GM-CSF, IL-2 Antigen presentation		Microbial products Immune complexes, GM-CSF, IL-2 Antigen presentation
Functions	Immunoregulation, with multiple actions on B- and T-cells via induction of IL-2, IL-4, IL-6, IL-8, and TNFα Fever, acute-phase response, hypotension, sleep, bone/cartilage resorption, hypotension, induction of PGE, collagenase and proteolytic enzymes and nitric oxide synthetase.		Inhibits biological activities of IL-1 by competing for binding to IL-1 receptors

after appropriate stimulation. IL-1 was initially described as lymphocyte activating factor (LAF)[13] because of its co-stimulation (with lectins) of T-cell proliferation mediated by T-cell growth factors such as IL-2 and IL-4. The other original description of IL-1 is as endogenous pyrogen (EP), the macrophage product that elicits fever *in vivo.*[14]

IL-1 has wide-ranging biological actions, mediating the local and systemic responses to injury and infection.[15] Interleukin-1 is dealt with in detail in the following chapter, and only its involvement in immune activation will be discussed here.

The IL-1 family is composed of at least six proteins produced by three genes (see Table 2). IL-1α and IL-1β are produced as propeptides of 31 to 33 kDa that are cleaved at the cell membrane to 17-kDa mature proteins. Of the precursor proteins, only pro-IL-1α is biologically active. IL-1 receptor antagonist (IL-1ra) is produced either as a secretory peptide with a leader sequence or as an intracellular form based on alternative first exons. Both IL-1ra proteins bind to the IL-1r but have no agonist activity. IL-1ra is discussed in detail in Chapter 9.

The IL-1α, IL-1β, and IL-1ra genes all lie within a 430-kb region on the long arm of human chromosome 2.[16]

Transcription of IL-1 is activated in human monocytes by bacterial agents, other cytokines (IL-1 itself; IFN-γ; IL-2; TNF-α), and other agents such as silica or urate crystals (mRNA is low or absent in resting cells). IL-1 agonists (IL-1α and IL-1β) induce transcription of many cytokines, such as IL-2, IL-4, IL-6, IL-8, and TNF-α. This explains the general immunopotentiating activities of IL-1. Downregulation of the IL-1 receptor, release of the soluble type II receptor, or predominance of IL-1ra, all limit the actions of IL-1, underlying the self-limitation of acute inflammation.

The IL-1 stimulation of IL-2 in T-cells seems to be mediated by nuclear factor AP-1 binding to the IL-2 promoter in T_H1 cells,[17] while stimulation of IL-6 production occurs via induction of NF-IL-6 and its binding to the IL-6 promoter.[18] Following antigen-TCR interaction, IL-2 is produced by T_H1 cells and IL-4 by T_H2 cells. Rapid induction of IL-6 and IL-4 also appears to be responsible for IL-1's extensive range of activities on B-cells (including proliferation and antibody production).

However, blocking IL-1 actions *in vitro* with anti-type I receptor antibodies[19] or high levels of IL-1ra[20] had no effect on antigen-driven T-cell activation. The relevance of the immunopotentiating role of IL-1 observed in most *in vitro* studies is therefore in some doubt, and the importance of *in vivo* studies will presumably be assessable in IL-1 knock-out mice.

Because of its radioprotective effects, particularly if administered with GM-CSF, IL-1 has been tested for recovery from radioactivity-produced lesions.[21] IL-1 has also been proposed as an adjuvant in vaccination, and an aid to WBC repopulation following chemotherapy. Antagonists of IL-1 find possible applications in the area of chronic inflammatory diseases (reviewed in Chapter 9).

III. INTERLEUKIN-2

In the mid-1970s tissue culture supernatants from human peripheral blood lymphocytes stimulated with mitogens were shown to support the growth of T-lymphocytes *in vitro*.[22] This activity was named T-cell growth factor (TCGF-I), and was associated with a protein of MW of 19 to 22 kDa by gel filtration and 14 to 16 kDa in SDS-PAGE, with a pI range from 6.5 to 8.3. Other transferrable activities had been described in activated T-cell supernatants, including thymocyte stimulation factor (TSF) and killer helper factor (KHF).

The molecular cloning of TCGF as IL-2[23] demonstrated that most of these activities could be attributed to IL-2 itself. The human IL-2 sequence codes for a 153-aa polypeptide, and cleavage of the 20-aa leader peptide produces a 133-aa protein of MW 15,420. Two of the three cysteines present in the protein (at 58 and 105) form a stable disulfide bond which is necessary for full biological activity. The apparent size heterogeneity is explained by variable glycosylation at a threonine residue at position 3.

IL-2 is coded by a single gene on human chromosome 4-q26, with four exons and three introns. The gene structure is similar to that of IL-4 and IL-5, although these other genes are clustered on Chr. 5.

Interleukin-2 gene transcription in T-cells is activated by two signals provided by antigen-presenting cells (APC): antigen (in the context of MHC) and cytokines such as IL-1. Several inducible factors bind to the promoter region of IL-2 causing transcriptional activation, including AP-1[17] and NF-κB[24] during antigen presentation, and the retroviral TAX protein during infections.[25] A high proportion of regulation is at the level of mRNA stabilization, where triggering of CD28 on the T-cell surface by B7 protein[26] stabilizes the intrinsically unstable mRNA.[27]

IL-2 is rapidly released, and exerts its effects through a saturable specific receptor found primarily on T-cells, B-cells, LAK cells, and NK cells.[28] The receptor comprises a 55-kDa α-chain (also called Tac antigen), a 70 to 75 kDa β-subunit, and a γ-chain of 64 kDa. The last two are expressed on resting T-cells, while the α-chain is induced during cellular activation. Signaling from the high-affinity α/β/γ receptor occurs with IL-2 in the picomolar range. Isolated α- and β-chains bind IL-2 with at least 1000 times lower affinity than the heterotrimer; the γ-chain does not seem to bind IL-2. The γ-subunit appears to be necessary for internalization of IL-2, while the β-chain, with a large cytoplasmic domain, is needed for signaling. The β-chain is intimately associated with at least two further intracellular proteins, each responsible for one specific signaling pathway. One is $p56^{lck}$, a member of the *src*-family of protein tyrosine kinases. This mediates $p21^{ras}$ activation and induction of *c-fos/c-jun* genes. A different cytoplasmic protein associated with the β-chain brings about activation of *c-myc* and cellular proliferation.[29]

The α-chain of IL-2r can be cleaved on the outer membrane, releasing a soluble form of the IL-2 receptor[30] that retains binding affinity and seem to be a marker of T-cell activation in inflammatory and autoimmune diseases.[31]

Interleukin-2 produced by activated T-cells is needed for expansion of cytotoxic T-lymphocytes and LAK and NK cell differentiation. In the adaptive immune response, production of IL-2 by T-lymphocytes is needed to initiate and support lymphocyte proliferation. IL-2 is the only cytokine produced by primitive T helper cells (T_HP) in response to antigen, and by T_H1 cells following activation. Response to IL-2 enables responsive cells to evolve to a further stage of differentiation, in due time becoming responsive to other differentiation-specific cytokines such as T_H1 and T_H2 cytokines. For this reason, many of the previously described actions of IL-2 on lymphocytes are in fact indirect, i.e., mediated by downstream cytokines. These include T-cell clonal expansion in the presence of antigen, augmentation of B-cell growth and Ig production, induction of IFN-γ, and IL-4 production and modulation of histamine release from basophils.

Recombinant IL-2 chimeric molecules fused to biological toxins have been used to downregulate T-cell responses in transplantation programs or to target IL-2r-bearing lymphoma cells.

Other important activities of IL-2, where it has been used as antitumoral agent, include the activation of NK cells for growth and tumoricidal activity and the generation of lymphokine-activated killer cells (LAK cells) from resting T-lymphocytes. Interleukin-2, like IFN-γ, is also reported to activate antitumoral cell cytotoxicity and other effector mechanisms in mononuclear phagocytes. This is compatible with the presence of IL-2 receptors on monocytes and, together with the report of IL-2 receptors on oligodendrocytes, confirms that IL-2, the first TCGF, is not solely a lymphocytotropic cytokine.

IV. INTERLEUKIN-3 AND COLONY STIMULATION FACTORS (CSFs)

Colony stimulating factors (CSFs) are a family of cytokines that control proliferation and differentiation of stem cells. Based on the types of colonies induced in stem-cell assays, four CSFs were originally described.[32] GM-CSF induces the differentiation from stem cells to macrophages or granulocytes, G-CSF causes proliferation of granulocyte precursors, and M-CSF induces monocytic differentiation. The fourth CSF is mCSF (multi-CSF or Interleukin-3) which induces colonies of multiple lineages, but also influences activities of mature cells. The relative importance of colony stimulating factors is still largely debated, but their role in hematopoiesis is likely to be more crucial than that in immunoregulation. Some of the characteristics of CSFs are reported in Table 3, and Interleukin-3 is here discussed in more detail.

Interleukin-3,[33] initially described as a colony stimulating factor, was soon renamed IL-3 because of its specific action in inducing 20α-hydroxysteroid-dehydrogenase in T-cells, which confers resistance to corticosteroids. Other biological activities from activated T-cells — described as mast cell growth factor (MCGF-II), multi-colony stimulating factor (mCSF), and burst-promoting activity (BPA), are also mediated by this cytokine.[34]

The existence of human IL-3 was under discussion for several years because its activity was elusive in human cultures, and mouse cDNA failed to cross-hybridize in human libraries. Gibbon IL-3 was then purified from MLA-144 cells (a T-cell line producing high levels of BPA activity) and the resulting cDNA was successfully used to

TABLE 3
Colony Stimulating Factors

	M-CSF	GM-CSF	G-CSF	Multi-CSF
Protein size	14.5 kDa	18–24 kDa	19 kDa	14.6 kDa
Molecular form	223 or 158 aa	127 aa	174 aa	133 aa
Gene location	5q33.1	5q23.1	17q11.2-21	5q23-31
Gene structure	8 Exons	4 Exons	5 Exons	5 Exons
Source	T-cells Neutrophils Macrophages Fibroblasts Endothelial cells	T-cells Neutrophils Macrophages Fibroblasts Endothelial cells Mast cells Eosinophils	T-cells Neutrophils Macrophages Fibroblasts Endothelial cells	T-cells Epithelial cells Keratinocytes Neuronal cells Mast cells
Stimuli	LPS, IL-1	LPS, IL-1, TNF	LPS, IL-1	Antigen presentation
Functions	Macrophage colonies Induces: PGE_2 IL-1, TNF, IFN gamma, antiviral	Granulocyte, eosinophils, Macrophage colonies IL-1, TNF Enhancement of eosinophil and neutrophil function Protection from bacterial and parasitic infection	Granulocyte colonies Terminal differentiation of myeloid cells Enhancement of neutrophil function	Multiple colonies of each lineage Enhancement of eosinophil functions

isolate hIL-3 cDNA from a human T-cell library.[35] The gene product is a 0.9-kb transcript whose translation yields a 152-aa precursor subsequently cleaved to a mature peptide of 133 aa (14.6 kDa). Differential glycosylation underlies the gel filtration appearance as a 15, 22, 28, and 36 kDa peptide. The gene, organized in 5 exons, maps to hChr. 5-q23 in the CSF cluster that includes GM-CSF, IL-4, and IL-5.[36]

Interleukin-3 is produced by epidermal cells, activated mast cells, and activated NK cells, but primarily by activated T-cells. Astrocyte and myelomonocytic cell lines have also been reported to produce IL-3.

The IL-3 receptor is a multisubunit complex, which binds IL-3 in the picomolar range. The 70-kDa α-subunit has a very short cytoplasmic domain while the 130-kDa β-subunit has an extended intracellular tail and is involved in signaling. Both receptors are part of the cytokine-receptor superfamily,[37] with two or four immunoglobulin-like domains and the WS motif (Trp-Ser-X-Trp-Ser). The β-chain also is a shared component of the IL-5 and GM-CSF receptor system and is probably responsible for the conflicting reports about IL-3, IL-5, and GM-CSF cross stimulation. Similarly to the IL-2 receptor, signaling involves both *c-jun* induction via a PKC-dependent pathway and *c-ras* protooncogene activation.

There is good evidence that multilineage colonies become less sensitive to IL-3 through differentiation, confirming its primary developmental role. Biological action of IL-3 includes stimulation of the differentiation of immature hematopoietic cells into macrophage, neutrophils, mast cell megakaryocyte, and erythroid colonies (mCSF activity); increased proliferation of cloned mast cells (MCGF activity); activation of granulocyte phagocytosis and monocyte production of IL-1, IL-6, and TNF-α; and downregulation of T-cell apoptosis.[38]

Therapeutic downregulation of IL-3 could be of use in IL-3-dependent lymphomas. Administration of IL-3 induces hematopoiesis[39] in the course of chemotherapy or radiation-induced leukopenias.

V. INTERFERON-γ

Interferons (IFNs) are cytokines identified as mediators of viral interference,[40] i.e., the ability of supernatants from virus-infected cells to protect other cells from viral infection. The family includes IFN-α (fibroblast IFN), IFN-β (leukocyte interferon), and IFN-γ (immune IFN). Interferon-γ is produced by activated T-cells, while IFN-α and IFN-β are produced by virtually any cell type; the actions of IFN-α and IFN-β within the innate immune system (enhancement of phagocytic and tumoricidal activities) and the adaptive immune system (modulation of MHC expression) seem to be redundant in the cytokine network. We describe, therefore, the principal characteristics of all interferons in Table 4, and discuss in more detail only IFN-γ, the immune interferon.

Interferon-γ was initially described as a new interferon (and called type II interferon) because of its higher MW (20 to 25 kDa by gel filtration) and its instability at pH 2.[41] IFN-γ cDNA[42] codes for a propeptide of 166 aa (17.1 kDa), 20 of which constitute the hydrophobic leader sequence. The two N-glycosylation sites at positions 38 and 100 explain the variable MW of the native IFN-γ. The IFN-γ gene has four exons and has been mapped to hChr. 12.

Production of IFN-γ by activated T-cells (both CD4+ and CD8+) is antigen-restricted. In the presence of macrophages, B-cells, or dendritic cells, IFN-γ synthesis by T_H1 lymphocytes is enhanced by IL-1 and IL-2. Natural killer cells (NK) also produce IFN-γ in response to IL-2 or activated B-cells.

Secreted IFN-γ, but not IFN-α or IFN-β, binds with intermediate affinity to a specific receptor of 90 kDa. This protein[43] probably represents the binding component of a higher-

TABLE 4
Interferons

	IFNα	IFN-β1	IFN-γ
Protein size	19 kDa	19 kDa	17 kDa
Molecular form	165 aa	166 aa	146 aa
Gene location	Chr. 9 (22 loci)	Chr. 9 p21	Chr. 12 q24.1
Gene structure	No introns	No introns	4 Exons
Source	Fibroblasts B-cells Macrophages T-cells	Fibroblasts B-cells Macrophages T-cells	T_H cells NK cells
Stimuli	Viruses, bacteria	Viruses, bacteria	Viruses, alloantigens, mitogens
Functions	Antiviral Antimitotic MHC class II (+) NK activity (+) Fever	Antiviral Antimitotic MHC class II (+)	Antiviral Antimitotic Synergy with TNF Class II MHC expression Enhancement of monocytic cytotoxicity Inhibits actions of IL-4 on B-cells Differentiative on T_H1 cells

affinity receptor complex, the signaling chain being as yet undescribed. Following interaction on the membrane, IFN-γ is rapidly internalized and it is apparently transported to the nucleus.

The biological activities of IFN-γ in the immune system are powerful and diverse. Apart from its antiviral activity, IFN-γ induces cytolytic activity in LAK and CTL cells and augments NK cytotoxicity, hence the early interest in the molecule as an anticancer agent. MHC class I and class II expression is upregulated by IFN-γ in most cell types (resulting in an accelerated immune response)[44,45] as well as IgA, IgM, and IgG Fc receptors on neutrophils (causing increased phagocytosis). Production of IFN-γ by T_H1 cells appears to be necessary for the induction of macrophage effector activities, possibly via induction of TNF receptors. TNF-α actions (including cytotoxicity) and IL-2-mediated polyclonal activation of B-lymphocytes are enhanced by IFN-γ, while IL-4 actions on B-cells and IL-1 actions on connective tissue catabolism are inhibited. IFN-γ emerges as a powerful immunopotentiator with a special ability to modulate cell-mediated immune responses.

Because of its immunopotentiator and tumoricidal activities, IFN-γ has been used in clinical trials for the treatment of certain types of cancer.[46,47] Other potential fields of application include states of immune suppression and systemic intracellular infections (parasitic and viral).

VI. INTERLEUKIN-4

Proliferation and differentiation of B-cells are partially driven by cytokines from activated T-cells. Engagement of B-cell surface IgM (by antigen, dextran, or anti-IgM ab) drives B-cells into the G1 phase, but then progression factors are needed to go into the S phase and T-cell help is needed to enter G2 and mitosis. The first B-cell growth factor, described (BCGF-1)[48] or B-cell stimulating factor-1 (BSF-1), was purified and characterized as IL-4.[49] The second growth factor (BCGF-2) was characterized as IL-5.

Interleukin-4 cDNA codes for a 153-aa protein, containing a leader sequence of 24 aminoacids.[50] The 129-aa (15 kDa) secreted protein has a pI of 10.4, and two potential glycosylation sites (native IL-4 is 18 to 20 kDa). The human IL-4 gene spans about 10 kb and has four exons. It has been mapped to the long arm of hChr. 5 in the same region as IL-3, GM-CSF, IL-5, and in close proximity to the ECGF, IL-9, MCSF genes and PDGF-rβ and M-CSF-r.[51] Such a clustering of cytokine genes is not unique: similarly, the IL-1 family of cytokines is grouped on Chr. 2 and the chemokine family on Chr. 4.

Interleukin-4 is an exquisitely lymphocytic cytokine, being produced by activated T_H2 and CD8+ cells. Targets for its actions are predominantly T- and B-lymphocytes, but also monocytes/macrophages and mast cells.

The IL-4 receptor appears to be a three-subunit complex. The first is a 135-kDa receptor which is involved in binding, the second is a membrane-bound extracellular protein, and the third chain is structurally highly homologous to other cytokine receptors with Ig-like domains, conserved cysteine residues, and the WS motif.[37] There is a soluble form of IL-4 receptor, a product of proteolytic cleavage, that retains its binding properties in solution (like the IL-2 receptor and others in this family).

IL-4 has powerful and diverse biological actions which include:

1. Supporting proliferation and viability of resting T-lymphocytes — during antigen presentation IL-4 induces the IL-1 type I receptor and, consequently, enhancement of atg presentation and T-cell activation (TCGF-II activity).
2. Enhancement of CTL and LAK cell development.
3. Induction of progression of B-cells and enhancement of IgG_1, IgE, and soluble IgM production (BCGF-I and BCDFε, BCDFγ activities).[52]
4. Upregulation of the IgE Fc receptor and MHC class II expression on B-cells.[53]
5. Stimulation of proliferation of the mast cell lines (MCGF-II activity).[54]
6. Co-stimulation of thymocyte maturation and bone marrow progenitors in the presence of other cytokines.
7. Induction of adhesion molecules on endothelial cells and tumoricidal activity in macrophages.
8. Inhibitory action on B-cell IgA secretion and monocyte cytokine expression (IL-1 and TNF).

Many of these actions may in fact be redundant *in vivo,* but the biological activities within the T- and B-cell compartments seem to be of a special importance to determine the final humoral and cellular response.

Potential therapeutic manipulation include suppression of IL-4 action to decrease production of IgE (type I hypersensitivity) or its administration in immunodeficiencies of hematopoietic and differentiating agents. Because of its inhibitory activities on monocyte cytokines, its use has been proposed in chronic inflammatory diseases.

VII. INTERLEUKIN-5

It was an early observation that supernatants from activated T-cells could substitute for the presence of T-lymphocytes in B-cell activation for T-dependent antigens (T-cell replacing factor activity — TRF).[55] TRF action on B-cells is not MHC restricted, but activity occurred only in the presence of MHC-compatible APC. The cytokine responsible for this action was purified, cloned, and named IL-5.[56,57] Its B-cell growth factor activity (BCGF-2) and BCDFα-BCDFμ activity (as IgA and IgM enhancers) were distinct from IL-4. In independent experiments, this cytokine had also been isolated and cloned as eosinophil differentiating factor (EDF). This activity was demonstrable in activated T-cell supernatant and in its absence (i.e., absence of T-cell help) eosinophilia and resistance to parasites could not be achieved *in vivo.*

Human IL-5 mRNA is translated as a 134-aa precursor (12.8 kDa),[56] presenting a clear stretch of hydrophobic amino acids as a signal peptide. The 115-aa mature protein has two N-glycosylation sites and two cysteines. In solution, native IL-5 is present as a dimer of 45 to 50 kDa. IL-5 is an exclusive product of T_H2 cells, where it is produced during cellular activation.[58] The gene of IL-5 is in human Chr. 5 q23–31, clustered with IL-3, IL-4 and GM-CSF. All of these genes have four exons, and the overall structure suggests some degree of coevolution.

The IL-5 receptor is formed from at least two chains.[59] A 60-kDa (415 aa) α-subunit is able to bind IL-5 at low affinity, but complexing with a β-chain (gp130) which is involved in signal transduction forms the high-affinity IL-5 receptor complex. This β-chain is in common with the GM-CSF, IL-3, and the IL-6 receptors. The IL-5 receptors are almost exclusively expressed on B-cells, granulocytes, and thymocytes.

The induction of eosinophil proliferation and eosinophilia *in vivo* by IL-5[60] is a key factor in the defense against parasites. Transgenic mice expressing high levels of IL-5 develop dramatic eosinophilia, and the mortality rate is multiplied in parasite-infected mice treated with anti-IL-5 antibodies.

Direct activities on B-cells by IL-5 appear to be limited to the stimulation of IgA and IgM secretion, and synergy with IL-4 for IgE production. However, the ability to enhance IL-2r expression brings about synergy with IL-2 on its own BCGF and BCDF activities. This same phenomenon is responsible for the synergy with IL-2 in inducing differentiation of NK cells and LAK cells.[61]

The actions on eosinophil maturation and growth can lead to IL-5 therapy to enhance protection from parasites; blocking of IL-5 could be tested in hypereosinophilic pathologies.[62]

VIII. INTERLEUKIN-6

Interleukin-6 was inadvertently cloned well before the discovery of its biological activities. In the cloning of IFN-β, an inducible 1.3-kb mRNA was isolated.[63] The protein product was called IFN-β_2 because, although structurally different, it was immunologically cross-reactive with IFN-β. When B-cell stimulatory factor-2 (BSF-2) was cloned,[64] not only was the same cytokine shown to be responsible for hepatocyte stimulating factor (HSF)[65] and hybridoma/plasmacytoma growth factor (HPGF) activity,[66] but the cDNA sequence was identical to the one deposited as IFN-β_2. All these activities were clearly associated with the same protein entity named Interleukin-6.

Interleukin-6 is a 184-aa (20.7 kDa) protein, resulting from the cleavage of its 28-aa leader sequence. Resting monocytes or T-cells do not produce IL-6. Bacterial or viral products, as well as IL-1 and TNF-α, elicit IL-6 transcription in many cell types; macrophages and activated T-cells are its major sources.

The IL-6 receptor[67] consists of a binding chain of 80 kDa and a signal-transducing protein (gp130). Human resting B-cells only express receptors upon stimulation, while resting T-cell receptor expression is independent from activation. Receptor activation brings tyrosine kinase and protein kinase activation of *jun* B and *tis* 11 gene transcription.[68] IL-6-induced DNA-binding proteins have been described in the rat (IL-6DBP) and in the human system (NF-IL6)[69] which bind the promoter of IL-6 responsive genes such as those of various acute-phase proteins.

The biological actions of IL-6 include proliferative effects on many cell types, such as B- and T-lymphocytes, keratinocytes, and mesangial cells, as well as differentiative effects, such as maturation of megakaryocytes and differentiation of neuronal cells and IL-3-dependent hematopoietic precursors.[70]

Interleukin-6 is also central in the induction of acute-phase response proteins from the liver during inflammation. While the function of some of these (such as fibrinogen or complement constituents) are of obvious importance in host defense, others (like SAA or CRP) are of more elusive advantageous consequences (see Chapter 3).

In the immune system, IL-6 has unique biological activities such as stimulation of IgG secretion in the absence of cellular proliferation (BCDF-γ activity).[71] In synergy with IL-1, IL-6 also promotes IgA and IgM production. Induction of antibody synthesis by IL-6 is independent of class switching, and suggests that IL-6 mediates terminal differentiation of B-cells into plasma cells.

IL-6 supports mature T-cell clonal expansion[72] in the presence of lectins or anti-CD3 antibodies (via induction of IL-2), induces CTL differentiation, and synergizes with IL-1 in its proliferative action on thymic T-cells. *In vitro,* IL-6 promotes the growth of hybridoma cells and plasmocytoma cell lines.[73]

Interleukin-6 seems to be a major inhibitor of myeloma cell growth, and anti-IL-6 therapy is being tested in myeloma and in mesangial proliferative glomerulonephritis cases.

IX. INTERLEUKIN-7

In comparison to the attention that has historically been given to lymphocyte differentiation and activation, early events in lymphopoiesis and commitment of lymphocytes are relatively unknown. Following the establishment of stromal marrow cells long-term culture[74] it has been possible to establish cultures of B-cell progenitors and understand their cellular and humoral requirements.

Interleukin-7 was described as a cytokine of 25 kDa produced by stromal cells and able to support growth of B-cell progenitors (and for this was initially named lymphopoietin-1; LP-1).[75] Human IL-7 cDNA[76] codes for a 177-aa precursor, cleaved to a 152-aa mature peptide (17.4 kDa), with three N-glycosylation sites and at least two disulfide links which are essential for activity. Injection of LPS or IL-1 in mice increases IL-7 gene expression, with accumulation of a 2.4-kb mRNA in the spleen, thymus, fetal liver cells, and bone marrow. Of interest, there is some evidence of alternative splicing, as liver tissue also presents a 1.8-kb mRNA species. The human gene has been mapped to Chr. 8q12–13.[77]

The IL-7 receptor has been described as a single chain protein belonging to the cytokine receptor superfamily (two conserved cysteine residues, a WS motif, and a cytoplasmic domain capable of signaling). Other associated proteins are likely to have a role in forming a high-affinity complex. Of interest, similarly to the IL-2r, the IL-1r type II, the IL-4r, and the TNFr p55, the IL-7 receptor exists in a soluble form that retains binding ability.[78]

IL-7 supports the growth of both pre-B-cells and pro-B-cells[79] from fresh bone marrow (an effect inhibited by TGF-β), but does not seem to mediate any maturation effect or induce Ig gene rearrangement. Other soluble mediators are probably needed for full B-cell maturation and commitment.

Interleukin-7 also has direct proliferative effects on thymocytes, in particular the early CD4- and CD8- cells,[80] which are inhibited by TGF-β. IL-7 has been recently demonstrated to play a role in T-cell maturation, by inducing the expansion of immature thymocytes until they become IL-4 and IL-2 responsive.[81] Thymic epithelial IL-7 has recently been reported to have a major role as a cofactor of TCR-β V(D)J rearrangement, with induction of RAG-1 and RAG-2 gene expression.[82] Confirming the importance of IL-7 in lymphocyte growth and differentiation, the IL-7r knock-out mouse lacks both

TABLE 5
Chemokines

	C-X-C Family	C-C Family
Gene location	4q12–21	17q11–32
Cytokine (human)	IL-8 GRO/MGSA PF-4 βTG IP10	MCAF LD-78 ACT-2 RANTES I-309
(mouse)	? MIP-2 ? ? CRG-2	JE MIP-1α MIP-1β ? TCA-3
Molecular form	8–10 kDa Dimer has 3D structure of MHC class I	8–10 kDa
Receptor	Only partially shared (type II)	Different receptors from IL-8
Source	Neutrophils, fibroblasts, monocytes/macrophages, T-cells, chondrocytes, epithelial cells, endothelial cells, mesangial cells, B-cells, Renal epithelium, smooth muscle cells, hepatocytes, platelets (PF-4 and βTG), glioblastoma cells, keratinocytes	
Stimuli	IL-1, TNF, phorbols, lectins, silica and urate crystals, IL-3, viruses, LPS, platelet activation for PF-4 and βTG	
Downregulators	Glucocorticoids, TGFβ, IL-4, vitamin D3, cyclosporin A	

T- and B-lymphocytes.[83] Other actions on mature T-cells includes the generation of LAK cells and CTL cells from T-lymphocyte cultures.

X. INTERLEUKIN-8 AND THE "CHEMOKINE" SUPERFAMILY

A group of peptide cytokines with chemotactic and proinflammatory activity has recently been described. They have a certain degree of structural homology and they are all small (8 to 10 kDa) secreted basic polypeptides. Because of its complexity, the family is referred to as "chemokines"[84] or "intercrines"[85] but the term IL-8 refers only to the oldest member of the family (see Chapter 3).

The chemokine family can be divided in two subfamilies, the C-C and C-X-C families (Table 5). The C-X-C family, so named because it contains four cysteine residues, two of which are separated by one amino acid, includes human IL-8, GRO/MGSA, PF-4, β-TG, and IP-10 proteins. All these are coded by single independent genes that cluster on Chr. 4 (q12–21).

Interleukin-8 was originally described in the course of experiments on IL-1-mediated effects on neutrophils, as monocyte-derived neutrophil chemotactic factor (MDNCF).[86] IL-8 had also been described as neutrophil activating peptide (NAP-1),[87] and monocyte- and lymphocyte-mediated neutrophil chemotactic peptide (MONAP and LYNAP).[88] IL-8 is a single protein product of 99 aa, processed with different N-terminus variants (69, 70, 72, and 77 aa each), giving bioactive products of about 8.2 kDa. In solution, the native protein is a homodimer, with a 3D structure resembling MHC class I. Upon stimulation by TNF-α or IL-1, IL-8 is secreted by a large variety of cell types (Table 5).

The IL-8 receptor is composed of at least two chains. The type I receptor only binds IL-8, while type II also binds the other C-X-C chemokines. Both receptors have seven transmembrane domains and are coupled to G proteins.[89] Following receptor interaction, Interleukin-8 is the most powerful activator of neutrophils (but not for free-radical production) and chemoattractant for monocytes and neutrophils. Within the adoptive immune response, it can be considered of importance because of its activity on lymphocyte migration[90] from the bloodstream into the site of action.

The C-C family includes other peptides such as MCAF/mcp.1, LD-78/PAT 464, ACT-2/PAT 744, RANTES, and I-309. All of these proteins are powerful chemotactic agents, their genes are closely linked on human Chr. 17, and their characteristics have been recently reviewed.[91] A special mention is needed for RANTES (Regulated upon Activation, Normal T Expressed and Secreted), which is inactive on neutrophils and seem to cause selective migration of memory T-cells. It is also noteworthy that this cytokine is a T-cell product, but its production is downregulated following T-cell activation.[92]

XI. INTERLEUKIN-9

Interleukin-9, or P40,[93] was initially described as a factor inducing megakaryocytoma growth, but its major actions have been demonstrated as T-cell growth factor (TCGF-III), mast cell activity enhancer (MEA), and for supporting growth of erythroid colonies.[94,95]

IL-9 cDNA codes for a 126-aa (17.8 kDa)[96] mature peptide containing four N-glycosylation sites and no cysteine residues. Activated T-cells produce IL-9, secreted as a glycosylated monomer of 20, 22, and 26 kDa.

The gene is organized in five exons in Chr. 5 q31–35, adjacent to the M-CSF gene and in proximity of the other CSF genes.[97] Such clustering with the CSF genes together with its biological actions reinforces the understanding of IL-9 as a colony stimulating factor on Chr. 5.

XII. INTERLEUKIN-10

This cytokine was discovered as a T_H2 helper cell product that inhibits development, function, and proliferation of T_H1 cells.[98] The factor was called cytokine synthesis inhibitor factor (CSIF), and when mouse and human cDNA clones were isolated the cytokine acquired the name of Interleukin-10.[99] Human IL-10 cDNA encodes for an 18-kDa propeptide of 178 aa with a clear hydrophobic sequence.[100] Processing at Ser19 leaves a mature peptide of 159 aa. The glycosylated protein is 19 to 21 kDa and, as IL-10 is present in solution as homodimer, the protein size in solution is about 40 kDa.

Human IL-10 is coded by a single gene on chromosome 1, with five exons and four introns. The cDNA is highly homologous (71%) to BCRF1,[101] an open reading frame in the EBV genome. Joining the list of cytokine sequences shared between human and viral genomes, the protein expressed *in vitro* by BCRF1 has recently been shown to have IL-10-like actions and is called "viral IL-10" or vIL-10.[102]

IL-10 is produced by T_H2 and T_H0 $CD4^+$ cells in the mouse system,[101,103] but also by activated mast cells, macrophages, keratinocytes, and activated B-cells.

Major inhibitory activities of IL-10 include suppression of antigen-presenting cell functions, with inhibition of T_H1 cell proliferation and T_H1 cytokine release. Antimacrophage actions of IL-10 also include downregulation of class II gene expression, and suppression of the production of IL-1α, IL-6, IL-8, GM-CSF, G-CSF, and TNF-α[104] from mononuclear phagocytes. Interleukin-10 is to be considered, therefore, the most potent immunosuppressor of macrophage-dependent responses[105] and type IV hypersensitivity.[106] IL-10 augments B-cell proliferation and differentiation into plasma cells committed to abundant production of IgM, IgA, and IgG antibodies.[107]

Administration of anti-IL-10 antibodies (IgM or IgG) *in vivo* leads to a decrease of peritoneal B-1 cells, very high levels of serum IFN-γ and a marked decrease of serum IgM and IgA antibodies.[108]

Its ability to suppress cellular responses makes IL-10 a good candidate for use in chronic inflammation, autoimmunity, and graft rejection. Phase II clinical trials have started, but no preliminary results have been reported to date. The biology and therapeutic potential of IL-10 are reviewed in Chapter 15.

XIII. INTERLEUKIN-11

Interleukin-11 was initially described as a mitogenic activity for the IL-6-dependent murine plasmacytoma cell line T1165, released from PU-34 — an immortalized, primate bone marrow-derived stromal cell line.[109] The cloning of the human cDNA encoding this cytokine by functional expression cloning was reported in 1990,[110] and most of the work since then on IL-11 has been directed towards its key role as a growth factor in hematopoiesis.

Mature human IL-11 is a 23-kDa, 178-residue protein that contains no potential glycosylation sites and, interestingly, no cysteine residues — a feature not found in any other cytokine gene. The human IL-11 genomic sequence is 7 kb in length and consists of five exons and four introns.[111] The IL-11 gene has been localized to the long arm of Chr. 19 at band 19q13.3-q13.4. The gene contains a second polyadenylation site in the 3′ untranslated region of exon 5 which encodes a short 1.5 kb IL-11 transcript instead of the 2.5 kb transcript observed in IL-1-induced human stromal and fibroblast cell lines. However, both transcripts have been shown to encode the same functional protein.[112]

A specific IL-11 binding protein has been identified in a number of cell lines as a 151-kDa protein[113] and recent studies have demonstrated that IL-11 shares the common signal transducer, gp130, with a number of other cytokines including IL-6, with which it shares a variety of functions.[114]

The principal sources of IL-11 are mesenchymal-derived adherent cells such as stromal fibroblasts, fetal lung fibroblasts, and trophoblasts. The principal biological functions of IL-11 are involved in hematopoiesis, specifically, augmenting IL-3-dependent development of megakaryocyte colonies in human bone marrow cultures[115] and bringing pluripotent stem cells out of the G_0 stage of the cell cycle to respond to other growth factors such as IL-3 and GM-CSF.[116] In addition, IL-11 plays various accessory roles in the immune system, primarily stimulating the T-cell-dependent development of immunoglobulin-producing B-cells *in vitro* and *in vivo*[117] and the synthesis of several acute-phase plasma proteins *in vitro*.[118] Additionally, IL-11 has been shown to act as an adipogenesis inhibitory factor in mouse preadipocytes.[112] IL-11 therefore represents another multipotential cytokine exerting various biological activities in the cytokine network.

XIV. INTERLEUKIN-12

A cytokine which is currently attracting much interest due to its essential role in the generation of T_H1 cells and its potent IFN-γ stimulatory properties is Interleukin-12. First described in 1989 as natural killer cell stimulatory factor (NKSF) isolated from three Epstein-Barr virus (EBV)-transformed cell lines,[119] IL-12 is unique amongst the interleukins in being a heterodimeric protein, consisting of two covalently linked, glycosylated chains. The p35 and p40 subunits are encoded by separate genes and the cDNAs for both subunits were cloned in 1991.[120] The p35 light chain comprises 197 amino acids and has limited homology with IL-6 and G-CSF. In common with most other cytokines, it also has an alpha-helix-rich structure.[121] By contrast, the 306-aa p40 heavy chain is not homologous to any other known cytokine but belongs to the hematopoietin receptor family, most closely resembling the IL-6 receptor and ciliary neurotrophic factor (CNTF) receptor. This has raised the possibility that the covalently linked IL-12 heterodimer derived from a primordial cytokine of the IL-6/CNTF family that is covalently linked to the soluble form of its own receptor.

Both IL-12 genes are thought to be single copy although no genomic sequence is available as yet. However, the subunits have been mapped to 5q31–33 and 3p12-q13.2 for the p40 and p35 subunits, respectively,[122] and p40 is thought to contain at least 5 exons spanning more than 20 kb. An IL-12 receptor has been identified on activated NK and T-cells as a 110-kDa membrane protein.[123] However, biological functions of IL-12 can be observed on both resting and activated NK and T-cells.

IL-12 is produced by cell types with accessory or antigen-presenting ability including monocytes/macrophages and B-cells, mainly in response to bacteria, bacterial products, or parasites. NK and T-cells, on which the biological activity of IL-12 and the presence of IL-12 receptor have been unequivocally demonstrated,[124] do not appear to be capable of producing IL-12.[125] Interestingly, several cell types, including NK and T-cells, express p35 mRNA but not p40 mRNA, the expression of which is highly regulated and present only in cell types able to produce the biologically active IL-12 heterodimer.

One of the primary functions of IL-12 is the facilitation of T_H1 cell generation, for which IL-12 production by monocytes is necessary.[126] Recent data have also established that IL-12 directly initiates T_H1 cell development in native CD4+ T-cells derived from mice, whereas the same cells develop into T_H2 cells in the presence of IL-4.[127] The effect of IL-4 appears to be dominant over IL-12, leading to the hypothesis that the dominant cytokines responsible for the balance between T_H1 and T_H2 cells are IL-12 and IL-4 for Th1 and Th2 differentiation, respectively, in a ying-yang type of equilibrium.[128]

In addition to its key role in T_H1 cell development, IL-12 has recently been shown to play a major role in the development of the cellular immune response to microbial pathogens, being the co-stimulatory factor (with IL-1/TNF-α) produced from stimulated macrophages required for the T-cell-independent release of IFN-γ from NK cells. This, in turn, feeds back and activates the macrophages to a microbicidal state.[129] In addition, macrophage-derived IL-12 conditions native helper T-cells to differentiate to the T_H1 pathway and so generate further IFN-γ from T_H1 cells to maximize macrophage microbicidal responses.[130] Thus, IL-12 serves to link the innate and adoptive cellular immune systems together.

Recent *in vivo* experiments have demonstrated that treatment with recombinant IL-12 results in prolonged survival of (T- and B-cell deficient) SCID mice following infection with an intracellular protozoan[131] and, in similar experiments, BALB/c mice, which are unable to mount a T_H1 response, have been shown to be protected from *Leishmania* infection by IL-12 treatment, leading to a T_H1 reaction.[132] Early investigations using PBMCs from asymptomatic seropositive HIV individuals, which are usually unresponsive, have demonstrated that IL-12 can restore HIV-specific cell-mediated immunity *in vitro,* and suggest a potential use of IL-12 in augmenting the diminished immunological functions associated with HIV infection.[133] Thus, the possibilities for using IL-12 therapeutically offer promising new strategies which are just beginning to be investigated.

XV. INTERLEUKIN-13

Interleukin-13 was originally identified as an induction-specific cDNA clone, termed P600, isolated from a mouse T_H2 cDNA library.[134] This was cloned in 1989,[135] and a 400 bp DNA fragment from P600 cDNA was used as a probe to identify the human homologue, which was subsequently cloned and designated IL-13.[136] At the same time, IL-13 was independently cloned by another group using an activated T-cell cDNA subtractive hybridization approach.[137]

Human IL-13 is a T-cell-specific cytokine secreted by activated T_H2 cells as a 132-aa protein which is mainly unglycosylated and has a molecular weight of 10 kDa. There is a 30% protein homology between human IL-13 and IL-4 and a 58% homology with murine IL-13. The gene gives rise to a 1.3-kb mRNA transcript, and the recently cloned 4.6-kb gene[138] has revealed that four exons constitute the mature mRNA encoding amino acids 1–44, 45–62, 63–97, and 98–132, respectively. Interestingly, IL-13 has been mapped to Chr. 5q31, in the same region that several other immunoregulatory and hematopoietic growth factors, i.e., IL-3, IL-4, IL-5, and GM-CSF have been mapped.

The principal functions of IL-13 *in vitro* appear to be the stimulation of B-cell proliferation and differentiation, the induction of several immunoglobulins, principally IgE, IgM, and IgG4 by B-cells, and the inhibition of proinflammatory cytokine production by monocytes and macrophages.[139] In these respects, its actions are very similar to those of IL-4 in that both cytokines display a strong T_H2-like response characterized by their anti-inflammatory effects and humoral response stimulation. The isotype switching properties of IL-13 have been shown to be IL-4-independent, suggesting that IL-13 has a distinct receptor subunit, although recent data from work involving a mutant IL-4 ligand and receptor binding studies have demonstrated that both IL-4 and IL-13 share a novel receptor component that functions in signal transduction.[140] This is probably the gamma chain known to be shared between IL-2, IL-4, and IL-7.

No *in vivo* studies have yet been performed for IL-13 so its therapeutic potential remains to be assessed. However, recent studies with IL-13 in HIV infection have shown that it inhibits production of the virus in primary blood-derived human macrophages *in vitro*[141] and that this property is unique to IL-13. Also, two alternatively spliced forms of the human IL-13 protein have been identified. However, although they differ by a single amino acid residue (glutamine 98) as a result of alternative splicing at the 5′ end of exon 4, they display identical biological activity.

XVI. INTERLEUKIN-14

Interleukin-14 has been studied for several years and is a high molecular weight B-cell growth factor (HMW-BCGF), a 60-kDa factor produced by a human B-lymphoma cell line. HMW-BCGF was purified to homogeneity, shown to bind specifically to activated human B-cells, and a monoclonal antibody was raised against it.[142] Using this monoclonal antibody, a receptor for HMW-BCGF was identified in 1988[143] and was subsequently determined to be 90 kDa in size.[144] Further studies examining the subpopulation of human B-cells responding to HMW-BCGF revealed that it also inhibits antibody secretion.[145] The cDNA for HMW-BCGF was finally cloned and published in July, 1993 and renamed IL-14.[146]

Results to date indicate that IL-14 is produced by normal T-cells (in addition to malignant B- and T-cells) as a 468-aa monomer with a predicted molecular weight of 53 kDa. The native protein is glycosylated and three potential sites have been identified. The gene encodes a 1.8-kb mRNA with a coding sequence of 1492 bases.

The principal *in vitro* functions of IL-14 are the induction of proliferation of activated B-cells and the inhibition of immunoglobulin secretion of mitogen-stimulated B-cells. Although it shares some functional activities with IL-4, IL-14 has a distinct receptor and intracellular signaling is thought to be quite different.[146]

XVII. INTERLEUKIN-15

The latest addition to the interleukin family is a cytokine that stimulates T-cell growth and shares a subunit of the IL-2 receptor.[147] This property was identified as a soluble factor from the simian kidney epithelial line CV-1/EBNA capable of supporting proliferation of the IL-2-dependent line CTLL-2. A full-length simian cDNA was isolated and used to screen a human library from bone marrow stromal cells (IMTLH).

Following LPS stimulation, human IL-15 mRNA rapidly accumulates in PBMCs, but not in activated peripheral blood T-cells. Placenta and skeletal muscle also express IL-15 mRNA, and at lower levels it is also present in heart, lung, liver, and kidney.

The expressed protein interacts with the β-chain of the IL-2 receptor. However, cell lines responsive to IL-2 and expressing the full receptor do not respond to IL-15, suggesting that a different chain is needed for its function and possibly mediates specific IL-2-independent effects.

XVIII. TUMOR NECROSIS FACTOR-α

Tumor necrosis factor-alpha (TNF-α), or cachectin, is a monocytic cytokine with major biological functions in the areas of inflammation, tissue metabolism and repair, immunity, cell growth, and differentiation.[155] This peptide cytokine has a fascinating history of nearly 100 years for its experimental antitumoral effects[156] and as a component of the endogenous pyrogen activity[157] (see also Chapters 3 and 10).

Because of its formidable actions in host defense and inflammation, this cytokine is examined in more detail in subsequent chapters of this book. Within the adaptive immune response, tumor necrosis factor has been reported to upregulate MHC class I expression in several cell types.[158] This effect is not as powerful as the effects of IFN-γ, and it is thought to be mediated by induction of IFN-β. TNF-α synergy with IFN-γ is evident in expression of class II MHC,[159] where the combination of the two cytokines maximally increases class II-dependent antigen presentation by nonprofessional APC such as astrocytes, endothelial cells, and pancreatic islet cells. Macrophage activation by T-cells is reported to be mainly due to secretion of IFN-γ and expression of TNF-α (in its soluble or membrane form) by T_H1 cells.

Both TNF-α and TNF-β have been tested, with varied success to activate TIL together with IL-2, and for their antitumoral effects *in vivo*.[160] Recently, the technique of the isolated perfused limb has allowed the administration of high doses of TNF without having systemic lethal effects. Major successes have been achieved with fibrosarcomas and in some types of melanomas.

Anti-TNF strategies are being tested in septic shock, graft rejection, and in chronic inflammatory diseases such as rheumatoid arthritis and Crohn's disease (reviewed by Bodmer and Foulkes in Chapter 10).

XIX. LYMPHOTOXIN (TNF-β)

This cytokine was originally described as a T-cell product (cytotoxicity factor, lymphotoxin) which was toxic to L929 cells or embryonic fibroblasts.[148,149] Following molecular characterization, which indicated significant homology to cachectin/TNF-α,[150] and because lymphotoxin and TNF-α share common receptors and therefore many biological activities, this molecule is also referred to as TNF-β.[151]

Although TNF-α and TNF-β genes are physically associated on Chr. 6 within the class III region of the MHC locus, and they are likely to be the result of a duplicating event, TNF-β is a lymphocytic cytokine and acts within the adaptive immune response, while TNF-α is mainly produced by activated macrophages and its roles in basic host defense outcast its immune functions.[152]

Lymphotoxin is a 20- to 25-kDa glycosylated protein that forms multimers in solution. Its cDNA codes for a 205-aa peptide cleaved to a mature 171-aa protein that is easily secreted by both CD4+ and CD8+ activated human lymphocytes.

TNF-β mediates CTL, NK, and LAK cell cytotoxicity. These cell types produce lymphotoxin following antigen presentation or in response to high levels of IL-2.[153] Other agents also released in this process include proteinases and perforin, that complement lymphotoxin-induced apoptosis.[154] Apoptosis in this case seem to be the result of inhibition of intracellular self-protective proteins. When available in sufficient concentrations, TNF-β also shares other actions of TNF-α such as induction of MHC, antiviral activities, and other actions on endothelial cells, neutrophils, and mesenchymal cells.

XX. TRANSFORMING GROWTH FACTOR-β

Transforming growth factor-β (TGF-β) is a cytokine produced by mesenchymal cells such as chondrocytes, myoblasts, fibroblasts, bone cells, and astrocytes, but platelets are a very important source of preformed TGF-β.[161] The protein is a homo- or heterodimer of a five-isoform chain, β1 to β5 (11 to 13 kDa for each chain). The most common form is the β1 homodimer, (25 kDa). Five different receptors types have been reported to bind TGF-β.[162]

TGF-β has important actions in the metabolism of connective tissue,[163] with a series of activities that result in connective tissue apposition and fibrosis. In the immune system, TGF-β enhances T-helper cell functions as effector or memory cells;[164] but the inhibitory effects on the immune system are the reason why TGF-β has received very early attention. TGF-β inhibits antibody production from B-cells (but stimulates IgA secretion),[165] inhibits lymphocyte growth, and antagonizes action or production of many cytokines.[166]

Because of this action, TGF-β promises to be a useful immunosuppressant,[167] while blocking TGF-β can be useful in the course of fibrotic diseases. The therapeutic potential of TGF-β is covered in Chapter 14.

XXI. ONCOSTATIN M

This cytokine[168] is a product of activated macrophages and T-cells. It belongs to a cytokine subfamily which includes IL-6, leukemia inhibitory factor (LIF), ciliary neurotropic factor (CNTF),[169] and G-CSF. The family has been defined by the common antiproliferative effects of these factors, by the distant structural similarities, and because their receptors share the β-chain (gp130) which is involved in signaling but not binding of the ligand. It derives that although not binding to the same receptors, many biological actions will be shared by the members of this family.

Oncostatin upregulates LDL receptors on liver cells, increasing their LDL uptake.[170] As the name implies, this cytokine has been originally described for its ability to inhibit growth of tumoral lines, but not their untransformed counterparts. This action is synergized by TGF-β, but not IFN-γ. Other important activities are the modulation of endothelial cell growth, IL-6 production, and fibrinolytic activity.[171] Administration of oncostatin has been proposed for experimental cancer therapy and in the control of cholesterol levels.

XXII. LEUKEMIA INHIBITORY FACTOR (LIF)

This cytokine was isolated as a factor able to modulate differentiation of the murine leukemia cell lines into macrophages or granulocytes.[172] Many other descriptive names for LIF, such as differentiation-inhibiting factor (DIA), human interleukin for DA cells

(HILDA), and cholinergic neuronal differentiation factor (CNDF), point to its actions on cellular growth and differentiation.

LIF is expressed by many resting cell types, but bone marrow cells produce very high levels of LIF in response to IL-1 and TNF,[173] and endometrial glands produce LIF to sustain embryonic development.[174]

LIF biological actions, for the reasons outlined for oncostatin M, are very similar to IL-6 (acute-phase response), GM-CSF (regulation of differentiation in hematopoietic lines and support of germ cell growth), and Oncostatin M (endothelial and myoblast proliferation). LIF has been shown to inhibit lipoprotein-lipase,[175] and its use has been proposed in the control of atherosclerosis. Because of its growth-arresting properties, its administration could be used for the therapy of myeloid leukemia and other myeloproliferative disorders.

XXIII. INTERLEUKIN-16

The International Cytolaine Society (ICS) Committee on Standardization and Nomenclature has recently recommended the name Interleukin-16 (IL-16) to be used to designate the cytokine Lymphocyte Chemoattractant Factor (LCF).[177] This product of CD8+ve T cells, endothelial cells, and eosinophils induces IL-2 receptor expression and T-lymphocyte migration through interaction with CD4 on T cells (and activation of $p56^{lck}$). Of particular interest for Type 1 reactions, histamine and serotonin are powerful inducers of LCF/IL-16 gene expression.

XXIV. CONCLUSIONS

The knowledge of cytokine physiology has been growing steadily in the last two decades. Studies on the gene regulation, production, and functions of cytokines *in vivo* have multiplied, and the importance of different cytokines in the pathogenesis of many human diseases has been established.

Following the first promising steps of cytokine and anti-cytokine therapies, and the unrelenting efforts in the dissection of immunoregulatory networks, targeted manipulation of cytokine pathways is now emerging as a true opportunity in modern medicine.

ACKNOWLEDGMENTS

Mr. John B. Mee is supported by a grant from the British Psoriasis Association. Our work is funded by the Arthritis and Rheumatism Council for Research (U.K.).

REFERENCES

1. **Paul, W. E. and Seder, R. A.,** Lymphocyte responses and cytokines. *Cell,* 1994; 76:241–251.
2. **Oppenheim, J. J. and Gery, I.,** From lymphodrek to Interleukin-1 (IL-1). *Immunol. Today,* 1993; 14:232–234.
3. **di Giovine, F. S., Symons, J., Manson, J., and Duff, G. W.,** Soluble mediators of Immunity: interleukins. In: Dick, C., Goodacre, J., Eds., *Immunopathogenic Mechanisms of Arthritis.* Lancaster: MTP Press, 1988:101–121.

4. **Callard, R. E. and Gearing, A. J. H.,** *The Cytokine Facts Book.* New York: Academic Press, 1994.
5. **Mosmann, T. R., Cherwinski, H., Bond, M. W., Giedlin, M. A., and Coffman, R. L.,** Two types of murine helper T cell clone. I. Definition according to profiles of lymphokine activities and secreted proteins. *J. Immunol.,* 1986; 136:2348–2354.
6. **Maggi, E., Delprete, G., Macchia, D., et al.,** Profiles of lymphokine activities and helper function for IgE in human T-cell clones. *Eur. J. Immunol.,* 1988; 18:1045–1050.
7. **Gajewski, T. F., Joyce, J., and Fitch, F. W.,** Antiproliferative effects of IFN-gamma in immune regulation. III. Differential selection of TH1 and TH2 murine helper lymphocyte-T clones using recombinant IL-2 and recombinant IFN-gamma. *J. Immunol.,* 1989; 143:15–22.
8. **Legros, G., Ben-Sasson, S. Z., Seder, R., Finkelmann, F. D., and Paul, W. E.,** Generation of IL-4 producing cells *in vivo* and *in vitro.* I1–2 and IL-4 are required for *in vitro* generation of IL-4 producing cells. *J. Exp. Med.,* 1990; 172:921–930.
9. **Fiorentino, D. F., Zlotnik, A., Vieira, P., et al.,** IL-10 acts on the antigen-presenting cell to inhibit cytokine production by TH1 cells. *J. Immunol.,* 1991; 146:3444–3451.
10. **Gajewski, T. F. and Fitch, F. W.,** Anti-proliferative effects of IFN-gamma in immune regulation. I. IFN-gamma inhibits the proliferation of TH2 but not TH1 murine helper lymphocyte T-clones. *J. Immunol.,* 1988; 140:4245–4252.
11. **Coffman, R. L. and Carty, J.,** A T-cell activity that enhances polyclonal ILE production and its inhibition by IFN-gamma. *J. Immunol.,* 1986; 136:949–954.
12. **Dinarello, C. A.,** Interleukin-1: aminoacid sequences, multiple biological activities and comparison with tumor necrosis factor (cachectin). *Year Immunol.,* 1986; 2:68–89.
13. **Gery, I., Gershon, R. K., and Waksman, B. H.,** Potentiation of the T-lymphocyte response to mitogens. I. The responding cell. *J. Exp. Med.,* 1972; 136:128–142.
14. **Atkins, E.,** Pathogenesis of fever. *Physiol. Rev.,* 1960; 40:580–592.
15. **di Giovine, F. S. and Duff, G. W.,** Interleukin-1, the first interleukin. *Immunol. Today,* 1990; 11:13–20.
16. **Nicklin, M. J. H., Weith, A., and Duff, G. W.,** A physical map of the region encompassing the human Interleukin-1α, Interleukin-1β, and Interleukin Receptor Antagonist genes. *Genomics,* 1994; 19:382–384.
17. **Muegge, K., Williams, T. M., Kant, J., et al.,** Interleukin-1 costimulatory activity on the interleukin-2 promoter via AP-1. *Science,* 1989; 246:249–251.
18. **Zhang, Y., Lin, J. X., and Vilcek, J.,** Interleukin-6 induction by TNF and IL-1 in human fibroblasts involves activation of a nuclear factor binding to a kappa-B-like sequence. *Mol. Cell. Biol.,* 1990; 10:3818–3823.
19. **McIntyre, K. W., Stepan, G. J., Kolinsky, K. D., et al.,** Inhibition of IL-1 binding and bioactivity in vitro and modulation of acute inflammation in vivo by IL-1 receptor antagonist and anti IL-1 receptor monoclonal antibody. *J. Exp. Med.,* 1991; 173:931–939.
20. **Nicod, L. P., Elhabre, F., and Dayer, J. M.,** Natural and recombinant Interleukin-1 receptor antagonist does not inhibit human T-cell proliferation induced by mitogens. *Cytokine,* 1992; 4:29–35.
21. **Neta, R., Sztein, M., Oppenheim, J. J., Gillis, S., and Douches, S. D.,** The *in vivo* effects of IL-1. I. Bone marrow cells are induced to cycle after administration of Interleukin-1. *J. Immunol.,* 1987; 139:1861–1866.
22. **Morgan, D. A., Ruscetti, F. W., and Gallo, R.,** Selective *in vitro* growth of T lymphocytes from normal bone marrows. *Science,* 1976; 193:1007–1008.
23. **Taniguchi, T., Matsui, H., Fujita, T., et al.,** Structure and expression of a cloned cDNA for human Interleukin-2. *Nature,* 1983; 302:305–310.
24. **Leung, K. and Nabel, G. J.,** HTLV-I transactivator induces Interleukin-2 receptor expression through an NF-K-B like factor. *Nature,* 1988; 333:776–778.
25. **Kronke, M., Leonard, W. J., Depper, J. M., and Greene, W. C.,** Deregulation of Interleukin-2 receptor gene-expression on HTLV-I induced adult T-cell leukemia. *Science,* 1985; 228:1215–1217.
26. **Thompson, C. B., Lindsten, T., Ledbetter, J. A., et al.,** CD28 activation pathway regulates the production of multiple T-cell derived lymphokine cytokines. *Proc. Natl. Acad. Sci. U.S.A.,* 1989; 86:1333–1337.

27. **Shaw, G. and Kamen, R.,** A conserved AU sequence from the 3′ untranslated region of GM-CSF messenger-RNA mediates selective messenger-RNA degradation. *Cell,* 1986; 46:659–667.
28. **Greene, W. C. and Leonard, W. J.,** The human IL-2 receptor. *Annu. Rev. Immunol.,* 1986; 4:69–96.
29. **Minami, Y., Kono, T., Miyazaki, T., and Taniguchi, T.,** The IL-2 receptor complex: its structure, function and target genes. *Ann. Rev. Immunol.,* 1993; 11:245–267.
30. **Treiger, B. F., Leonard, W. J., Svetlik, P., Rubin, L. A., Nelson, D. L., and Greene, W. C.,** A secreted form of the human IL-2 receptor. *J. Immunol.,* 1986; 136:4099–4105.
31. **Symons, J. A., Wood, N. C., di Giovine, F. S., and Duff, G. W.,** Soluble interleukin-2 receptor in rheumatoid arthritis: correlation with disease activity, Interleukin-1 and interleukin-2 inhibition. *J. Immunol.,* 1988; 141:2612–2618.
32. **Metcalf, D.,** The molecular biology and functions of the colony stimulating factors. *Blood,* 1986; 67:257–267.
33. **Ihle, J. N., Pepersack, L., and Rebar, L.,** Regulation of T-cell differentiation. *In vitro* induction of 20-alpha hydroxysteroid dehydrogenase in splenic lymphocytes from athymic mice by a unique lymphokine. *J. Immunol.,* 1981; 126:2184–2189.
34. **Schrader, J. W.,** The panspecific hemopoietin of activated lymphocyte-T (IL-3). *Annu. Rev. Immunol.,* 1986; 4:205–230.
35. **Otsuka, T., Miyajima, A., Brown, N., et al.,** Isolation and characterization of an expressible cDNA encoding human IL-3-induction of IL-3 mRNA in human T cell clones. *J. Immunol.,* 1988; 140:2288–2295.
36. **Pettenati, M. J., Lebeau, M. M., Lemons, R. S., et al.,** Assignment of CSF-1 to 5q33.1. Evidence for clustering of genes regulating hematopoiesis and for their involvement in the deletion of the long arm of chromosome 5 in myeloid disorders. *Proc. Natl. Acad. Sci. U.S.A.,* 1987; 84:2970–2974.
37. **Cosman, D., Lyman, S. D., Idzerza, R. L., et al.,** A new cytokine receptor superfamily. *TIBS,* 1990; 15:265–270.
38. **Collins, M. K. L., Marvel, J., Malde, P., and Lopezrivas, A.,** Interleukin-3 protects murine bone marrow cells from apoptosis induced by DNA damaging agents. *J. Exp. Med.,* 1992; 176:1043–1051.
39. **Groopman, J. E., Molina, J. M., and Scadden, D. T.,** Hematopoietic growth factors. Biology and clinical applications. *N. Engl. J. Med.,* 1989; 321:1449–1459.
40. **Isaac, A. and Lindemann, J.,** The phenomenon of viral interference. *Proc. R. Soc. (London) Ser. B,* 1957; 147:258–273.
41. **Farrar, M. A. and Schreiber, R. D.,** The molecular cell biology of interferon gamma and its receptor. *Annu. Rev. Immunol.,* 1993; 11:571–611.
42. **Gray, P. W. and Goeddel, D. V.,** Cloning and expression of murine immune interferon cDNA. *Proc. Natl. Acad. Sci. U.S.A.,* 1983; 80:5842–5846.
43. **Gray, P. W., Leong, S., Fennie, E. H., et al.,** Cloning and expression of the cDNA for the murine interferon-gamma receptor. *Proc. Natl. Acad. Sci. U.S.A.,* 1989; 86:8497–8501.
44. **Wong, G. H. W., Clarklewis, I., Mckimmbreschkin, J. L., Harris, A. W., and Schrader, J. W.,** Interferon-gamma induces enhanced expression of Ia-antigens and H-2 antigens on B-lymphoid, macrophage and myeloid cell lines. *J. Immunol.,* 1983; 131:788–793.
45. **Todd, I., Pujolborrell, R., Hammond, L. J., Bottazzo, G. F., and Feldmann, M.,** Interferon-gamma induces HLA-DR expression by thyroid epithelium. *Clin. Exp. Immunol.,* 1985; 61:265–273.
46. **Krown, S. E.,** Interferons in malignancy: biological products or biological response modifiers. *J. Natl. Cancer Inst.,* 1988; 80:306–312.
47. **Garnick, M. B., Reich, S. D., Maxwell, B., Covalgoldsmith, S., Richie, J. P., and Rudnick, S. A.,** Phase I/II study of recombinant interferon gamma in advanced renal-cell carcinoma. *J. Urol.,* 1988; 139:251–255.
48. **Howard, M., Farrar, J., Hilfiker, M., et al.,** Identification of a T-cell derived B cell growth factor distinct from interleukin-2. *J. Exp. Med.,* 1982; 155:914–923.
49. **Ohara, J., Lahet, S., Inman, J., and Paul, W. E.,** Partial purification of murine B cell stimulatory factor (BSF-)-1. *J. Immunol.,* 1985; 135:2518–2523.

50. **Yokota, T., Otsuka, T., Mosmann, T. R., et al.,** Isolation and characterisation of a mouse interleukin cDNA clone, homologous to mouse B-cell stimulatory factor-1, that express B cell and T cell stimulating activity. *Proc. Natl. Acad. Sci. U.S.A.,* 1986; 83:5894–5898.
51. **Lebeau, M. M., Lemons, R. S., Espinosa, R., Larson, R. A., Arai, N., and Rowley, J. D.,** IL-4 and IL-5 map to human chromosome 5 in a region encoding growth factors and receptors and are deleted in myeloid leukemias with a del(5q). *Blood,* 1988; 73:647–653.
52. **Paul, W. E.,** Interleukin-4: a prototypic immunoregulatory lymphokine. *Blood,* 1991; 77:1859–1870.
53. **Snapper, C. M., Hooley, J. J., Atasoy, U., Finkelman, F. D., and Paul, W. E.,** Differential regulation of murine B cell FcγRII expression by CD4+ T helper subsets. *J. Immunol.,* 1989; 143:2133–2138.
54. **Mosmann, T. R., Bond, M. W., Coffman, R. L., Ohara, J., and Paul, W. E.,** T cell and mast cell lines respond to B cell stimulatory factor-1. *Proc. Natl. Acad. Sci. U.S.A.,* 1986; 83:5654–5659.
55. **Schimpl, A. and Wecker, E.,** Replacement of T cell function by a T cell product. *Nature,* 1972; 237:15–17.
56. **Azuma, C., Tanabe, T., Konishi, M., et al.,** Cloning of cDNA for human T-cell replacing factor (interleukin-5) and comparison murine homologue. *Nucl. Acids Res.,* 1986; 14:9149–9158.
57. **Takatsu, K., Harada, N., Hara, Y., et al.,** Purification and physiochemical characterisation of murine T cell replacing factor (TRF). *J. Immunol.,* 1985; 134:382–389.
58. **Cherwinski, H., Schumacher, J. H., Brown, K. D., and Mosmann, T. R.,** Two types of mouse helper T cell clone. III. Further differences in lymphokine synthesis between Th1 and Th2 clones revealed by RNA hybridisation, functionally monospecific bioassays, and monoclonal antibodies. *J. Exp. Med.,* 1987; 166:1229–1238.
59. **Devos, R., Plaetinck, G., Vanderheyden, J., et al.,** Molecular basis of a high affinity murine Interleukin-5 receptor. *EMBO J.,* 1991; 10:2133–2137.
60. **Sanderson, C. J., Warren, D. J., and Strath, M.,** Identification of a lymphokine that stimulates eosinophil differentiation *in vitro*. *J. Exp. Med.,* 1985; 162:60–79.
61. **Aoki, T., Kikuchi, H., Mitayake, S., et al.,** Interleukin-5 enhances Interleukin-2 mediated lymphokine-activated killer activity. *J. Exp. Med.,* 1989; 170:583–588.
62. **Tai, P. C., Ackerman, S. J., Spry, C. J. F., Olsen, E. G. J., Dunnette, S., and Gleich, G. J.,** Deposits of eosinophil granule proteins in cardiac tissues of patients with eosinophilic endomyocardial disease. *Lancet,* 1987; 1:643–647.
63. **Weisenbach, J., Chernajovsky, Y., Zeevi, M., et al.,** Two interferon mRNA in human fibrolasts: in vitro translation and *Escherichia coli* cloning studies. *Proc. Natl. Acad. Sci. U.S.A.,* 1980; 77:7152–7156.
64. **Hirano, T., Yasukawa, K., Harada, H., et al.,** Complementary cDNA for a novel human interleukin (BSF-2) that induces B lymphocytes to produce immunoglobulin. *Nature,* 1986; 324:73–76.
65. **Gauldie, J., Richards, C., Harnish, D., Landsorp, P., Baumann, H.,** Interferon-β2/BSF-2 shares identity with monocyte-derived hepatocyte stimulating factor (HSF) and regulates the major acute phase protein response in liver cells. *Proc. Natl. Acad. Sci. U.S.A.,* 1987; 84:7251–7255.
66. **Van Snick, J., Vink, A., Cayphas, S., and Uyttenhove, C.,** Interleukin-HP1, a T-cell-derived hybridoma growth factor that supports the *in vitro* growth of murine plasmacytomas. *J. Exp. Med.,* 1987; 165:641–649.
67. **Hirano, T.,** Interleukin-6 (IL-6) and its receptor: their role in plasma cell neoplasias. *Int. J. Cell Cloning,* 1991; 9:166–184.
68. **Nakajima, K., and Wall, R.,** Interleukin-6 signals activating JunB and Tis11 gene transcription in a B-cell hybridoma. *Mol. Cell. Biol.,* 1991; 11:1409–1418.
69. **Isshiki, H., Akira, S., Tanabe, O., et al.,** Constitutive and IL-1 inducible factors interact with the IL-1 responsive element in the IL-6 gene. *Mol. Cell. Biol.,* 1990; 10:2757–2764.
70. **Van Snick, J.,** Interleukin-6: an overview. *Annu. Rev. Immunol.,* 1990; 8:253–278.
71. **Hirano, T., Teranishi, T., and Onoue, K.,** Human helper T-cell factor(s). III. Characterization of B-cell differentiation factor-I (BCDF-I). *J. Immunol.,* 1984; 132:229–234.

72. **Lotz, M., Jirik, F., Kabouridis, P., et al.,** B-cell stimulating factor-2/interleukin-6 is a costimulant for human thymocytes and lymphocytes-T. *J. Exp. Med.,* 1988; 167:1253–1258.
73. **Kawano, M., Hirano, T., Matsuda, F., et al.,** Autocrine generation and requirement of BSF-2/IL-6 for human multiple myelomas. *Nature,* 1988; 332:83–85.
74. **Whitlock, C. A., and Witte, O. N.,** Long-term culture of lymphocytes-B and their precursors from murine bone-marrow. *Proc. Natl. Acad. Sci. U.S.A.,* 1982; 79:3608–3612.
75. **Namen, A. E., Schmierer, A. E., March, C. J., et al.,** B-cell precursor growth-promoting activity. Purification and characterisation of a growth factor active on lymphocyte precursors. *J. Exp. Med.,* 1988; 167:988–1002.
76. **Namen, A. E., Lupton, S., Hjerrild, K., et al.,** Stimulation of B-cell progenitors by cloned murine Interleukin-7. *Nature,* 1988; 333:571–573.
77. **Sutherland, G. R., Baker, E., Fernandez, K. E. W., et al.,** The gene for human interleukin-7 is at 8q12–13. *Hum. Genet.,* 1989; 82:371–372.
78. **Goodwin, R. G., Friend, D., Ziegler, S. F., et al.,** Cloning of the human and murine Interleukin-7 receptors. Demonstration of a soluble form and homology to a new receptor superfamily. *Cell,* 1990; 60:941–951.
79. **Lee, G., Namen, A. E., Gillis, S., Ellingsworth, L. R., and Kincade, P. W.,** Normal B-cell precursors responsive to recombinant murine IL-7 and inhibition of IL-7 activity by transforming growth factor-beta. *J. Immunol.,* 1989; 142:3875–3883.
80. **Okazaki, H., Ito, M., Sudo, T., et al.,** IL-7 promotes thymocyte proliferation and maintains immunocompetent thymocytes bearing alpha-beta-T-cell or gamma-delta-T-cell receptors *in vitro.* Synergism with IL-2. *J. Immunol.,* 1989; 143:2917–2922.
81. **Chantry, D., Turner, M., and Feldmann, M.,** Interleukin-7 (murine pre-B-cell growth factor lymphopoietin-1) stimulates thymocyte growth. Regulation by transforming growth factor. *Eur. J. Immunol.,* 1989; 19:783–786.
82. **Muegge, K., Vila, M. P., and Durum, S. K.,** Interleukin-7, a cofactor for V(D)J rearrangement of the T-cell receptor beta-gene. *Science,* 1993; 261:93–95.
83. **Alderson, M. R., Sassenfeld, H. M., and Widmer, M. B.,** Interleukin-7 enhances cytolytic lymphocyte-T generation and induces lymphokine-activated killer cells from human peripheral blood. *J. Exp. Med.,* 1990; 172:577–587.
84. **Miller, M. D. and Krangel, M. S.,** Biology and biochemistry of the chemokines — a family of chemotactic and inflammatory cytokines. *Crit. Rev. Immunol.,* 1992; 12:17–46.
85. **Oppenheim, J. J., Zachariae, C. O. C., Mukaida, N., and Matsushima, K.,** Properties of the novel proinflammatory supergene "intercrine" cytokine family. *Ann. Rev. Immunol.,* 1991; 9:617–648.
86. **Matsushima, K., Morishita, K., Yoshimura, T., et al.,** Molecular cloning of a human monocyte-derived neutrophil chemotactic factor (MDNCF) and the induction of MDNCF mRNA by Interleukin-1 and tumor necrosis factor. *J. Exp. Med.,* 1988; 167:1883–1893.
87. **Baggiolini, M., Walz, A., and Kunkel, S. L.,** Neutrophil activating peptide-1 Interleukin-8, a novel cytokine that activates neutrophils. *J. Clin. Inv.,* 1989; 84:1045–1049.
88. **Schroder, J. M., Mrowietz, U., and Christophers, E.,** Identification of different charged species of a human monocyte derived neutrophil activating peptide (MONAP). *Biochem. Biophys. Res. Commun.,* 1988; 152:277–284.
89. **Kelvin, D. J., Michiel, D. F., Johnston, J. A., et al.,** Chemokines and serpentines. The molecular biology of chemokine receptors. *J. Leuk. Biol.,* 1993; 54:604–612.
90. **Larsen, C. G., Anderson, A. D., Appella, E., Oppenheim, J. J., and Matsushima, K.,** The neutrophil-activating protein (NAP-1) is also chemotactic for lymphocytes-T. *Science,* 1989; 243:1464–1466.
91. **Schall, T. J.,** Biology of the RANTES/SIS cytokine family. *Cytokine,* 1991; 3:165–183.
92. **Schall, T. J., Bacon, K., Toy, K. J., and Goeddel, D. V.,** Selective attraction of monocytes and lymphocytes-T of the memory phenotype by cytokine RANTES. *Nature,* 1990; 347:669–671.
93. **Donahue, R. E., Yang, Y. C., and Clark, S. C.,** Human P40 T-cell growth factor (Interleukin-9) supports erythroid colony formation. *Blood,* 1990; 75:2271–2275.
94. **Druez, C., Coulie, P., Uyttenhove, C., and Van Snick, J.,** Functional and biochemical characterization of mouse P40/IL-9 receptors. *J. Immunol.,* 1990; 145:2494–2499.

95. **Renauld, J. C., Houssiau, F., Druez, C., Uyttenhove, C., Vink, A., and Van Snick, J.,** Interleukin-9. *Int. Rev. Exp. Pathol.,* 1993; 34:99–109.
96. **Yang, Y. C., Ricciardi, S., Ciarletta, A., Calvetti, J., Kelleher, K., and Clark, S. C.,** Expression cloning of a cDNA encoding a novel hematopoietic growth-factor. Human homologue of murine T-cell growth factor P40. *Blood,* 1989; 74:1880–1884.
97. **Mock, B. A., Krall, M., Kozak, C. A., et al.,** IL-9 maps to mouse chromosome 13 and human chromosome 5. *Immunogenetics,* 1990; 31:265–270.
98. **Fiorentino, D. F., Bond, M. W., and Mosmann, T. R.,** Two types of mouse helper T cell. IV. Th2 clones secrete a factor that inhibits cytokine production by Th1 clones. *J. Exp. Med.,* 1989; 170:2081–2095.
99. **Moore, K. W., O'garra, A., de Waal Malefyt, R., Vieira, P., and Mosmann, T. R.,** Interleukin-10. *Ann. Rev. Immunol.,* 1993; 11:165–190.
100. **Vieira, P., de Waal Malefyt, R., Dang, M. N., et al.,** Isolation and expression of human cytokine synthesis inhibitory factor (CSIF/IL-10) cDNA clones: homology to Epstein-Barr virus open reading frame BCRF1. *Proc. Natl. Acad. Sci. U.S.A.,* 1991; 88:1172–1176.
101. **Moore, K. W., Rousset, F., and Banchereau, J.,** Evolving principles in immunopathology: interleukin-10 and its relationship to Epstein-Barr virus protein BCRF1. *Springer Semin. Immunopathol.,* 1991; 13:157–166.
102. **Hsu, D. H., de Waal, Malefyt, R., Fiorentino, D. F., et al.,** Expression of IL-10 activity by Epstein-Barr virus protein BCRF1. *Science,* 1990; 250:830–832.
103. **Yssel, H., de Waal Malefyt, R., Roncarolo, M. G., et al.,** Interleukin-10 is produced by subsets of human CD4+ T cell clones and peripheral blood T cells. *J. Immunol.,* 1992; 149:2378–2384.
104. **Fiorentino, D. F., Zlotnik, A., Mosmann, T. R., Howard, M., and O'garra, A.,** IL-10 inhibits cytokine production by activated macrophages. *J. Immunol.,* 1991; 147:3815–3822.
105. **Bogdan, C., Vodovotz, Y., and Nathan, C.,** Macrophage deactivation by IL-10. *J. Exp. Med.,* 1991; 174:1549–1555.
106. **de Waal Malefyt, R., Haanen, J., Spits, H., et al.,** IL-10 and viral IL-10 strongly reduce antigen-specific human T cell proliferation by diminishing the antigen presenting capacity of monocytes via down-regulation of class II MHC expression. *J. Exp. Med.,* 1991; 174:915–924.
107. **Rousset, F., Garcia, E., Defrance, T., et al.,** IL-10 is a potent growth and differentiation factor for activated human B lymphocytes. *Proc. Natl. Acad. Sci. U.S.A.,* 1992; 89:1890–1893.
108. **Ishida, H., Hastings, R., Kearny, J., and Howard, M.,** Continuous anti-I1–10 antibody administration depletes mice of CD5 B cells but not conventional B cells. *J. Exp. Med.,* 1992; 175:1213–1220.
109. **Nordan, R. P. and Potter, M.,** A macrophage-derived factor required by plasmacytomas for survival and proliferation *in vitro. Science,* 1986; 233:566–569.
110. **Paul, S. R., Bennett, F., Calvetti, J. A., et al.,** Molecular cloning of a cDNA encoding interleukin-11, a stromal cell-derived lymphopoietic and haematopoietic cytokine. *Proc. Natl. Acad. Sci. U.S.A.,* 1990; 87:7512–7516.
111. **McKinley, D., Wu, Q., Yang-Feng, T., and Yang, Y.,** Genomic sequence and chromosomal location of human interleukin-11 gene (IL-11). *Genomics,* 1992; 13:814–819.
112. **Kawashima, I., Ohsumi, J., Mita-Honjo, K., et al.,** Molecular cloning of cDNA encoding adipogenesis inhibitory factor and identity with interleukin-11. *FEBS Lett.,* 1991; 283:199–202.
113. **Yin, T., Miyazawa, K., and Yang, Y.-C.,** Characterization of interleukin-11 receptor and protein tyrosine phosphorylation induced by interleukin-11 in mouse 3T3-L1 cells. *J. Biol. Chem.,* 1992; 267:8327–8351.
114. **Yin, T., Taga, T., Tsang, M. L.-S., Yasukawa, K., Kishimoto, T., and Yang, Y.-C.,** Involvement of IL-6 sigmal transducer gp 130 in IL-11-mediated signal transduction. *J. Immunol.,* 1993; 151:2555–2561.
115. **Bruno, E., Briddell, R. A., Cooper, R. J., and Hoffman, R.,** Effects of recombinant interleukin-11 on human megakaryocyte progenitor cells. *Exp. Haematol.,* 1991; 19:378–381.
116. **Musashi, M., Yang, Y.-C., Paul, S. R., Clark, S. C., Sudo, T., and Ogawa, M.,** Direct and synergistic effects of interleukin-11 on murine haemopoiesis in culture. *Proc. Natl. Acad. Sci. U.S.A.,* 1991; 88:765–769.
117. **Yin, T., Schendel, P., and Yang, Y.-C.,** Enhancement of *in vitro* and *vivo* antigen-specific antibody responses by interleukin-11. *J. Exp. Med.,* 1992; 175:211–216.

118. **Baumann, H. and Schendel, P.,** Interleukin-11 regulates the hepatic expression of the same plasma protein genes as interleukin-6. *J. Biol. Chem.,* 1991; 266:20424–20427.
119. **Kobayashi, M., Fitz, L., Ryan, M., et al.,** Identification and purification of natural killer cell stimulatory factor (NKSF), a cytokine with multiple biologic effects on human lymphocytes. *J. Exp. Med.,* 1989; 170:827–845.
120. **Wolf, S. F., Temple, P. A., Kobayashi, M., et al.,** Cloning of cDNA for natural killer cell stimulatory factor, a heterodimeric cytokine with multiple biologic effects on T and natural killer cells. *J. Immunol.,* 1991; 146:3074–3081.
121. **Merberg, D. M., Wolf, S. F., and Clark, S. C.,** Sequence similarity between NKSF and the IL-6/G-CSF family. *Immunol. Today,* 1992; 13:77–78.
122. **Sieburth, D., Jabs, E. W., Warrington, J. A., et al.,** Assignment of genes encoding a unique cytokine (IL-12) composed of two unrelated subunits to chromosomes 3 and 5. *Genomics,* 1992; 14:59–62.
123. **Chizzonite, R., Truitt, T., Desai, B. P., et al.,** IL-12 receptor. I. Characterisation of the receptor on phytohemagglutinin-activated human lymphoblasts. *J. Immunol.,* 1992;148:3117–3124.
124. **Chan, S. H., Perussia, B., Gupta, J. W., et al.,** Induction of interferon-gamma production by natural-killer-cell stimulatory factor — characterisation of the responder cells and synergy with other inducers. *J. Exp. Med.,* 1991; 173:869–879.
125. **D'Andrea, A., Rengaraju, M., Valiante, N. M., et al.,** Production of natural killer cell stimulatory factor (interleukin-12) by peripheral blood mononuclear cells. *J. Exp. Med.,* 1992; 176:1387–1398.
126. **Manetti, R., Parronchi, P., Giudizi, M. G., et al.,** Natural-killer-cell stimulatory factor (interleukin-12 (IL-12)) induces T-helper type I (Th1)-specific immune-responses and inhibits the development of IL-4-producing Th cells. *J. Exp. Med.,* 1993; 177:1199–1204.
127. **Hseih, C. S., Macatonia, S. E., Tripp, C. S., Wolf, S. F., O'garra, A., and Murphy, K. M.,** Development of Th1 CD4+ T-cells through IL-12 produced by Listeria-induced macrophages. *Science,* 1993; 260:547–549.
128. **Trinchieri, G.,** Interleukin-12 and its role in the generation of TH1 cells. *Immunol. Today,* 1993; 14:335–338.
129. **Tripp, C. S., Wolf, S. F., and Unanue, E. R.,** Interleukin-12 and tumor necrosis factor α are costimulators of interferon γ production by natural killer cells in severe combined immunodeficiency mice with listeriosis, and interleukin-10 is a physiologic antagonist. *Proc. Natl. Acad. Sci. U.S.A.,* 1993; 90:3725–3729.
130. **Locksley, R. M.,** Interleukin-12 in host defence against microbial pathogens. *Proc. Natl. Acad. Sci. U.S.A.,* 1993; 90:5879–5880.
131. **Gazzinelli, R. T., Hieny, S., Wynn, T. A., Wolf, S., and Sher, A.,** Interleukin-12 is required for the T-lymphocyte-independent induction of interferon γ by an intracellular parasite and induces resistance in T-cell-deficient hosts. *Proc. Natl. Acad. Sci. U.S.A.,* 1993; 90:6115–6119.
132. **Heinzel, F. P., Schoenhaut, D. S., Rerko, R. M., Rosser, L. E., and Gately, M. K.,** Recombinant interleukin-12 cures mice infected with *Leishmania major. J. Exp. Med.,* 1993; 177:1505–1509.
133. **Clerici, M., Lucey, D. R., Berzofsky, J. A., et al.,** Restoration of HIV-specific cell-mediated immune responses by interleukin-12 *in vitro. Science,* 1993; 262:1721–1724.
134. **Cherwinski, H. M., Schumacher, J. H., Brown, K. D., and Mosmann, T. R.,** Two types of mouse helper T cell clone. III. Further differences in lymphokine synthesis between Th1 and Th2 clones revealed by RNA hybridisation, functionally monospecific bioassays and monoclonal antibodies. *J. Exp. Med.,* 1987; 166:1229–1244.
135. **Brown, K. D., Zurawski, S. M., Mosmann, T. R., and Zurawski, G.,** A family of small inducible proteins secreted by leukocytes are members of a new superfamily that includes leukocyte and fibroblast-derived inflammatory agents, growth factors and indicators of various activation processes. *J. Immunol.,* 1989; 142:679–687.
136. **McKenzie, A. N. J., Culpepper, J. A., de Waal Malefyt, R., et al.,** Interleukin-13, a T-cell-derived cytokine that regulates human monocyte and B-cell function. *Proc. Natl. Acad. Sci. U.S.A.,* 1993; 90:3735–3739.
137. **Minty, A., Chalon, P., Derocq, J. M., et al.,** Interleukin-13 is a new human lymphokine regulating inflammatory and immune responses. *Nature,* 1993; 362:248–250.

138. **McKenzie, A. N. J., Li, X., Largaespada, D. A., et al.,** Structural comparison and chromosomal localisation of the human and mouse IL-13 genes. *J. Immunol.,* 1993; 150:5436–5444.
139. **Punnonen, J., Aversa, G., Cocks, B. G., et al.,** Interleukin-13 induces interleukin-4-independent IgG4 and IgE synthesis and CD23 expression by human B cells. *Proc. Natl. Acad. Sci. U.S.A.,* 1993; 90:3730–3734.
140. **Zurawski, S. M., Vega, F., Huyghe, B., and Zurawski, G.,** Receptors for interleukin-13 and interleukin-4 are complex and share a novel component that functions in signal transduction. *EMBO J.,* 1993; 12:2663–2670.
141. **Montaner, L. J., Doyle, A. G., Collin, M., et al.,** Interleukin-13 inhibits Human Immunodeficiency Virus type I production in primary blood-derived human macrophages *in vitro. J. Exp. Med.,* 1993; 178:743–747.
142. **Ambrus, J. L., Jurgensen, C. H., Brown, E. J., and Fauci, A. S.,** Purification to homogeneity of a high molecular weight human B cell growth factor; demonstration of specific binding to activated B cells and development of a monoclonal antibody to the factor. *J. Exp. Med.,* 1985; 162:1319–1335.
143. **Ambrus, J. L., Jurgensen, C. H., Brown, E. J., McFarland, P., and Fauci, A. S.,** Identification of a receptor for High Molecular Weight Human B Cell Growth Factor. *J. Immunol.,* 1988; 141:861–869.
144. **Uckun, F. M., Fauci, A. S., Chandan-Langlie, M., Myers, D. E., and Ambrus, J. L.,** Detection and characterisation of Human High Molecular Weight B Cell Growth Factor receptors on leukemic B cells in chronic lymphocytic leukemia. *J. Clin. Invest.,* 1989; 84:1595–1608.
145. **Ambrus, J. L., Chesky, L., Stephany, D., McFarland, P., Mostowski, H., and Fauci, A. S.,** Functional studies examining the sub-population of human B lymphocytes responding to high molecular weight B cell growth factor. *J. Immunol.,* 1990; 145:3949–3955.
146. **Ambrus, J. L., Pippin, J., Joseph, A., et al.,** Identification of a cDNA for a human high-molecular-weight B-cell growth factor. *Proc. Natl. Acad. Sci. U.S.A.,* 1993; 90:6330–6334.
147. **Grabstein, K. H., Eisenman, J., Shanebeck, K., et al.,** Cloning of a T cell Growth factor that interacts with the b chain of the Interleukin-2 receptor. *Science,* 1994; 264:965–968.
148. **Ruddle, N. H. and Waksman, B. H.,** Cytotoxycity mediated by soluble antigen and lymphocytes in delayed hypersensitivity. III. Analysis of mechanism. *J. Exp. Med.,* 1968; 128:1267–1279.
149. **Granger, G. A. and Williams, T. W.,** Lymphocyte cytotoxycity *in vitro*. Activation and release of a cytotoxic factor. *Nature,* 1968; 218:1253–1254.
150. **Gray, P. W., Aggarwal, B. B., Benton, C. V., et al.,** Cloning and expression of cDNA for human lymphotoxin, a lymphokine with tumour necrosis activity. *Nature,* 1984; 312:721–724.
151. **Shalaby, M. R., Aggarwal, B. B., Rinderknecht, E., Svedersky, L. P., Finkle, B. S., and Palladino, M. A.,** Activation of human polymorphonuclear neutrophil functions by interferon gamma and tumour necrosis factor. *J. Immunol.,* 1985; 135:2069–2073.
152. **Paul, N. L. and Ruddle, N. H.,** Lymphotoxin. *Ann. Rev. Immunol.,* 1988; 6:407–438.
153. **Tite, J. P., Powell, M. B., and Ruddle, N. H.,** Protein antigen specific Ia-restricted cytolytic T-cells. Analysis of frequency, target-cell susceptibility, and mechanism of cytolysis. *J. Immunol.,* 1985; 135:25–33.
154. **Schmid, D. S., Hornung, R., McGrath, K. M., Paul, N., and Ruddle, N. H.,** Target cell DNA fragmentation is mediated by lymphotoxin and tumor necrosis factor. *Lymph. Res.,* 1987; 6:195–202.
155. **Beutler, B. and Cerami, A.,** Cachectin and tumor necrosis factor as two sides of the same biological coin. *Nature,* 1986; 320:584–588.
156. **Old, L. J.,** Tumor necrosis factor (TNF). *Science,* 1985; 230:630–632.
157. **Atkins, E.,** Fever: its history, cause and function. *Yale J. Biol. Med.,* 1982; 55:283–289.
158. **Selmaj, K. W., Farooq, M., Norton, W. T., Raine, C. S., and Brosnan, C. F.,** Proliferation of astrocytes *in vitro* in response to cytokines. A primary role for tumor necrosis factor. *J. Immunol.,* 1990; 144:129–135.
159. **Watanabe, Y. and Jacob, C. O.,** Regulation of MHC class II antigen expression. Opposing effects of tumor necrosis factor-alpha on IFN-gamma induced HLA-DR and Ia expression depends on the maturation and differentiation stage of the cell. *J. Immunol.,* 1991; 146:899–905.

160. **Vassalli, P.,** The pathophysiology of tumor Necrosis factors. *Annu. Rev. Immunol.,* 1992; 10:411–452.
161. **Sporn, M. B., Roberts, A., Wakefield, L. M., and Assoian, R. K.,** Transforming growth factor-β: biological function and chemical structure. *Science,* 1986; 233:532–534.
162. **O'Grady, P., Kuo, M. D., Baldassare, J. J., Huang, S. S., and Huang, J. S.,** Purification of a new type high-molecular weight receptor (type V receptor) of transforming growth factor beta (TGF-β) from bovine liver. Identification of the type V TGF-β receptor in cultured cells. *J. Biol. Chem.,* 1991; 266:8583–8589.
163. **Mustoe, T. A., Pierce, G. F., Thomason, A., Gramates, P., Sporn, M. B., and Deuel, T. F.,** Accelerated healing of incisional wounds in rates induced by transforming growth factor beta. *Science,* 1987; 237:1333–1336.
164. **Swain, S. L., Huston, G., Tonkonogy, S., and Weinberg, A.,** Transforming growth factor-beta and IL-4 cause Helper T-cell precursors to develop into distinct effector helper-cells that differ in lymphokine secretion pattern and cell-surface phenotype. *J. Immunol.,* 1991; 147:2991–3000.
165. **Kehrl, J. H., Thevenin, C., Rieckmann, P., and Fauci, A. S.,** Transforming growth factor beta suppresses human lymphocyte-B Ig production by inhibiting synthesis and the switch from the membrane form to the secreted form of Ig mRNA. *J. Immunol.,* 1991; 146:4016–4023.
166. **Stoeck, M., Miescher, S., Barras, C., and Vonfliedner, V.,** Transforming growth factor beta-1 and beta-2 interfere with IL-2 driven and IL-4 driven proliferation of lymphocyte-T. *Immunobiology,* 1994; 178:85, (abstract).
167. **Wahl, S. M., Hunt, D. A., Wong, H. L., et al.,** Transforming growth factor-beta is a potent immunosuppressive agent that inhibits Il-1 dependent lymphocyte proliferation. *J. Immunol.,* 1988; 140:3026–3032.
168. **Brown, T. J., Lioubin, M. N., and Marquardt, H.,** Purification and characterisation of cytostatic lymphokines produced by activated human lymphocytes-T. Synergistic antiproliferative activity of TGF-β1, IFN-γ, and Oncostatin M for human melanoma cells. *J. Immunol.,* 1987; 139:2977–2983.
169. **Rose, T. M. and Bruce, A. G.,** Oncostatin M is a member of a cytokine family that includes leukemia inhibitory factor, G-CSF and Interleukin-6. *Proc. Natl. Acad. Sci. U.S.A.,* 1991; 88:8641–8645.
170. **Grove, R. I., Mazzucco, C. E., Radka, S. F., Shoyab, M., and Kiener, P. A.,** Oncostatin M upregulates low-density lipoprotein receptors in HepG2 cells by a novel mechanism. *J. Biol. Chem.,* 1991; 266:18194–18199.
171. **Brown, T. J., Rowe, J. M., Liu, J. W., and Shoyab, M.,** Regulation of Il-6 expression by Oncostatin-M. *J. Immunol.,* 1991; 147:2175–2180.
172. **Tomida, M., Yamamototiyamaguchi, Y., and Mozumi, M.,** Purification of a factor inducing differentiation of mouse myeloid leukemic M1-cells from conditioned medium of mouse fibroblast L929-cells. *J. Biol. Chem.,* 1984; 259:978–9482.
173. **Wetzler, M., Talpaz, M., Lowe, D. G., Baiocchi, G., Guttermann, J. U., and Kurzrock, R.,** Constitutive expression of leukemia inhibitory factor RNA by human bone-marrow stromal cells and modulation by IL-1, TNF-α and TGF-β. *Exp. Haematol.,* 1991; 19:347–351.
174. **Bhatt, H., Brunet, L. J., and Stewart, C. L.,** Uterine expression of leukemia inhibitory factor coincides with the onset of blastocyst implantation. *Proc. Natl. Acad. Sci. U.S.A.,* 1991; 88:11408–11412.
175. **Metcalf, D., Hilton, D., and Nicola, N. A.,** Leukemia inhibitory factor can potentiate murine megakaryocyte production *in vitro. Blood,* 1991; 77:2150–2153.
176. **Mosmann, T. R. and Moore, K. W.,** The role of IL-10 in cross regulation of TH1 and TH2 responses, *Immunoparasitology Today,* A49–A53, 1991.
177. **Center, D. M., Berman, J. S., Kornfeld, H., Theodore, A. C., and Cruickshank, W. W.,** The lymphocyte chemoattractant factor, *J. Lab. Clin. Med.,* 125, 167–172, 1995.

Chapter 3

CYTOKINES IN INFLAMMATION: AN OVERVIEW

E. Roy Pettipher

CONTENTS

I. INTRODUCTION

Cytokines, a large group of non-antibody proteins which include interleukins, interferons, growth factors, and colony stimulating factors (see Chapter 1), have many diverse and overlapping biological activities which are relevant to all aspects of the inflammatory response. Cytokines can have both local and systemic effects on the host, but differ from classical hormones in the diversity of their cellular sources and the multiplicity of their biological effects. Much research has focused on the cells involved in the immune response as important sources of cytokines (reviewed in Chapter 2). Tissue macrophages and circulating monocytes can produce cytokines such as IL-1, TNFα, and IL-8, while T lymphocytes produce distinct patterns of lymphokines depending on their phenotype, i.e., Th1 cells, which mediate delayed-type hypersensitivity, produce IL-2 and interferon-γ, while Th2 cells produce IL-4 and IL-5 and are associated with allergic disease. It should be emphasized, however, that nonimmune cells such as endothelial cells, fibroblasts, and epithelial cells also represent important sources of cytokines.

0-8493-8381-1/96/$0.00+$.50

The inflammatory effects of cytokines range from stimulation of the acute phase response to activation of leukocytes and connective tissue cells. The systemic overproduction of cytokines can lead to physiological effects such as fever and acute phase protein production, or pathological effects such as sustained hypotension and widespread increased vascular permeability, such as occurs in septic shock. Local production of cytokines can lead to activation of the endothelium, leukocyte infiltration, angiogenesis, and connective tissue remodeling.

It is unlikely that any particular cytokine acts in isolation at sites of inflammation and it is becoming clear that cytokines act synergistically to bring about various inflammatory events. For example, systemic administration of combinations of IL-1 and TNFα are much more toxic than the sum of the effects of either cytokine alone. As the combination of many pro-inflammatory cytokines acting in concert is likely to have a devastating effect on the integrity of host tissues, it is not surprising that many systems have evolved to regulate cytokine production and activity. These inhibitory systems include endogenous receptor antagonist proteins, soluble cytokine receptors, and anti-inflammatory cytokines which can decrease pro-inflammatory cytokine synthesis.

The role of cytokines in the systemic response to infection such as fever, the acute phase response, and septic shock has been dealt with elsewhere.[1] This review will focus on the local actions of the cytokines which mediate tissue injury in chronic inflammatory diseases.

II. EFFECTS OF CYTOKINES ON LEUKOCYTE EMIGRATION

Neutrophil adhesion to postcapillary venules and subsequent emigration into tissues are cellular hallmarks of acute inflammation which can be elicited by the action of cytokines. The process of neutrophil infiltration into tissues involves a number of discrete steps — firstly, neutrophils begin to roll on the endothelium of postcapillary venules, dramatically reducing their velocity and increasing their contact with the endothelium. Then, if the neutrophils are activated by contact with chemotactic factors, they firmly adhere to the endothelium and subsequently migrate through gaps in the endothelium to infiltrate the inflamed tissue.

A. AN *IN VITRO* MODEL FOR LEUKOCYTE-ENDOTHELIAL CELL INTERACTIONS

Our understanding of the molecular interactions underlying these various steps of leukocyte adhesion and diapedesis have increased dramatically in the last few years based largely on *in vitro* experiments. Leukocyte rolling has been modeled *in vitro* and shown to be mediated by lectin-sugar interactions.[2] The expression of the lectin E-selectin, formerly known as ELAM-1, can be induced on endothelial cells by the action of cytokines IL-1 or TNFα.[3] Purified preparations of E-selectin can support leukocyte rolling *in vitro*[4] and antibodies against E-selectin can inhibit neutrophil binding to cytokine-activated endothelial cells.[3,5] The neutrophil ligand(s) for E-selectin are believed to be structures expressing the sialylated tetrasaccharides sialyl Lewis X[6] or sialyl Lewis A[7] which are constitutively expressed.

However, neutrophils do not flatten and firmly adhere to endothelial cells until they interact with a chemotactic factor. Chemotactic factor activation of neutrophils leads to a rapid increase in the affinity/avidity of the CD11/CD18 integrin complexes without increased surface expression of these molecules.[8] Chemotactic cytokines or chemokines produced from endothelial cells are likely to be important in mediating the initial activation and firm adhesion of leukocytes. In the case of neutrophils, the α-chemokine or C-X-C family of cytokines, which includes IL-8, NAP-2, and GROα, are likely to be involved. Interleukin-8 has been shown to be produced from endothelial cells *in vitro* in response to IL-1 and TNFα[9] and is selectively chemotactic for neutrophils.[10] As IL-8 is a heparin-binding protein, it has been postulated that it can bind to glycosaminoglycan-containing matrix surrounding the endothelial cell, which could serve as a mechanism to present chemotactic factors to neutrophils rolling on the endothelium of postcapillary venules.[11] The importance of IL-8 in mediating this process has been demonstrated by the ability of neutralizing antibodies to block transendothelial migration of neutrophils in an *in vitro* model of the blood vessel wall.[12]

Similar mechanisms exist for monocyte and lymphocyte adhesion, although members of the β-chemokine or C-C family are responsible for the activation of mononuclear cells. β-Chemokines include RANTES, MCP-1, and MIP1α and β and these have have been shown to be chemotactic for monocytes and some lymphocyte subpopulations[13] (see Chapter 2). The adhesion molecules expressed on endothelial cells which bind mononuclear cells are also known and have been designated ICAM-1[14] and VCAM-1,[15] both members of the immunoglobulin superfamily. VCAM-1 can be induced on endothelial cells by the addition of IL-1, TNFα, or IL-4 and binds to VLA-4 expressed on monocytes, lymphocytes, and eosinophils, but not neutrophils.[16,17] ICAM-1 is constitutively expressed and can be further upregulated by incubation of endothelial cells with IL-1 or TNFα.[14] ICAM-1 can bind both mononuclear cells and neutrophils.

These experiments highlight the importance of cytokines in activation of both endothelial cells and leukocytes in *in vitro* models of leukocyte-endothelial cell interactions.

B. CYTOKINES AND LEUKOCYTE EMIGRATION *IN VIVO*

The cytokines interleukin-1 and tumor necrosis factor cause accumulation of neutrophils and monocytes when injected at low concentrations into tissues of laboratory animals.[18,19] In fact, IL-1 is the most potent inflammatory mediator so far described for eliciting leukocyte infiltration, with IL-1α and β being equipotent in this regard.[20] The effects are particularly striking when IL-1 and TNFα are combined, leading to a synergistic increase in leukocyte infiltration.[21] The mechanism for the synergistic interaction between IL-1 and TNFα has not been elucidated but is likely to be physiologically relevant and extend to other cytokine interactions. Although IL-1 and TNFα induce leukocyte accumulation *in vivo,* they are not directly chemotactic and, unlike chemotactic factors, act by a process requiring *de novo* protein synthesis.[22,23] These results are consistent with newly synthesized proteins such as E-selectin and the chemokine IL-8, being required for the initiation of leukocyte infiltration in response to IL-1 and TNFα. In fact, the time course of neutrophil infiltration in response to IL-1 and TNFα is delayed compared to that of chemotactic factors.[19] It is not yet proven that the *in vivo* responses to IL-1 and TNFα involve E-selectin, although it is known that antibodies to E-selectin can inhibit leukocyte infiltration in response to immune complex-induced lung injury,[24] a model where IL-1 and TNFα have been implicated by *in vivo* antibody-blocking experiments.[25,26] Investigation of the role of IL-8 in neutrophil infiltration in experimental models of inflammation is hampered by the findings that rats and mice produce a homolog of human GROα rather than IL-8 in response to cytokine stimulation. However, the production of both IL-8 and GROα has been demonstrated in the rabbit, and antibodies against IL-8 have recently been shown to inhibit lung reperfusion injury[27] and IL-1- and LPS-induced synovitis[28] in this species.

The effects of cytokines on mononuclear cell emigration into tissues have been studied less extensively than those on neutrophils. However, infiltration of monocytes in response to IL-1 and TNFα has been noted.[20] As IL-8 is not chemotactic for monocytes it is possible that members of the β-chemokine family such as MCP-1, MIP1α/β, and RANTES are induced by IL-1 and TNFα and are responsible for the observed monocytic infiltrate.[13] It has recently been demonstrated that injection of human RANTES in dog skin can induce a rich monocytic infiltrate, while the numbers of infiltrating neutrophils were low.[29] Interestingly, RANTES also stimulated the accumulation of large numbers of eosinophils in these experiments, which raises the possibility that RANTES may participate in the cellular component of allergic disease.

Generally, it has proven difficult to elicit lymphocyte infiltration in response to single mediators. However, TNFα has been found capable of eliciting lymphocyte infiltration *in vivo,* while IL-1α or TNFβ are less active under similar conditions.[30] Interferon-γ is also effective in these assays and combinations of some of these cytokines are synergistic in respect to stimulating lymphocyte influx *in vivo.* Although TNFα and interferon-γ are known to upregulate endothelial ligands for lymphocyte adhesion such as ICAM-I and VCAM-1, these cytokines are not directly chemotactic for lymphocytes. Members of the β-chemokine family are candidates for the secondary lymphocyte chemotactic factors, but further studies need to be conducted to determine whether they can fulfill this role *in vivo.* The α-chemokine, IL-8, can cause the accumulation of CD4+ T cells when applied subcutaneously by a mechanism dependent on infiltrating neutrophils.[31]

III. PRIMING EFFECTS OF CYTOKINES

In addition to the direct effects of cytokines on leukocyte and endothelial cell function, cytokines can prime cells to respond to other mediators. GM-CSF and TNFα can prime neutrophils such that responses to chemotactic factors are greatly enhanced. These exaggerated responses include generation of reactive oxygen species and production of lipoxygenase products of arachidonic acid metabolism.[32–34] One priming response that has been studied at the molecular level is the increased production of LTB_4 by neutrophils in response to GM-CSF. The priming effect is mediated both at the level of availability of substrate and activation of 5-lipoxygenase.[34] Thus, GM-CSF can increase the production of the 5-lipoxygenase protein by a posttranscriptional mechanism[35] and GM-CSF and TNFα have both been shown to increase the expression of message for the 5-lipoxygenase activating protein (FLAP), a membrane-bound protein which serves as a docking site for 5-lipoxygenase.

The mechanisms for the priming effects of cytokines *in vivo* are less well defined but can be particularly dramatic, e.g., GM-CSF can convert a normally nontoxic dose of LPS into a lethal one and is accompanied by striking elevations in circulating levels of both TNFα and IL-6.[36] This phenomenon is not related to increases in circulating or resident macrophages and is extremely rapid. This property is not shared by G-CSF, which in fact reduces TNFα production *in vivo.*[36] As GM-CSF is present in rheumatoid joints[37] these priming effects may be of importance in the pathogenesis of rheumatoid arthritis.

In allergic asthma, eosinophils are considered to play an important role as they are present in allergic airways and can release cationic proteins which damage the epithelial cells. Interleukin-5 is a differentiation and growth factor for eosinophils[38] that can directly activate the effector functions of mature eosinophils[39] and can increase the responsiveness of eosinophils to other mediators.[40] Interleukin-5 can also prime basophils to release exaggerated amounts of histamine and leukotrienes, mediators which contribute to allergic bronchospasm and congestion in asthma.[41]

Thus it can be envisaged that the direct actions of cytokines and other mediators can be amplified by the priming actions of other cytokines, and these cellular synergistic responses provide a mechanism whereby vanishingly small quantities of one particular cytokine, which alone would not provoke significant biological effects, can make a dramatic contribution to the host's response to injury. The role of 5-lipoxygenase in the control of cytokine synthesis is briefly reviewed in Chapter 5.

IV. INDUCTION OF PROSTAGLANDIN SYNTHESIS

It has been recognized for some time that the oxidative metabolism of arachidonic acid is increased in inflamed tissues and that some of the cyclooxygenase products such as prostaglandin E_2 (PGE_2) and prostacyclin (PGI_2) are potent inflammatory mediators.[42] Prostaglandins E_2 and I_2 are potent vasodilators that can also increase the sensitivity of nerve endings to chemical and mechanical stimuli and by these mechanisms contribute to hyperemia, edema, and pain. Indeed, the anti-inflammatory action of nonsteroidal anti-inflammatory drugs (NSAIDs) can be explained by their ability to inhibit prostaglandin synthesis in inflamed tissues.[43] Recently, it has been discovered that there are two distinct forms of cyclooxygenase — a constitutive form present in normal tissues (COX 1) and an inducible form (COX 2). COX 2 is induced by the action of cytokines and its synthesis is regulated by corticosteroids.[44] The rapid production of prostaglandins, as occurs in response to bradykinin in many cells, is believed to be due to the activity of COX 1. This enzyme is postulated to have a physiological role in the production of vasodilator prostaglandins to maintain blood flow to the gastric mucosa and kidney. In contrast, prostaglandin production in response to cytokines such as IL-1 and TNFα is delayed and is due to the synthesis of COX 2 in inflamed tissues. The recent findings that COX 2 is responsible for prostaglandin synthesis in inflamed tissues gives hope that agents can be discovered which selectively inhibit COX 2 but do not compromise prostaglandin synthesis in the gastric mucosa or kidney, and hence may be free of mechanism-related side effects.

Although prostaglandins can enhance pain responses to chemical and mechanical stimuli and increase blood flow to sites of inflammation, these arachidonic acid metabolites can also inhibit leukocyte function, an effect which is mediated by elevation of intracellular cAMP. Interestingly, responses which are regulated in this way include the production of some cytokines. Low ng/ml concentrations of PGE_2 can suppress TNFα production at the level of the message,[45] while the production of other cytokines such as IL-1 or IL-8 are not inhibited by prostaglandin treatment. Furthermore, inhibition of prostaglandin formation *in vitro* or *in vivo* enhances TNFα production, which may lead to exacerbation of tissue damage. In a model of endotoxic shock in mice, both indomethacin and ibuprofen enhance TNFα production and increase mortality at concentrations which inhibit prostaglandin formation (Pettipher and Wimberly, in press). This negative feedback mechanism has also been shown to occur in humans where ibuprofen further elevated plasma levels of TNFα in volunteers infused with low doses of LPS.[46,47] NSAIDs have also been shown to increase TNFα production locally at sites of inflammation.[48] which may explain the exacerbation of cartilage degradation seen with NSAIDs in some models of arthritis.[49,50] Thus, although prostaglandins are involved in mediating the pain and increased blood flow associated with acute inflammation, they also have important modulatory effects on cytokine production and, consequently, inhibition of their production by NSAID treatment may exacerbate cytokine-mediated tissue injury.

V. INDUCTION OF NITRIC OXIDE SYNTHASE

Nitric oxide (NO) is synthesized by mammalian cells and is involved in a variety of physiological functions including the regulation of blood pressure, neurotransmission, and microbicidal activity. Arginine is the biochemical source of NO which, in endothelial and neuronal cells, is converted to citrulline by a calcium-dependent constitutive enzyme which releases small quantities of NO as a co-product. This pathway operates under physiological conditions and is important in processes such as the control of vascular tone and in neurotransmission in central and peripheral nerves. However, at sites of infection or inflammation bacterial products or cytokines can induce the synthesis of a calcium-independent NO synthase in macrophages, smooth muscle cells, and other cells, which can generate large quantities of NO. Increased nitrite (a stable breakdown product of NO) has been detected at sites of inflammation such as in RA synovial fluid,[51] and the presence of the inducible NO synthase has been demonstrated biochemically and immunohistologically in inflammatory bowel disease and asthma.[52,53] Increased NO production, by elevating cGMP in the smooth muscle of arterioles and causing vasodilatation, is likely to contribute to the increased blood flow to inflamed tissue which, in turn, will potentiate leakage of plasma from the lumen of blood vessels into tissue spaces. In addition to the vasoactive properties, NO can react with the superoxide radical to generate peroxynitrite which can cause tissue injury by decomposing to the hydroxy radical. It is not yet known whether the peroxynitrite pathway is operative *in vivo,* but under physiological conditions *in vitro* Kupffer cells can be induced to generate NO-dependent chemiluminescence (an indirect measurement of OH· production).[54] Both TNFα and IL-1 have been shown to be important in the induction of NO synthase *in vivo* as demonstrated by the use of blocking antibodies and receptor antagonists.[55,56]

Inhibitors of the arginine-NO pathway, such as monomethyl-L-arginine, have been tested *in vivo* and have been found to reduce inflammation in various models including immune complex-induced lung injury,[57] acetic acid-induced colitis,[58] and experimental arthritis.[59] To date, the compounds tested have been nonselective inhibitors of both the constitutive and inducible enzymes. It would be of interest to test a specific inhibitor of the inducible NO synthase since inhibition of the constitutive NO synthase may compromise basal blood flow to tissues independently of the inflammatory reaction.

VI. CARTILAGE DEGRADATION

Interleukin-1 and TNFα can stimulate living cartilage to undergo resorption, a process which involves degradation of both its collagen and proteoglycan components. The process is presumed to involve the production of proteases which has been demonstrated in isolated chondrocytes. The induction of matrix metalloproteinases (MMPs) by IL-1 and TNFα has been particularly well studied, the combined action of which can degrade all known constituents of the articular cartilage extracellular matrix at neutral pH. In fact, interstitial collagenase (MMP 1) is the only known mammalian enzyme which can degrade native triple-helical collagen, although type II collagen is a poor substrate for this enzyme compared to type I collagen. Other metalloproteinases such as stromelysin (MMP 3) and gelatinases (MMPs 2 and 9) are also induced by IL-1 and TNFα in chondrocytes and can degrade cartilage proteoglycan. These metalloproteinases cleave aggrecan between the G1 and G2 domains at the Asn 341-Phe 342 bond, while in IL-1-stimulated cartilage aggrecan is cleaved at the Glu 373-Ala 374 bond.[60] The difference in the aggrecan degradation pattern between the metalloproteinases and IL-1 casts doubt on the role of stromelysin in cartilage proteoglycan loss induced by IL-1. The activity in cartilage which cleaves proteoglycan in response to IL-1 has been called "aggrecanase" but the identity of the enzyme has not yet been established.

Interleukin-1 can stimulate articular cartilage proteoglycan loss *in vivo* after intra-articular injection,[18,61–63] an effect which is independent of the infiltrating neutrophils.[63,64] This suggests that IL-1 causes proteoglycan loss by a direct action on the chondrocytes in cartilage, a view supported by findings of elevated stromelysin levels in joint fluid after intra-articular injection of IL-1.[63] The levels of stromelysin correlated with the levels of proteoglycan fragments in joint fluid[63] although this does not prove a causal relationship. Furthermore, the vast majority of the stromelysin detected in biological systems is in an inactive proform and it has proven difficult to detect stromelysin in a catalytically active form either *in vitro* or *in vivo*.

Although *in vitro* experiments suggest that neutrophils can digest cartilage proteoglycan by releasing elastase and cathepsin G, several studies indicate infiltrating neutrophils do not contribute to cytokine-mediated cartilage degradation. Firstly, cartilage proteoglycan loss after cytokine injection is not reduced in neutropenic animals[64] or in animals where neutrophil accumulation is blocked by treatment with anti-CD18 antibodies.[63] Secondly, injection of low doses of LPS can elicit equivalent neutrophil infiltration to that induced by IL-1, but unlike IL-1, LPS does not stimulate cartilage breakdown acutely.[18] Similar observations have been made with IL-8, which stimulates neutrophil accumulation but no cartilage proteoglycan loss after intra-articular injection.[65] However, LPS also causes the accumulation of monocytes/macrophages and if these cells are further activated with a second dose of LPS, cartilage degradation proceeds rapidly (Pettipher and Henderson, unpublished observations). The second LPS stimulus is likely to cause massive release of IL-1 as occurs *in vitro* when monocytes are given two consecutive LPS stimulations.[66] Indeed, infiltrating monocytes have been shown to be a source of IL-1 in rabbit joints injected with LPS, and blocking the activity of IL-1 can reduce LPS-induced cartilage degradation.[67] Although it has often been assumed that neutrophils are responsible for cartilage degradation in septic arthritis, it has been demonstrated experimentally that even when a suppurative arthritis was induced in rabbit knee joints by the intra-articular injection of live *Haemophilus influenzae,* blockade of neutrophil infiltration did not reduce cartilage degradation.[68]

In summary, the cytokines IL-1 and TNFα stimulate cartilage degradation by an action which is independent of infiltrating leukocytes. The importance of these cytokines in the cartilage degradation that occurs in inflammatory and degenerative joint disease will be determined by the use of selective antagonists or blocking antibodies.

VII. CYTOKINE INHIBITORS

It is clear that cytokines have many potent pro-inflammatory effects and their importance as endogenous mediators of inflammatory responses can be inferred from the existence of several effective mechanisms for their neutralization and removal. Various mechanisms to regulate cytokine production and activity are discussed below.

A. CYTOKINE RECEPTOR ANTAGONISTS

Presently, the only known cytokine receptor antagonist is the interleukin-1 receptor antagonist, IL-1ra. IL-1ra is a 22-kDa protein which was originally isolated from monocytes activated with adherent immune complexes or IgG[69] and from the urine of patients with fever or myelocytic leukemia.[70] IL-1ra has sinced been cloned and expressed,[71] shown to share 30% amino acid homology with IL-1β and 19% with IL-1α, and to displace IL-1 from both the type I and type II receptor with a similar potency to that of IL-1 itself.[72,73] IL-1ra can effectively antagonize the effects of IL-1 *in vitro* and *in vivo,* although 100-fold molar excess of IL-1ra is required to inhibit most biological responses because only a low degree of receptor occupation is required for IL-1 to elicit a maximal response. Monocytes can be activated to produce IL-1ra and during the process of aging into macrophages begin to constitutively express IL-1ra, a phenomenon which is enhanced in the presence of GM-CSF.[74] Neutrophils can also be induced to produce IL-1ra by the addition of TNFα or GM-CSF and IL-1ra represents the major *de novo* synthesized protein secreted by neutrophils.[75] Although neutrophils produce approximately 20× less IL-1ra than monocytes on a per cell basis, the large numbers of neutrophils present at inflamed sites gives this cellular source biological relevance. Indeed, IL-1ra has been detected in high concentrations in RA synovial fluid, where the levels correlated with the synovial fluid neutrophil count and where IL-1ra synthesis by synovial fluid neutrophils was demonstrated *ex vivo.*[76]

In addition to RA, IL-1ra has been detected in the plasma of septic shock patients and is rapidly produced after endotoxin infusion in normal volunteers.[77] Thus, IL-1ra is produced in response to a variety of inflammatory insults and serves to downregulate the tissue-damaging potential of IL-1. Consequently, a deficiency in IL-1ra production may cause an imbalance between these pro- and anti-inflammatory cytokines and may predispose to chronic inflammatory disease. This has been nicely demonstrated in Lyme's disease patients where a low level of IL-1ra relative to IL-1 was a poor prognostic factor, while high levels of IL-1ra predicted recovery.[78] The role of IL-1ra as a therapeutic agent is discussed in detail in Chapter 9.

B. SOLUBLE RECEPTORS

The shedding of cytokine receptors upon cell activation was first noted with IL-2 receptors.[79] Symons et al.[80] later showed that the levels of soluble IL-2 receptors in synovial fluid correlated with disease activity in RA and suggested that these receptors may have anti-inflammatory activity. Soluble TNFα receptors of both types (p55 and p75) are released during inflammatory reactions[81] and have been detected in high concentrations in RA synovial fluid.[82] Since these receptors can serve as TNFα inhibitory proteins, their production may regulate TNFα-mediated inflammatory responses. Of particular interest is the observation that activated neutrophils release soluble TNFα receptors[83] and neutrophil elastase can liberate a ligand-binding portion from the 75-kDa TNFα receptor.[84] Since neutrophils are by far the most abundant cell-type in RA synovial fluid, these cells may represent the most relevant source of TNFα inhibitor. A soluble IL-1 receptor has also been detected in RA synovial fluid[80] and its ability to bind IL-1β selectively suggests it may be the type II IL-1 receptor. As neutrophils express a nonfunctionally coupled type II IL-1 receptor on their cell surfaces, neutrophils could possibly also be a major source of IL-1 inhibitory proteins in inflamed tissues. Further details of the therapeutic potential of soluble cytokine receptors are to be found in Chapter 10.

C. ANTI-INFLAMMATORY CYTOKINES

Several cytokines and growth factors possess *in vitro* activities which characterize them as "anti-inflammatory" cytokines. The cytokines IL-4, IL-10, and IL-13 have been reported to inhibit the synthesis of IL-1 and TNFα and other pro-inflammatory cytokines by monocytes.[85–87] Interleukin-4 also reduces prostaglandin production by monocytes.[85]

Production of collagenase and stromelysin can also be reduced by some growth factors such as TGFβ.[88] Anti-inflammatory cytokines can also act via the induction of natural inhibitors — IL-1ra production is increased by both IL-4 and TGFβ[89,90] while the synthesis of the protease inhibitor, TIMP, is stimulated by IL-6 and TGFβ.[91,92]

However, the *in vitro* activities of these factors cannot always be extrapolated to the whole organism. For example, despite the *in vitro* activities described above, TGFβ stimulates synovitis and cartilage degradation when injected into the joints of animals.[93,94] Detailed discussions of the therapeutic potential of 'antagonist' cytokines are to be found in Chapter 14 and 15.

VIII. SUMMARY

Cytokines represent an ill-defined group of proteins ranging from interleukins, through hematopoietic colony stimulating factors, to growth factors. Cytokines are diverse in structure and function and are probably the most potent endogenous inflammatory mediators so far discovered. The roles of many cytokines such as TNFα, IL-1, and IL-8 have been well defined by the use of blocking antibodies and natural antagonists, while the precise roles of the more recently discovered cytokines such as RANTES have yet to be established. Cytokines are particularly important in mediating the tissue injury and remodeling associated with chronic aspects of inflammatory diseases. The plethora of mechanisms which have evolved to inhibit cytokine production and action probably reflect the destructive potential of cytokines. Many of these anti-inflammatory mechanisms involve neutrophils, i.e., neutrophils are major sources of IL-1ra, soluble type II IL-1 receptors, and soluble TNFα receptors. In addition to the well-characterized role of neutrophils in the host defense against infection, neutrophils may also serve an important function in limiting cytokine-mediated tissue injury in chronic inflammation.

REFERENCES

1. **Dinarello, C. A.,** Anti-cytokine strategies, *Eur. Cytokine Networks,* 3, 7, 1992.
2. **Lasky, L. A.,** Selectins: interpreters of cell-specific carbohydrate information during inflammation, *Science,* 258, 964, 1992.
3. **Bevilacqua, M. P., Pober, J. S., Mendrick, D. L., Cotran, R. S., and Gimbrone, M. A.,** Identification of an inducible endothelial-leukocyte adhesion molecule, *Proc. Natl. Acad. Sci. U.S.A.,* 84, 9238, 1987.
4. **Lawrence, M. B. and Springer, T. A.,** Neutrophils roll on E-selectin, *J. Immunol.* 151, 6338, 1993.
5. **Luscinskas, F. W., Brock, A. F., Arnaout, M. A., and Gimbrone, M. A.,** Endothelial-leukocyte adhesion molecule-1-dependent leukocyte (CD11/CD18)-dependent mechanisms contribute to polymorphonuclear leukocyte adhesion to cytokine-activated human vascular endothelium, *J. Immunol.,* 142, 2257, 1989.
6. **Phillips, M. L., Nudelman, E., Gaeta, F. C. A., Perez, M., Singhal, A. K., Hakomori, S.-I., and Paulson, J. C.,** ELAM-1 mediates cell adhesion by recognition of a carbohydrate ligand, sialyl LeX, *Science,* 250, 1130, 1990.
7. **Berg, E. L., Robinson, M. K., Mansson, O., Butcher, E. C., and Magnani, J. L.,** A carbohydrate domain common to both sialyl Lewis a and sialyl Lewis x is recognized by the endothelial cell leukocyte molecule ELAM-1, *J. Biol. Chem.,* 266, 14869, 1991.

8. **Vedder, N. B. and Harlan, J. M.,** Increased surface expression of CD11b/CD18 (Mac-1) is not required for stimulated neutrophil adherence to cultured endothelium, *J. Clin. Invest.,* 81, 676, 1988.
9. **Strieter, R. M., Kunkel, S. L., Showell, H. J., Remick, D. G., Phan, S. H., Ward, P. A., and Marks, R. M.,** Endothelial cell gene expression of a neutrophil chemotactic factor by TNF-alpha, LPS and IL-1 beta, *Science,* 243, 1467, 1989.
10. **Yoshimura, T., Matsushima, K., Oppenheim, J. J., and Leonard, E. J.,** Neutrophil chemotactic factor produced by lipopolysaccharide (LPS)-stimulated human blood mononuclear leukocytes: partial characterization and separation from interleukin-1 (IL-1), *J. Immunol.,* 139, 788, 1987.
11. **Rot, A.,** Endothelial cell binding of NAP-1/IL-8: role in neutrophil emigration, *Immunol. Today* 13, 291, 1992.
12. **Huber, A. R., Kunkel, S. L., Todd, R. F., and Weiss, S. J.,** Regulation of transendothelial neutrophil migration by endogenous interleukin-8, *Science,* 254, 99, 1991.
13. **Schall, T. J.,** Biology of the RANTES/SIS cytokine family, *Cytokine,* 3, 165, 1991.
14. **Springer, T. A.,** Adhesion molecules of the immune system, *Nature,* 346, 425, 1990.
15. **Osborn, L., Hession, C., Tizard, R., Vassallo, C., Luhowskyj, S., Chi-Rosso, G., and Lobb, R.,** Direct expression cloning of vascular cell adhesion molecule 1, a cytokine-induced endothelial protein that binds to lymphocytes, *Cell,* 59, 1203, 1989.
16. **Bochner, B. S., Luscinskas, F. W., Gimbrone, M. A., Newman, W., Sterbinsky, S. A., Derse-Anthony, C. P., Klunk, D., and Schleimer, R. P.,** Adhesion of human basophils, eosinophils, and neutrophils to interleukin-1-activated human vascular endothelial cells. Contribution of endothelial cell adhesion molecules, *J. Exp. Med.,* 173, 1553, 1991.
17. **Schleimer, R. P., Sterbinsky, S. A., Kaiser, J., Bickel, C. A., Klunk, D. A., Tomioka, K., Newman, W., Luscinskas, F. W., Gimbrone, M. A., McIntyre, B. W., and Bochner, B. S.,** IL-4 induces adherence of human eosinophils and basophils but not neutrophils to endothelium. Association with the expression of VCAM-1, *J. Immunol.,* 148, 1086, 1992.
18. **Pettipher, E. R., Higgs, G. A., and Henderson, B.,** Interleukin-1 induces leukocyte infiltration and cartilage proteoglycan degradation in the synovial joint, *Proc. Natl. Acad. Sci. U.S.A.,* 83, 8749, 1986.
19. **Cybulsky, M. I., Chan, M. K. W., and Movat, H. Z.,** Acute inflammation and microthrombosis induced by endotoxin, interleukin-1, and tumor necrosis factor and their implication in Gram-negative infection, *Lab. Invest.,* 58, 365, 1988.
20. **Henderson, B. and Pettipher, E. R.,** Comparison of the inflammatory activities after intraarticular injection of natural and recombinant IL-1α and IL-1β in the rabbit, *Biochem. Pharmacol.,* 37, 4171, 1988.
21. **Henderson, B. and Pettipher, E. R.,** Arthritiogenic actions of recombinant IL-1 and tumour necrosis factor α in the rabbit: evidence for synergistic interactions between cytokines *in vivo, Clin. Exp. Immunol.,* 75, 306, 1989.
22. **Rampart, M. and Williams, T.J.,** Evidence that neutrophil accumulation induced by interleukin-1 requires both local protein synthesis and neutrophil CD18 antigen expression *in vivo*, *Br. J. Pharmacol.,* 94, 1143, 1988.
23. **Cybulsky, M. I., McComb, D. J., and Movat, H. Z.,** Protein synthesis dependent and independent mechanisms of neutrophil emigration. Different mechanisms of inflammation in rabbits induced by interleukin-1, tumor necrosis factor alpha or endotoxin versus leukocyte chemoattractants, *Am. J. Pathol.,* 135, 227, 1989.
24. **Mulligan, M. S., Varani, J., Dame, M. K., Lane, C. L., Smith, C. W., Anderson, D. C., and Ward, P. A.,** Role of endothelial-leukocyte adhesion molecule 1 (ELAM 1) in neutrophil-mediated lung injury in rats, *J. Clin. Invest.,* 88, 1396, 1991.
25. **Warren, J. S., Yabroff, K. R., Remick, D. G., Kunkel, S. L., Chensue, S. W., Kunkel, R. G., Johnson, K. J., and Ward, P. A.,** Tumor necrosis factor participates in the pathogenesis of acute immune complex alveolitis in the rat, *J. Clin. Invest.,* 84, 1873, 1989.
26. **Warren, J. S.,** Intrapulmonary interleukin-1 mediates acute immune complex alveolitis in the rat, *Biochem. Biophys. Res. Commun.,* 175, 604, 1991.
27. **Sekido, N., Mukaida, N., Harada, A., Nakanishi, I., Watanabe, Y., and Matsushima, K.,** Prevention of lung reperfusion injury in rabbits by a monoclonal antibody against interleukin-8, *Nature,* 365, 654, 1993.

28. **Akahoshi, T., Endo, H., Hosaka, S., Kondo, H., Kashiwazaki, S., and Matsushima, K.,** Essential involvement of IL-8 on neutrophil recruitment and joint destruction in experimental acute arthritis, Combined Meet. 8th Lymphokine Workshop and 4th Workshop on Cytokines, Osaka, October 1993, 231, 1993.
29. **Meurer, R., Van Riper, G., Feeney, W., Cunningham, P., Hora, J., D., Springer, M. S., MacIntyre, D. E., and Rosen, H.,** Formation of eosinophilic and monocytic intradermal inflammatory sites in the dog by injection of human RANTES but not human Monocyte Chemoattractant Protein 1, human macrophage inflammatory protein 1 α, or human interleukin-8, *J. Exp. Med.,* 178, 1913, 1993.
30. **Issekutz, T. B. and Stoltz, J. M.,** Stimulation of lymphocyte migration by endotoxin, tumor necrosis factor and interferon, *Cell. Immunol.,* 120, 165, 1989.
31. **Kudo, C., Araki, A., Matsushima, K., and Sendo, F.,** Inhibition of IL-8-induced W3/25 + (CD 4+) T lymphocyte recruitment into subcutaneous tissues of rats by selective depletion of *in vivo* neutrophils with a monoclonal antibody, *J. Immunol.,* 147, 2196, 1991.
32. **Dahinden, C. A., Zingg, J., Maly, F. E., and de Weck, A. L.,** Leukotriene production in human neutrophils primed by recombinant human granulocyte/macrophage-stimulating factor and stimulated with the complement component C5a and FMLP as second signals, *J. Exp. Med.,* 167, 1281, 1988.
33. **DiPersio, J. F., Billing, P., Williams, R., and Gasson, J. C.,** Human granulocyte-macrophage colony-stimulating factor and other cytokines prime human neutrophils for enhanced arachidonic acid release and leukotriene B_4 synthesis, *J. Immunol.,* 140, 4315, 1988.
34. **McColl, S. R., Krump, E., Naccache, P. H., Poubelle, P. E., Braquet, P., Braquet, M., and Borgeat, P.,** Granulocyte-macrophage colony-stimulating factor increases the synthesis of leukotriene B_4 by human neutrophils in response to platelet-activating factor. Enhancement of both arachidonic acid availability and 5-lipoxygenase activation, *J. Immunol.,* 146, 1204, 1991.
35. **Pouliot, M., McDonald, P. P., Khamzina, L., Borgeat, P., and McColl, S. R.,** Granulocyte-macrophage colony-stimulating factor enhances 5-lipoxygenase levels in human polymorphonuclear leukocytes, *J. Immunol.,* 152, 851, 1994.
36. **Gorgen, I., Hartung, T., Leist, M., Niehorster, M., Tiegs, G., Weitzel, F., and Wendel, A.,** Granulocyte colony stimulating factor treatment protects rodents against lipopolysaccharide-induced toxicity via suppression of systemic tumor necrosis factor, *J. Immunol.,* 149, 918, 1992.
37. **Xu, W. D., Firestein, G. S., Taetle, R., Kaushansky, K., and Zvaifler, N. J.,** Cytokines in chronic inflammatory arthritis. II. Granulocyte-macrophage colony-stimulating factor in rheumatoid synovial effusions, *J. Clin. Invest.,* 83, 876, 1989.
38. **Clutterbuck, E. J., Hirst, E. M., and Sanderson, C. J.,** Human interleukin-5 regulates the production of eosinophils in human bone marrow cultures: comparison and interaction with IL-1, IL-3, IL-6, and GM-CSF, *Blood,* 73, 1504, 1989.
39. **Lopez, A. F., Sanderson, C. J., Gamble, J. R., Campbell, H. D., Young, I. G., and Vadas, M. A.,** Recombinant human recombinant interleukin-5 is a selective activator of human eosinophil function, *J. Exp. Med.,* 167, 219, 1988.
40. **Sehmi, R., Wardlaw, A. J., Cromwell, O., Kurihara, K., Waltmann, P., and Kay, A. B.,** Interleukin-5 selectively enhances the chemotactic response of eosinophils obtained from normal but not eosinophilic subjects, *Blood,* 79, 2952, 1992.
41. **Bischoff, S. C., Brunner, T., de Weck, A. L., and Dahinden, C. A.,** Interleukin-5 modifies histamine release and leukotriene generation by human basophils in response to diverse agonists, *J. Exp. Med.,* 172, 1577, 1990.
42. **Salmon, J. A. and Higgs, G. A.,** Prostaglandins and leukotrienes as inflammatory mediators, *Br. Med. Bull.,* 43, 285, 1987.
43. **Vane, J. R.,** Inhibition of prostaglandin synthesis as a mechanism of action for aspirin-like drugs, *Nature,* 231, 232, 1971.
44. **Masferrer, J. L., Seibert, K., Zweifel, B., and Needleman, P.,** Endogenous glucocorticoids regulate an inducible cyclooxygenase enzyme, *Proc. Natl. Acad. Sci. U.S.A.,* 89, 3917, 1992.
45. **Scales, W. E., Chensue, S. W., Otterness, I. G., and Kunkel, S. L.,** Regulation of monokine gene expression. Prostaglandin E2 suppresses tumor necrosis factor but not interleukin-1 α or β-mRNA and cell-associated bioactivity, *J. Leukocyte Biol.,* 45, 416, 1989.

46. **Spinas, G. A., Bloesch, D., Keller, U., Zimmerli, W., and Cammisuli, S.,** Pretreatment with ibuprofen augments circulating tumor necrosis factor-α, interleukin-6 and elastase during acute endotoxinemia, *J. Infect. Dis.,* 163, 89, 1991.
47. **Martich, G. D., Danner, R. L., Ceska, M., and Suffredini, A. F.,** Detection of interleukin-8 and tumor necrosis factor in normal humans after intravenous endotoxin: the effect of anti-inflammatory agents, *J. Exp. Med.,* 173, 1021, 1991.
48. **Ferrandiz, M. L. and Foster, S. J.,** Tumour necrosis factor production in a rat airpouch model of inflammation. Role of eicosanoids, *Agents Actions,*32, 289, 1991.
49. **Bottomly, K. M. K., Griffiths, R. J., Rising, T. J., and Steward, A.,** A modified mouse air pouch model for evaluating the effects of compounds on granuloma-induced cartilage degradation, *Br. J. Pharmacol.,* 93, 627, 1988.
50. **Pettipher, E. R., Henderson, B., Edwards, J. C. W., and Higgs, G. A.,** Effect of indomethacin on swelling, lymphocyte influx and cartilage proteoglycan depletion in experimental arthritis, *Ann. Rheum. Dis.,* 48, 623, 1989.
51. **Farrell, A. J., Blake, D. R., Palmer, R. M. J., and Moncada, S.,** Increased concentrations of nitrite in synovial fluid and serum samples suggest increased nitric oxide synthesis in rheumatic diseases, *Ann. Rheum. Dis.,* 51, 1219, 1992.
52. **Boughton-Smith, N. K., Evans, S. W., Hawkey, C. J., Cole, A. T., Balsitis, M., Whittle, B. J. R., and Moncada, S.,** Nitric oxide synthase activity in ulcerative colitis and Crohn's disease, *Lancet,* 342, 338, 1993.
53. **Kobzik, L., Bredt, D. S., Lowenstein, C. J., Drazen, J., Gaston, B., Sugarbaker, D., and Stamler, J. S.,** Nitric oxide synthase in human and rat lung: Immunocytochemical and histochemical localization, *Am. J. Respir. Cell. Mol. Biol.,* 9, 371, 1993.
54. **Wang, J.-F., Komarov, P., Sies, H., and de Groot, H.,** Contribution of nitric oxide synthase to luminol-dependent chemiluminescence generated by phorbol-ester-activated Kupffer cells, *Biochem. J.,* 279, 311, 1991.
55. **Thiemermann, C., Wu, C.-C., Szabo, C., Perretti, M., and Vane, J. R.,** Role of tumour necrosis factor in the induction of nitric oxide synthase in a rat model of endotoxin shock, *Br. J. Pharmacol.,* 110, 177, 1993.
56. **Szabo, C., Wu, C.-C., Gross, S. S., Thiemermann, C., and Vane, J. R.,** Interleukin-1 contributes to the induction of nitric oxide synthase by endotoxin *in vivo*, *Eur. J. Pharmacol.,* 250, 157, 1993.
57. **Mulligan, M. S., Moncada, S., and Ward, P. A.,** Protective effects of inhibitors of nitric oxide synthase in immune complex-induced vasculitis, *Br. J. Pharmacol.,* 107, 1159, 1992.
58. **Miller, M. J. S., Sadowska-Krowicka, H., Chotinaruemol, S., Kakkis, J. L., and Clark, D. A.,** Amelioration of chronic ileitis by nitric oxide synthase inhibition, *J. Pharmacol. Exp. Ther.,* 264, 11, 1993.
59. **McCartney-Francis, N., Allen, J. B., Mizel, D. E., Albina, J. E., Xie, Q.-W., Nathan, C. F., and Wahl, S. M.,** Suppression of arthritis by an inhibitor of nitric oxide synthase, *J. Exp. Med.,* 178, 749, 1993.
60. **Sandy, J. D., Flannery, C. R., Neame, P. J., and Lohmander, L. S.,** The structure of aggrecan fragments in human synovial fluid. Evidence for the involvement in osteoarthritis of a novel proteinase which cleaves the Glu373-Ala 374 bond of the interglobular domain, *J. Clin. Invest.,* 89, 1512, 1992.
61. **Arner, E. C., DiMeo, T. M., Ruhl, D. M., and Pratta, M. A.,** *In vivo* studies on the effects of human recombinant interleukin-1β on articular cartilage, *Agents Actions,* 27, 254, 1989.
62. **O'Byrne, E. M., Blancuzzi, V., Wilson, D. E., Wong, M., and Jeng, A. Y.,** Elevated substance P and accelerated cartilage degradation in rabbit knees injected with interleukin-1 and tumor necrosis factor, *Arthr. Rheum.,* 33, 1023, 1990.
63. **McDonnell, J., Hoerrner, L. A., Lark, M. W., Harper, C., Dey, T., Lobner, J., Eiermann, G., Kazazis, D., Singer, I., and Moore, V. L.,** Recombinant human interleukin-1b-induced increase in levels of proteoglycans, stromelysin and leukocytes in rabbit synovial fluid, *Arthr. Rheum.,* 35, 799, 1992.
64. **Pettipher, E. R., Henderson, B., Moncada, S., and Higgs, G. A.,** Leucocyte infiltration and cartilage proteoglycan loss in immune arthritis in the rabbit, *Br. J. Pharmacol.,* 95, 169, 1988.

65. **Endo, H., Akahoshi, T., Takagishi, K., Kashiwazaki, S., and Matsushima, K.,** Elevation of interleukin-8 (IL-8) levels in joint fluids of patients with rheumatoid arthritis and the induction by IL-8 of leukocyte infiltration and synovitis in rabbit joints, *Lymphokine Cytokine Res.,* 10, 245, 1991.
66. **Chin, J. and Kostura, M. J.,** Dissociation of IL-1β synthesis and secretion in human blood monocytes stimulated with bacterial cell wall products, *J. Immunol.,* 151, 5574, 1993.
67. **Matsukawa, A., Ohkawara, S., Maeda, T., Takagi, K., and Yoshinaga, M.,** Production of IL-1 and IL-1 receptor antagonist and the pathological significance in lipopolysaccharide-induced arthritis in rabbits, *Clin. Exp. Immunol.,* 93, 206, 1993.
68. **Jafari, H. S., Saez-Llorens, X., Paris, M., Rinderknecht, S., Friedland, I., Ehrett, S., Severien, C., Olsen, K. D., Burns, D. K., Harper, C. F., Lark, M. W., Thonar, E. J.-M., and McCracken, G. H.,** Dexamethasone attenuation of cytokine-mediated articular cartilage degradation in experimental lapine Haemophilus arthritis, *J. Inf. Dis.,* 168, 1186, 1993.
69. **Arend, W. P., Joslin, F. G., Thompson, R. C., and Hannum, C. H.,** An interleukin-1 inhibitor from human monocytes: production and characterization of biological properties, *J. Immunol.,* 143, 1851, 1985.
70. **Seckinger, P., Lowenthal, J. W., Williamson, K., Dayer, J.-M., and MacDonald, H. R.,** A urine inhibitor of interleukin-1 activity that blocks ligand binding, *J. Immunol.,* 139, 1546, 1987.
71. **Eisenberg, S., Evans, R. J., Arend, W. P., Verderber, E., Brewer, M. T., Hannum, C. H., and Thompson, R. C.,** Primary structure and functional expression from complementary DNA of a human interleukin-1 receptor antagonist, *Nature,* 343, 341, 1990.
72. **Hannum, C. H., Wilcox, C. J., Arend, W. P., Joslin, F. G., Dripps, D. J., Heimdal, P. L., Armes, L. G., Sommer, A., Eisenberg, S. P., and Thompson, R. C.,** Interleukin-1 receptor antagonist activity of a human interleukin-1 inhibitor, *Nature,* 343, 336, 1990.
73. **Granowitz, E. V., Clark, B. D., Mancilla, J., and Dinarello, C. A.,** Interleukin-1 receptor antagonist competitively inhibits the binding of interleukin-1 to the type II interleukin-1 receptor, *J. Biol. Chem.,* 266, 14147, 1991.
74. **Janson, R. W., Hance, K. R., and Arend, W. P.,** Production of IL-1 receptor antagonist by human *in vitro*-derived macrophages. Effects of lipopolysaccharide and granulocyte-macrophage colony-stimulating factor, *J. Immunol.,* 147, 4218, 1991.
75. **McColl, S. R., Paquin, R., Menard, C., and Beaulieu, A. D.,** Human neutrophils produce high levels of the interleukin-1 receptor antagonist in response to granulocyte/macrophage colony-stimulating factor and tumor necrosis factor α, *J. Exp. Med.,* 176, 593, 1992.
76. **Malyak, M., Swaney, R. E., and Arend, W. P.,** Levels of synovial fluid interleukin-1 receptor antagonist in rheumatoid arthritis and other arthropathies, *Arthr. Rheum.,* 36, 781, 1993.
77. **Fischer, E., Van Zee, J., Marano, M. A., Rock, C. S., Kenney, J. S., Poutsiaka, D. D., Dinarello, C. A., Lowry, S. F., and Moldawer, L. L.,** Interleukin-1 receptor antagonist circulates in experimental inflammation and in human disease, *Blood,* 79, 2196, 1992.
78. **Miller, L. C., Lynch, E. A., Isa, S., Logan, J. W., Dinarello, C. A., and Steere, A. C.,** Balance of synovial fluid IL-1β and IL-1 receptor antagonist and recovery from Lyme arthritis, *Lancet,* 341, 146, 1993.
79. **Rubin, L. A., Kurman, C. C., Fritz, M. E., Biddison, W. E., Boutin, B., Yarchoan, R., and Nelson, D. L.,** Soluble interleukin-2 receptors are released from activated human lymphoid cells *in vitro, J. Immunol.,* 135, 3172, 1985.
80. **Symons, J. A., Eastgate, J. A., and Duff, G. W.,** Purification and characterization of a novel soluble receptor for interleukin-1, *J. Exp. Med.,* 174, 1251, 1991.
81. **Spinas, G. A., Keller, U., and Brockhaus, M.,** Release of soluble receptors for tumor necrosis factor (TNF) in relation to circulating TNF during experimental endotoxinemia, *J. Clin. Invest.,* 90, 533, 1992.
82. **Cope, A. P., Aderka, D., Doherty, M., Engelmann, H., Gibbons, D., Jones, A. C., Brennan, F. M., Maini, R. N., Wallach, D., and Feldmann, M.,** Increased levels of soluble tumour necrosis factor receptors in sera and synovial fluid of patients with rheumatic diseases, *Br. J. Rheumatol.,* 31, 211, 1992.
83. **Porteu, F. and Nathan, C. F.,** Shedding of tumor necrosis factor receptors by activated neutrophils, *J. Exp. Med.,* 172, 599, 1990.

84. **Porteu, F., Brockhaus, M., Wallach, D., Engelmann, H., and Nathan, C. F.,** Human neutrophil elastase releases a ligand-binding fragment from the 75-kDa tumor necrosis factor (TNF) receptor, *J. Biol. Chem.,* 266, 18846, 1991.
85. **Hart, P. H., Vitti, G. F., Burgess, D. R., Whitty, G. A., Piccoli, D. S., and Hamilton, D. A.,** Potential anti-inflammatory effects of interleukin-4. Suppression of human monocyte tumor necrosis factor alpha, interleukin-1, and prostaglandin E2, *Proc. Natl. Acad. Sci. U.S.A.,* 86, 3803, 1989.
86. **Fiorentino, D. F., Zlotnik, A., Mosmann, T. R., Howard, M., and O'Garra, A.,** IL-10 inhibits cytokine production by activated macrophages, *J. Immunol.,* 147, 3815, 1991.
87. **Minty, A., Chalon, P., Derocq, J.-M., Dumont, X., Guillemot, J.-C., Kaghad, M., Labit, C., Leplatois, P., Liauzun, P., Miloux, B., Minty, C., Casellas, P., Loison, G., Lupker, J., Shire, D., Ferrara, P., and Caput, D.,** Interleukin-13 is a new human lymphokine regulating inflammatory and immune responses, *Nature,* 362, 248, 1993.
88. **Chandrasekhar, S. and Harvey, A. K.,** Transforming growth factor-β is a potent inhibitor of IL-1 induced protease activity and cartilage proteoglycan degradation, *Biochem. Biophys. Res. Commun.,* 157, 1352, 1988.
89. **Vannier, E., Miller, L. C., and Dinarello, C. A.,** Coordinated anti-inflammatory effects of interleukin-4. Interleukin-4 suppresses interleukin-1 production but upregulates gene expression and synthesis of interleukin-1 receptor antagonist, *Proc. Natl. Acad. Sci. U.S.A.,* 89, 4076, 1992.
90. **Turner, M., Chantry, D., Katsikis, P., Berger, A., Brennan, F. M., and Feldmann, M.,** Induction of the interleukin-1 receptor antagonist protein by transforming growth factor- β, *Eur. J. Immunol.,* 21, 1635, 1991.
91. **Sato, T., Ito, A., and Mori, Y.,** Interleukin-6 enhances the production of tissue inhibitor of metalloproteinases (TIMP) but not that of matrix mettaloproteinases by human fibroblasts, *Biochem. Biophys. Res. Commun.,* 170, 824, 1990.
92. **Wright, J. K., Cawston, T. E., and Hazleman, B. L.,** Transforming growth factor beta stimulates the production of the tissue inhibitor of metalloproteinases (TIMP) by human synovial and skin fibroblasts, *Biochim. Biophys. Acta,* 1094, 207, 1991.
93. **Allen, J. R., Manthey, C. L., Hand, A. R., Ohura, K., Ellingsworth, L., and Wahl, S. M.,** Rapid onset synovial inflammation and hyperplasia induced by transforming growth factor β, *J. Exp. Med.,* 171, 231, 1990.
94. **Elford, P. R., Graeber, M., Ohtsu, H., Aeberhard, M., Legendre, B., Wishart, W. L., and MacKenzie, A. R.,** Induction of swelling, synovial hyperplasia and cartilage proteoglycan loss upon intra-articular injection of transforming growth factor β-2 in the rabbit, *Cytokine,* 4, 232, 1992.

Chapter 4

OVERVIEW OF CYTOKINE MODULATION

Brian Henderson

CONTENTS

I. INTRODUCTION

From the preceding chapters it is clear that cytokines play a role in cell, tissue, and organismal homeostasis as well as in the induction and perpetuation of widely diverse states of tissue and systemic pathology ranging from boils to cancers. On this basis, most of the world's pharmaceutical industries have research, development, and clinical programs aimed at blocking the synthesis or action of one or another cytokine. The main problem for the drug industry is that the growth in our understanding of cytokine biology generally outstrips the managerial capacity to recognize relevant targets. This lag phase is a major problem in areas of rapidly expanding knowledge, as the time taken to develop inhibitors of a particular mediator can be measured in decades. In the early 1980s the leukotrienes and other products of lipoxygenase action on arachidonate were described and their biological actions were perceived to make them important therapeutic targets in diseases such as asthma and arthritis. Fifteen years or more later, bioactive inhibitors of 5-lipoxygenase are only now poised to enter the clinical arena for the treatment of asthma. The lag time from the recognition that these mediators were key therapeutic targets to their entering the clinic has to some extent been lengthened by periods of uncertainty about the role of these lipid mediators in the diseases with which they were first perceived to be involved. A similar problem can now be seen with the role of pro-inflammatory cytokines in the pathogenesis of septic shock (see Chapter 9) as none of the cytokine-inhibiting agents tested in clinical trial have yet produced clear clinical benefits.

Before discussing the various methods available to block cytokines it is important to state that there is mounting evidence to suggest that cytokines have important roles in organismal homeostasis, e.g., in the bone marrow-derived production of leukocytes. As has been described in Chapter 1, the generation of transgenic mice in which particular genes have been inactivated by homologous recombination can give clues to the role

0-8493-8381-1/96/$0.00+$.50

played by the knock-out gene product in the normal state. Such studies have now shown that the suppression of the transcription of cytokines such as IL-2,[1] IL-10,[2] TGFβ,[3] and the p55 TNFα receptor[4] can render the transgenic animal susceptible to life-threatening inflammatory states. The possibility that prolonged therapy with a cytokine-blocking agent could produce tissue pathology must therefore be considered and examined before such agents reach the clinic. Of course to some extent this will depend on the risk/benefit ratio produced by any cytokine-modulating therapeutic. The analogy in this case is with the nonsteroidal anti-inflammatory drugs which produce many side effects, the most prominent being gastropathy. In spite of this these drugs are widely used in the treatment of the intractable pain of arthritis, the patients generally feeling that the risk/benefit ratio is in their favor.

There is overwhelming evidence that cytokines are major therapeutic targets in some, if not most, human diseases. In this brief chapter the possible ways that cytokines can be modulated for therapeutic effect will be discussed, largely to signpost the rest of the volume where these individual targets will be discussed in greater detail. The first faltering attempts to develop anti-cytokine agents started only a short time ago in the mid-1980s, but through the growth in the biopharmaceutical industry in the last decade we currently have a number of anti-cytokine agents in clinical trials. The methods used to modulate cytokines will be briefly dealt with and interested readers should refer to recent reviews for more details.[5–7]

II. THERAPEUTIC MODULATION OF CYTOKINES

Four major approaches can be taken to the modulation of cytokines, namely: (1) inhibition of cytokine synthesis, (2) inhibition of cytokine release, (3) inhibition of cytokine action, or (4) inhibition of cytokine intracellular signaling pathways. The various points of intervention are detailed in Table 1 and will be dealt with in turn.

A. INHIBITION OF CYTOKINE SYNTHESIS

This can be achieved at a number of levels. The first is the blockade of the intracellular signaling pathways which result in the transcription of particular cytokine genes. In the past few years it has been demonstrated that the potent immunosuppressants — cyclosporine and FK506 — inhibit the synthesis of lymphokines such as IL-2, IL-3, IL-4, and γ-IFN in this manner, by forming a complex with specific cellular binding proteins which then inhibits the calmodulin-dependent protein phosphatase, calcineurin. The function of calcineurin is believed to be to dephosphorylate a subunit of the lymphokine transcription factor NFAT (nuclear factor of activated T cells) which can then enter the nucleus and form the active transcriptional complex to switch on the genes transcribing IL-2 and related cytokines. The formation of the drug-binding protein complex inhibits this dephosphorylation and thus the cell fails to generate the transcriptional complex.[8] This mechanism is discussed in detail in Chapter 6. One of the curious clinical findings is that cyclosporine is surprisingly ineffectual in treating rheumatoid arthritis (RA), a disease until recently considered to be an autoimmune condition driven by antigen-specific T cells. The relative inefficaciousness of cyclosporine supports a growing belief that lymphokines play a minor role in the pathology of established RA.[9] As outlined in Chapter 6, the understanding of the mechanism of action of these and other novel immunosuppressants should help in the development of selective agents for blocking postreceptor/pretranscriptional intracellular signaling pathways involved in cytokine synthesis.

TABLE 1
Methods Used for Therapeutic Modulation of Cytokines

1. Inhibit cellular activation pathways resulting in cytokine gene transcription
2. Directly inhibit transcription of cytokine gene
3. Inhibit the splicing/metabolism of cytokine mRNA
4. Inhibit cytokine mRNA translation by oligonucleotide-based agents
5. Inhibit the proteolytic processing of cytokines
6. Inhibit specific cytokine release pathways
7. Block action of soluble cytokines by neutralizing antibodies or soluble cytokine receptors
8. Antagonize binding of cytokines to their receptors
9. Inhibit cytokine-specific intracellular signaling pathways
10. Block cytokine-induced cellular activation by antagonist cytokines

Oligonucleotide-based drugs — There is enormous interest in agents which can selectively inhibit gene transcription or mRNA translation. Such agents are generally modified oligonucleotides with nuclease-resistant bonds. Antisense oligonucleotides designed to hybridize with the AUG translation initiation codon have been the most popular choice and inhibition of the synthesis of a wide range of proteins including cytokines and cytokine receptors *in vitro* has been reported[10] (see Chapter 8), and one group has shown that antisense probes to the murine type I IL-1 receptor have anti-inflammatory activity *in vivo.*[11] This is a very active area of pharmaceutical research although one with a large number of pharmacokinetic and pharmacodynamic problems to overcome. The use of such agents to selectively inhibit cytokine synthesis is reviewed in detail in Chapter 8.

Low molecular mass inhibitors — There are a growing number of low molecular mass compounds reported to inhibit the synthesis of cytokines (Table 2). Most are active in the micromolar range and are therefore potential therapeutic agents. Most have an unknown mechanism of action. Tenidap is one of the compounds that has been claimed to inhibit IL-1 synthesis. This agent is now in clinical trial in RA and has been reported to show clinical effects consistent with the inhibition of cytokine synthesis.[12] The only tenuous common association shown by most of these compounds is the ability to inhibit arachidonic acid oxidation. It is not clear, however, if this property is related to inhibition of cytokine synthesis. Some rather interesting compounds have been reported to inhibit cytokine synthesis. These are discussed in detail in Chapter 5. Of interest are reports that vitamins such as vitamin D and vitamin K have the capacity to inhibit cytokine synthesis. This conjures up a possible additional role for these vitamins.

B. INHIBITION OF CYTOKINE RELEASE

If it does not prove possible to block the transcription or translation of a cytokine it may prove possible to block the release of the cytokine from the producer cell. This can now be achieved in at least two ways. The first is to block the proteases which activate certain cytokines such as IL-1β and TNFα, and the second probably relies on the fact that the release of certain cytokines is via a novel pathway.

Inhibitors of cytokine processing — Certain cytokines such as the key pro-inflammatory molecules IL-1β and TNFα require proteolytic processing to produce the active form. This processing is achieved by the action of specific proteases. The enzyme responsible for the conversion of the 31-kDa IL-1β precursor to the active 17-kDa form is termed IL-1β converting enzyme (ICE) and a number of companies including MSD, Sterling Winthrop, and Vertex have developed potent inhibitors and reported activity *in*

TABLE 2
Low Molecular Mass Inhibitors of Cytokine Synthesis

Compound	Cytokine Inhibited
Glucocorticoids	Many cytokines
Prostanoids	IL-1, TNF
Taurolidine	IL-1
SK&F 105809	IL-1
SK&F 86002	TNF
Tenidap (Pfizer)	IL-1
Pentamidine	IL-1
IX (Sandoz)	IL-1
Antioxidants	IL-1, IL-6, TNF
Substituted chalcones	IL-1, TNF
E5090 (Tsukuba)	IL-1
E3330 (Tsukuba)	TNF
SA 3443 (Santen)	TNF
RP54754	IL-1
RPR102194	IL-1
TA383	IL-6
Immunosuppressants	Lymphokines
Sodium aurothiomalate	IL-8
Anti-arthritic compounds	Cytokines/lymphokines
Vitamin D_3	IL-8
Vitamin K	IL-6

vivo[13,14] (also Chapter 7). The recent determination of a high resolution structure of an ICE-inhibitor complex should aid the development of selective inhibitors of this enzyme.[14] This paper also reports that the antirheumatic drugs Auranofin and gold thiomalate are noncompetitive inhibitors of ICE, adding yet another feather in the cap of these multifarious and toxic drugs. It has also been established recently that the processing of the TNFα precursor is catalyzed by a metalloproteinase[15] and that inhibitors of the matrix metalloproteinases (matrixins: involved in the remodeling of the connective tissues and implicated as key mediators of tissue breakdown in arthritis, cancer, corneal ulceration, etc.) can block this processing.[15–17] These inhibitors have also been shown to have anti-TNF activity *in vitro*. This suggests the attractive possibility that any single metalloproteinase inhibitor could not only block the enzymes causing tissue damage in diseases such as cancer metastasis or rheumatoid arthritis, but could also block the production of TNFα, a cytokine with pro-inflammatory properties and the capacity to stimulate metalloproteinase gene transcription.

Inhibition of cytokine release — In addition to requiring processing, IL-1 and a few other cytokines lack signal sequences required for export and are released from cells by a novel pathway which could represent a selective therapeutic target.[18] A few compounds have been reported to inhibit IL-1 release including pentamidine isothiocyanate, an antiprotozoal agent, and IX 207,887, a compound developed by Sandoz.[6] In animal models IX demonstrates anti-inflammatory, antipyretic, antinociceptive and anti-arthritic activities.[19] A double blind clinical trial in patients with RA, IX demonstrated significant anti-inflammatory and anti-arthritic activity, although the trial was too short to determine if there was any inhibition of the progression of joint damage.[20] Sandoz has also reported the synthesis of tri- and tetrapeptides

TABLE 3
Actions of Cytokine-Neutralizing Antibodies

Cytokine Neutralized	Disease Model
IL-1	Antigen-induced arthritis, endotoxin shock
TNFα	Many and varied including endotoxin shock EAA[a], contact hypersensitivity, arthritis, Schwartzmann reaction
IL-3	Helminth-induced mastocytosis
γ-IFN	Schwartzmann reaction, endotoxin shock, arthritis, cerebral malaria
IL-3 and GM-CSF	Cerebral malaria
IL-4	Leishmaniasis
IL-5	Helminth-induced eosinophilia
IL-6	Endotoxin shock
PDGF	Atherosclerosis
TGFβ	Fibrosis

[a] EAA — experimental allergic encephalomyelitis. References can be found in Reference 6.

which inhibit the release of IL-1β with *in vitro* IC_{50} values of 0.1μ*M*. The action of these compounds is described in more detail in Chapter 5.

C. INHIBITION OF CYTOKINE ACTION

The inhibition of cytokine synthesis and release is still proving difficult to achieve and few compounds have reached clinical trial. In contrast, there are a number of agents now in clinical trial whose function is to bind and inactivate cytokines which have been produced by cells and released into the extracellular fluid, and ultimately prevent their binding to cellular receptors. Two major forms of therapeutic modality are under investigation: (1) agents such as neutralizing monoclonal antibodies and soluble cytokine receptors which bind to the cytokine in the fluid phase and sterically inhibit its binding to the receptor, and (2) antagonists of cytokine binding to its cell surface receptor. In addition, a third modality — the use of 'antagonist' cytokines — is under experimental investigation.

Cytokine-neutralizing monoclonal antibodies — Monoclonal antibodies (Mabs) have a number of advantages as therapeutics:

1. They can be raised to any cytokine target
2. Their natural milieu is the blood, extracellular fluids, and mucosal surfaces
3. They have long circulating half-lives
4. Their physical properties of size and affinity can be modified to the therapeutic demand

However, they have one major disadvantage — they are raised in rodents and produce a strong immune response in recipients. This has now been overcome by the generation of chimeric antibodies (mouse variable region and human Fc) and humanized antibodies, in which only the antigen-binding hypervariable regions are murine.[21] There is now a significant body of experimental evidence to support the idea that antibody-based neutralization of cytokines can inhibit tissue pathology (Table 3). Anti-TNFα Mabs have been assessed in patients with septic shock[22] and the results are discussed in detail in Chapter 10. One open trial of a chimerized anti-TNFα Mab produced by Centocor for the treatment of RA has produced a great deal of excitement,[23] with clinical benefits being reported, and a placebo-controlled double blind trial has confirmed the initial results. Other centers in London are testing the Celltech anti-TNFα Mab in double blind dose-ranging studies. Again, these studies are discussed in Chapter 10.

A note of caution should be introduced at this point with regard to the physical mass and size of the IgG molecule, in particular with diseases such as rheumatoid arthritis where tissue pathology is localized to particular tissues. The neutralization of TNFα in rodent models of arthritis such as adjuvant and collagen-induced arthritis inhibits synovitis and tissue damage,[24] giving confidence that neutralization of this cytokine in humans should produce clinical benefits. In contrast, the neutralization of this cytokine fails to inhibit tissue damage in antigen-induced arthritis in mice and rabbits.[25] The discrepancy between these two sets of results may simply reflect the well-known differences in response of these experimental lesions to nonsteroidal anti-inflammatory drugs (NSAIDs). Adjuvant and collagen arthritis are largely cured by NSAIDs while these agents have no effect on the progressive damage to articular cartilage and bone damage in antigen-induced arthritis models — a reflection of the clinical experience with these drugs. It has long been established that nonsteroidal drugs such as aspirin or ibuprofen have no ability to inhibit the destruction of the joint tissues, which is the major feature of rheumatoid arthritis. As TNFα is a potent inducer of cyclooxygenase II and thus stimulates prostanoid synthesis, the neutralization of this cytokine would be likely to inhibit prostanoid formation and this could be responsible for some of the reported effects. In the author's laboratory it has been found that a neutralizing Mab to rabbit TNFα has a profound anti-inflammatory effect in animals with antigen-induced arthritis, with complete inhibition of joint swelling with 1 week of treatment — an effect seen with no other drug tested. However, these animals showed no inhibition of cartilage proteoglycan loss.[25] This substantiates the growing belief that tissue damage and inflammation in the rheumatoid joint may not be causally linked.[26] Another explanation for the failure to inhibit damage to the articular cartilage in rabbits with arthritis may lie in the cellular source of the cytokines. Our experience was that the chondrocytes of the articular cartilage produced much of the TNFα seen in the joint.[25] Now, articular cartilage is a peculiar tissue which cannot be penetrated by macromolecules of molecular weight >60 kDa. Thus, and this is a general point, if the agent used therapeutically to block cytokine function cannot reach the cytokine then it will be unable to induce any therapeutic benefit.

The size of antibodies is a definite problem as is their non-oral route of administration. Another potential problem with the fully humanized antibodies that are planned to be employed clinically is that prolonged administration of a large amount of a particular idiotope could induce an anti-idiotypic response. Of particular concern is the possibility that some of these anti-idiotypic antibodies could mimic TNFα and thus exacerbate the disease being treated. One exciting development which could overcome these various problems is the work of Saragovi's group, who have reported the synthesis of low molecular weight nonpeptide mimetics of antibody-combining sites.[27] If this proves to be a universally applicable strategy then antibodies would simply become the first step in the development of low molecular weight antagonists.

Soluble cytokine receptors — A second approach to the inhibition of cytokines in the extracellular milieu is to use soluble forms of cytokine receptors. This is one strategy developed by Mother Nature to control cytokines. Such receptors have a higher binding affinity than antibodies but suffer from having a short circulating half-life. However, by linking an antibody Fc region to soluble receptors the circulating half-life can be considerably extended[28] and a number of investigators are producing such hybrid molecules with multiple binding sites which should enhance cytokine binding. Soluble receptors have been shown to suppress a range of experimental models of tissue pathology. Soluble

IL-1 receptor was shown to inhibit local allergen-induced inflammation in human volunteers, with no side effects being reported.[29] Hoffman La Roche is currently in early phase clinical trials in RA with a soluble TNF receptor-IgG Fc complex that is apparently showing clinical efficacy (Fenner, personal communication). The potential use of soluble cytokine receptors is discussed in more detail in Chapter 10.

Cytokine antagonists — When pharmacologists think about inhibiting the action of an agonist their first thought is to develop a receptor antagonist. Such ideas initiated the drive to develop anti-cytokine therapeutics and in spite of extensive and expensive research none of the world's pharmaceutical companies has yet devised a cytokine receptor antagonist. The reasons for this probably relate to the enormously high affinities of binding between cytokines and their receptor and the multipoint attachment of these two cognate proteins. However, Mother Nature has had many eons of pharmaceutical experimentation and has developed an antagonist of IL-1. This protein, interleukin-1 receptor antagonist (IL-1ra), shares only 20 to 30% homology with the IL-1s but is now regarded as the third member of the IL-1 family and binds to the functional type I IL-1 receptor without stimulating an agonist response. It is therefore a useful probe of the residues in IL-1ra which trigger receptor-mediated cellular activation.[30] First described in the late 1980s and cloned and expressed in 1990,[31] during the past four years IL-1ra has been shown to be a potent inhibitor of many experimental models of human disease, in particular septic shock (reviewed in Reference 6), and has been tested in clinical trials for various conditions. IL-1ra is in clinical trial for a number of diseases including septic shock, rheumatoid arthritis, and inflammatory bowel disease. Most attention has focused on the use of IL-1ra for the treatment of septic shock, and these clinical trials are described in Chapter 9, but is unlikely to be of therapeutic benefit in RA as it has a very short half-life in the circulation. Unfortunately, at the time of writing, Synergen, who have pioneered IL-1ra under the trade name *Antril,* have suffered a huge setback with the failure of their second phase III trial of Antril for septic shock.[32] It is the author's view that a novel approach to agonist receptor interactions will be needed to develop clinically useful cytokine antagonists.

Antagonist cytokines — Cytokines are paracrine and autocrine mediators and as such form complex interacting networks capable of both stimulating or inhibiting cellular activity. The capacity for both actions probably depends on the initial state of the system. It has emerged in the last few years that certain cytokines act to downregulate the cellular activation induced by other cytokines. Molecules such as IL-4, IL-10, and TGFβ have been shown to be able to inhibit experimental pathologies including inflammatory lesions and experimental models of arthritis.[6,33] The actions of TGFβ are reviewed in Chapter 14 and those of IL-10 in Chapter 15. The problem inherent in the clinical use of this therapeutic approach is that the administration of these cytokines to healthy animals can, by itself, induce pathology. Much more information is needed to determine if it is possible to use such antagonist cytokines therapeutically in chronic diseases such as RA.

Inhibition of cytokine-induced intracellular signaling — The effects of cytokines are due to specific upregulation of receptor-mediated intracellular signaling pathways. Blocking these pathways will inhibit the action of the cytokine. We have already discussed the mechanism of action of the immunosuppressants cyclosporine and FK506 and this is reviewed in Chapter 6. Many pharmaceutical companies have therapeutic programs to develop inhibitors of cytokine-induced intracellular signaling. Perhaps the most advanced is the program to develop inhibitors of tyrosine kinases (TKs) which are increasingly seen to be involved in cytokine-generated cellular activation. A number of bacterial metabolites such as genistein and erbstatin are known inhibitor of TKs. Erbstatin has been used to develop a family of rationally designed TK inhibitors termed Tyrphostins,[34] which have been shown to inhibit experimental septic shock[35] and are currently in clinical trial for psoriasis.[34] The recent report from Parke-Davis of a selective inhibitor of epidermal growth factor receptor tyrosine kinase with a Ki value of 5 p*M* suggests the possibility that the goal of therapeutic inhibition of growth factor receptors is a not-too-distant one.[36] Other compounds, such as Lisofylline, an inhibitor of phosphatidic acid which blocks cell activation by pro-inflammatory cytokines, are also in early phase clinical trials.

III. CONCLUSIONS

The last decade has seen enormous strides in our understanding of the biology of cytokines and their role in human disease. We are already seeing the introduction into the clinic of therapeutic cytokines such as γ-interferon, α-interferon, erythropoietin, and colony-stimulating factors. The next decade, hopefully, should see the clinical use of cytokine modulators which downregulate the action of selected cytokines. The greatest efforts in this respect have been devoted to the therapy of septic shock, an acute condition believed to be caused by the endotoxin-driven production of pro-inflammatory cytokines such as IL-1 and TNFα. The acute nature of this disease and the experimental finding that septic shock in rodents, lagomorphs, and primates could be blocked by single anti-endotoxin or anti-cytokine agents gave the biopharmaceutical industry confidence that it could produce effective therapeutics. The experience thus far is cautionary. As described by Wright,[32] a number of biopharmaceutical companies have now become unstuck in their efforts to develop clinically effective therapies for this condition. This includes well-known companies such as Centocor, Synergen, Xoma, and Immunex. Given the enormous complexity of the pathology of chronic diseases such as rheumatoid arthritis, which is the next major target for anti-cytokine therapies, our expectations should remain at a sensible, low level that by blocking single cytokines by complex agents such as cloned chimerized antibodies or soluble receptor-antibody complexes we can inhibit tissue pathology. Only time will tell if it is possible to inhibit complex cytokine-induced pathology by neutralizing or inhibiting the synthesis of single cytokines. Having introduced this note of caution, it is clear that if one wishes to block the pathology of humankind's major diseases this will only be achieved by targeting cytokines.

REFERENCES

1. **Sadlack, B., Merz, H., Schorle, H. et al.,** Ulcerative colitis-like disease in mice with a disrupted interleukin-2 gene. *Cell,* 75, 263–274, 1993.
2. **Kuhn, R., Lohler, J., Rennick, D. et al.,** Interleukin-10-deficient mice develop chronic enterocolitis. *Cell,* 75, 263–274, 1993.
3. **Shull, M. M., Ormsby, A. B., Kier, S. et al.,** Targeted disruption of the mouse transforming growth factor β, gene results in multifocal inflammatory disease. *Nature,* 359, 693–699, 1992.
4. **Pfeffer, K., Matsuyama, T., Kundig, T. M. et al.,** Mice deficient for the 55 kDa tumor necrosis factor receptor are resistant to endotoxic shock, yet succumb to *L. monocytogenes* infection. *Cell,* 73, 457–467, 1993.
5. **Henderson, B. and Blake, S.,** Therapeutic potential of cytokine manipulation. *Trends Pharmacol. Sci.,* 13, 145–152, 1992.
6. **Henderson, B. and Poole, S.,** Modulation of cytokine function: therapeutic applications. *Adv. Pharmacol.,* 25, 53–115, 1994.

7. **Henderson, B.,** Cytokines and proteinase inhibitors in the modulation of connective tissue destruction in rheumatoid arthritis, Prospects for therapy. *Clin. Immunother.* 2, 167–174, 1994.
8. **Liu, J.,** FK506 and cyclosporin: molecular probes for studying intracellular signal transduction. *Trends Pharmacol. Sci.,* 14, 182–188, 1993.
9. **Firestein, G. S. and Zvaifler, N. J.,** How important are T cells in chronic rheumatoid synovitis?. *Arthritis Rheum.,* 33, 768–773, 1990.
10. **Erickson, R. P. and Izant, J. G. (Eds.)** *Gene Regulation: Biology of Antisense RNA and DNA,* Raven Press, New York, 1992.
11. **Burch, R. M. and Mahan, L. C.,** Oligonucleotide antisense to the interleukin-1 receptor mRNA blocks the effects of interleukin-1 in cultured murine and human fibroblasts and in mice. *J. Clin. Invest.,* 88, 1190–1196, 1991.
12. **Loose, L. D., Sipe, J. D., and Kirby, D. S.,** Reduction of acute phase proteins with Tenidap sodium, a cytokine-modulating anti-rheumatic drug. *Br. J. Rheumatol.,* 33 (Suppl. 3), 19–25, 1993.
13. **Thornberry, N. A., Bull, H. G., Calaycay, J. R. et al.,** A novel heterodimeric cysteine protease is required for interleukin-1β processing in monocytes. *Nature (London),* 356, 768–774, 1992.
14. **Wilson, M., Black, J. F., Thomson, J. A. et al.,** Structure and mechanism of interleukin-1β converting enzyme. *Nature (London),* 370, 270–275, 1994.
15. **Mohler, K. M., Sleath, P. R., Fitzner, J. N. et al.,** Protection against a lethal dose of endotoxin by an inhibitor of tumour necrosis factor processing. *Nature (London),* 370, 218–220, 1994.
16. **Gearing, A. J. H., Beckett P., Christodouiou M. et al.,** Processing of tumour necrosis factor-α precursor by metalloproteinases. *Nature (London),* 370, 555–557, 1994.
17. **McGeehan, G. M., Becherer, J. D., Bast, R. C. et al.,** Regulation of tumour necrosis factor-α processing by a metalloproteinase inhibitor. *Nature (London),* 370, 558–561, 1994.
18. **Rubartelli, A. and Sitia, R.,** Interleukin-1β and thioredoxin are secreted through a novel pathway of secretion. *Biochem. Soc. Trans.,* 19, 255–259, 1991.
19. **MacKenzie, A. R., Bollinger, P., Cooper, P. H., Del Pozo, E., Elford, P., Graeber, M., Gubler, H.-U., Leutwiler, A., Payne, T., Reich, W., Schnyder, J., and Weuthrich, H.-J.,** Pharmacology of a new anti-arthritic agent IX 207–887. *Br. J. Rheumatol.,* 30 (Suppl. 2), 75–78, 1991.
20. **Dougados, M., Combe, B., Beveridge, T., Bourdeux, I., Lallemand, A., Armor, B., and Sany, J.,** IX 207–887 in rheumatoid arthritis. A double-blind clinical trial. *Arthritis Rheum.,* 35, 999–1006, 1992.
21. **Adair, J.,** Engineering antibodies for therapy. *Immunol. Rev.,* 130, 1–36, 1992.
22. **Fisher, C. J., Opal, S. M., Dhainaut, J.-F. et al.,** Influence of an anti-tumour necrosis factor monoclonal antibody on cytokine levels in patients with sepsis. *Crit. Care Med.,* 21, 317–328, 1993.
23. **Elliot, M. J., Maini, R. N., Feldmann, M. et al.,** Treatment of rheumatoid arthritis with chimaeric monoclonal antibodies to tumour necrosis factor-α. *Arthritis Rheum.,* 36, 1681–1690, 1993.
24. **Williams, R. O., Feldmann, M., and Maini, R. N.,** Anti-tumour necrosis factor ameliorates joint disease in murine collagen-induced arthritis. *Proc. Natl. Acad. Sci. U.S.A.,* 89, 9784–9788, 1993.
25. **Lewthwaite, J. C., Blake, S. M., Vetterlein, O. et al.,** The effect of anti-TNFα monoclonal antibodies on the induction of antigen-induced arthritis in the rabbit. *Trans. Orthoped. Res. Soc.,* 19, 342, 1993.
26. **Henderson, B., Hardingham, T., Blake, S., and Lewthwaite, J.,** Experimental animal models in the study of the mechanisms of articular cartilage loss in rheumatoid arthritis. In, *Joint Destruction in Arthritis and Osteoarthritis,* van den Berg, V., Ed., Basel:Birkhauser Verlag, 15–26, 1993.
27. **Saragovi, H. U., Fitzpatrick, D., Raktabutr, A., Nakanishi, H., Kahn, M., and Greene, M. I.,** Design and synthesis of a mimetic from an antibody complementarity-determining region. *Science,* 253, 792–795, 1991.
28. **Capon, D. J., Chamow, S. M., Mordenti, J. et al.,** Designing CD4 immunoadhesins for AIDS therapy. *Nature (London),* 337, 525–531, 1989.
29. SCRIP, Immunex IL-1 receptor results 1870, 24, 1993.
30. **Arend, W. P.,** Interleukin-1 receptor antagonist. A new member of the interleukin-1 family. *J. Clin. Invest.,* 88, 1694–1697, 1991.
31. **Hannum, C. H., Wilcox, C. J., Arend, W. P. et al.,** Interleukin-1 receptor antagonist activity of a human interleukin-1 inhibitor. *Nature (London),* 343, 336–340, 1990.

32. **Wright, L.,** In shocking Synergen, sepsis tallies sixth victim. *Bio/Technology,* 12, 855, 1994.
33. **Thorbecke, G. J., Shah, R., Leu, C. H., Kuruvilla, A. P., and Palladino, M. A.,** Involvement of endogenous tumour necrosis factor α and transforming growth factor β during induction of collagen type II arthritis in mice. *Proc. Natl. Acad. Sci. U.S.A.,* 89, 7375–7379, 1992.
34. **Levitski, A.,** Tyrphostins: tyrosine kinase blockers as novel antiproliferative agents and dissectors of signal transduction. *FASEB J.,* 6, 3275–3262, 1992.
35. **Novogrodsky, A., Vanichin, A., Patya, M., Gazit, A., Osherov, N., and Levitzki, A.,** Prevention of lipopolysaccharide-induced lethal cytotoxicity by tyrosine kinase inhibitors. *Science,* 264, 1319–1322, 1994.
36. **Fry, D. W., Kraker, A. J., and McMichael, A.,** A specific inhibitor of the epidermal growth factor receptor tyrosine kinase. *Science,* 265, 1093–1096, 1994.

Low Molecular Mass Cytokine Inhibitors

Chapter 5

LOW MOLECULAR WEIGHT INHIBITORS OF CYTOKINE SYNTHESIS

Andrew Roland MacKenzie

CONTENTS

0-8493-8381-1/96/$0.00+$.50

I. INTRODUCTION

A large number of agents have been described which inhibit the production or function of cytokines. Some of these, such as the glucocorticoids and retinoids, have well-described multiple hormone actions and cannot be considered specific cytokine inhibitors. More interesting are the increasingly large number of substances described as inhibiting cytokine production or action by unknown mechanisms. To restrict the scope of the current chapter, only these newer molecules will be considered, chiefly in the context of their effects on IL-1α and IL-1β. References to effects on other cytokines will be largely restricted to indications of selectivity of the effects.

II. GENERAL CONSIDERATIONS

In spite of the considerable progress and interest in the use of recombinant proteins in the therapy of human diseases, there can be little doubt that the therapeutic modality of choice is a low molecular weight substance which can be conveniently given enterally, topically, or by inhalation. Considerable effort is currently being made within both academic laboratories and the pharmaceutical industry to identify compounds which can inhibit the production or action of those cytokines thought to play a role in the pathology of a variety of diseases. For the most part, the low molecular weight inhibitors so far discovered have been identified empirically, often with no clear understanding of their molecular mechanisms of action. Most of the compounds discussed in this chapter fall into this category.

Figure 1 shows a typical cytokine cascade in which a stimulus induces cell activation leading to the transcription of one or usually several cytokine genes. The newly synthesized gene products may be secreted immediately, stored for later release, or released following posttranslational processing of a pro form of the cytokine. In some cases the pro form may be secreted intact and is processed outside the cell. Once the mature cytokine is released it binds to a specific receptor on the responding cell where intracellular signal transduction occurs, finally leading to the physiological response of the cell. The response of the target cell may be stimulation or inhibition of function and may even involve the synthesis of further cytokines. Achievement of cytokine inhibition can be theoretically envisaged in several different ways, as depicted in Figure 1.

The most direct way to inhibit a cytokine is to antagonize its action by preventing binding to its specific receptor. However, cytokines are proteins with molecular weights ranging from 8,000 to about 50,000 kDa and bind to their receptors with affinities generally in the range of 10 to 100 p*M*. This high affinity is in part due to the large surface area of interaction between the cytokine and its receptor. On thermodynamic grounds it therefore seems highly unlikely that low molecular weight receptor antagonists for cytokines can be produced with affinities high enough for them to be of therapeutic value.

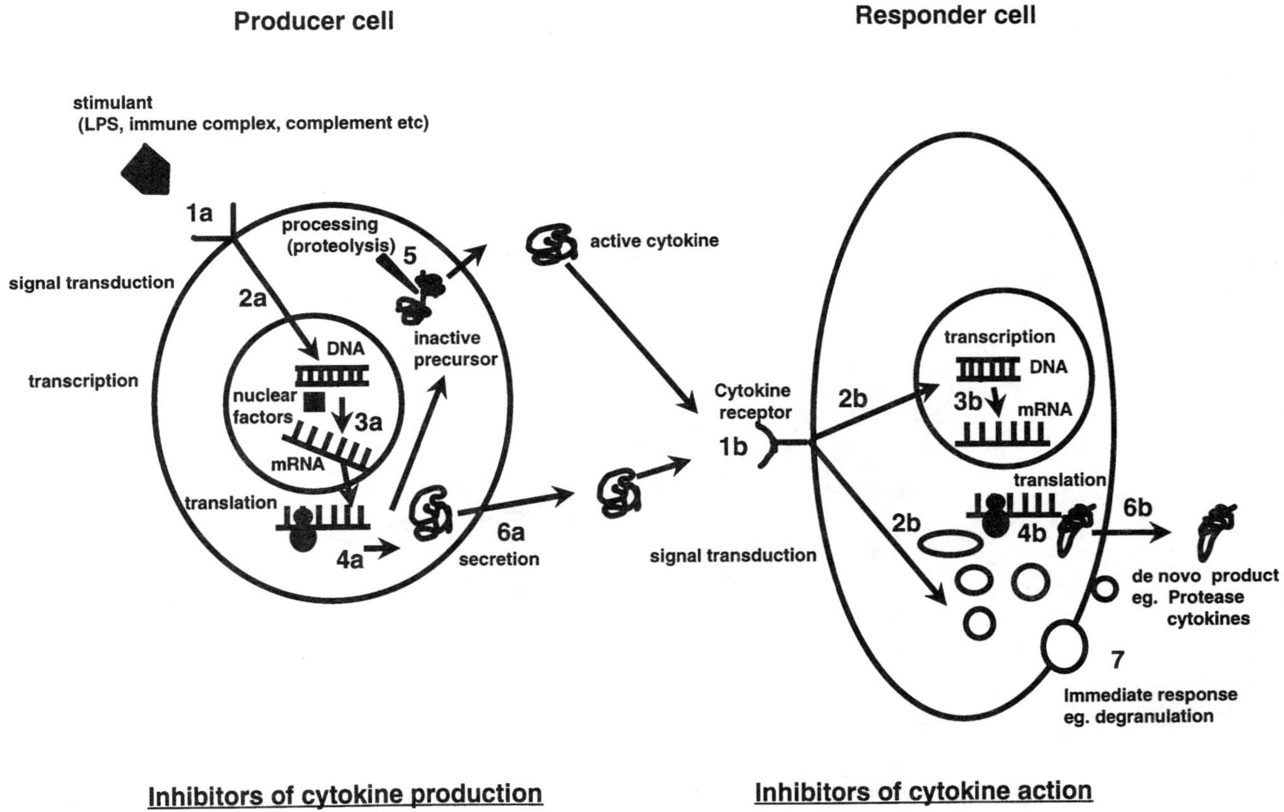

FIGURE 1. Scheme of cytokine production and action showing possible sites for pharmacological intervention. *Inhibitors of cytokine production:* 1a: receptor antagonist for stimulant of cytokine production; 2a: inhibitor of signal transduction for the stimulant; 3a: inhibitor of transcription, e.g., transcription factor binding; 4a: inhibitor of translation; 5: inhibitor of processing, e.g., proteolytic cleavage of an inactive precursor; 6a: inhibition of secretion. *Inhibitors of cytokine action:* 1b: receptor antagonist for the cytokine. Physiological antagonism through 2b: inhibition of signal transduction; 3b: inhibition of transcription; 4b: inhibition of translation; 6b: inhibition of secretion; 7: inhibition of effector mechanism, e.g., degranulation, phagocytosis; 8: inhibition of product action, e.g., proteolytic degradation.

The binding affinities for low molecular weight ligands of other physiological receptors are generally around 1 to 10 n*M*. For example, dexamethasone binds to the glucocorticoid receptor with a Kd of circa 1 n*M* and PGI_2 binds to receptors on platelets with a Kd of 12 n*M*. However, a few rare examples of low molecular weight ligands with extraordinarily high affinities do exist, such as biotin which binds to streptavidin with a Km of $10^{-15}M$, which suggests that the possibility of low molecular weight receptor antagonists for cytokines cannot be totally eliminated.

Physiological antagonism of the effects of cytokines may be much easier to achieve by interference with the signal transduction pathways within the cell. Such approaches, however, are likely to lack specificity because of the commonality of many signal transduction pathways both for different receptors within a cell and for different cell types. Nevertheless, experience tells us that such approaches can lead to usable, even if not ideal, drugs.

Inhibition of cytokine synthesis or release is another potential way to control cytokine effects. Control may be exerted by

1. Preventing the delivery of the stimulating signal to the producer cell
2. Preventing gene expression (transcription)
3. Preventing translation
4. Preventing secretion
5. Preventing posttranslational processing (see Figure 1)

The first of these suffers from the same problems associated with the inhibition of cytokine action. Many stimulants of cytokine production are themselves high molecular weight mediators (and may indeed be other cytokines), making receptor antagonism difficult to achieve. Blockade of signal transduction may be possible and in some circumstances may be surprisingly selective. Recent studies have shown that Type IV phosphodiesterase inhibitors selectively inhibit TNFα production from monocytes.[1]

Control of cytokine expression by regulation of transcription (or transcription factors) has been achieved with low molecular weight compounds. Probably the best examples are the cyclosporins and macrolides which prevent the secretion primarily of T-cell cytokines, e.g., IL-2, IL-4, γIFN, and GM-CSF. In addition, some monocyte cytokines may also be inhibited under some circumstances the actions of these two classes of compounds will be described in Chapter 6 and the use of transcription/translation inhibitors in Chapter 8).

Prevention of secretion of cytokines is possible, but once again specificity may present considerable problems. However, prevention of posttranslational processing of those cytokines which require it, should provide a high level of specificity and may in fact prevent secretion as a consequence of the failure to correctly process the precursor. IL-1β requires the action of a highly specific protease to produce the mature active 17-kDa form. Inhibition of the protease promises not only efficacy but a high level of selectivity (see Chapter 7).

The majority of the cytokine-production inhibitors that will be described in this chapter work through largely undefined mechanisms. Furthermore, many of these compounds have additional properties which makes it difficult to assess the relative importance of the cytokine inhibitory effects in *in vivo* models in which indirect methods of assessment are often used.

Finally, cytokine action may be indirectly inhibited by inducing the production of other cytokines which inhibit the action of the target cytokine. Prominent amongst such cytokines are the IL-1 receptor antagonist (IL-1ra) and TGFβ.

III. INHIBITORS OF IL-1 PRODUCTION

A. CYCLOOXYGENASE/5-LIPOXYGENASE INHIBITORS

For the greater part, classical cyclooxygenase inhibitors, which are usually described as nonsteroidal anti-inflammatory drugs (NSAIDs), have no effects on cytokine production from blood monocytes.[2] However, it has been previously reported that 5-lipoxygenase inhibitors BW-755c and 5,8,11,14-eicosatetraynoic acid (ETYA) inhibit IL-1 synthesis by human monocytes.[3] These inhibitors are relatively nonspecific in their actions and other more selective 5-lipoxygenase inhibitors do not have this effect.[4] Over the past few years, however, a number of compounds have emerged which interfere with eicosanoid production but also inhibit IL-1 production, primarily from monocytes. Interestingly, most if not all of these compounds were discovered from a series of compounds having combined cyclooxygenase/5-lipoxygenase inhibitory activity. In addition, a number of other antioxidant-type compounds have shown effects on IL-1 generation, which has led to the suggestion that 5-lipoxygenase activity is involved in the regulation of IL-1 biosynthesis.[4,5] Subsequent findings, however, show that there is no clear correlation between 5-lipoxygenase inhibition and effects on IL-1 production.[2,6,7] While it therefore seems unlikely that compounds in this class exert their effects on IL-1 via 5-lipoxygenase inhibition, possible effects on other lipoxygenase pathways such as those of 12 and 15 lipoxygenases as an explanation have not been completely eliminated.

1. SK&F 86002

a. In Vitro

The first such compound to be described was SK&F 86002[4,8] (Figure 2). SK&F 86002 inhibits cyclooxygenase with an IC_{50} of 120 μM and 5-lipoxygenase at 10 μM. By contrast, in whole cells PGE_2 production is inhibited with an IC_{50} of 1 μM. Inhibition of IL-1β release from human monocytes occurs at about the same concentration (1 to 2 μM). SK&F 86002 is not specific for IL-1, but the effects on other cytokines occur at generally higher concentrations. TNFα is inhibited at 5 to 8 μM, and at even higher concentrations (20 μM) interferon-α, GM-CSF, and IL-6 are also inhibited. Monocytes/macrophages are not the only cells affected by SK&F 86002. IL-8 production from human endothelial cells is also inhibited, with an IC_{50} of around 7 μM.[9] These results suggest that the effects on cytokine production are not due to a general downregulation of protein synthesis, but represent a selective action on certain secretory proteins from monocytes.

Since there is no evidence that pure cyclooxygenase inhibitors inhibit cytokine production, it is tempting to conclude that the inhibition of the 5-lipoxygenase pathway is responsible for the action on cytokines. However, other 5-lipoxygenase inhibitors including BW755c and phenidone,[10] were examined and found in this study (unlike the study

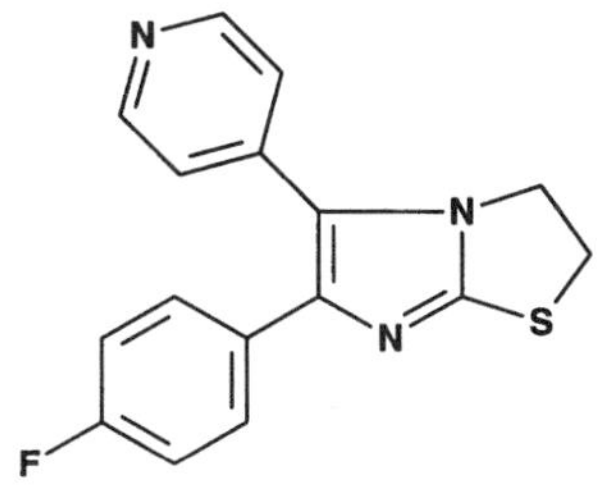

FIGURE 2. Structure of SK&F 86002.

of Dinarello et al.[3]) to have a weak or no effect on cytokine production. The lack of correlation with 5-lipoxygenase/cyclooxygenase inhibition is further reinforced by studies in which a series of compounds related to SK&F 86002 showed a clear dissociation of IL-1 inhibition and effects on arachidonic acid metabolism.[10]

Although some effects of SK&F 86002 on steady state mRNA levels were reported,[10] these occur only at concentrations tenfold higher than those necessary to inhibit IL-1 release at the protein level. At concentrations between 1 and 5 μM, no inhibition of mRNA for IL-1β was observed in either THP-1 cells or human monocytes.[9] Neither have effects been seen on the rates of induction nor decay in the case of TNFα. When levels of precursor 31 kDa IL-1β were measured, inhibition was seen with an IC_{50} of 5 μM, suggesting that the primary action is at the translational or posttranslational level.

SK&F 86002 appears to have no effects on neutrophil function[11] (expression of LFA-1 [CD11b/CD18], fMLP-induced aggregation, and adherence to human endothelium and PMA-induced chemiluminescence). This is in contrast to another inhibitor of IL-1 (IX 207–887 [Section III.B.1.] which inhibits granule release and O_2^- production).

b. In Vivo

SK&F 86002 has protective effects in two murine models of endotoxic shock, which is consistent with the view that the compound inhibits cytokine production *in vivo.*[12] In the first model, SK&F 86002 at 100 mg/kg p.o. protected mice from the lethal effects of endotoxin in D-galactosamine-primed mice. By contrast, the combined cyclooxygenase/lipoxygenase inhibitor phenidone and cyclooxygenase inhibitors gave only partial protection, suggesting that protection by SK&F 86002 was not because of its effects on eicosanoids. In the second model, LPS administration to *Propionibacterium acnes*-sensitized mice induces a shock condition. In this model, SK&F 86002 also afforded protection and, in contrast to naproxen and indomethacin, caused a decrease in circulating TNFα levels. Similar observations were also made in an endotoxic shock model in the rat, where dose-related inhibition of TNFα levels was seen with an ED_{50} of 32 mg/kg p.o. IL-1 is not normally found in the circulation even following endotoxic shock, so it has not been possible to show that IL-1 is inhibited in a similar way.

In collagen arthritis in mice, SK&F 86002 inhibits at the relatively high dose of 50 mg/kg, giving 62% inhibition of clinical manifestations and 55% inhibition of serum amyloid P (SAP) levels.[13] There are no published studies in other models of inflammatory arthritis, but the compound appears to be inactive in adjuvant arthritis (Graeber, M., personal communication).

2. SK&F 105809

a. In Vitro

SK&F 105809 (Figure 3), another dual cyclooxygenase/5-lipoxygenase inhibitor which is structurally related to SK&F 86002, possesses anti-inflammatory activity in rodent models. The active metabolite is the sulfide (SK&F 105561) which markedly inhibits IL-1 production from LPS-stimulated human monocytes with an IC_{50} of 2 μM, which is similar to its efficacy as an inhibitor of LTB_4 production from A23187-stimulated monocytes but 20-fold weaker than its effects on seminal vesicle cyclooxygenase.[14] By contrast, in whole cells PGE_2 production is inhibited with an IC_{50} of only 0.1 μM. With the recent description of two forms of cyclooxygenase, it is tempting to speculate that SK&F 105809 might be a selective Type II cyclooxygenase inhibitor.

In the context of its effects on cytokines the compound is not selective, showing an inhibition of TNFα and IL-8 production at similar or slightly higher concentrations than those necessary for IL-1.[15]

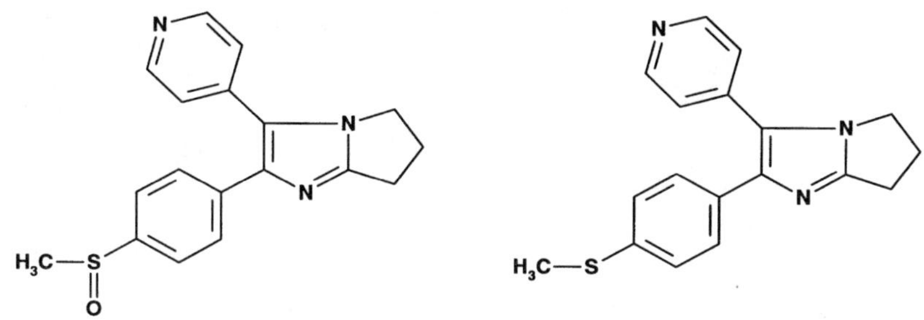

FIGURE 3. Structures of SK&F 105809 and its active metabolite SK&F 105561.

b. In Vivo

Studies in animal models of inflammation suggest that SK&F 105809 has properties that differentiate it from classical cyclooxygenase-inhibiting NSAIDs. Polymorphonuclear cell accumulation was inhibited in a variety of models including arachidonic acid ear edema, and monosodium urate and carrageenan-induced peritonitis. ED_{50} values in the range 44 to 72 mg/kg p.o. were observed.[16] These studies do not clearly indicate whether cytokine inhibition is responsible for some of the anti-inflammatory effect or whether it is entirely accounted for by the combination of cyclooxygenase and 5-lipoxygenase inhibition. The observation that PGE_2 and LTB_4 levels in the inflamed ear exudates are inhibited with ED_{50}s of 15 and 41 mg/kg, respectively, leaves the question open since it cannot be assumed that there is a direct relationship between prostanoid concentration and edema.

A problem associated with the study of the *in vivo* effects of SK&F 105809 and its active metabolite (SK&F 105561) are their very short half lives (14 to 18 min) which probably accounts for the lack of anti-arthritic activity seen in adjuvant arthritis when given once or twice daily (MacKenzie, A. R., unpublished). However, activity in collagen arthritis in the mouse has been reported[13] but a high dose of 60 mg/kg t.i.d. p.o. was required to give 89% inhibition of clinical disease and 67% inhibition of serum amyloid P (SAP). Since this dose is much higher than that required to inhibit PGE_2 and LTB_4 *in vivo,* SK&F 105809 appears to have additional actions which would be consistent with its *in vitro* effects on cytokine production.

Evidence that the compound does inhibit cytokines *in vivo* comes from studies in murine endotoxic shock models.[15] In the D-g galactosamine-enhanced shock model, SK&F 105809 produced a marked fall in plasma TNFα levels (40 to 90%) at doses of 10 to 100 mg/kg p.o. In the *P. acnes*-sensitized model, which produces much higher concentrations of TNFα in the plasma, 30 mg/kg p.o. produced a 90% inhibition. In addition, these doses protected animals from the lethal effects of endotoxin. In view of evidence from other sources of the importance of TNFα in shock, it seems probable that, in these models at least, the effects of the compound on cytokine production are important. Interestingly, although details have so far not been published in full, analogues derived from SK&F 105809 have been synthesized which inhibit cytokine production while having no significant effects on arachidonic acid metabolism.

3. Tenidap Sodium (CP 66,248–2)

a. In Vitro

Tenidap (Figure 4) is another compound which is claimed to inhibit both cyclooxygenase and 5-lipoxygenase pathways, but which also appears to inhibit IL-1 production. In

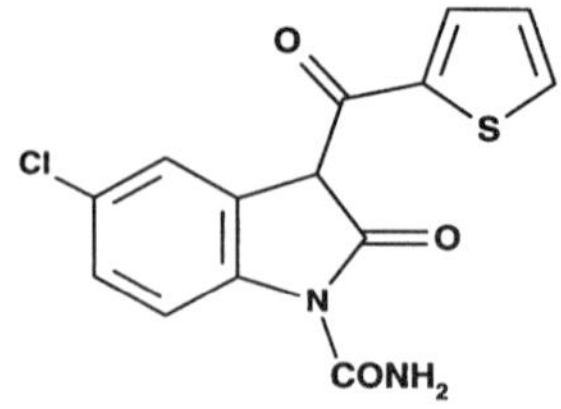

FIGURE 4. Structure of Tenidap.

zymosan-stimulated mouse peritoneal macrophages, Tenidap at 30 μ*M* completely inhibited PGE_2, LTC_4 and IL-1 production in the absence of serum. In the presence of 3% fetal calf serum, although PGE_2 and IL-1 were still markedly inhibited, LTC_4 was unaffected.[17] These results support the conclusion that the inhibition of IL-1 is not due indirectly to an effect on 5-lipoxygenase.

The effect of Tenidap on IL-1 production appears to be independent of the stimulus used, similar inhibition being achieved using either zymosan or LPS, (IC_{50} of 3 μ*M*). Both IL-1α and IL-1β appear to be inhibited by Tenidap, and Western blotting shows that the appearances of the 34- and 36-kDa pro forms of both cytokines are blocked at 3 μ*M*, suggesting that synthesis is blocked. The effect, however, is selective since general protein synthesis as measured by ^{35}S-methionine incorporation is only affected at higher concentrations (>10 μ*M*). Studies reported to date do not indicate whether IL-1 transcription is inhibited.

The effects of Tenidap on cytokines are not restricted to IL-1. Indeed, in human peripheral blood mononuclear cells the rank order of potency for cytokine inhibition is IL-6 > TNFα > IL-1, the IC_{50} for the former being 2 μg/ml.[18] Importantly, the concentrations of Tenidap achieved therapeutically are about 20 μg/ml,[18] i.e., substantially higher than the effective *in vitro* concentrations in this system.

To determine whether Tenidap could act as a physiological antagonist of IL-6, it was tested on the hepatoma cell line (HEPG2) and stimulated with conditioned medium containing IL-6. No effects were seen at concentrations up to 30 μg/ml. This finding contrasts with the down-regulation of IL-1 receptors seen in human cartilage-derived chondrocytes treated with Tenidap, which probably also accounts for the inhibition of metalloproteinase secretion seen by these authors.[19] Tenidap therefore also appears to act as a physiological antagonist of IL-1- and TNFα-induced bone resorption[20] by a mechanism which is independent of prostaglandin inhibition.

Protective effects of Tenidap against the effects of IL-1 on intact cartilage explants has also been reported.[21] Tenidap, at 5 to 10 μ*M*, prevented the catabolic effects of IL-1α (loss of glycosaminoglycans [GAGs]) and to a lesser extent prevented the reduced synthesis of GAGs as measured by incorporation of $^{35}SO_4$. At higher concentrations, however, Tenidap itself had some inhibitory effects on GAG synthesis.

Thus, Tenidap behaves as a cytokine release inhibitor and also in some circumstances as an inhibitor of cytokine action.

b. In Vivo

Direct demonstration of an effect on cytokine production *in vivo* in experimental animals has not been reported. A number of *in vivo* effects consistent with cytokine inhibition are, however seen and a number of clinical studies have been reported which demonstrate the inhibition of cytokine production. In one study, synovial fluid levels of IL-1 in rheumatoid patients were reduced by Tenidap.[22] Two further studies of RA patients receiving Tenidap showed that IL-6 levels in plasma were significantly decreased.

The effect was similar to that seen for piroxicam plus hydroxychloroquine (but not piroxicam alone), but not with diclofenac plus auranofin which were compared in the same studies.[23,24] A significant reduction in acute phase proteins (CRP) as well as ESR was also reported in these studies, which is consistent with an inhibitory effect on cytokine production. Reductions in acute phase proteins are not seen when classical NSAIDs are used to treat RA patients.

4. Antioxidants

a. In Vitro

Several antioxidants have been reported to inhibit IL-1 production from human and murine monocytes and macrophages. One of the first reports showed that quinoline antibiotics inhibit IL-1 production by human peripheral blood monocytes.[25] Furthermore, the hypocholesterolemic agent, probucol,[26,27] was also shown to inhibit IL-1β production. In a more recent study, Eugui et al.[28] showed that a range of antioxidants of moderate lipophilicity inhibited cytokine formation (IL-1β, TNFα and IL-6) at low micromolar concentrations. These included butylated hydroxyanisole (BHA), tetrahydropapaveroline (THP), apomorphine, norapomorphine, nordihydroguaiaretic acid (NDGA), and mepacrine. Not all antioxidants showed this effect and several hydrophilic antioxidants had insignificant activity in the range of 50 to 200 μ*M*. Compounds of the latter type which were tested included ascorbic acid, α-tocopherol, mannitol, trolox, butylated hydroxytoluene, quercetin, *N,N′*-diphenyl-*p*-phenylenediamine, and zileuton. The effects on cytokines were not due to a general effect on protein synthesis, as determined by measuring (^{3}H)leucine incorporation into U-937 cells, which was only affected by 10 to 20-fold higher concentrations. Analysis of the protein profiles of PBMCs after incorporation of (^{3}H)methionine on sizing gels showed that only a few bands, corresponding in size to cytokines, disappeared under the influence of THP. Interestingly, this effect seems to be cell type restricted. IL-6 production from fibroblasts stimulated with IL-1β was unaffected by THP and a range of other antioxidants which inhibited IL-6 from PBMCs.

Since these antioxidants reduce mRNAs for IL-1β, IL-1α, and IL-6 to nonstimulated levels, the site of action appears to be at the transcriptional level or at a stage in the activation signal transduction pathway prior to transcription. Nuclear run-on assays showed that for IL-1β transcription is markedly inhibited. By contrast, transcription of c-jun in the same cells was unaffected.

When nuclear transcription factors NF-κB and AP-1 were examined, their levels were found to be markedly reduced by THP if the intact cells were treated with the compound. No direct effects could be seen when THP was added directly to the nuclear extract oligonucleotide mixture in the gel-shift assay. The effects of NF-κB and AP-1 were selective since several other transcription factors (SP-1, CRE, CTF/NF-1, and OCT) were unaffected. Exactly how antioxidants influence the levels of NF-κB and AP-1 is unclear, but earlier studies have shown that redox status influences both constitutive and induced NF-κB translocation.[29] However, other data suggest that reducing conditions *favor* cytokine transcription. Malter and Hong[30] have shown that in PMA-stimulated Jurkat cells, an adenosine/uridine binding protein (AUBF), is upregulated. AUBF binds to (and is thought to protect) mRNAs containing the AUUUA motif which is associated with the inherent instability of most cytokine mRNAs. Agents which block thiols inhibit the binding of AUBF to mRNA, which would be expected to lead to a reduction of the amount of cytokine protein produced. It follows from this that antioxidants would be expected to favor stability of mRNA (and hence increased cytokine production) rather than the observed inhibitory effects. However, the exact function of AUPFs is not well understood and these two observations may yet prove reconcilable.

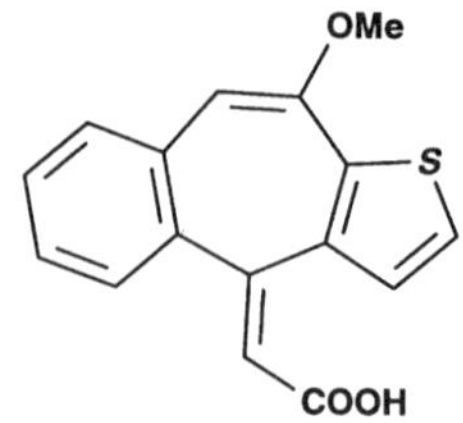

FIGURE 5. Structure of IX 207–887.

b. In Vivo

In the study of Eugui et al.[28] the *in vivo* inhibitory effects of apomorphine (150 mg/kg) were seen on IL-1β levels in the plasma of mice given a lethal dose of LPS (200 μg). When IL-1 was measured in lysates of LPS-elicited mouse macrophages prepared directly after harvesting, inhibition of IL-1β was also observed. Relatively high doses (50 to 100 mg/kg) of THP, apomorphine, and norapomorphine were required, with predosing twice before LPS and again 2 h after LPS.

In addition, other antioxidants have been reported to reduce mortality or symptoms in endotoxemia. 2-Octadecyl ascorbic acid is active in a mouse model of endotoxemia[31] and the 21-aminosteroid tirilazad mesylate reduced symptoms of endotoxemia in calves.[32] While in these cases the correlation with cytokine inhibition is not available, it seems likely that this mechanism could at least partly be responsible for the efficacy.

B. NOVEL CLASSES OF COMPOUNDS INHIBITING IL-1

1. IX 207–887

a. In Vitro

Unlike the compounds described above, IX 207–887 (Figure 5) and its known metabolites are devoid of effects on seminal vesicle cyclooxygenase and human neutrophil 5-lipoxygenase and yet they inhibit IL-1 production from mouse and human[33] monocytes. In human peripheral blood monocytes stimulated with endotoxin, IX 207–887 caused a concentration-related inhibition of IL-1β release with an IC_{50} of 60 μ*M*.[33] Although this is a high concentration, it should be borne in mind that in clinical trials in rheumatoid arthritis, doses of 800 mg/day gave trough plasma levels of 30 μ*M* and peak levels of 220 μ*M*, indicating that the concentrations required for IL-1 inhibition do occur at effective antirheumatic doses. Measurement of cell-associated IL-1β suggested that the synthesis of IL-1β was unaffected and that the effect is to prevent secretion. Consistent with this finding is the observation that there was no reduction in mRNA levels for IL-1β when analyzed by Northern blotting. It is not clear whether IX 207–887 inhibits IL-1α in humans because of the low levels of this cytokine produced by monocytes. In the mouse, however, which produces predominantly IL-1α, IX 207–887 showed a significant but incomplete inhibition of total IL-1 as measured by bioassay.

In all the *in vitro* studies reported, IX 207–887 does not show evidence of cytotoxicity at cytokine inhibitory levels (as measured by lactate dehydrogenase release) nor is constitutive lysozyme secretion affected. In contrast to IL-1, IL-6 and TNFα are only weakly affected or remain unaffected by IX 207–887.

b. In Vivo

Several effects have been reported which are consistent with inhibition of IL-1 release. IX 207–887 inhibits endotoxin-induced fever in rats with an ED_{50} of 20 mg/kg. In the

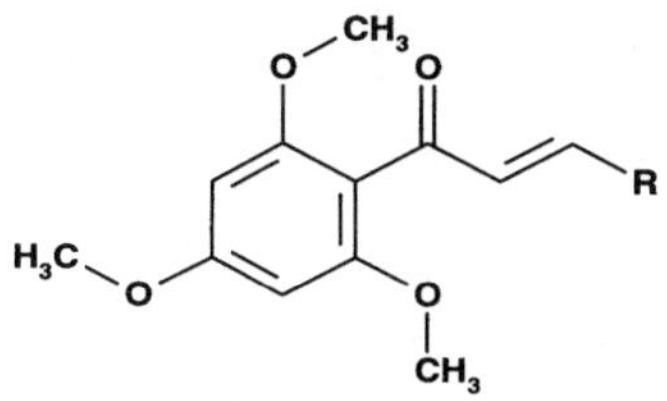

FIGURE 6. General structure of 2′-substituted chalcones.

absence of cyclooxygenase inhibition such an effect could be due to inhibition of IL-1 release, which has been shown by use of the IL-1 receptor antagonist[34] to play a critical role in the development of pyrexia.

The acute phase response is also dependent upon the cytokine cascade and it is known that exogenous IL-1 can directly or indirectly induce this response. In this context it is interesting to note that IX 207–887 inhibits α2-macroglobulin responses to endotoxin with an ED_{50} of about 20 mg/kg.[35] This appears to differentiate IX 207–887 from NSAIDs and is again consistent with an effect on IL-1 release.

In several models of arthritis, IX 207–887 inhibits with an ED_{50} of 5 to 10 mg/kg. While the antiarthritic effect cannot be conclusively attributed to inhibition of IL-1, it is known that several of these models are dependent on IL-1 production, which can be blocked with neutralizing antibodies or the IL-1 receptor antagonist. In the absence of effects on cyclooxygenase *in vitro,* no effect on arachidonic acid metabolism in platelets, vascular endothelium, or gastric mucosa, and only weak effects on prostaglandin levels in inflammatory lesions occur. Inhibition of IL-1 appears to be responsible for the antiarthritic effects.

2. 2′-Substituted Chalcone Derivatives

a. In Vitro

A series of 2′-substituted chalcone derivatives with the structure shown in Figure 6 have been reported to possess inhibitory effects on cytokine biosynthesis.[36] A wide range of compounds were reported which inhibit IL-1β at low micromolar concentrations. The compounds were not selective for IL-1 as TNFα was also inhibited, but generally at three- to tenfold higher concentrations. The mechanism of action of these compounds is not understood, however, an extensive structure-activity relationship study suggests that their activity is not due to the antioxidant properties of these compounds, although some in the series may have additional activity deriving from this property (see above). Batt et al.[36] argue that the activity also is not due to nonspecific alkylation of biological nucleophiles through the electrophilic enone system they contain. Firstly, IL-1 inhibition occurs at concentrations that are not overtly cytotoxic. The enone does, however, appear to be necessary for activity, but the substitution pattern suggests that the molecules must fit a specific receptor, which is then alkylated by the enone.

b. In Vivo

A number of chalcones have been tested for effects on LPS-induced lethal shock in mice. When given every 4 h these compounds were able to protect against mortality, but only at the very high dose of 600 mg/kg.

The most potent of these compounds inhibited IL-1 from elutriated human monocytes stimulated by LPS with an IC_{50} of 0.22 μ*M.* The effects of at least some of these

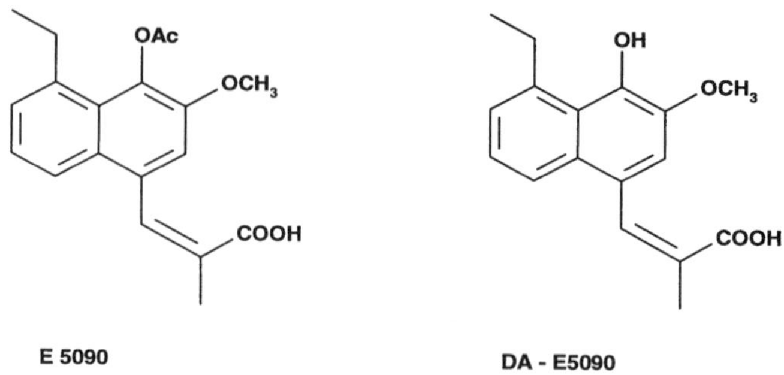

FIGURE 7. Structures of E5090 and its active metabolite DA-E5090.

compounds appear to be specific since cytotoxicity as measured by MTT production did not occur at concentrations effective at inhibiting IL-1β. In a septic shock model in the mouse induced by the i.p. administration of LPS, survival was increased from 0 to about 50% by very high doses of selected derivatives (600 mg/kg p.o.). The mechanism of action of these compounds is unknown, but does not appear to be related to effects on arachidonic acid metabolism. Slightly weaker effects were also seen on TNFα so the assumption is that the effects are at an early pretranscriptional level. No data, however, have been reported on mRNA levels to support this view.

It is of interest that of the many reported effects of chalcones, one of the more prominent is anti-inflammatory activity.[37,38]

3. E5090

E5090 is a prodrug which requires deacetylation to the active form, DA-E5090. The structures of both are shown in Figure 7. *In vitro* studies have been performed with DA-E5090 rather than the parent compound. Neither the parent nor metabolite inhibits ram seminal vesicle cyclooxygenase although they are said to inhibit prostaglandin and leukotriene production from whole cell systems at 3 to 14 μ*M*, depending on the model used. Unfortunately, these important data appear not to have been published in full, but have been reported only in the form of meetings abstracts.

a. In Vitro

DA-E5090 shows inhibitory effects on both IL-1α and IL-1β production from LPS-stimulated human monocytes — the IC_{50}s lying between 3 and 10 μ*M*. Both cell-associated and released IL-1 were inhibited, suggesting that the effect was on the synthesis of the cytokines. The effects are not due to nonspecific inhibition of protein synthesis, as shown by (^{35}S)-methionine incorporation studies.[39] The effect is mediated primarily at a pretranslational step since Northern blot analysis showed that DA-E5090 suppressed mRNA levels for both IL-1α and IL-1β at 3 and 10 μ*M*, while having no effect on β-actin mRNA.

The effect appears to be independent of the stimulus, suggesting that the site of action is not an early signal transduction mechanism.

b. In Vivo

Unlike most of the compounds discussed in this chapter, there is direct evidence that E5090 can inhibit IL-1 production *in vivo*.[40] Using a rat air-pouch model in which a cellular exudate is induced by carboxymethyl cellulose (CMC), which is then stimulated

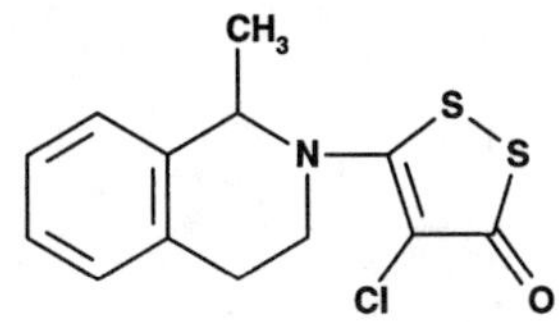

FIGURE 8. Structure of RP 54745.

in vivo by the injection of LPS into the pouch, E5090 was shown to inhibit IL-1 levels in the exudate at 25 to 200 mg/kg. In the absence of a specific ELISA for rat IL-1β, the Lymphocyte Activating Factor (LAF) assay was used to quantify total IL-1. LAF activity was found mostly in the cellular compartment of the exudate and is therefore probably mostly IL-1α. However, the results from bioassays (particularly those depending on proliferation as a read-out) must be considered with caution. It is not possible to conclude that IL-1 production is inhibited since the loss of LAF activity could equally well be explained by an elevation of IL-1-receptor antagonist (IL-1ra) levels or soluble IL-1 receptor.

IL-6 and PGE_2 levels were also inhibited in this model, but no effect was seen on TNFα. By contrast, indomethacin at 3 mg/kg inhibited only PGE_2 production. In a variant of this model in which animals were preimmunized with azobenzenearsonate-conjugated acetyl bovine serum albumin and then subsequently challenged with this antigen after induction of the CMC exudate, E5090 again inhibited IL-1 and PGE_2 levels and weakly affected TNFα levels. In this case IL-6 was unaffected. Thus E5090 seems to be a relatively selective inhibitor of IL-1 amongst the cytokines. It is not clear if the effects on PGE_2 production are related. It is unlikely that the effect on IL-1 is a consequence of PGE_2 inhibition since indomethacin at 3 mg/kg gives a greater inhibition of PGE_2 than even 200 mg/kg E5090 but, nevertheless fails to have any effect on IL-1 synthesis.

Several other *in vivo* effects are consistent with an effect on IL-1 production. E5090 inhibits acute inflammation (carrageenan edema)[41] at high doses (>100 mg/kg). The compound also possesses antipyretic activity against LPS- and yeast-induced pyrexia at 50 and 30 mg/kg, respectively, in the rat. Finally, E5090 has shown good antiarthritic effects in adjuvant arthritis and collagen arthritis in rats.[41] Paw swelling was inhibited at 25 mg/kg p.o. and effects were also seen on the erythrocyte sedimentation rate (ESR) and peripheral blood leukocyte number. The latter two features are also suppressed by 5 mg/kg p.o. prednisolone but not by 1 mg/kg p.o. indomethacin. It seems, therefore, that even if some of the anti-inflammatory effects of E5090 can be explained by the inhibition of prostaglandin production, the compound also appears to have additional activities which are consistent with an effect on IL-1 production.

4. RP 54745

RP 54745 (Figure 8) is a further compound which lacks cyclooxygenase inhibitory activity at the relatively high concentration of 50 μ*M* and only slightly reduces 5-lipoxygenase at 5 μ*M*.[42] In this case the enzymes used were derived from sonicated RBL_1 cells. The effects on IL-1 reported below therefore appear to be independent of effects on arachidonic acid metabolism.

a. *In Vitro*

RP 54745 inhibits the production of total IL-1 activity from LPS-stimulated murine peritoneal macrophages as measured by LAF assay.[42] Full concentration response data are

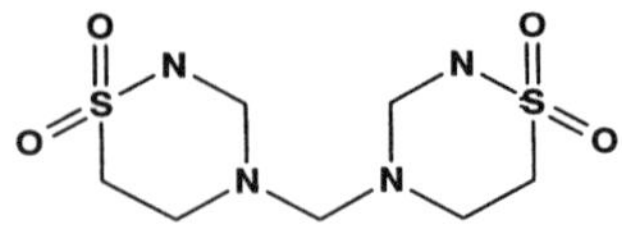

FIGURE 9. Structure of taurolidine.

not shown but 1 μ*M* caused between 80 and 55% inhibition of IL-1 activity, depending on the time after stimulation with LPS. When mRNA levels were examined by Northern blot analysis both IL-1α and IL-1β message levels were significantly reduced by 3 μ*M* RP 54745. By contrast the mRNA for TNFα was hardly affected. Thus RP 54745 appears to differ from most of the IL-1 inhibitors described so far by having a selective effect. Unfortunately, no data have been published on a range of other cytokines to determine exactly how selective this is.

Although apparently selective for particular cytokines, RP 54745 does have other effects on LPS-stimulated macrophages. The hexosemonophosphate shunt is inhibited with an IC_{50} of 3 μ*M* and release of the lysosomal enzyme β-glucuronidase was inhibited 70% by 2.5 μ*M*. Since these events are early and rapid responses to zymosan, it seems likely that the site of action of RP 54745 is at an early stage in signal transduction. Unfortunately, the cytokine release studies were performed using a different stimulus so it is not possible to compare the potency of the compound on the different effects and to conclude whether or not the effects are likely to occur through a common mechanism.

Cytotoxicity data are not presented in this study, but the selectivity of the effect for IL-1 suggests that this is not a toxic phenomenon.

b. In Vivo

When mice were treated with RP 54745 followed by *ex vivo* stimulation of peritoneal macrophages by LPS, cell-associated IL-1 production was inhibited in a dose-dependent fashion over the range of 5 to 25 mg/kg p.o.[42] At the lower dose, the inhibition could be overcome by increasing the concentration of LPS. By contrast, extracellular IL-1 was inhibited only at 25 mg/kg. When the macrophages were stimulated *in vivo* with LPS, both cell-associated and extracellular IL-1 appeared equally sensitive. At the mRNA level it was confirmed that 25 mg/kg/day substantially inhibited IL-1β message levels (70%) and only slightly affected TNFα (22%). No studies have been reported so far on inflammatory models in which IL-1 is thought to play a role.

5. Taurolidine

Taurolidine (Figure 9) is an analogue of the amino acid, taurine, and is used in humans as an adjunct therapy in the treatment of infections, both as an antimicrobial agent and as an antitoxin.[43,44] Taurolidine is formed by the condensation of two molecules of taurine with three molecules of formaldehyde. When in aqueous solution it undergoes hydrolysis to form free methylol groups which cross-link proteins. It is the cross-linking of bacterial proteins that is thought to bestow the antibacterial effects.[45] It has also been found that large doses of taurolidine given to patients with endotoxemia caused a fall in LPS levels, suggesting that it may also inactivate LPS.[44]

a. In Vitro

In view of the apparent ability to cross-link and inactivate LPS it is not too surprising after preincubation of LPS with taurolidine for 60 min prior to addition to peripheral blood mononuclear cells (PBMC) that both IL-1β and TNFα levels were inhibited.[46] Somewhat surprisingly, the efficacy of taurolidine did not appear to depend on the molar ratio of LPS to taurolidine, which would be expected if the effect is primarily on the cross-linking of LPS. In a further series of experiments,[46] PBMCs were preincubated with taurolidine at concentrations of 10 to 100 μg/ml for 30 min at 37°C. When the cells were subsequently stimulated with LPS (without washing the cells) over 90% reductions in TNFα and IL-1β reactivity were observed at concentrations above 40 μg/ml. The continued presence of taurolidine does not appear necessary for the inhibition of cytokine production. As short a time as 30 min incubation with 80 μg/ml, followed by washing, was as effective as the continuous presence of taurolidine.

The molecular mechanism is unclear at the cellular level. Cross-linking of cell surface proteins could lead to either the inactivation of receptors or molecules involved in signal transduction or, alternatively, might lead to activation of certain receptors. Bedrosian et al.[46] suggest that elevation of cAMP might account for the effects on cytokine production, and cite as evidence the bell-shaped concentration response curve for taurolidine which shows a slight stimulation of cytokine production at low concentrations of taurolidine. Such responses are common for substances which raise cAMP. This explanation, however, is inconsistent with other findings that the elevation of cAMP levels by phosphodiesterase inhibitors such as rolipram leads to the selective inhibition of TNFα but not IL-1β.[47]

b. *In Vivo*

A number of *in vivo* effects have been reported for taurolidine which are consistent with an effect on cytokine inhibition. Mortality in mice receiving LPS is reduced by taurolidine[48] as is LPS-induced fever in rabbits.[49] To what extent these effects are accounted for by direct effects on cytokine-producing cells, rather than through the inactivation of endotoxin, is not clear.

IV. SLOW-ACTING ANTIRHEUMATIC DRUGS

Slow-acting antirheumatic drugs (SAARDs) are a diverse range of agents which have been found empirically to give benefit in rheumatoid arthritis. At one time such agents were considered to be disease modifying and were referred to as disease modifying antirheumatic drugs (DMARDs). It is now generally accepted that these drugs do not influence the progression of erosive disease in rheumatoid patients. The mechanism of action of these drugs is unknown and, not surprisingly, over the past few years attention has been paid to the possibility that modulation of cytokine expression might play a role in their action. Although highly diverse, they will be considered here as a group.

A. *IN VITRO*

The first SAARDs to be shown to inhibit cytokine production were chloroquine and gold salts. IL-1 levels were found to be inhibited both in mixed cell culture systems[50,51] and in purified monocyte cultures.[52–55] Sulfasalazine has also been reported to inhibit IL-1 production.[55,56] By contrast, d-penicillamine and bucillamine have no effects on IL-1. The question of selectivity has been addressed in relatively few studies. Evans and Zuckerman[57] showed that auranofin inhibited TNFα production and Danis et al.[58] subsequently made a careful study in which they compared the effects of a range of SAARDs on IL-1α, IL-1β, TNFα, and IL-6 using specific ELISA assays. Using elutriated human peripheral blood monocytes stimulated with LPS, Danis et al.[58] showed that auranofin,

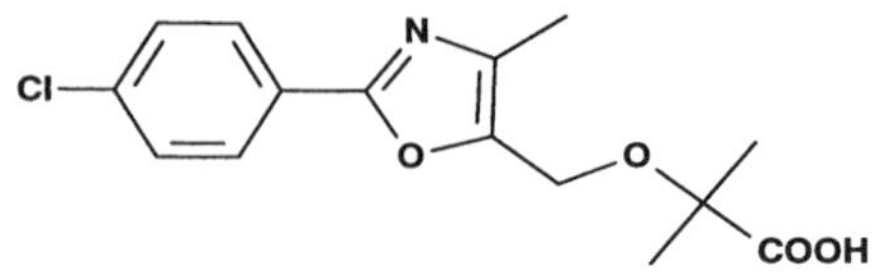

FIGURE 10. Structure of Romazarit.

hydroxychloroquine (HCQ), and sulfasalazine inhibited all four cytokines at therapeutically relevant concentrations. Some differential effects were seen. Sulfasalazine, for instance, inhibited IL-1 secretion less well than the levels of cell-associated IL-1. In addition, sulfasalazine was generally less effective on IL-6 production. HCQ showed a different profile, having better effects on secretion than on cell-associated IL-1. In marked contrast, gold sodium thiomalate (GST), d-penicillamine, and the metabolites of sulfasalazine, sulfapyridine, and 5-aminosalicylic acid (5-ASA), showed no significant effects on cytokine production at up to 200 μg/ml.

The effects were not restricted to LPS stimulation. Sulfasalazine and HCQ were shown to also inhibit IL-1α and IL-1β from monocytes stimulated with either γ-interferon + IL-2 or γ-interferon + GM-CSF over a similar concentration range (12.5 to 50 μg/ml).

Unlike the other compounds tested, GST and auranofin showed a clear *stimulation* of both IL-1 and TNFα at low concentrations (1 to 50 ng/ml) with the inhibitory effects of auranofin only becoming apparent at higher concentrations. IL-6 behaved differently, however, and was not significantly elevated by either GST or auranofin.

As a group, therefore, it seems that cytokine inhibition is not a common mechanism which accounts for the activity of SAARDs in rheumatoid arthritis. Nevertheless, in some cases, effects on cytokines occurring at therapeutically relevant concentrations may contribute to the mode of action. The fact is that SAARDs (DMARDs) are defined as agents having inhibitory effects on a number of biological parameters of rheumatoid arthritis (in particular the acute phase protein, CRP) and their mechanism of action is certainly consistent with an effect on cytokine production, but it is impossible to determine if this is direct or indirect.

V. INHIBITORS OF IL-1 ACTION

To date, no low molecular weight receptor antagonists for IL-1 have been described. However, a few compounds have been reported to act as physiological antagonists of IL-1.

A. ROMAZARIT

Romazarit (Figure 10) is an agent with anti-inflammatory and antiarthritic activities[59] which has no significant effects on arachidonic acid metabolism. Against sheep seminal vesicle cyclooxygenase it is inactive at 6.5 m*M*. It does not inhibit arachidonate release, LTB_4 production, or 6-keto-PGF_1 production by ionophore-stimulated adherent rat peritoneal cells. Metabolic conversion to a substance with inhibitory effects on arachidonate metabolism seems unlikely, since after treating rats for 1 week with 100 mg/kg no effects were seen on urinary PGE_2 production.

Romazarit does, however, inhibit PGE_2 release from human synovial fibroblasts in culture stimulated with IL-1 with an IC_{50} of 150 μ*M*.[60] In rheumatoid arthritis patients receiving 350 mg of Romazarit, plasma levels of greater than 100 μg/ml are achieved,[61]

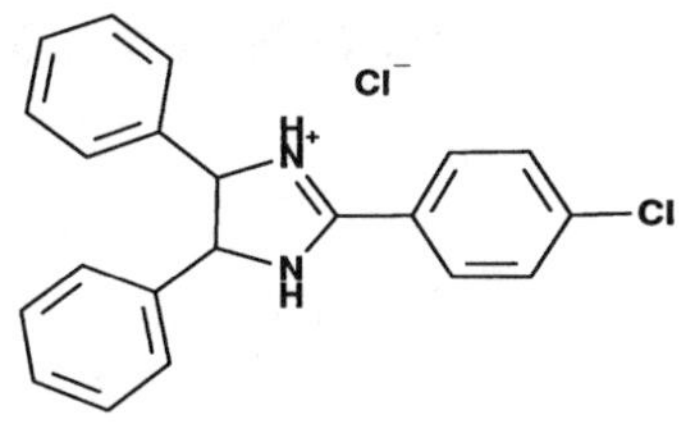

FIGURE 11. Structure of TA-383.

suggesting that the effect on PGE_2 production could occur *in vivo.* One possible explanation for the effects of Romazarit on PGE_2 production is that it selectively inhibits Type II cyclooxygenase in fibroblasts. This, however, is made less likely by the observations in a further study which show inhibitory effects on IL-1-induced collagenase synthesis in the dermal fibroblast line CCD-43 over a concentration range similar to those needed for effects on PGE_2.[60]

Additional studies have shown that Romazarit does not affect either the binding of IL-1 to its receptor, nor its internalization as a receptor/ligand complex.[60] Romazarit therefore behaves as a physiological antagonist of IL-1.

An *in vivo* observation which suggests that Romazarit may possess inhibitory properties against IL-1 *in vivo* comes from a study in which IL-1-induced serum amyloid P was measured. Romazarit at 100 mg/kg substantially, but not completely, inhibited the response.

B. TA-383

TA-383 (Figure 11) is an immunomodulatory compound described by Ueno et al.[62] In a recent publication it was shown that TA-383 inhibits IL-6 production from mouse synovial fibroblasts stimulated with IL-1β.[63] The effects occurred between 0.1 to 10 μ*M* but did not completely inhibit IL-6 production even at the highest level. The effect was not due to a general depression of protein synthesis, as determined by ^{3}H-leucine uptake. It was also reported that TA-383 did not affect the expression of the Type I IL-1 receptor on these cells. No direct evidence was reported to show whether or not TA-383 directly inhibited IL-1 binding.

In vivo, TA-383 has been shown to inhibit collagen-induced arthritis in the mouse.[62] In parallel, it was also demonstrated that serum IL-6 levels were dramatically reduced, which could reflect the observed action on IL-1-induced IL-6 production *in vitro.* So far, no studies on other actions of IL-1 have been reported so it is difficult to determine if this is a specific effect on IL-6 synthesis or truly a physiological inhibition of IL-1 action.

VI. CONCLUDING REMARKS

It is clear that an increasingly large number of compounds (including many which have been reported only in the form of meetings abstracts and are not described here) inhibit IL-1 *in vitro.* For most of these compounds, studies to examine directly the effects on IL-1 *in vivo* are urgently needed, particularly for those compounds which inhibit IL-1 release. It is to be hoped that in the near future studies will be carried out to help elucidate which of these compounds really owe their *in vivo* anti-inflammatory activity to an effect on IL-1.

REFERENCES

1. **Molnar-Kimber, K., Yonno, L., Heaslip, R., et al.,** Modulation of TNFα and IL-1β from endotoxin-stimulated monocytes by selective PDE isoenzyme inhibitors, *Agents Actions,* 39, Spec. Conf. Issue C77–79, 1993.
2. **Hartman, D. A., Ochalski, S. J., and Carlson, R. P.,** The effects of anti-inflammatory and antiallergic drugs on the release of IL-1β and TNFα in the human whole blood assay, *Agents Actions,* 39, Spec. Conf. Issue, C70, 1993.
3. **Dinarello, C. A., Bishai, I., Rosenwasser, L. J., et al.,** Influence of lipoxygenase inhibitors on the *in vitro* production of human leukocytic pyrogen and lymphocyte activating factor (interleukin-1), *Int. J. Immunopharmacol.,* 6, 43, 1984.
4. **Lee, J. C., Rebar, L., and Layden, J. T.,** Effect of SK&F 86002 on cytokine production by human monocytes, *Agents Actions,* 27, 277, 1989.
5. **Bender, P. E. and Lee, J. C.,** Pharmacological modulation of interleukin-1, *Annu. Rep. Med Chem.,* 25, 25, 1989.
6. **Liu, D. S., Liew, F. Y., and Rhodes, J.,** Immunoregulatory properties of novel specific inhibitors of 5-lipoxygenase, *Immunopharmacology,* 17, 1, 1989.
7. **Parkar, B. A., McCormick, M. E., and Foster, S. J.,** Leukotrienes do not regulate interleukin-1 production by activated macrophages, *Biochem. Biophys. Res. Commun.,* 169, 422, 1989.
8. **Lee, J. C., Votta, B., Griswold, D. E., et al.,** Inhibitory effects of SK&F 86002 on monocyte IL-1 production, *Agents Action,* 27, 280, 1989.
9. **Lee, J. C., Badger, A. M., Griswald, D. E., et al.,** Bicyclic imidazoles as a novel class of cytokine biosynthesis inhibitor, *Ann. N.Y. Acad. Sci.,* 696, 149, 1993.
10. **Lee, J. C., Votta, B., Dalston, B. J., et al.,** Inhibition of human monocyte IL-1 production by SKF-86002, *Int. J. Immunother.,* 6, 1, 1990.
11. **Shappel, S. B., Taylor, A. A., Hughes, H., et al.,** Comparison of antioxidant and nonantioxidant lipoxygenase inhibitors of neutrophil function. Implication for pathogenesis of myocardial reperfusion injury, *J. Pharmacol. Exp. Ther.,* 252, 531, 1990.
12. **Badger, A. M., Olivera, D., Talmadge, J. E., et al.,** Protective effect of SK&F 86002, a novel dual inhibitor of arachidonic acid metabolism, in murine models of endotoxin shock: inhibition of tumor necrosis factor as a possible mechanism of action, *Circ. Shock,* 27, 51, 1989.
13. **Griswold, D. E., Hillegass, P. C., Meunier, P. C., et al.,** Effect of inhibitors of eicosanoid metabolism in murine collagen-induced arthritis, *Arthritus Rheum.,* 31, 1406, 1988.
14. **Marshall, P. J., Griswold, D. E., Breton, J., et al.,** Pharmacology of the pyrrolimidazole SKF-105809 I. Inhibition of inflammatory cytokine production and of 5-lipoxygenase-mediated and cyclooxygenase-mediated metabolism of arachidonic acid, *Biochem. Pharmacol.,* 42, 813, 1991.
15. **Olivera, D. L., Esser, K. M., and Lee, J. C.,** Beneficial effects of SK&F 105809, a novel cytokine suppressive agent, in murine models of endotoxin shock, *Circ. Shock,* 37, 301, 1992.
16. **Griswold, D. E., Marshall, P. J., Lee, J. C., et al.,** Pharmacology of the pyrrolimidazole SKF-105809 II. Anti-inflammatory activity and inhibition of mediator production *in vivo, Biochem. Pharmacol.,* 42, 825, 1991.
17. **Otterness, I. G., Bliven, M. L., Downs, T. D., et al.,** Inhibition of interleukin-1 synthesis by tenidap: a new drug for arthritis, *Cytokine,* 3, 277, 1991.
18. **Sipe, J. D., Bartie, L. M., and Loose, L. D.,** Modification of proinflammatory cytokine production by the antirheumatic agents tenidap and naproxen. A possible correlate with clinical acute phase response, *J. Immunol.,* 148, 480, 1992.
19. **Martel-Pelletier, J., McCollum, R., Tardif, G., et al.,** Tenidap, a new antirheumatic drug reduces IL-1R expression and IL-1 induced expression of metalloproteinases in cartilage human chondrocytes, *Arthritis Rheum.,* 35 (Suppl. 9), s131, 1992.
20. **Al Humidan, A. M., Reilly, K. M., Leeming, M. R. G., et al.,** Tenidap inhibits bone resorption by PTH, 1,25 Vit D3, IL-1, TNF and PGE_2 *in vitro* by mechanisms independent of inhibition of prostaglandin synthesis, *Arthritis Rheum.,* 34 (Suppl. 9), s192, 1991.
21. **Dingle, J. T., Leeming, M. R. G., and Martindale, J. J.,** Effect of tenidap on cartilage integrity *in vitro, Ann. Rheum. Dis.,* 52, 292, 1993.
22. **McDonald, B., Loose, L., and Rosenwasser, L. J.,** Synovial fluid IL-1 in RA patients receiving a novel arachidonate inhibitor, CP-66248, *Arthritis Rheum.,* 31 (Suppl. 4), s52, 1988.

23. **Leeming, M. R. G.,** A double-blind randomised comparison of tenidap versus auranofin plus diclofenac in early rheumatoid arthritis, *Rev. Esp. Rheumatol.,* 20 (Suppl. 1), Th189, 1993.
24. **Shanahan, W. R., Kraska, A. R., Kirby, D. S., et al.,** Comparison of tenidap, piroxicam and the combination of piroxicam and hydroxychloroquine in the treatment of rheumatoid arthritis, *Rev. Esp. Rheumatol.,* 20 (Suppl. 1), Th80, 1993.
25. **Roche, Y., Fay, M., and Gougerot-Pocidalo, M. A.,** Effects of quinolines on interleukin-1 production *in vitro* by human monocytes, *Immunopharmacology,* 13, 99, 1987.
26. **Ku, G., Doherty, N. S., Schmidt, L. F., et al.,** *Ex vivo* lipopolysaccharide-induced interleukin-1 secretion from murine peritoneal macrophages inhibited by probucol, a hypocholesterolemic agent with anti-oxidant properties, *FASEB J.,* 4, 1645, 1990.
27. **Akeson, A. L., Woods, C. W., Mosher, L. B., et al.,** Inhibition of IL-1β expression in THP-1 cells by probucol and tocopherol, *Atherosclerosis,* 86, 261, 1991.
28. **Eugui, E. M., DeLustro, B., Rouhafza, S., et al.,** Some antioxidants inhibit, in co-ordinate fashion, the production of tumour necrosis factor-α, IL-1β and IL-6 by human peripheral blood mononuclear cells, *Int. Immunol.,* 6, 409, 1994.
29. **Israel, N., Gougerat-Pocidalo, M.-A., Aillet, F., et al.,** Redox status influences constitutive or induced NFκB translocation and HIV long terminal repeat activity in human T and monocytic cell lines, *J. Immunol.,* 149, 3386, 1992.
30. **Malter, J. S. and Hong, Y.,** A redox switch and phosphorylation are involved in the posttranslational up-regulation of the adenosine-uridine binding factor by phorbol ester and ionophore, *J. Biol. Chem.,* 266, 3167, 1991.
31. **Nonaka, A., Manabe, T., and Tobe T.,** Effect of a new synthetic free radical scavenger, 2-octadecyl ascorbic acid, on the mortality in mouse endotoxaemia, *Life Sci.,* 47, 1933, 1990.
32. **Braughler, J. M., Pregenzer, J. F., Chase, R. L., et al.,** Novel 21-aminosteroids as potent inhibitors of iron dependent lipid peroxidation, *J. Biol. Chem.,* 262, 10438, 1987.
33. **Schnyder, J., Bollinger, P., and Payne, T.,** Inhibition of interleukin-1 release by IX 207–887, *Agents Actions,* 30, 350, 1990.
34. **Coceani, F., Lees, J., Redford, J., and Bishai, I.,** Interleukin-1 receptor antagonist. Effectiveness against interleukin-1 fever, *Can. J. Physiol. Pharmacol.,* 70, 1590, 1992.
35. **Mackenzie, A. R.,** Pharmacological properties of IX 207–887, in *The Non-Steroidal Anti-inflammatory Drugs. Mechanisms and Clinical Uses,* 2nd ed., Lewis, A. J. and Furst, D. E., Eds., Marcel Dekker, New York, 1994, 391.
36. **Batt, D. G. Goodman, R., Jones, D. G., et al.,** 2′-Substituted chalcone derivatives as inhibitors of interleukin-1 biosynthesis, *J. Med. Chem.,* 36, 1434, 1993.
37. **Rao, M. N. A., Naidoo, L., and Ramaman, P.N.,** Antiinflammatory activity of phenyl styryl ketones, *Pharmazie,* 46, 542, 1991.
38. **Singh, G. B., Leach, G. D. H., and Atal, C. K.,** Antiinflammatory actions of methyl and phenyl 3-methoxy-4-hydroxystyryl ketones, *Arzneim. Forsch.,* 37, 435, 1987.
39. **Goto, M., Chiba, K., Hashida, R., et al.,** A novel inhibitor of IL-1 generation, E5090. *In vitro* inhibitory effects on the generation of IL-1 by human monocytes, *Agents Actions,* Suppl. 32, 225, 1991.
40. **Chiba, K., Goto, M., and Shirota, H.,** A novel inhibitor of IL-1 generation, E5090. *In vivo* inhibitory effect on the generation of IL-1-like factor and on granuloma formation in an air-pouch mode, *Agents Actions,* Suppl. 32, 231, 1991.
41. **Shirota, H., Chiba, K., Goto, M., et al.,** Antiinflammatory properties of E5090. A novel orally active inhibitor of IL-1 generation, *Agents Actions,* Suppl. 32, 219, 1991.
42. **Folliard, F., Bousseau, A., and Terlain, B.,** RP 54745, a potential antirheumatic compound. I. Inhibitor of macrophage stimulation and interleukin-1 production, *Agents Actions,* 36, 119, 1992.
43. **Browne, M. K., Leslie, G. B., and Pfirrmann, R. W.,** Taurolin: a new chemotherapeutic agent, *J. Appl. Bacteriol.,* 41, 363, 1976.
44. **McCartney, A. C. and Browne, M. K.,** Clinical studies on administration of taurolin in severe sepsis. A preliminary study, in *Bacterial Endotoxins: Pathophysiological Effects, Clinical Significance and Pharmacological Control,* Levein, J., Bullen, H. R., ten Gate, J. W., van Deventer, S. J. H., and Sturk, A., Eds., Alan R. Liss, New York, 1988, 361.
45. **Browne, M. K., Leslie, G. B. Pfirrmann, R. W., et al.,** The *in vitro* and *in vivo* activity of taurolin against anaerobic pathogenic organisms, *Surg. Gynecol. Obstet.,* 145, 842, 1977.

46. **Bedrosian, I., Sofia, R. D., Wolff, S. M., et al.,** Taurolidine, an analogue of the amino acid taurine, suppresses interleukin-1 and tumor necrosis factor synthesis in human peripheral blood mononuclear cells, *Cytokine,* 3, 568, 1991.
47. **Molnar-Kimber, K., Yonno, L., Heaslip, R., et al.,** Modulation of TNFα and IL-1β from endotoxin-stimulated monocytes by selective PDE isoenzyme inhibitors, *Agents Actions,* 39, Spec. Conf. Issue C77–79, 1993.
48. **Pfirrmann, R. W. and Leslie, G. B.,** The anti-endotoxic activity of taurolin in experimental animals, *J. Appl. Bacteriol.,* 46, 97, 1979.
49. **Pfirrmann, R. W.,** Taurolidin: Ein neues Konzept zur antimikrobiellen Chemotherapie chirugischer Infektionen Einführung und Uebersicht, in *Taurolin,* Bruckner, W. L. and Pfirrmann, R. W., Eds., Urben and Schwarzenberg, Munich, p. 3, 1985.
50. **Salmeron, G. and Lipskey, P. P.,** Immunosuppressive potential of antimalarials, *Am. J. Med.,* Suppl. 75, 19, 1983.
51. **Lewis, G. P. and Gordon, D.,** Cytokines as potential targets for anti-rheumatic therapy: a pharmacologist's view, *Br. J. Rheumatol.,* 24 Suppl. 1, 29, 1985.
52. **Barrett, M. L. and Lewis, G. P.,** Unique properties of auranofin as a potential anti-rheumatic drug, *Agents Actions,* 19, 109, 1985.
53. **Rordorf-Adam, C., Lazdins, J., Woods-Cook K., et al.,** An assay for the detection of interleukin-1 synthesis inhibitors: effects of antirheumatic drugs, *Drugs Exp. Clin. Res.,* 15, 355, 1989.
54. **Danis, V. A., Kulesz, A. J., Nelson, D. S., et al.,** The effect of gold sodium thiomalate and auranofin on lipopolysaccharide-induced interleukin-1 production by blood monocytes *in vitro:* variation in healthy subjects and patients with arthritis, *Clin. Exp. Immunol.,* 79, 335, 1990.
55. **Okamoto, M., Sasano, M., Goto, M., et al.,** Suppressive effect of anti-rheumatic drugs on interleukin-1β release from human peripheral blood monocytes, *Int. J. Immunopharmacol.,* 12, 39, 1990.
56. **Remvig, L., Andersen, B.,** Salicylazosulfapyridine (salazopyrin) effect on endotoxin-induced production of interleukin-1-like factor from human monocytes *in vitro, Scand. J. Rheumatol.,* 19, 11, 1990.
57. **Evans, G. F., and Zuckerman, H.,** Pharmaocologic modulation of TNF production by endotoxin stimulated macrophages: *in vitro* and *in vivo* effects of auranofin, and other chryso-therapeutic compounds, *Agents Actions,* 26, 329, 1989.
58. **Danis, V. A., Rathjen, D. A., and Brooks, P. M.,** The effect of slow acting antirheumatic drugs on the production of cytokines by human monocytes, *Inflammopharmacology,* 1, 315, 1992.
59. **Bloxham, D. P., Bradshaw, D., Cashin, C. H., et al.,** Biologic properties of romazarit (Ro 31–3948), a potential disease-modifying antirheumatic drug, *J. Pharmacol. Exp. Ther.,* 252, 1331, 1990.
60. **Bradshaw, D., and Williams, P. E. O.,** Romazarit, in *The Non-Steroidal Antiinflammatory Drugs. Mechanisms and Clinical Uses,* 2nd ed., Lewis, A. J. and Furst, D. E. Eds., Marcel Dekker, New York, 1994, 317.
61. **Williams, P. E. O., Bird, H. A., Minty, S., et al.,** A pharmacokinetic and tolerance study of romazarit in patients with rheumatoid arthritis, *Biopharm. Drug Dispos.,* 13, 119, 1992.
62. **Ueno, M., Sugita, T., Furukawa, O., et al.,** Immunopharmacological studies on a new antirheumatic drug, TA-383, *Jpn. Pharmacol.,* 58 (Suppl. 1), 123, 1992.
63. **Sugita, T., Ueno, M., Furukawa, O., et al.,** Effect of a novel anti-rheumatic drug, TA-383, on type II collagen-induced arthritis — suppressive effect of TA-383 on interleukin-6 production, *Int. J. Immunopharmacol.,* 15, 515, 1993.

Chapter 6

IMMUNOSUPPRESSANTS AS CYTOKINE INHIBITORS

Jerzy A. Schmidt and Robert V. Bundick

CONTENTS

0-8493-8381-1/96/$0.00+$.50

I. INTRODUCTION

Once the technical problems of transplant surgery had been overcome in the late 1950s the discovery of immunosuppressive agents was essential for this life-saving technique to progress as a routine procedure. The first generation of immunosuppressive drugs were selected on their antiproliferative properties. These drugs, including alkylating agents, cytostatic agents, and antimetabolites, were primarily developed to control the proliferation of neoplastic cells and considered as potential immunosuppressives by virtue of their antiproliferative properties on immune cells. Although these agents acted as potent immunosuppressives, they were nonspecific, inhibiting any cell in mitosis and resulting in undesirable side effects. The next generation of agents, including the glucocorticosteroids and antisera directed toward lymphocyte cell surface proteins, had more selectivity toward immune cells. These have been most successful when used in combination with other immunosuppressives but still retain undesirable side effects. Presently there is a new group of immunosuppressives derived from microbial broths that is receiving a great deal of attention; this group includes cyclosporin A, FK 506, and rapamycin. These new-generation compounds are not only proving to have exciting potential in the clinic but are also allowing immunologists to elucidate entirely new intracellular signaling pathways.

Cytokine biology is a relatively recent addition to our knowledge of the interactions of the immune response, in fact, the cloning and sequencing of Interleukin-2 (IL-2) has been achieved only in the last decade. Previously, the cytokines were defined as factors present in conditioned media and were generally referred to as "lymphokines" or "monokines". As understanding of the immune response progressed, it was apparent that many immunosuppressive drugs acted on the production or action of certain cytokines. Furthermore, understanding the mechanisms of these drugs has also helped us to elucidate the role lymphokines play in orchestrating the immune response.

This chapter will endeavor to review a selection of immunosuppressive drugs and focus on how their activity can be explained by their effects on cytokine biology. We will not simply present a list of compounds which act by inhibiting cytokine production and action, nor give a comprehensive review on all aspects of each immunosuppressive compound, but rather discuss how immunosuppressive drugs have evolved and how their activities relate to each other and to cytokine biology. We will also discuss the capacity of immunosuppressive drugs to selectively inhibit individual cytokines in light of recent observations showing the imbalance of T helper cell subsets, defined by their cytokine repertoire, in the pathology of various immune dysfunctions. Finally, we will speculate on the kind of profile of cytokine production or action that might be desirable in immunosuppressive drugs of the future. We will limit our discussion to pharmacological agents. Other immunosuppressive approaches such as naturally occurring cytokine inhibitors, cytokines themselves, or antibodies are covered in detail in other chapters. For a wider consideration of some of the immunosuppressive drugs we consider here, the reader is directed to several recent and comprehensive reviews.[1–3]

II. FIRST-GENERATION IMMUNOSUPPRESSIVES

By virtue of their well-defined antiproliferative activity directed at the level of disrupting the nuclear integrity of the cell, the antimetabolites, cytostatic agents, and alkylating agents are not considered to have a direct effect on cytokine production or action. The glucocorticosteroids (for example, prednisolone) however, have a more subtle mechanism of action and there are a number of reports showing that they inhibit the production of monocyte and T-cell-derived cytokines *in vitro*.[4–10] The cytokines inhibited include

IL-1, IL-2, IL-3, IL-6, and interferon gamma (IFNγ). These findings have recently been extended *in vivo.*[11] Inhibition is at the level of transcription. Glucocorticoid-responsive elements have been characterized within sensitive genes and these are thought to regulate cytokine transcription. In the case of IL-2 this regulation is specific, and glucocorticosteroids have no effect on other transcription factors which also control the gene.[12] The role of the lymphokines in controlling T and B lymphocyte production and function is outlined in Chapter 2.

III. CYCLOSPORIN A — A NEW CLASS OF IMMUNOSUPPRESSIVE DRUG

The discovery of cyclosporin A revolutionized the field of organ transplantation. It was first isolated in 1970 and is one of a family of cyclic undecapeptides produced by *Tolypocladium infantum Gans* which possesses weak antifungal activity but strong immunosuppressive properties. Cyclosporin A is a potent inhibitor of lymphocyte mitogenic responses *in vitro* and inhibits both humoral and cell-mediated immune responses *in vivo.*[13–15] When introduced into the clinic its exact mechanism of action was unknown, but it was clear that the drug was not steroid-like. It had already been successfully used in transplant patients when the first reports of its effects on cytokine production appeared. Bunjes et al.[16] found that the drug inhibited the release of IL-1 and IL-2 and proposed this as the mechanism of inhibition of primary cytotoxic T cell responses thought to be important in graft rejection. Further work indicated that these were not the only cytokines whose production was inhibited, and by the mid-1980s it had been established that inhibition was at the level of mRNA production and a family of genes in T lymphocytes were inhibited.[17–19] These genes encoded for cytokines produced by T helper cells only after activation and included those encoding for IL-2, IFNγ, TNFα, IL-3, and IL-4. This family of genes is coordinately expressed early after T cell activation.

Inhibition of these activation-induced cytokines explained many of the observed effects of cyclosporin A. What was still unclear was whether the drug also had other modes of action in addition to those described above. Helper T cells appear to have a second cyclosporin A-sensitive step after the inhibition of IL-2 production. Exogenously added IL-2 cannot replace the inhibited cytokine and overcome the effects of the drug on T cell activation.[20] This suggests that IL-2 action is also a target. However, the drug has no effect on IL-2-induced proliferation of preactivated cells or IL-2-dependent CTLL cells used in the standard IL-2 bioassay.[16,21] This pointed to an effect on IL-2 receptor expression and many groups have attempted to clarify this point. However, conflicting data were reported (reviewed by DiPadova[22]) and it was argued that inhibition is not a direct effect but rather a consequence of inhibition of earlier events in T cell activation. Foxwell et al.[21] attempted to resolve this point and found that surface expression of both the p55 and p75 chains of the IL-2 receptor is inhibited, but that inhibition can be partially reversed by exogenous IL-2.

As with the question of the IL-2 receptor, it is still controversial whether cyclosporin A has effects on non-T-cell-derived cytokines. IL-1 production by macrophages has been reported to be inhibited[16] but this has been contradicted.[23] Cyclosporin A was found to have no effect on monokine message levels, however, Remick et al.[24] reported that, as in T cells, TNF production by macrophages could also be inhibited. This was in the absence of any suppression of TNF mRNA levels or the expression of cell-associated TNF activity, suggesting another mode of action at the posttranslational level. The mechanism of this inhibition is unknown, but it may be related to the modulation of the antigen-presenting capacity of these cells by cyclosporin A.[25,26] It should be noted that doses of cyclosporin A required to show effects on antigen presentation are higher than needed for inhibition of T cell cytokine gene expression. A number of compounds have now been reported to inhibit cytokine synthesis at the posttranslational level (see Chapter 5).

Having identified that T cell cytokine gene expression was a target for cyclosporin A, the next problem was to identify the sequence of intracellular events affected by the drug. Much of the progress in this area was linked with studies of the mode of action of a second potent immunosuppressive drug, FK 506, identified in the late 1980s, and these two drugs along with rapamycin, a compound related to FK 506, are best discussed together.

IV. FK 506 — A SUCCESSOR TO CYCLOSPORIN A?

In the mid-1980s researchers at Fujisawa Pharmaceuticals in Japan identified FK 506, a novel immunosuppressive drug, as a result of a random screening program. FK 506 was shown to be a potent inhibitor of the mixed lymphocyte reaction.[27–29] The compound is a naturally occurring cyclic macrolide isolated from the spent culture medium of the filamentous bacterium *Streptomyces tsukubaensis.* It is structurally unrelated to cyclosporin A (Figure 1) but possesses very similar immunosuppressive properties both *in vitro* and *in vivo.* Interest in this compound stemmed mainly from the fact that it was at least 50- to 100-fold more active than cyclosporin A *in vitro* and about 10- to 30-fold more active *in vivo.*

Progress in understanding the mechanism of action of FK 506 was rapid as it was compared directly to cyclosporin A. Similar immunosuppressive properties were not surprising in relatively nonspecific assays such as the MLR. However, a surprising result was the activation of the same set of immediate early genes in T cells being inhibited by both cyclosporin A and FK 506. Tocci et al.[30] showed that mRNA levels for IL-2, IL-3, IL-4, TNFα, and IFNγ were not elevated on activation of T cells even though immediate early genes (C-fos, C-myc) or late genes (IL-2 receptor) were unaffected. This was the same pattern as seen previously for cyclosporin A. Thus, rather than FK 506 being a new kind of immunosuppressive drug, it appeared to act in an identical manner to cyclosporin A. Numerous studies confirmed this view, as reviewed elsewhere.[1] Experience in the clinic with FK 506 indicated that this similarity extended to its unwanted side effects. Both drugs are similarly nephrotoxic and diabetagenic, these being the principal side effects limiting use in humans.[31] These side effects, therefore, are likely to be a consequence of this class of drug and it may be difficult to eliminate them in modified analogues. However, not all side effects are common to both drugs. FK 506 is neither hypertensive nor causes hirsutism nor gingival hyperplasia.[31,32]

V. FK 506 AND CYCLOSPORIN A HAVE DISTINCT CELLULAR RECEPTORS BUT A COMMON TARGET

Despite having the same effects on the appearance of immediate early T cell activation genes, studies with radiolabeled analogues of FK 506 and cyclosporin A showed that each bound to a distinct cellular receptor. The principal receptor for cyclosporin A is cyclophilin,[33] a 17-kDa protein, whereas the main FK 506 receptor FKBP-12 (*FK* 506 *B*inding *P*rotein) is a distinct 12-kDa protein encoded by a separate gene.[34–37] Both receptors are, however, abundant cellular components, comprising between 0.1 and 1% of total cellular protein. They are not the only intracellular receptors for these drugs and other receptors will be discussed later.

A breakthrough came with the discovery that cyclophilin possessed peptidyl prolyl cis-trans isomerase activity.[38,39] Further studies showed that FKBP-12 also had this activity,[40,41] providing a plausible explanation for the similar activities of the two drugs via inhibition of this enzyme activity although how this translated into inhibition of gene expression was not clear. Deletion experiments in lower eukaryotes established that cyclophilin was a functional receptor, as loss of the gene or loss of the ability to bind to cyclosporin A correlated well with resistance to the antifungal properties of the drug.[42] Similar experiments selecting for resistance to FK 506 gave a surprising result: resistance mapped to a locus distinct from FKBP.[43] Disruption of the FKBP gene in yeast did, however, confer resistance to rapamycin, an antifungal and immunosuppressive agent related to FK 506 (see later), indicating that FKBP is a functional receptor[44–46] although other loci are also involved.

Establishing a role for cyclophilin or FKBP in the immunosuppressive actions of the compounds had to be by more indirect means. A good but not wholly convincing correlation between inhibition of the mixed lymphocyte reaction and cyclophilin binding for a number of cyclosporin A analogues was found.[47] Similar studies with FK 506 were more positive. A good correlation with binding to the major FKBP and inhibition of IL-2 production was obtained[48] (Figure 2). Most surprisingly, the data suggested that FK 506 was not acting as an inhibitor of a cellular process but rather as an agonist, and in some way *inducing* the downregulation of IL-2 production. This was based on a number of points from the correlation studies. Firstly, most analogues caused 50% inhibition of IL-2 production at concentrations well below that required to give 50% occupancy of the receptor. Thus complete inhibition of IL-2 production occurred before all receptors were occupied. This is clearly incompatible with the drug acting simply to inhibit an activity inherent to FKBP-12 such as the peptidyl prolyl cis-trans isomerase. However, the coefficient of regression for the data was close to unity, strongly suggesting that the receptor was involved. We were very conscious of trivial explanations for this difference between receptor binding and functional properties and investigated whether compound losses in either of the assays were responsible, but these were discounted.[48] Secondly, we found a small number of analogues of FK 506 which were able to bind to the receptor but were not inhibitors of IL-2 production. This may have indicated either that the receptor being measured was not the functionally correct one or that these particular compounds could not enter the cell. Neither, however, was the case as all of these inactive analogues were able to antagonize the actions of FK 506 (Figure 3). This supported our view that the receptor we were measuring was the true receptor and that the active analogues were acting as agonists. Again, it is difficult to rationalize these data with a simple activity where FK 506 inhibits the enzyme activity of the receptor. Other groups have also described inactive analogues, either produced synthetically from the putative binding domain of FK 506[49] or else as a structural analogue of the whole molecule.[50,51]

These same considerations can be applied to similar data obtained with cyclosporin. Discrepancies in the correlation can be at least partly explained by an agonist action similar to FK 506. Thus very quickly after the common enzymic activity of the two receptors was described, it was clear to us that inhibition of peptidyl prolyl isomerase activity was not responsible for the immunosuppressive properties and it was not the meeting point of the mechanisms of the two drugs.

If the common enzyme activity was not involved in immunosuppression, then an alternative explanation had to be found. It was still possible that there was a common target for these receptors. This was explored by Liu et al.[52] who found that the complexes of either FK 506 or cyclosporin A with their relevant receptors inhibited the calmodulin-dependent protein phosphatase 2B, also known as calcineurin. Importantly, analogues of FK 506 capable of binding to the receptor but which were not immunosuppressive did not recognize calcineurin. The cyclosporin-cyclophilin complex was, however, able to displace

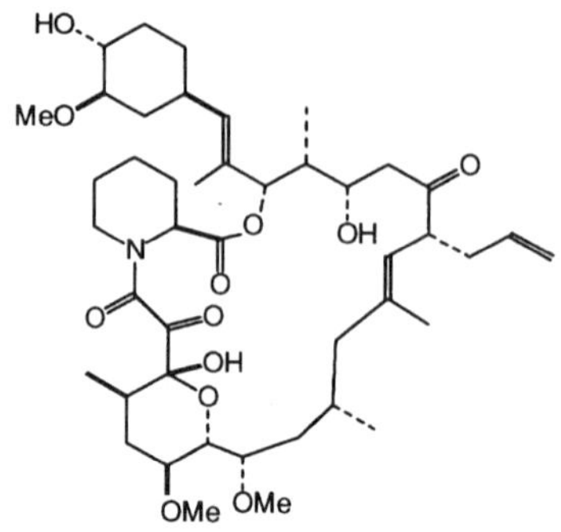

FK 506

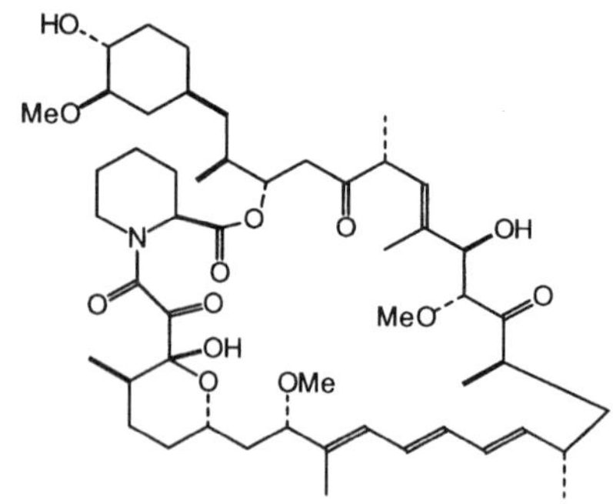

Rapamycin

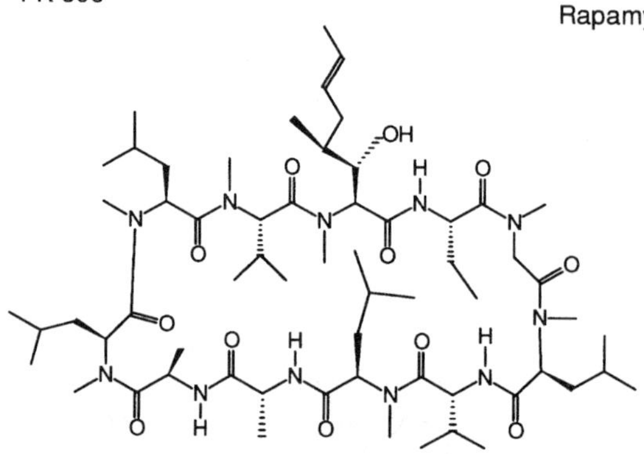

Cyclosporin A

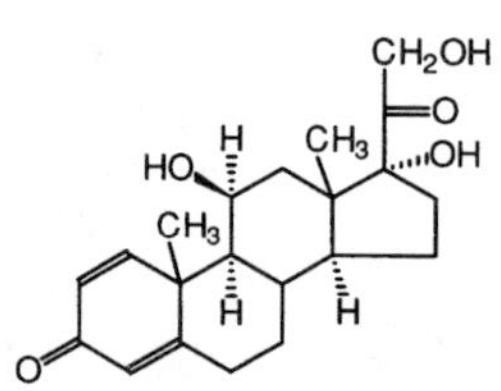

Prednisolone

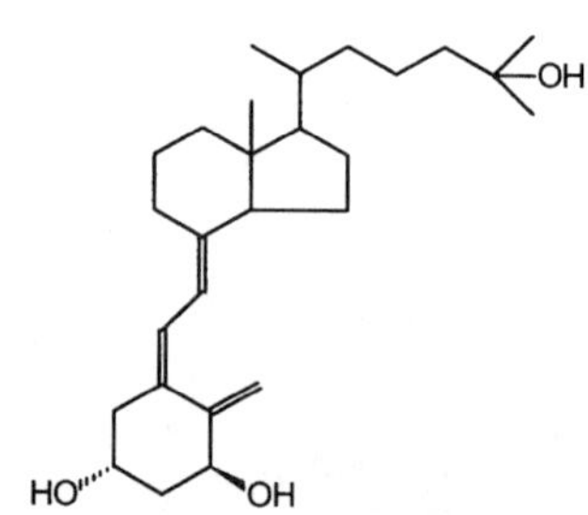

1,25 - Dihydroxy vitamin D_3

FIGURE 1. Structures of the immunosuppressive compounds discussed.

the FK 506-FKBP-12 complex from binding to calcineurin. Further evidence linking calcineurin with IL-2 gene expression was subsequently found. The enzyme is present in T cells and is inhibited by similar concentrations of the drug as is used to inhibit IL-2 production.[53] Furthermore, experiments where the calcineurin had been overexpressed in T cells made the cells better able to produce IL-2 and, more importantly, more resistant to inhibition by FK 506.[54,55] This target was compatible with our view of FK 506 acting as an agonist. The level of calcineurin in cells is much lower than FKBP-12, thus not all FKBP-12 molecules need to be bound to inhibit all calcineurin phosphatase activity. In addition, inactive analogues, such as rapamycin (see below), could not form the complex when bound to FKBP-12,[52] yet their ability to displace FK 506 from FKBP-12 accounted

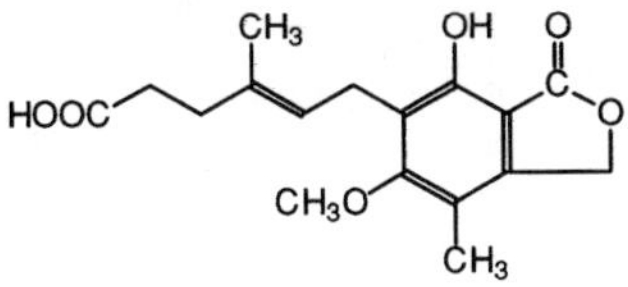

Mycophenolic Acid

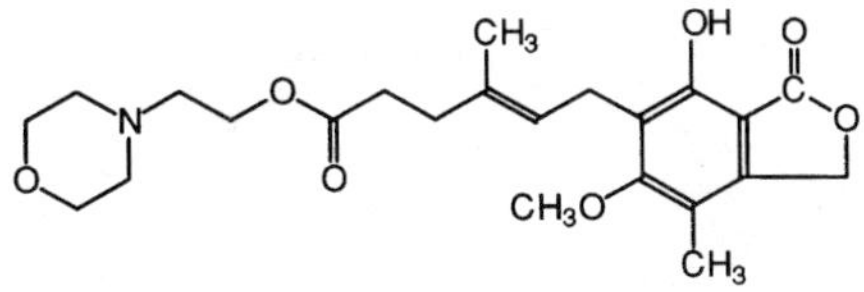

RS-61443

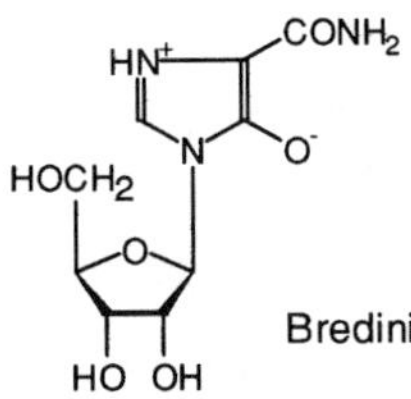

Bredinin

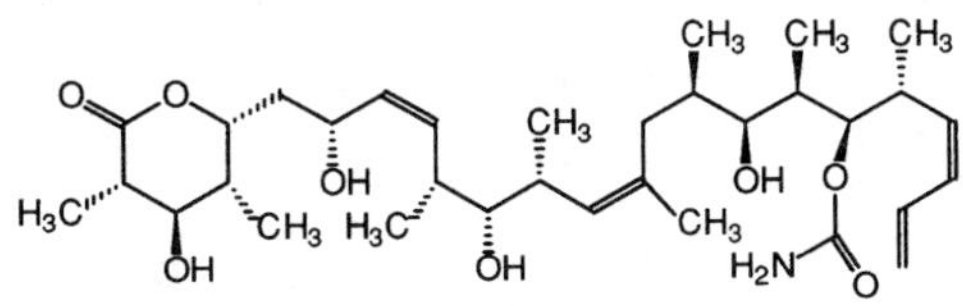

Discodermolide

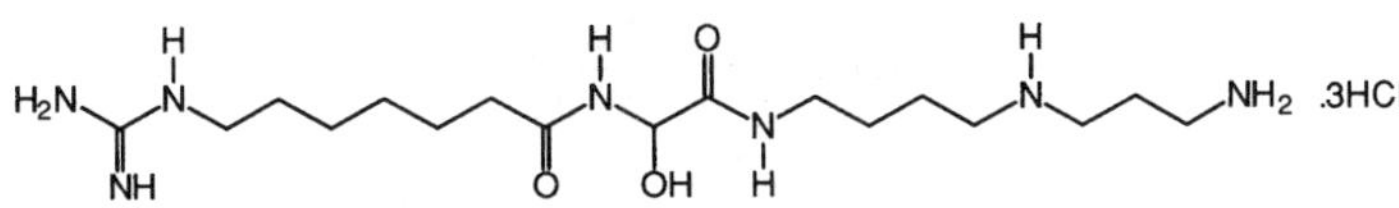

15-Deoxyspergualin

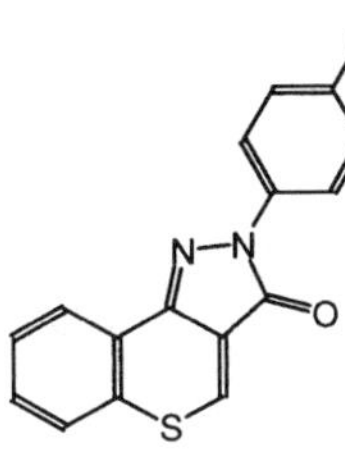

CP-17,193

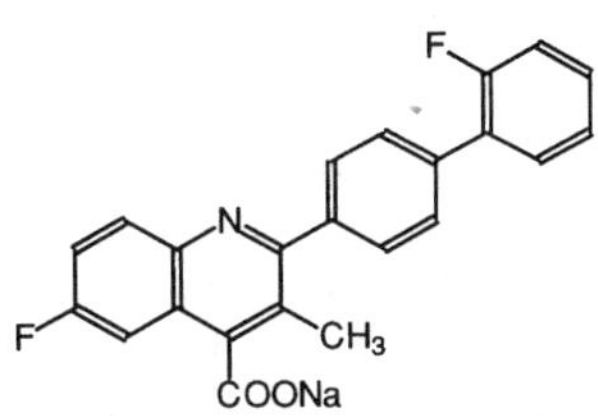

Brequinar Sodium

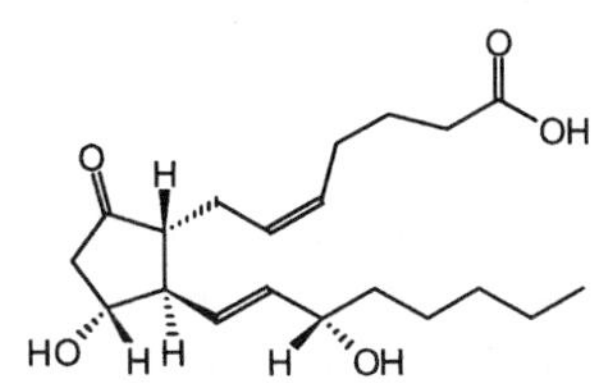

Prostaglandin E_2

FIGURE 1. (continued)

for their antagonistic properties. Finally, agonists with different apparent efficacy could be explained as analogues which, when bound to FKBP-12, had different affinities for calcineurin than the FK 506-FKBP-12 complex.

VI. HOW IS INHIBITION OF CALCINEURIN PHOSPHATASE ACTIVITY RELATED TO INHIBITION OF CYTOKINE GENE EXPRESSION?

Until its identification as a target for both FK 506 and cyclosporin A, the phosphatase activity of calcineurin had not featured in any scheme of events involved in T cell

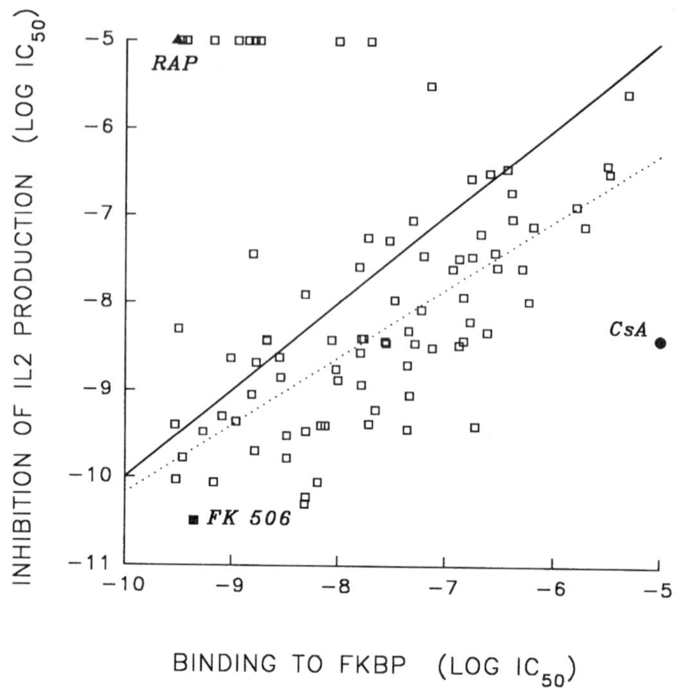

FIGURE 2. Correlation of the inhibition of IL-2 release from Jurkat cells with binding to the rat splenocyte receptor for analogues of FK 506. Each data point (□) represents a structurally modified analogue of FK 506 and the point corresponding to FK 506 is indicated. Also marked are cyclosporin A (●) and rapamycin (▲) and nine compounds that bind but are inactive (these are not included in the regression analysis). The broken line shows linear regression analysis of the data, the solid line indicates the line of identity between the two assay systems. (From Bundick, R. V., Donald, D. K., Eady, R. P., Hutchison, R., Keogh, R. W., Schmidt, J. A. and Wells, E., *Transplantation,* 53, 1150, William and Wilkins, 1992. With permission.)

activation. The main question now was where did it fit? Early studies had shown that both FK 506 and cyclosporin A did not prevent the rise in intracellular calcium or any of the other changes to known second messengers occurring immediately on ligation of the T cell receptor.[56–60] However, it was clear that both drugs specifically inhibited only those events which followed triggering through Ca^{2+}-associated signal transduction pathways. Activation of IL-2 production and proliferation through other pathways not involving mobilization of intracellular Ca^{2+}, such as can be achieved by triggering T cells with anti-CD28 and PMA, was insensitive to inhibition.[61,62] This placed calcineurin after these early events.

The transcription factors controlling expression of the IL-2 gene have mainly been elucidated by Crabtree and co-workers.[63] Both FK 506 and cyclosporin A specifically inhibited the appearance of two of these factors — NFAT and OAP/OCT-1 (also called NFIL2A)[12,63–65] — again confirming their similar mode of action. Neither drug inhibited the transcription of the gene when added to nuclear run-on assays,[30] suggesting no direct effect on the transcription process itself. These transcription factors are mainly restricted to activated T cells and control the expression of more than just the IL-2 gene. Motifs (although generally not identical to that in the IL-2 gene) for binding these transcription factors can be found upstream of all the immediate early genes inhibited by these drugs as well as a number of other genes such as the serine esterases characteristically found

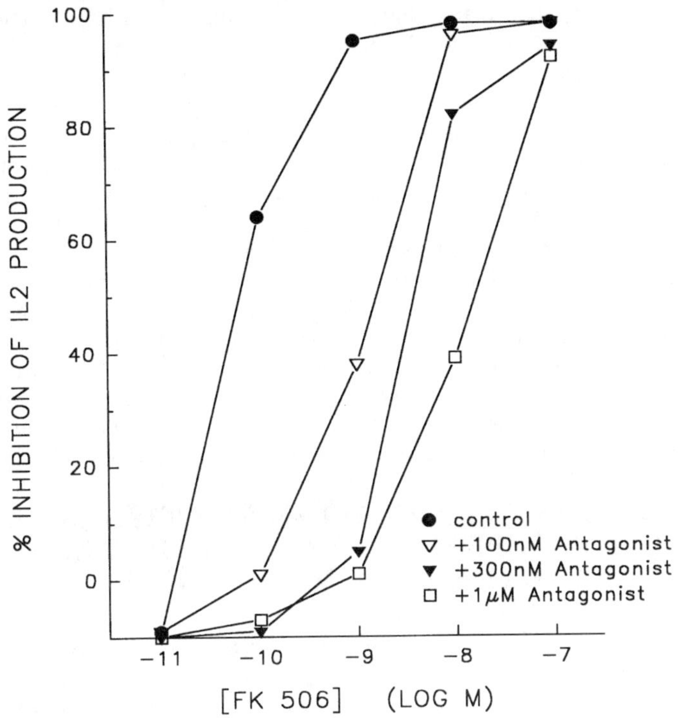

FIGURE 3. Antagonism of the inhibition of IL-2 production from Jurkat cells by FK 506 with an analogue which is able to bind to the receptor but shows no activity of its own, such as depicted in Figure 2. Inhibition of IL-2 production was as described by Henderson et al.[80] except that the antagonist was added at the same time as FK 506 at the concentrations stated.

in cytotoxic T cells (results of the authors' searches of the EMBL database). Interestingly, the maturation of cytotoxic T cells is another process inhibited by the same concentrations of both FK 506 and cyclosporin A as are active on cytokine gene expression.[66–69] Furthermore, perforin and granzyme B mRNA expression in $CD8^+$ T cells is sensitive to cyclosporin A.[70]

The dose response for inhibition of the appearance of functional NFAT or OAP/OCT-1 closely paralleled that for inhibition of target cytokine gene expression,[65] suggesting that these are the critical targets. Another transcription factor also controlling the IL-2 gene is the more ubiquitous NFκβ. The appearance of functional NFκβ is also inhibited by both FK 506 and cyclosporin A, but much larger doses of drug are required.[66] This is consistent with the relatively T-cell-specific action of these drugs, but does not discount that some of the side effects of the drug may be related to inhibition of the more ubiquitous transcription factor. Recently, a second step at which FK 506 and rapamycin may control cytokine mRNA levels has been described.[71] Both drugs were found to selectively enhance the degradation of IL-2 and GM-CSF mRNA but to a different extent. Whether this mode of action is related to inhibition of calcineurin phosphatase activity has yet to be established.

At the time of writing, it has not been resolved exactly how calcineurin is involved in the production of active transcription factors. Much attention has been focused on the generation of active NFAT. This transcription factor, in common with many others, has more than one subunit in its active state. These have different cellular locations and it is thought that the active transcription factor is generated when a preexisting T-cell-specific component present in nonactivated cells is brought together with an induced nuclear component.[72] The inducible nuclear component is the fos/jun (AP-1) transcription factor, and both the preexisting and inducible subunits recognize nucleotide sequences within the promoter site.[73] An increase in intracellular Ca^{2+} triggers translocation of the preexisting unit into the nucleus,[72] whereas the activation of protein kinase C induces fos/jun.[74] Since FK 506 and cyclosporin A only appear to inhibit Ca^{2+}-mediated events and the appearance of active AP-1 transcription factor itself is not affected by either drug,[65] the preexisting NFAT subunit is the most likely target for calcineurin. This may be either direct or indirect through a phosphorylation cascade. Recent data[75] suggest that the preexisting NFAT subunit is indeed phosphorylated and, furthermore, that it can be directly dephosphorylated by calcineurin. This would be consistent with other known targets of the protein phosphatase 2B. These include the regulatory subunits of the cyclic AMP dependent protein kinase phosphatase and other similar regulatory subunits. In these cases, dephosphorylation by calcineurin activates the subunit (reviewed by Cohen[76]).

VII. FK 506 AND RAPAMYCIN ARE TWO DISTINCT IMMUNOSUPPRESSIVE DRUGS THAT SHARE THE SAME RECEPTOR

The naturally occurring macrolide, rapamycin, was isolated by researchers at Ayerst Laboratories in the mid-1970s, well before the discovery of FK 506. Structurally, it is very similar to FK 506 (Figure 1), particularly on the left-hand side as shown in the figure, from the pipecolate through to the lactone. Originally it was described as an antifungal antibiotic, but its development was limited by potent immunosuppressive side effects![76] Subsequently, and in the light of the discovery of its structural similarities to FK 506, it too has been shown to be a potent immunosuppressive compound able to prolong transplanted allografts (reviewed by Morris[3]).

A major surprise was that rapamycin was not active in all of the *in vitro* assays which were inhibited by both FK 506 and cyclosporin A and was able to inhibit responses unaffected by the latter two drugs. Like FK 506 and cyclosporin A, rapamycin was a potent inhibitor of the MLR and the proliferative responses of T cells to mitogens and other antigens.[48,78] In these assays, its activity was as potent as FK 506, with IC_{50}s in the 0.1 to 0.8 n*M* range. However, it was also able to inhibit the proliferative response of PBMC to LPS,[79] unlike FK 506 or cyclosporin A, again with subnanomolar activity (Schmidt and Bundik, unpublished data), but it was unable to inhibit IL-2 production.[48,78] At the mRNA level no inhibition of cytokine gene expression was seen,[78] neither were there any effects on the appearance of the NFAT or OAP/OCT-1 transcription factors.[12,80] Rapamycin was, however, able to bind to FKBP-12 with essentially the same affinity as FK 506. This prompted us to investigate whether rapamycin was similar to other inactive FK 506 analogues we had identified (as depicted in Figures 2 and 3). As described above, these analogues bound to FKBP-12 and could be used as antagonists of FK 506. Figure 4a shows that rapamycin is also an antagonist of FK 506 and causes a right shift of the dose response curve. Schild analysis of this antagonism indicates that it is competitive in nature. We have also identified a convenient *in vitro* assay which is sensitive to rapamycin and insensitive to FK 506.[48] This involves PMA stimulation of murine CTLL cell proliferation and the response can be completely inhibited by rapamycin at subnanomolar levels. Using this assay, we have been able to show that the converse is also true, namely that FK 506 is an antagonist of rapamycin (Figure 4b). Again, this antagonism was

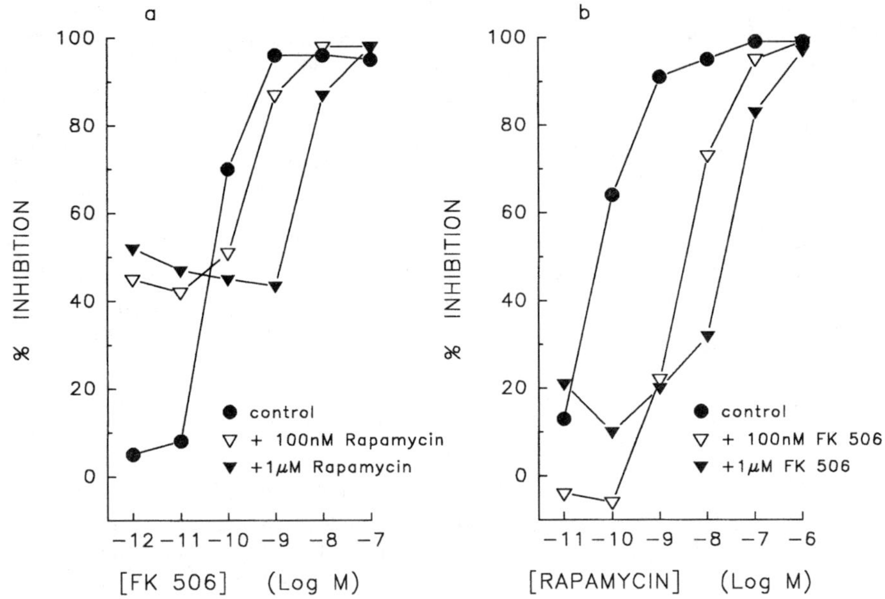

FIGURE 4. Mutual antagonism of FK 506 and rapamycin. Panel (a) shows inhibition of IL-2 production from Jurkat cells by FK 506 in the presence of increasing concentrations of rapamycin. Note that rapamycin itself inhibits IL-2 production at the concentrations used, giving at most 50% inhibition. When combined with FK 506 it is clearly able to reverse IL-2 inhibition but only to the level seen with rapamycin alone. Panel (b) shows inhibition of the PMA-stimulated proliferation of murine CTLL cells by rapamycin in the presence of increasing concentrations of FK 506. (From Bundick, R. V., Donald, D. K., Eady, R. P., Hutchison, R., Keogh, R. W., Schmidt, J. A. and Wells, E., *Transplantation,* 53, 1150, William and Wilkins, 1992. With permission.)

competitive. This mutual antagonism between FK 506 and rapamycin in T cells has been documented by others[81,82] and also occurs in other systems, such as inhibition of mediator release from neutrophils, basophils, and mast cells.[83–85] It is important to note that no antagonism of cyclosporin A has been seen with rapamycin. This is consistent with these two classes of drugs having distinct cellular binding proteins.

Rapamycin, like FK 506, is therefore a potent immunosuppressive compound but acts by a different mechanism. The antagonism studies give us the important message that whatever the mechanism of rapamycin, it is through the same functional receptor as FK 506. We therefore view FK 506 as an inactive analogue of rapamycin and rapamycin as an inactive analogue of FK 506.

What is the mechanism of action of rapamycin? The compound has been reported to be an inhibitor of cytokine action rather than production. It is able to inhibit the proliferation of IL-2- and IL-4-dependent T cells[77] (although inhibition is not complete), IL-4-induced production of IgE[86] as well as IL-2, and *S. aureus* induction of IgG in pure B cell populations.[79] These latter activities may explain the early observation that rapamycin is able to inhibit the generation of an IgE response to antigens/allergens.[76] The mutual antagonism tells us that these activities are mediated by binding to the same FKBP as FK 506, but the rapamycin-FKBP-12 complex is unable to inhibit calcineurin activity.[52] Rapamycin activity is not restricted to Ca^{2+}-mediated events since, unlike FK 506 and cyclosporin A, it can inhibit anti-CD28/PMA-stimulated T cell proliferation.[78,82] Cells are also inhibited at a later point in the cell cycle compared to those inhibited by FK 506 or cyclosporin A.[88]

The direct target for the rapamycin-FKBP-12 complex is not yet known, but a consequence of rapamycin treatment of cells is an inhibition of cellular p70 S6 protein kinase activity.[89–92] This kinase phosphorylates the 40S ribosomal protein S6 which is necessary for protein synthesis. Activation of the S6 kinase itself is dependent on phosphorylation and this event is blocked in rapamycin-treated cells. This is true for a number of growth-promoting stimuli in many cells types including IL-2 stimulation of T cells,[91–92] and occurs at concentrations comparable to those blocking S phase entry of T cells. Other T cell kinases such as the erk encoded MAP kinases or the rsk encoded S6 kinases, are unaffected.[89] The link between the rapamycin-FKBP complex and inhibition of p70 S6 kinase awaits to be discovered. A recent candidate is phosphatidyl inositol-3-kinase. The *TOR 2* gene which confers rapamycin sensitivity in yeast encodes a homolog of the mammalian enzyme.[93]

VIII. THE PROBLEM OF MULTIPLE RECEPTORS

Thus far, we have only considered the major receptors for FK 506/rapamycin and cyclosporin A, namely FKBP-12 and cyclophilin. Other unrelated proteins are able to bind either drug and are candidates as mediators of certain noncytokine-related effects. These are reviewed by Sigal and Dumont.[1] More interesting are the cyclophilin- and FKBP-12-related proteins found both in T cells and other cell types. Cyclophilin and FKBP-12 are each members of a large family of related proteins. What is unknown is exactly how binding to any of these other members relates to the immunosuppressive properties of the drugs. To date, most of the information we have relates only to the binding affinities of the drugs for the proteins. However, a number of interesting points can be raised.

A second cyclophilin has been described[94] which appears to be a membrane-associated form. Both cyclophilins can mediate inhibition of bovine brain calcineurin when complexed with cyclosporin A.[95] Interestingly, the IC_{50} for inhibition depended on the substrate used and varied between the two cyclophilin forms. In the case of the FKBPs, at least three other forms have been reported. FKBP-13 is closely related to FKBP-12 except that, like the cyclophilins, its sequence predicts a protein anchored to the endoplasmic reticulum.[96] It binds FK 506 with lower affinity than FKBP-12 (Ka of 3×10^{-8} *M* vs. 3×10^{-10} *M* for FKBP-12[97]). This affinity is too low to explain the very potent activity of FK 506, suggesting that this receptor does not transduce FK 506 effects on cytokine gene expression. To support this, FKBP-13 when bound to FK 506 is unable to make a complex with calcineurin.[98] Rapamycin binds to FKBP-13 with about tenfold greater affinity than FK 506.[97] Even so, this is still too low to explain its potent *in vitro* immunosuppressive properties. Furthermore, Schild analysis of the mutual antagonism between FK 506 and rapamycin predicts that each compound binds with approximately the same affinity to the common functional receptor (Schmidt and Bundick, unpublished data, and consideration of the mutual antagonism presented by Dumont et al.[81]) eliminating the involvement of a differential binding receptor. This important point can be used to discount a functional role of a rapamycin-selective FKBP of approximately 25 kDa. This receptor has been reported to bind rapamycin with nanomolar affinity but binds FK 506 approximately 200-fold less well.[99,100] If this receptor was responsible for the immunosuppressive action of rapamycin, this would predict that FK 506 would be a much poorer antagonist of rapamycin than rapamycin is of FK 506, which is clearly not the case.

Not enough information is yet available on the other FKBPs to determine whether they play a functional role in FK 506 or rapamycin action. Interestingly, one of these is hsp 56,[101] which can form a complex with the glucocorticoid receptor and hsp 90 in the absence of the drug. It is still controversial whether macrolide binding has any effects on the functioning of the glucocorticoid receptor. Hutchison et al.[102] reported no effect on receptor folding or function, whereas Ning and Sanchez[103] found that rapamycin and FK 506 potentiated glucocorticoid receptor signaling. In view of their reciprocal antagonism, the finding that both drugs similarly affect steroid receptor function discounts this as the primary mode of immunosuppressive action. However, it does provide a link between two pathways of immunosuppression and may have bearing on the combined use of steroids with macrolides.

Good antagonists of cyclosporin A which would allow a similar comparison of the relationship between functional receptors and binding proteins for the cyclophilin family have not been described.

IX. FK 506, CYCLOSPORIN A, AND CYTOKINES FROM NON-T CELLS

Cyclosporin A and FK 506 have actions on several other cell types. In B cells, both drugs are potent inhibitors of activation induced by some signals, for example, anti IgM, but not of other stimuli such as LPS.[78,104–106] Again, both drugs only affect Ca^{2+}-mediated events. Interestingly, T-cell-independent immunoglobulin production induced by IL-6 in B cell lines is sensitive to cyclosporin A but not FK 506.[107] This is one of the few instances where cyclosporin A action is not the same as FK 506. Granulocytic cells are another target and both drugs are able to inhibit degranulation and mediator release from mast cells, basophils, and neutrophils.[83–85,108] Besides blocking the immediate events of degranulation, cyclosporin A was also found to block MIP-1α and MIP-1β gene expression in mast cell lines induced by IgE.[109] This same signal induces other cytokines in mast cells, such as IL-4 and IL-5, and it would be surprising if these products were not similarly inhibited although no reports have yet been published. However, as with T cells, only events which involve Ca^{2+}-associated signal transduction pathways are inhibited whereas Ca^{2+}-independent degranulation is unaffected.[110]

The effects of cyclosporin A and FK 506 on cytokine production from macrophages is still an uncertain area. Cyclosporin A was discussed earlier and the effects of FK 506 appear similar. It has generally been accepted that neither drug has marked effects on IL-1 production from normal monocytes or macrophages however stimulated, but Keicho et al.[111] found that by using a specific ELISA, they could measure a significant inhibition of IL-1α production by FK 506 in U937 cells, peripheral blood monocytes, and alveolar macrophages. Inhibition, though, was not complete and IA_{50}s of between 1 and 10 n*M* indicated that higher concentrations of the drug were required than for comparable effects on T cell cytokines. This puts in question the physiological role of this inhibition. Furthermore, human monocytes generally produce more IL-1β than IL-1α and this may explain why so few effects had been seen by others on the overall IL-1 production.

While evidence is pointing to FK 506 and cyclosporin A being inhibitors of Ca^{2+}-mediated events in hematopoietic cells, they are not general inhibitors of all Ca^{2+}-triggered responses. For example, MIP is an inducible gene in a number of cells, but induction of the gene in macrophages is not inhibitable with cyclosporin A.[112,113] Thus, these drugs appear to define a subtype of a Ca^{2+}-mediated event which is sensitive to the drug.

X. OTHER IMMUNOSUPPRESSIVE DRUGS

In recent years a number of new immunosuppressive drugs have been described along with novel approaches to immunosuppression, such as the use of antibodies to cytokines or lymphocyte surface antigens, soluble receptors to cytokines, or the cytokines themselves as immunomodulators of the cytokine network. We will not deal with the latter approaches as they are considered elsewhere in this volume. However, a few of the emerging low molecular weight drugs deserve mention due mainly to their demonstrable efficacy *in vivo* or interesting *in vitro* properties. Many immunomodulatory compounds are not included as we have limited our discussions to novel compounds which are either showing promise in the clinic or have well-documented effects on cytokine production or action. See the previous chapter for a consideration of low molecular weight inhibitors of cytokine synthesis.

A. DEOXYSPERGUALIN

Spergualin analogues, particularly 15-deoxyspergualin (Figure 1), are potent immunosuppressive and antiallergic agents. 15-Deoxyspergualin inhibits both cellular and humoral responses *in vivo*[114–116] and has proved effective in organ transplantation in humans.[117] Encouraging results with xenografts have also been obtained.[118] It appears most promising when used in combination therapy with other immunosuppressive drugs.[119] So far, its exact mechanism of action is unknown. Like FK 506 and cyclosporin A, it inhibits Con A-induced T cell proliferation and the MLR.[120,121] Unlike FK 506, it also inhibits cytotoxic T cell killing[122] as well as LPS-induced blastogenesis,[120] suggesting it is generally antiproliferative. These properties are most likely due to inhibition of polyamine biosynthesis[123] which would explain general antiproliferative properties. It does not seem to act by inhibiting T cell cytokine production.[121] Nemoto et al.[120] reported a stimulation rather than inhibition of IL-1 and IL-2 production from rat spleen cells.

B. MYCOPHENOLIC ACID/BREDININ/RS-61443

These compounds are purine synthesis inhibitors[124–126] and it has long been known that agents of this type are immunosuppressive due to the importance of this biochemical pathway in lymphocyte proliferation. Their advantage over earlier purine synthesis inhibitors such as methotrexate and azathioprine is that they appear to be more selective for T and B cells. These compounds are potent inhibitors of mitogenic responses[126,127] and, like rapamycin, are not selective for Ca^{2+}-dependent pathways of activation. Inhibition of proliferation by these agents is at a relatively late stage of cell activation as cytokine production (IL-1, IL-2) is unaffected. RS-61443, a morpholinoethyl ester of mycophenolic acid, is the most promising compound to date.

C. BREQUINAR SODIUM

Brequinar sodium is currently receiving attention as a promising immunosuppressive agent.[128] However, rather than its activity being due to any selective effect on cytokine biology, its known properties as a potent inhibitor of dihydroorotate acid dehydrogenase, the fourth enzyme along the *de novo* pyrimidine biosynthesis pathway,[129,130] explain its immunosuppressive action. In this regard, it is very similar to the purine synthesis inhibitors described above.

D. VITAMIN D_3

Vitamin D_3 exerts its biological properties through a major metabolite, 1,25-dihydroxyvitamin D_3, and has diverse activities on many cells and tissues.[131] It has a unique receptor through which its many activities occur. This receptor is a member of the superfamily of steroid receptors[132] and, like other steroid receptors, recognizes specific regulatory sites on DNA when complexed with the ligand. In T cells, the IL-2 gene is one of the vitamin D_3-sensitive genes and the drug acts to suppress transcription.[133] It also suppresses expression of other T cell activation genes such as IL-3,[134] IFNγ,[135] and GM-CSF.[136] These are the same genes which are sensitive to cyclosporin A and FK 506, but unlike the latter two drugs, vitamin D_3 enhances TNFα expression.[137] These activities would account for why the compound is immunosuppressive, but its more widespread effects make it difficult to predict exactly what other immunomodulatory actions it would have. Vitamin D_3 has been found to enhance IL-1 and TGFβ transcription from cultured macrophages and rat calvarial cells, respectively.[138,139] If these effects occur *in vivo,* the enhanced IL-1 and TNFα production might be expected to be pro-inflammatory, but the TGFβ would potentiate the immunosuppressive activity. Among the noncytokine genes suppressed is the cellular oncogene c-myc,[140] which is normally an immediate early T cell activation gene and the transferrin receptor.[133] Both of these genes are important in growth in T cells. Interestingly, their expression is unaffected by FK 506 and cyclosporin A. This activity would contribute to the immunosuppressive action of vitamin D_3 but other effects such as upregulation of MHC class II[141] and Fc receptor[134] expression could have a counter effect. Vitamin D_3 is therefore a hormone with pleiotropic actions, the sum of which is responsible for the overall activity.

E. PYRAZOLOQUINOLINES

Numerous pyrazoloquinoline derivatives and compounds with related structures have appeared in the patent literature as having immunomodulatory properties. Patent applications have been submitted by Pfizer,[142] American Home Products,[143] Carlo-Erba,[144,145] and Boots.[146,147] CP-17193 (Figure 1) is one such compound described by researchers at Pfizer in the early 1980s[148,149] that is clearly therapeutically effective in models of autoimmune disease such as the spontaneous lupus disease developing in NZB/NZW mice and in chronic relapsing experimental allergic encephalomyelitis in guinea pigs.[150,151] In the former, there is a reversal in the reduction of IL-2 production which normally accompanies the disease, as well as a reduction in other parameters such as glomerulonephritis and the appearance of anti-DNA antibodies.[150] The compound inhibits T cell blast formation *in vitro,*[152] but this requires relatively high doses (20 μ*M*). Our own studies have indicated that it is unlike FK 506, cyclosporin, or rapamycin with regard to its effects on IL-2 production or action (unpublished data). This compound, therefore, appears to be an example of a class of immunomodulants which has been extensively investigated by the pharmaceutical industry but for which no mechanism of action has been found.

F. PROSTAGLANDIN E

Prostaglandins of the E series, particularly PGE_2, have long been known to be immunosuppressive. Prostaglandin analogues prolong cardiac allografts in experimental transplants[153] and in the clinic, the bioavailable prostaglandin analogue, misoprostol, has been recently shown to reduce renal allograft rejection when given in combination with either cyclosporin A or prednisolone.[154] The effects of prostaglandins on T cells and macrophages has been extensively studied.

PGE_2 inhibits monocyte/macrophage IL-1 and TNF production, however, the mechanism may not be the same for each cytokine. Inhibition of IL-1 production appears to be at a posttranscriptional level[155,156] although not all groups can show potent inhibition of IL-1 production by prostanoids.[157] In contrast, inhibition of TNFα production is readily demonstrable and can be measured at the level of induction of gene transcription.[158,159] In addition, there appears to be species specificity in the effects of PGE_2. Haynes et al.[157] found that while a number of prostanoids including PGE_2 were potent inhibitors of IL-1 production from rat or mouse macrophages, they had little effect on human macrophages. The same study found that TNF production was inhibited in all species and, in contrast, IL-6 production was enhanced. As well as inhibiting IL-1 production, prostanoids are also potent inhibitors of the co-mitogenic activity of IL-1 on T cells.[157,160] These properties would be expected to contribute to the immunosuppressive activity of PGE_2.

PGE_2 also has potent activities directly on T cells. It inhibits T cell mitogenesis[161,162] and appears to act at several levels. In T helper cells, PGE_2 inhibits IL-2[163,164] and IFNγ[165] production and expression of IL-2 receptor[166] and transferrin receptor.[167] The inhibition of IL-2R expression may simply be as a consequence of inhibition of IL-2 production, as inhibition can be overcome with excess cytokine.[168] PGE_2 has no effect on IL-4 production.[165] The situation with IL-3 and GM-CSF is more complex. Initial reports suggested that GM-CSF production was enhanced by PGE_2.[169] Further studies showed that this was only seen in some T cell clones and both GM-CSF and IL-3 could be inhibited in other clones.[170] The pattern of T cell cytokine inhibition by PGE_2 is therefore unlike that seen with the other immunosuppressive drugs discussed here.

Numerous studies have implicated cyclic AMP as the critical second messenger involved in mediating the immunosuppressive properties of PGE_2.[156,166,171] Agents which artificially raise cAMP have similar effects. Permeant cAMP analogues, direct activators of adenylate cyclase (forskolin), inhibitors of phosphodiesterase and cholera toxin, which activates the stimulatory GTP binding protein, all have similar effects to PGE_2.[166,171,172] Cyclic AMP appears to act to block the activation signal. No mRNA for IL-2 appears in T cells activated in the presence of PGE_2.[171] The block is very early after the activation signal as PI_3 metabolism itself is inhibited by agents which raise cAMP,[173] as is the rise in intracellular Ca^{2+}.[173,174] Our own studies, using a series of prostaglandin analogues known to bind to distinct PG receptor subtypes, confirms that the inhibition of the murine MLR is mediated by the EP_2 receptor known to act by raising intracellular cyclic AMP (Figure 5). This also suggests that other prostaglandin receptor subtypes which activate different second messengers are not involved. It is not yet known exactly how elevation of cAMP is coupled to inhibition of the other second messengers.

Perhaps the most significant finding to date is the observation that PGE_2 is able to inhibit the cytokines produced by Th1 cells but not Th2 cells.[165] This is discussed in more detail below.

G. DISCODERMOLIDE

We have included this novel compound not because extensive literature exists nor because it possesses a profoundly unique property, but because initial reports describing its *in vitro* properties focused immediately on its effects on cytokine production and action. This reflects the increasing importance that an understanding of the effects of compounds on cytokine biology has in evaluating new immunosuppressive compounds. While useful in early screens, mitogenic assays alone will detect antiproliferative agents as well as new immunomodulants. A knowledge of any effects on cytokine biology is now crucial in progressing novel agents.

Discodermolide is a lactone isolated from the marine sponge *Discodermia dissoluta.* It inhibits the GVH response *in vivo*[175] and Con A-induced mitogenesis and the MLR *in vitro.*[176] While it appeared as active as cyclosporin A, it was unable to inhibit IL-2 production but had significant effects on the surface expression of the IL-2 receptor on both human and murine cells.[176] Thus, it appears different from both FK 506, cyclosporin, and rapamycin and warrants further study.

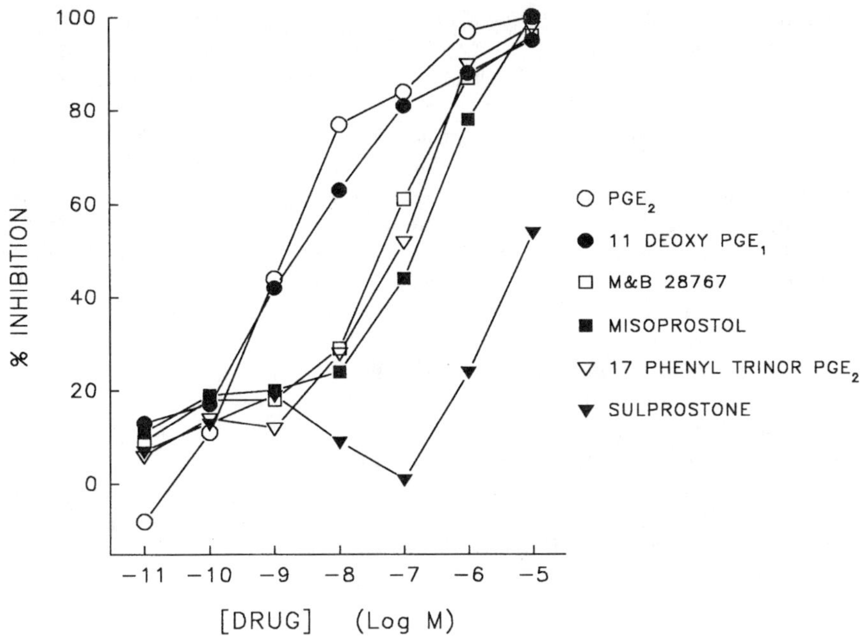

FIGURE 5. Inhibition of the murine mixed lymphocyte reaction with prostaglandin analogues. The murine mixed lymphocyte reaction was set up as described by Henderson et al.[80] in the presence of the prostaglandin analogues indicated. The classification of the receptor specificity of the prostaglandin analogues is as determined by Lawrence et al.[203] PGE_2 is equipotent on EP_1, EP_2 and EP_3 receptors. The potency order of the other analogues is as follows: 11-deoxy-PGE_1, $EP_2 > EP_3 > EP_1$; M & B 28767, $EP_3 \gg EP_2 = EP_1$; Misoprostol, $EP_3 > EP_2 > EP_1$; 17-phenyl-trinor PGE_2, $EP_1 > EP_3 > EP_2$; Sulprostone, $EP_3 > EP_1$ with virtually no binding to EP_2.

XI. IMMUNOSUPPRESSIVE DRUGS IN COMBINATION

In the clinic, immunosuppressive drugs are rarely given alone and are usually administered in combination. FK 506 and cyclosporin A are given with steroids, and even the new approaches using antibodies such as OKT3 almost invariably are given to patients already taking cyclosporin A. A number of studies have looked at the interactions *in vitro* on both the inhibition of mitogenic responses and inhibition of IL-2 production between FK 506, cyclosporin A, and steroids. No significant synergy was found between cyclosporin A and prednisolone, rather, the effects were additive.[177] However, this is still a useful finding clinically as it suggests that combined therapy could maintain immunosuppression while avoiding unwanted side effects. Others have reported a synergism between cyclosporin A and FK 506[66,178] but care has to be taken in analyzing the results as the relatively steep dose response curves for FK 506 and cyclosporin A can mislead the interpretation. More critical analysis of the two drugs in combination clearly shows only additive effects.[179]

Given their similar mechanism of action, we now understand why when used together, FK 506 and cyclosporin A would only have additive activity. However, none of the *in vitro* studies predicted the problems encountered when the drugs were tried in combination *in vivo*. Co-administration led to raised levels of circulating cyclosporin A.[180] This was due to FK 506 inhibiting the metabolism of cyclosporin A by cytochrome P450 mixed function oxidase.[181,182] Thus, rather than being beneficial, the raised cyclosporin A levels increased the risk of harmful side effects and required careful monitoring.

Combinations of cyclosporin A and rapamycin might prove more effective. This combination shows promise in prolonging grafts, as might be predicted from the differing actions of the drugs.[3] Although their actions are no more than additive *in vitro,*[88] combination of the two drugs may lead to an improvement of the therapeutic ratio *in vivo.* FK 506 and rapamycin might not at first seem to be such an effective combination due to the mutual antagonism of the compounds. However, given that both FK 506 and rapamycin behave as agonists and a large molar excess of each is required for effective antagonism, therapeutic doses of each may be useful in combination. When mixed as such, the effects of combined FK 506 and rapamycin appear additive.[88]

Combination therapy utilizing cyclosporin A and PGE_2 (and other prostaglandin analogues) is proving useful in reducing the frequency of acute renal allograft rejection.[160] This combination has been studied *in vitro* and has shown that beneficial effects may be in part due to marked synergism between the two agents in inhibiting macrophage procoagulant activity.[183] This combination is one of the few reported where synergism was clearly present. In contrast, cyclosporin A had only additive effects with 1,25-dihydroxyvitamin D_3 on inhibition of IL-2 production.[184]

XII. FUTURE PROSPECTS — SELECTIVE CYTOKINE INHIBITORS

The recent classification of human T helper cell subtypes into those resembling murine Th1 and Th2 cells has greatly helped our understanding of a number of immunologically mediated diseases in man.[185] Th1 cells produce a different set of cytokines to Th2 cells upon antigenic stimulation. Many of the cytokines produced are common to both cell types, but in human cells IL-4 and IL-5 production in the absence of IFNγ is characteristic of Th2 cells, whereas IFNγ production without IL-4 or IL-5 is seen only with Th1 cells. Th0 cells have characteristics of both cell types.[185] Whether a specific antigen triggers a Th1, Th2, or Th0 response depends in part on the antigen itself and on the individual responder.[186–189] It is now clear that in atopic individuals, the response is skewed towards a predominance of Th2 cells (reviewed by Ricci[190]). This is true not only for responses to allergens, which might be expected to give rise to a Th2 cell response, but also for antigens such as Purified Protein Derivative (PPD) which in normal individuals gives rise to a Th1 response. PBMC from severe atopics will produce antigen-specific Th0 or even Th2 cell clones in response to antigens which only give Th1 responses in normal individuals. It has also been recently proposed that a Th1 to Th2 switch is a critical step in the etiology of HIV infection.[191]

This new information suggests novel targets for therapeutic intervention. New therapies can be envisaged where drugs are used to modify the T helper cell response away from that which is causing pathology. Initially, this may be achieved using cytokines themselves to modify the response. High levels of IL-4 and IFNγ modify the response towards Th2 and Th1, respectively.[188] Other cytokines thought to play a role in determining the balance are IFNα, IL-12, and TGFβ.[192–194] IL-10 is controversial; in murine systems it clearly is a Th2 cell product which acts to inhibit Th1 cell cytokine production.[195] However, in humans it is not a T helper cell subtype-specific product and has more general effects, being an inhibitor of the generation of both Th1 and Th2 cells.[196,197]

Antibodies or soluble cytokine receptors may also prove to be an initial approach, but a long-term aim would be to develop therapies which can be given orally.

A number of groups have begun to look for selective effects in current immunosuppressive drugs. Most of the reports describe differential effects on the production of cytokines characteristic of Th1 or Th2 cells. Encouraging results have been obtained with cyclosporin A and PGE_2. Murine Th1 and Th2 clones differ in their responses to both drugs. Cyclosporin A shows about a ten-fold selectivity for inhibition of IL-2 production from Th1 clones over IL-4 production from Th2 clones.[198] However, this selectivity was not seen for cyclosporin A or FK 506 on human peripheral blood T cells analyzed at the single-cell level.[199] Additionally, this latter study also showed no selective effects on the expression of IFNγ, IL-5, or TNFα. Interestingly, while FK 506 and cyclosporin A are potent inhibitors of early T cells activation genes as discussed previously, neither of these two drugs, nor rapamycin, blocks Con A-induced IL-10 mRNA expression in a murine Th2 cell line.[200] This suggests that the IL-10 gene is under different control to other cytokine genes in T cells. However, the picture is not yet clear since Durez et al.[201] reported that induction of IL-10 by anti-CD3 was sensitive to cyclosporin A, whereas induction by LPS was not. PGE_2 is similar, and has been reported to inhibit the production of Th1 but not Th2 lymphokines.[165] Interestingly, in some Th0 clones, IL-2 and IL-4 production were differentially regulated by PGE_2, but in other clones both cytokines were equally inhibited. This is preliminary evidence for a further level in the regulation of T helper cell gene expression which might prove to be yet another target for intervention. Most of the data with PGE_2 have been obtained using murine T cell clones and little information is available for human cells. PGE_2 also promotes the differentiation of T helper cells towards Th2. Culturing either uncommitted T cell lines in the presence of PGE_2 or even committed Th0 cell clones for 1 week leads to a change in phenotype and the cells become more Th2-like in that, on restimulation, IL-4 production increases relative to IFNγ.[202]

XIII. CONCLUDING REMARKS

Immunosuppressive drugs have come a long way since they were first used in humans. Originally they were chosen for their antiproliferative properties, preferably with some selectivity for lymphocyte proliferation over other cell types. Our increased understanding of the role of lymphokines and cytokines in the immune response, however, has led to the quest for immunosuppressive drugs that have more selective effects on individual cytokine production or action. On the other hand, much of our progress in unraveling some of the processes occurring in T cell activation has been linked to the discovery of immunosuppressive drugs. Cyclosporin A, FK 506, and rapamycin have proved to be useful tools in this regard and many surprises may yet await us even with these relatively well-characterized drugs. In the future, new immunosuppressive drugs may provide both the academic immunologist and the physician with the tools they both need to further our understanding of the immune response and to develop novel and effective therapies for those autoimmune diseases for which no satisfactory treatment exists.

ACKNOWLEDGMENTS

We acknowledge the roles of our associates Drs. Eady, Keogh, Donald, Wells, and Hutchinson in the formulation of our views on the mechanism of action of FK 506 and rapamycin.

REFERENCES

1. **Sigal, N. H. and Dumont, F. J.,** Cyclosporin A, FK 506 and Rapamycin. Pharmacologic probes of lymphocyte signal transduction, *Annu. Rev. Immunol.,* 10, 519, 1992.
2. **Thomson, A. W., Ed.,** *Cyclosporin. Mode of Action and Clinical Application,* Kluwer Academic, Dordrecht, 1989.
3. **Morris, R. E.,** Rapamycins: Antifungal, antitumour, anti-proliferative and immunosuppressive macrolides, *Transplant. Rev.,* 6, 39, 1992.
4. **Knudsen, P. J., Dinarello, C. A., and Strom, T. B.,** Glucocorticoids inhibit transcriptional and post-transcriptional expression of Interleukin-1 in U937 cells, *J. Immunol.,* 139, 4129, 1987.
5. **Lee, S. W., Tsuo, A.-P., Chan, H., Thomas, J., Petrie, K., Eugui, E. M., and Allison, A. C.,** Glucocorticoids selectively inhibit the transcription of the interleukin-1 gene and decrease the stability of interleukin-1 mRNA, *Proc. Natl. Acad. Sci. U.S.A.,* 85, 1204, 1988.
6. **Reem, G. H. and Yeh, N.-H.,** Interleukin-2 regulates expression of its receptor and synthesis of gamma interferon by human T lymphocytes, *Science,* 225, 429, 1984.
7. **Arya, S. K., Wong-Staal, F., and Gallo, R. C.,** Dexamethasone mediated inhibition of human T cell growth factor and interferon gamma messenger RNA, *J. Immunol.,* 133, 273, 1984.
8. **Vacca, A., Martinotti, S., Screpanti, I., Maroder, M., Felli, M. P., Farina, A. R., Gismondi, A., Santoni, S., Frati, L., and Gulino, A.,** Transcriptional regulation of the interleukin-2 gene by glucocorticoid hormones, *J. Biol. Chem.,* 265, 8075, 1990.
9. **Culpepper, J. A. and Lee, F.,** Regulation of IL-3 expression by glucocorticoids in cloned murine T lymphocytes, *J. Immunol.,* 135, 3191, 1985.
10. **Zanker, B., Walz, G., Weider, K. J., and Strom, T. B.,** Evidence that the glucocorticosteroids block expression of the human interleukin-6 gene by accessory cells, *Transplantation,* 49, 183, 1990.
11. **Kunica, J. E., Talle, M. A., Denhardt, G. H., Brown, M., Prince, L.A., and Goldstein, G.,** Immunosuppression by glucocorticoids. Inhibition of production of multiple cytokines by *in vivo* administration of dexamethasone, *Cell. Immunol.,* 149, 39, 1993.
12. **Granelli-Piperno, A., Nolan, P., Inaba, K., and Steinman, R. M.,** The effects of immunosuppressive agents on the induction of nuclear factors that bind to sites on the Interleukin-2 promoter, *J. Exp. Med.,* 172, 1869, 1990.
13. **Borel, J. F., Feurer, C., Guberland, H. U., and Staehelin, H.,** Biological effects of cyclosporin A: a new antilymphocyte agent, *Agents Actions,* 6, 468, 1976.
14. **Borel, J. F.,** Comparative study of *in vitro* and *in vivo* drug effects on cell mediated cytotoxicity, *Immunology,* 31, 631, 1976.
15. **Borel, J. F., Feurer, C., Magnee, C., and Staehelin, H.,** Effects of the new antilymphocytic peptide cyclosporin A in animals, *Immunology,* 32, 1017, 1977.
16. **Bunjes, D., Hardt, C., Rollinghoff, M., and Wagner, H.,** Cyclosporin A mediates immunosuppression of primary cytotoxic T cell responses by impairing the release of Interleukin-1 and Interleukin-2, *Eur. J. Immunol.,* 11, 657, 1981.
17. **Kronke, M., Leonard, W. J., Depper, J. M., Arya, S. K., Wong-Staal, F., Gallo, R. C., Waldman, T. A., and Greene, W. C.,** Cyclosporin A inhibits T cell growth factor gene expression at the level of mRNA transcription, *Proc. Natl. Acad. Sci. U.S.A.,* 81, 5214, 1984.
18. **Granelli-Piperno, A., Andrus, L., and Steinman, R. M.,** Lymphokine and nonlymphokine mRNA levels in stimulated human T cells. Kinetics, mitogen requirements and effects of cyclosporin A, *J. Exp. Med.,* 163, 922, 1986.
19. **Espevik, T., Figari, I. S., Shalaby, M. R., Lackides, G. A., Lewis, G. D., Shepard, H. M., and Palladino, M. A.,** Inhibition of cytokine production by cyclosporin A and Transforming Growth Factor, *J. Exp. Med.,* 166, 571, 1987.
20. **Gelfand, E. W., Cheung, R. K., and Mills, G. B.,** The cyclosporins inhibit lymphocyte activation at more than one site, *J. Immunol.,* 138, 1115, 1987.
21. **Foxwell, B. M. J., Simon, J., Herrero, J.-J., Taylor, D., Woerly, G., Cantrell, D., and Ryffel, B.,** Anti-CD3 antibody induced expression of both p55 and p75 chains of the Interleukin-2 receptor on human T lymphocytes is inhibited by cyclosporin A, *Immunology,* 69, 104, 1990.
22. **DiPadova, F. E.,** Pharmacology of cyclosporin (Sandimmune). V. Pharmacological effects on immune function: *in vitro* studies, *Pharmacol. Rev.,* 41, 373, 1989.

23. **Hess, A. D., Tutschka, P. J., Pu, Z., and Santos, G. W.,** Effects of cyclosporin on human lymphocyte responses in vitro. IV. Production of T cell stimulatory growth factors and development of responsiveness to these growth factors in cyclosporin A treated primary MLR cultures, *J. Immunol.,* 128, 360, 1982.
24. **Remick, D. G., Nguyen, D. T., Eskandari, M. K., Streiter, R. M., and Kunkel, S. L.,** Cyclosporin A inhibits TNF production without decreasing TNF mRNA levels, *Biochem. Biophys. Res. Commun.,* 161, 551, 1989.
25. **Esa, A. H., Converse, P. J., and Hess, A. D.,** Cyclosporin inhibits soluble antigen and alloantigen presentation by human monocytes *in vitro, Int. J. Immunopharmacol.,* 9, 893, 1987.
26. **Snyder, D. S., Wright, C. L., and Ting, C.,** Inhibition of human monocyte antigen presentation but not HLA-DR expression by cyclosporin, *Transplantation,* 44, 407, 1987.
27. **Kino, T., Hatanaka, H., Hashimoto, M., Nishiyama, M., Goto, T., Okuhara, M., Kohsaka, M., Aoki, H., and Inamaka, H.,** FK 506, a novel immunosuppressant isolated from a streptomycetes. I. Fermentation, isolation and physicochemical and biological characteristics, *J. Antibiot.,* 40, 1249, 1987.
28. **Goto, T., Kino, T., Hatanaka, H., Nishiyama, M., Okuhara, M., Kohsaka, M., Aoki, H., and Inamaka, H.,** Discovery of FK 506, a novel immunosuppressant isolated from *Streptomycetes tsukubaensis, Transplant. Proc.,* 19 (Suppl. 6), 4, 1987.
29. **Kino, T., Hatanaka, H., Miyata, S., Inamura, N., Nishiyama, M., Yajima, T., Goto, T., Okuhara, M., Kohsaka, N., Aoki, H., and Ochiai, T.,** FK 506, a novel immunosuppressant isolated from a streptomycetes. II. Immunosuppressive effect of FK 506 *in vitro, J. Antibiot.,* 40, 1256, 1987.
30. **Tocci, M. J., Matkovich, D. A., Collier, K. A., Kwok, P., Dumont, F. J., Lin, S., Degudicibus, S., Siekierka, J. J., Chin, J., and Hutchinson, N. I.,** The immunosuppressant FK 506 selectively inhibits expression of early T cell activation genes, *J. Immunol.,* 143, 718, 1989.
31. **Fung, J. J., Alessiani, M., Abu-Elmagd, K., Todo, S., Shapiro, R., Tzakis, A., Van-Thiel, D., Armitage, J., Jain, A., McCauley, J., Selby, R., and Starzl, T. E.,** Adverse effects associated with the use of FK 506, *Transplant. Proc.,* 23, 3105, 1991.
32. **MacLeod, A. M. and Thomson, A. W.,** FK 506, an immunosuppressant for the 1990s, *Lancet,* 337, 25, 1991.
33. **Handschumacher, R. E., Harding, M. W., Rice, J., Drugge, R. J., and Speicher, D. W.,** Cyclophilin, a specific cytosolic binding protein for cyclosporin A, *Science,* 226, 544, 1984.
34. **Siekierka, J. J., Staruch, M. J., Hung, S. H., and Sigal, N. H.,** FK 506, A potent novel immunosuppressive agent binds to a cytosolic protein which is distinct from the cyclosporin A binding protein cyclophilin, *J. Immunol.,* 143, 1580, 1989.
35. **Haendler, B., Hofer, W. R., and Hofer, E.,** Complementary DNA for human T cell cyclophilin, *EMBO J.,* 6, 947, 1987.
36. **Standaert, R. F., Galat, A., Verdine, G. L., and Schreiber, S. L.,** Molecular cloning and overexpression of the human FK 506 binding protein, FKBP, *Nature,* 346, 671, 1990.
37. **Maki, N., Sekiguchi, F., Nishimaki, J., Miwa, K., Hayano, T., Takahashi, N., and Suzuki, M.,** Complementary DNA encoding the human T cell FK 506 binding protein, a peptidyl prolyl cis-trans isomerase distinct from cyclophilin, *Proc. Natl. Acad. Sci. U.S.A.,* 87, 5440, 1990.
38. **Takahashi, N., Hayano, T., and Suzuki, M.,** Peptidyl prolyl cis-trans isomerase is the cyclosporin A binding protein cyclophilin, *Nature,* 337, 473, 1989.
39. **Fischer, G., Wittman, L. B., Lang, K., Kiefhaber, T., and Schmid, F. X.,** Cyclophilin and peptidyl prolyl cis-trans isomerase are probably identical proteins, *Nature,* 337, 476, 1989.
40. **Siekierka, J. J., Hung, S. H., Poe, M., Lin, C. S., and Sigal, N. H.,** A cytosolic binding protein for the immunosuppressant FK 506 has peptidyl prolyl isomerase activity but is distinct from cyclophilin, *Nature,* 341, 755, 1989.
41. **Harding, M. W., Galat, A., Uehling, D. E., and Schreiber, S. L.,** A receptor for the immunosuppressant FK 506 is a cis-trans peptidyl prolyl isomerase, *Nature,* 341, 758, 1989.
42. **Tropschug, M., Barthelmess, I. B., and Neupert, W.,** Sensitivity to cyclosporin A is mediated by cyclophilin in *Neurospora crassa* and *Saccharomyces cerevisiae, Nature,* 342, 953, 1989.
43. **Brizuella, L., Chrebet, G., Bostian, K. A., and Parent, S. A.,** Anti-fungal properties of the immunosuppressant FK 506. Identification of an FK 506 responsive yeast gene distinct from FKBI, *Mol. Cell. Biol.,* 11, 4616, 1991.

44. **Weiderrecht, G., Brizuella, L., Elliston, K., Sigal, N. H., and Siekierka, J. J.,** FKBI encodes a non-essential FK 506 binding protein in *Saccharomyces cerevisiae* and contains regions suggesting homology to the cyclophilins, *Proc. Natl. Acad. Sci. U.S.A.,* 88, 1029, 1991.
45. **Heitman, J., Movva, N. R., Hiestand, P. C., and Hall, M. N.,** FK 506 binding protein proline rotamase is a target for the immunosuppressive agent FK 506 in *Saccharomyces cerevisiae, Proc. Natl. Acad. Sci. U.S.A.,* 88, 1948, 1991.
46. **Heitman, J., Movva, N. R., and Hall, M. N.,** Targets for cell cycle arrest by the immunosuppressant rapamycin in yeast, *Science,* 253, 905, 1991.
47. **Quesniaux, V. F., Schreier, M. H., Wenger, R. M., Hiestand, P. C., Harding, M. W., and Regenmortel, M. H. V.,** Molecular characteristics of cyclophilin-cyclosporin interaction, *Transplantation,* 46, 235, 1988.
48. **Bundick, R. V., Donald, D. K., Eady, R. P., Hutchinson, R., Keogh, R. W., Schmidt, J. A., and Wells, E.,** FK 506 as an agonist to induce inhibition of Interleukin-2 production, *Transplantation,* 53, 1150, 1992.
49. **Bierer, B. E., Somers, P. K., Wandless, T. J., Burakoff, S. J., and Schreiber, S. L.,** Probing immunosuppressant action with a non-natural immunophilin ligand, *Science,* 250, 556, 1990.
50. **Kawai, M., Lane, B. C., Hsieh, G. C., Mollison, K. W., Carter, G. W., and Luly, J. R.,** Structure activity profiles of macrolactam immunosuppressant FK 506 analogues, *FEBS Lett.,* 316, 107, 1993.
51. **Chen, S. S., Wickler, L. S., and Arison, B. H.,** New L-683590 microbial transformation product. EP 0403242 A, 1993.
52. **Liu, J., Farmer, J. D., Lane, W. S., Friedman, J., Weissman, I., and Schreiber, S. L.,** Calcineurin is a common target of cyclophilin-cyclosporin A and FKBP-FK 506 complexes, *Cell,* 66, 807, 1991.
53. **Fruman, D. A., Klee, C. B., Bierer, B. E., and Burakoff, S. J.,** Calcineurin phosphatase activity in T lymphocytes is inhibited by FK 506 and cyclosporin A, *Proc. Natl. Acad. Sci. U.S.A.,* 89, 3686, 1992.
54. **O'Keefe, S. J., Tamura, J., Kincaid, R. L., Tocci, M. J., and O'Neill, E. A.,** FK 506 and cyclosporin A sensitive activation of the Interleukin-2 promoter by calcineurin, *Nature,* 357, 692, 1992.
55. **Clipstone, N. A. and Crabtree, G. R.,** Identification of calcineurin as a key signalling enzyme in T lymphocyte activation, *Nature,* 357, 695, 1992.
56. **Metcalfe, S.,** Cyclosporine does not prevent cytosolic calcium changes associated with lymphocyte activation, *Transplantation,* 38, 161, 1984.
57. **Bijsterbosch, M. K. and Klauss, G. G.,** Cyclosporin does not inhibit mitogen induced inositol phospholipid degradation in mouse lymphocytes, *Immunology,* 56, 435, 1985.
58. **Fidelus, R. H. and Laughter, A. H.,** Protein kinase activation and the immunosuppressant cyclosporin, *Transplantation,* 41, 187, 1986.
59. **Fujii, Y., Fujii, S., and Kaneko, T.,** Effect of a novel immunosuppressive agent FK 506 on mitogen induced inositol phospholipid degradation in rat thymocytes, *Transplantation,* 47, 1081, 1989.
60. **Lin, C. S., Boltz, R. C., Siekierka, J. J., and Sigal, N. H.,** FK 506 and cyclosporin A inhibit highly similar signal transduction pathways in human T lymphocytes, *Cell Immunol.,* 133, 269, 1991.
61. **Bloemena, E., van Oers, R. H. J., Weinreich, S., Stilma-Meinesz, A. P., Schellekens, P. Th. A., and van Lier, R. A. W.,** The influence of cyclosporin A on the alternative pathways of human T cell activation, *Eur. J. Immunol.,* 19, 943, 1989.
62. **Bierer, B. E., Schreiber, S. L., and Burakoff, S. J.,** The effect of the immunosuppressant FK 506 on alternate pathways of T cell activation, *Eur. J. Immunol.,* 21, 439, 1991.
63. **Emmel, E. A., Verweij, C. L., Durand, D. B., Higgins, K. M., Lacy, E., and Crabtree, G. R.,** Cyclosporin A specifically inhibits function of nuclear proteins involved in T cell activation, *Science,* 246, 1617, 1989.
64. **Randak, C., Brabletz, T., Hergenrother, M., Sobotta, I., and Serfling, E.,** Cyclosporin A suppresses the expression of the Interleukin-2 gene by inhibiting the binding of lymphocyte specific factors to the IL-2 enhancer, *EMBO J.,* 9, 2259, 1990.

65. **Mattila, P. S., Ullman, K. S., Fiering, S., Emmel, E. A., McCutcheon, M., Crabtree, G. R., and Herzenberg, L. A.,** The actions of cyclosporin A and FK 506 suggest a novel step in the activation of T lymphocytes, *EMBO J.,* 9, 4425, 1990.
66. **Sawada, S., Suzuki, G., Kawase, Y., and Takahu, F.,** Novel immunosuppressive agent FK 506. *In vitro* effects on cloned T cell activation, *J. Immunol.,* 139, 1797, 1987.
67. **Heeg, K., Gillis, S., and Wagner, H.,** IL-4 bypasses the immune suppressive effect of cyclosporin A during the *in vitro* induction of murine cytotoxic T lymphocytes, *J. Immunol.,* 141, 2330, 1988.
68. **Thomas, J., Matthews, C., Carroll, R., Loreth, R., and Thomas, F.,** The immunosuppressive action of FK 506. *In vitro* induction of allogeneic unresponsiveness in human CTL precursors, *Transplantation,* 49, 390, 1990.
69. **Johansson, A. and Moller, E.,** Evidence that the immunosuppressive effects of FK 506 and cyclosporin are identical, *Transplantation,* 50, 1001, 1990.
70. **Lu, P., Garcia-Sanz, J. A., Lichtenheld, M. G., and Podack, E. R.,** Perforin expression in human peripheral blood mononuclear cells. Definition of an IL-2 independent pathway of perforin induction in $CD8^+$ T cells, *J. Immunol.,* 148, 3354, 1992.
71. **Hanke, J. H., Nichols, L. N., and Coon, M. E.,** FK 506 and rapamycin selectively enhance degradation of IL-2 and GM-CSF mRNA, *Lymphokine Cytokine Res.,* 11, 221, 1992.
72. **Flanagan, W. F., Corthesy, B., Bram, R. J., and Crabtree, G. R.,** Nuclear association of a T cell transcription factor blocked by FK 506 and cyclosporin A, *Nature,* 352, 803, 1991.
73. **Jain, J., McCaffrey, P. G., Valge-Archer, V. E., and Rao, A.,** Nuclear factor of activated T cells contains *fos* and *jun, Nature,* 356, 801, 1992.
74. **Jain, J., Valge-Archer, V. E., and Rao, A.,** Analysis of the AP-1 sites in the IL-2 promoter, *J. Immunol.,* 148, 1240, 1992.
75. **McCaffrey, P. G., Perrino, B. A., Soderling, T. R., and Rao, A.,** NF-ATp, a T lymphocyte DNA binding protein that is a target for calcineurin and immunosuppressive drugs, *J. Biol. Chem.,* 268, 3747, 1993.
76. **Cohen, P.,** The structure and regulation of protein phosphatases, *Annu. Rev. Biochem.,* 58, 453, 1989.
77. **Martel, R. R., Klicius, J., and Galet, S.,** Inhibition of the immune response by rapamycin, a new antifungal antibiotic, *Can. J. Physiol. Pharmacol.,* 55, 48, 1977.
78. **Dumont, F. J., Staruch, M. J., Koprak, S. L., Melino, M. R., and Sigal, N. H.,** Distinct mechanisms of suppression of murine T cell activation by the related macrolides FK 506 and rapamycin, *J. Immunol.,* 144, 251, 1990.
79. **Wicker, L. S., Boltz, R. C., Matt, V., Nichols, E. A., Peterson, L. B., and Sigal, N. H.,** Suppression of B cell activation by cyclosporin A, FK 506 and rapamycin, *Eur. J. Immunol.,* 20, 2277, 1990.
80. **Henderson, D. J., Naya, I., Bundick, R. V., Smith, G. M., and Schmidt, J. A.,** Comparison of the effects of FK 506, cyclosporin A and rapamycin on IL-2 production, *Immunology,* 73, 316, 1991.
81. **Dumont, F. J., Melino, M. R., Staruch, M. J., Koprak, S. L., Fischer, P. A., and Sigal, N. H.,** The immunosuppressive macrolides FK 506 and rapamycin act as reciprocal antagonists in murine T cells, *J. Immunol.,* 144, 1418, 1990.
82. **Bierer, B. E., Mattila, P. S., Standaert, R. F., Herzenberg, L. A., Burakoff, S. J., Crabtree, G. R., and Schreiber, S. L.,** Two distinct signal transmission pathways in T lymphocytes are inhibited by complexes formed between an immunophilin and either FK 506 or rapamycin, *Proc. Natl. Acad. Sci. U.S.A.,* 87, 9231, 1990.
83. **Hultsch, T., Albers, M. W., Schreiber, S. L., and Hohman, R. J.,** Immunophilin ligands demonstrate common features of signal transduction leading to exocytosis or transcription, *Proc. Natl. Acad. Sci. U.S.A.,* 88, 6229, 1991.
84. **De Paulis, A., Cirillo, R., Ciccarelli, A., Condorelli, M., and Marone, G.,** FK 506, a potent novel inhibitor of the release of pro-inflammatory mediators from human Fc_{ε} $R1^+$ cells, *J. Immunol.,* 146, 2374, 1991.
85. **De Paulis, A., Cirillo, R., Ciccarelli, A., De Crescenzo, G., Oriente, A., and Marone, G.,** Characterisation of the anti-inflammatory effect of FK 506 on human mast cells, *J. Immunol.,* 147, 4278, 1991.

86. **Luo, H., Chen, H., Daloze, P., Chang, J. Y., St. Louis, G., and Wu, J.,** *In vitro* IgE production by Interleukin-4 stimulated human peripheral blood mononuclear cells is suppressed by rapamycin, *Clin. Immunol. Immunopathol.,* 61, 410, 1991.
87. **Luo, H., Chen, H., Daloze, P., Chang, J. Y., St. Louis, G., and Wu, J.,** Inhibition of *in vitro* immunoglobulin production by rapamycin, *Transplantation,* 53, 1071, 1992.
88. **Metcalfe, S. M. and Richards, F. M.,** Cyclosporine, FK 506 and rapamycin. Some effects on early activation events in serum free mitogen activated mouse spleen cells, *Transplantation,* 49, 798, 1990.
89. **Price, D. J., Grove, J. R., Calvo, V., Avruch, J., and Bierer, B. E.,** Rapamycin induced inhibition of the 70 kilodalton S6 protein kinase, *Science,* 257, 973, 1992.
90. **Chung, J., Kuo, C. J., Crabtree, G. R., and Blenis, J.,** Rapamycin-FKBP specifically blocks growth dependent activation of and signalling by the 70 kD S6 protein kinases, *Cell,* 69, 1227, 1992.
91. **Calvo, V., Crews, C. M., Vik, T. A., and Bierer, B. E.,** Interleukin-2 stimulation of p70 S6 kinase activity is inhibited by the immunosuppressant rapamycin, *Proc. Natl. Acad. Sci. U.S.A.,* 89, 7571, 1992.
92. **Kuo, C. J., Chung, J., Fiorentino, D. F., Flanagan, W. M., Blenis, J., and Crabtree, G. R.,** Rapamycin selectively inhibits interleukin-2 activation of p70 S6 kinase, *Nature,* 358, 70, 1992.
93. **Kunz, J., Henriquez, R., Schneider, U., Deuter-Reinhard, M., Movva, N. R., and Hall, M. N.,** Target of rapamycin in yeast, TOR 2, is an essential phosphatidyl inositol kinase homolog required for G_1 progression, *Cell,* 585, 1993.
94. **Price, E. R., Zydowsky, L. D., Jin, M. J., Baker, C. H., McKeon, F. D., and Walsh, C. T.,** Human cyclophilin B: a second cyclophilin gene encodes a peptidyl prolyl isomerase with a signal sequence, *Proc. Natl. Acad. Sci. U.S.A.,* 88, 1903, 1991.
95. **Swanson, S. K.-H., Born, T., Zydowsky, L. D., Cho, H., Chang, H. Y., Walsh, C. T., and Rusnak, F.,** Cyclosporin mediated inhibition of bovine calcineurin by cyclophilins A and B, *Proc. Natl. Acad. Sci. U.S.A.,* 89, 3741, 1992.
96. **Jin, Y.-J., Albers, M. W., Lane, W. S., Bierer, B. E., Schreiber, S. L., and Burakoff, S. J.,** Molecular cloning of a membrane associated human FK 506 and rapamycin binding protein FKBP-13, *Proc. Natl. Acad. Sci. U.S.A.,* 88, 6677, 1991.
97. **Rosborough, S. L., Fleming, M., Nelson, P. A., Boger, J., and Harding, M. W.,** Identification of FKBP related proteins with antibodies of predetermined specificity and isolation by FK 506 affinity chromatography, *Transplant. Proc.,* 23, 2890, 1991.
98. **Bram, R. J., Hung, D. T., Martin, P. K., Schreiber, S. L., and Crabtree, G. R.,** Identification of the immunophilins capable of mediating inhibition of signal transduction by cyclosporin A and FK 506 — roles of calcineurin binding and cellular location, *Mol. Cell. Biol.,* 13, 4760, 1993.
99. **Galat, A., Lane, W. S., Standaert, R. F., and Schreiber, S. L.,** A rapamycin sensitive 25 kDa immunophilin, *Biochemistry,* 31, 2427, 1992.
100. **Jin, Y.-J., Burakoff, S. J., and Bierer, B. E.,** Molecular cloning of a 25 kDa high affinity rapamycin binding protein, FKBP 25, *J. Biol. Chem.,* 267, 10942, 1992.
101. **Yem, A. W., Tomasselli, A. G., Heinrikson, R. L., Zurcher-Neely, H., Ruff, V. A., Johnson, R. A., and Deibel, M. R.,** The hsp56 component of steroid receptor complexes binds to immobilised FK 506 and shows homology to FKBP 12 and FKBP 13, *J. Biol. Chem.,* 267, 2868, 1992.
102. **Hutchison, K. A., Scherrer, L. C., Czar, M. J., Ning, Y. M., Sanchez, E. R., Leach, K. L., Diebel, M. R., and Pratt, W. B.,** FK 506 binding to the 56 kilodalton immunophilin (hsp56) in the glucocorticoid receptor heterocomplex has no effect on receptor folding or function, *Biochemistry,* 32, 3953, 1993.
103. **Ning, Y. M. and Sanchez, E. R.,** Potentiation of glucocorticoid receptor mediated gene expression by the immunophilin ligands FK 506 and rapamycin, *J. Biol. Chem.,* 268, 6073, 1993.
104. **Walliser, P., Benzie, C. R., and Kay, J. E.,** Inhibition of murine B lymphocyte proliferation by the novel immunosuppressive drug FK 506, *Immunology,* 68, 434, 1989.
105. **Suzuki, N., Sakane, T., and Tsunematsu, T.,** Effects of a novel immunosuppressive agent on human B cell activation, *Clin. Exp. Immunol.,* 79, 240, 1990.
106. **Kay, J. E., Kromwel, L., Doe, S. E., and Denyer, M.,** Inhibition of T and B lymphocyte proliferation by rapamycin, *Immunology,* 72, 544, 1991.

107. **Stevens, C., Lempert, N., and Freed, B. M.,** The effects of immunosuppressive agents on *in vitro* production of human immunoglobulins, *Transplantation,* 51, 1240, 1991.
108. **Marone, G., Triggiani, M., Cirillo, R., Giacummo, A., Siri, L., and Condorelli, M.,** Cyclosporin A inhibits the release of histamine and peptide leukotriene C4 from human lung mast cells, *Ric. Clin. Lab.,* 18, 53, 1988.
109. **Burd, P. R., Rogers, H. W., Gordon, J. R., Martin, C. A., Jayaraman, S., Wilson, S. D., Dvorak, A. M., Galli, S. J., and Dorf, M. E.,** Interleukin-3 dependent and independent mast cells stimulated with IgE and antigen express multiple cytokines, *J. Exp. Med.,* 170, 245, 1989.
110. **Forrest, M. J., Jewell, M. E., Koo, G. C., and Sigal, N. H.,** FK 506 and cyclosporin A: selective inhibition of calcium ionophore induced polymorphonuclear leukocyte degranulation, *Biochem. Pharmacol.,* 42, 1221, 1991.
111. **Keicho, N., Sawada, S., Kitamura, K., Yotsumoto, H., and Takanu, F.,** Effects of an immunosuppressant FK 506 on interleukin-1 production by human macrophages and a macrophage-like cell line U937, *Cell. Immunol.,* 132, 285, 1991.
112. **Burd, P. R., Freeman, G. F., Wilson, S. D., Berman, M., De Kruyff, R., Billings, P. R., and Dorf, M. E.,** Cloning and characterisation of a novel T cell activation gene, *J. Immunol.,* 139, 3126, 1987.
113. **Kwon, B. S. and Weissman, S. M.,** cDNA sequences of two inducible T cell genes, *Proc. Natl. Acad. Sci. U.S.A.,* 86, 1963, 1989.
114. **Umezawa, H., Ishizuka, M., Takeuchi, T., Abe, F., Nemoto, K., Shibuya, K., and Nakamura, T.,** Suppression of tissue graft rejection by spergualin, *J. Antibiot.,* 38, 283, 1985.
115. **Walter, P., Thies, J., Harbauer, G., Dickneite, G., Sedlacek, H. H., and Vonnahme, F.,** Allogeneic heart transplantation in the rat with a new antitumoral drug 15-deoxyspergualin, *Transplant. Proc.,* 18, 1293, 1986.
116. **Nemoto, K., Hayashi, M., Abe, F., Nakamura, M., Ishizuka, M., and Umezawa, H.,** Immunosuppressive activities of 15-deoxyspergualin in animals, *J. Antibiot.,* 40, 561, 1987.
117. **Amemiya, H., Suzuki, S., Ota, K., Takahashi, K., Sonoda, T., Ishibashi, M., Omoto, R., Koyama, I., Dohi, K., Fukuda, Y., and Fukao, K.,** A novel rescue drug, 15-deoxyspergualin. First clinical trials for recurrent graft rejection in renal transplant recipients, *Transplantation,* 49, 337, 1990.
118. **Marchman, W., Araneda, D., DeMasi, R., Taylor, D., Larkin, E., Alqaisi, M., and Thomas, F.,** Prolongation of xenograft survival after combination therapy with 15-deoxyspergualin and total lymphoid irradiation in the hamster to rat cardiac xenograft model, *Transplantation,* 53, 30, 1992.
119. **Gannedahl, G., Wanders, A., Carlsson, M., and Tufveson, G.,** 15-Deoxyspergualin effects in rat heart allograft transplantation — relation to dose, timing and cyclosporine, *Transplantation,* 55, 455, 1993.
120. **Nemoto, K., Abe, F., Nakamura, T., Ishizaka, M., Takeuchi, T., and Umezawa, H.,** Blastogenic responses and the release of Interleukins 1 and 2 by spleen cells obtained from the rat skin allograft recipients administered with 15-deoxyspergualin, *J. Antibiot.,* 40, 1062, 1987.
121. **Fujii, H., Takada, T., Nemoto, K., Abe, F., Fujii, A., Talmadge, J. E., and Takeuchi, T.,** Deoxyspergualin, a novel immunosuppressant, markedly inhibits human mixed lymphocyte reaction and cytotoxic T lymphocyte activity *in vitro*, *Int. J. Immunopharmacol.,* 14, 731, 1992.
122. **Kerr, P. G. and Atkins, R. C.,** Deoxyspergualin inhibits cytotoxic T lymphocytes but not NK or LAK cells, *Immunol. Cell. Biol.,* 69, 177, 1991.
123. **Hibasami, H., Tsukada, T., Suzuki, R., Takano, K., Takaji, S., Takeuchi, T., Shirakawa, S., Murata, T., and Takashima, K.,** 15-Deoxyspergualin, an antiproliferative agent for human and mouse leukaemia cells shows inhibitory effects on the synthetic pathway of polyamines, *Anticancer Res.,* 11, 325, 1991.
124. **Franklin, T. J. and Cook, J. M.,** Inhibition of nucleic acid synthesis by mycophenolic acid, *Biochem. J.,* 113, 515, 1969.
125. **Mizuno, K., Tsujino, M., Takada, M., Hayashi, M., Atsumi, K., Asano, K., and Matsuda, T.,** Studies on bredinin, I. Isolation, characterisation and biological properties, *J. Antibiot.,* 24, 775, 1974.

126. **Morris, R. E., Hoyt, E. G., Murphy, M. P., Eugui, E. M., and Allison, A. C.,** Mycophenolic acid morpholinoethyl ester (RS61443) is a new immunosuppressant that prevents and halts heart allograft rejection by selective inhibition of T and B cell purine synthesis, *Transplant. Proc.,* 22, 1659, 1990.
127. **Zeevi, A., Woan, M., Yao, G.-Z., Ventkataramanan, R., Todo, S., Starzl, T. E., and Duquesnoy, R. J.,** Comparative in vitro studies on the immunosuppressive activities of mycophenolic acid, Bredinin, FK 506, cyclosporin A and rapamycin, *Transplant. Proc.,* 23, 2928, 1991.
128. **Makowka, L. and Cramer, D. V.,** Brequinar sodium: a new immunosuppressive drug for transplantation, *Transplant Sci.,* 2, 50, 1992.
129. **Chen, S.-F., Ruben, R. L., and Dexter, D. L.,** Mechanism of action of the novel anticancer agent 6-fluoro-2-(2′-fluoro-1,1′-biphenyl-4-yl)-3-methyl-4-quinolinecarboxylic acid sodium salt (NSC 368390): inhibition of *de novo* pyrimidine nucleotide biosynthesis, *Cancer Res.,* 46, 5014, 1986.
130. **Peters, G. J., Schwartsmann, G., Nadal, J. C., Laurensse, E. J., van Goeningen, C. J., van der Vijgh, W. J., and Pinedo, H. M.,** *In vivo* inhibition of the pyrimidine *de novo* enzyme dihydroorotic acid dehydrogenase by brequinar sodium (DUP-785; NSC 368390) in mice and patients, *Cancer Res.,* 50, 4644, 1990.
131. **Henry, H. L. and Norman, A. W.,** Vitamin D. Metabolism and biological actions, *Ann. Rev. Nutr.,* 4, 493, 1984.
132. **McDonnell, D. P., Mangelsdorf, D. J., Pike, J. W., Hussler, M. R., and O'Malley, B. W.,** Molecular cloning of complementary DNA encoding the avian receptor for vitamin D, *Science,* 235, 1214, 1987.
133. **Rigby, W. F. C., Denone, S., and Fanger, M. W.,** Regulation of lymphokine production and human T lymphocyte activation by 1,25-dihydroxyvitamin D_3, *J. Clin. Invest.,* 79, 1659, 1987.
134. **Abe, J., Moriya, Y., Saito, M., Sugawara, Y., Suda, T., and Nishii, Y.,** Modulation of cell growth, differentiation and production of IL-3 by 1,25-dihydroxyvitamin D_3 in the murine myelomonocytic leukaemia line WEHI-3, *Cancer Res.,* 46, 6316, 1986.
135. **Reichel, H., Koeffler, H. P., Tobler, A., and Norman, A. W.,** 1,25-Dihydroxyvitamin D_3 inhibits interferon synthesis by normal human peripheral blood lymphocytes, *Proc. Natl. Acad. Sci. U.S.A.,* 84, 3385, 1987.
136. **Tobler, A., Gasson, J., Reichel, H., Norman, A. W., and Koeffler, H. P.,** Granulocyte macrophage colony stimulating factor sensitive and receptor mediated regulation by 1,25-dihydroxyvitamin D_3 in normal human peripheral blood lymphocytes, *J. Clin. Invest.,* 79, 1700, 1987.
137. **Steffen, M., Cayre, Y., Manogue, K. R., and Moore, M. A.,** 1,25-Dihydroxyvitamin D_3 transcriptionally regulates tumour necrosis factor mRNA during HL60 cell differentiation, *Immunology,* 63, 43, 1988.
138. **Bhalla, A. K., Amento, E. P., and Krane, S. M.,** Differential effects of 1,25-dihydroxyvitamin D_3 on human lymphocytes and monocytes/macrophages: inhibition of interleukin-2 and augmentation of interleukin-1 production, *Cell. Immunol.,* 98, 311, 1986.
139. **Petkovich, P. M., Wrana, J. L., Grigordiadis, A. E., Heersche, J. N., and Sodek, J.,** 1,25-Dihydroxyvitamin D_3 increases epidermal growth factor receptors and transforming growth factor-like activity in a bone derived cell line, *J. Biol. Chem.,* 262, 13424, 1987.
140. **Simpson, R. U., Hsu, T., Begley, D. A., Mitchell, B. S., and Alizadeh, B. N.,** Transcriptional regulation of the c-myc protooncogene by 1,25-dihydroxyvitamin D_3 in HL60 promyelocytic leukaemia cells, *J. Biol. Chem.,* 262, 4104, 1987.
141. **Amatruda, T. T., Bohman, R., Ranyard, J., and Koeffler, H. P.,** Pattern of expression of HLA-DR and HLA-DQ antigens and mRNA in myeloid differentiation, *Blood,* 69, 1225, 1987.
142. **Lombardino, J. G., Otterness, I. G., and Muren, J. F.,** [1]Benzothiopyrano[4,3-c]pyrazoles as Immunoregulatory Agents, U.S. Patent 4,268,516, 1981.
143. **Skotnicki, J. S., Gilman, S. C., Steinbaugh, B. A., and Musser, J. H.,** Pyrazolo(4,3-c)quinolines, U.S. Patent 4,748,246, 1987.
144. **Doria, G., Isetta, A. M., Ferrari, M., and Trizio, D.,** Heteroaryl 3-oxo-propanenitrile derivatives and process for their preparation, Eur. Pat. 274,443, 1987.
145. **Doria, G., Isetta, A. M., Ferrari, M., and Trizio, D.,** Condensed pyrazole derivatives and process for their preparation, Eur. Pat. 278,603, 1987.

146. **Bowen, J. G., Hockley, M. H., Housley, J. R., Hunneyball, I. M., Titman, R. B., and Webber, D. G.,** Therapeutic [1]benzothiopyrano[4,3-c]-pyrazoles, Eur. Pat. 354,694, 1988.
147. **Titman, R. B., Hockley, M. H., and Gill, O. S.,** Therapeutic agents, WO 9,111,448A 1990.
148. **Otterness, I. G.,** Comparative activity of CP-17193 and five established immunosuppressive towards the antigens SRBC and EL4, *Clin. Exp. Immunol.,* 46, 332, 1981.
149. **Lombardino, J. G. and Otterness, I. G.,** Novel immunosuppressive agents. Potent immunological activity of some Benzothiopyrano-[4,3-c]-pyrazol-3-ones, *J. Med. Chem.,* 24, 830, 1981.
150. **Bundick, R. V. and Eady, R. P.,** The effects of CP 17193, an immunosuppressive pyrazoloquinoline, on the development of spontaneous lupus disease in NZBW F1 hybrid mice, *Clin. Exp. Immunol.,* 89, 179, 1992.
151. **Symons, J. A.,** D. Phil. Thesis, University of York, United Kingdom, 1987.
152. **Otterness, I. G. and Bliven, M. L.,** The mechanism of tumour allograft survival induced by CP-17193, *Immunopharmacology,* 4, 341, 1982.
153. **Kamei, T., Callery, M. P., and Flye, M. W.,** Intragraft delivery of 16,16-dimethyl PGE_2 induces donor specific tolerance in rat cardiac allograft recipients, *Transplantation,* 51, 242, 1991.
154. **Moran, M., Mozes, M. F., Maddux, M. S., Veremis, S., Bartkus, C., Ketel, B., Pollak, R., Wallemark, C., and Jonasson, O.,** Prevention of acute graft rejection by the prostaglandin E_1 analogue misoprostol in renal-transplant recipients treated with cyclosporine and prednisolone, *New Engl. J. Med.,* 332, 1183, 1990.
155. **Kunkel, S. L. and Chensue, S. W.,** Arachidonic acid metabolites regulate interleukin-1 production, *Biochem. Biophys. Res. Commun.,* 128, 892, 1985.
156. **Knudsen, P. J., Dinarello, C. A., and Strom, T. B.,** Prostaglandins posttranscriptionally inhibit monocyte expression of interleukin-1 activity by increasing intracellular cyclic adenosine monophosphate, *J. Immunol.,* 137, 3189, 1986.
157. **Haynes, D. R., Whitehouse, M. W., and Vernon-Roberts, B.,** The prostaglandin E_1 analogue, Misoprostol, regulates inflammatory cytokines and immune functions *in vitro* like the natural prostaglandins E_1, E_2 and E_3, *Immunology,* 76, 251, 1992.
158. **Kunkel, S. L., Wiggins, R. C., and Larrick, J.,** Regulation of macrophage tumour necrosis factor production by prostaglandin E_2, *Biochem. Biophys. Res. Commun.,* 137, 404, 1986.
159. **Scales, W. E., Chensue, S. W., Otterness, I., and Kunkel, S. L.,** Regulation of monokine gene expression. Prostaglandin E2 suppresses tumour necrosis factor but not interleukin-1 alpha or beta mRNA and cell associated bioactivity, *J. Leukocyte Biol.,* 45, 416, 1989.
160. **Otterness, I. G., Bliven, M. L., Eskra, J. D., Reinke, M., and Hanson, D. C.,** Regulation of interleukin-1 production. The role of prostaglandins, *Cell. Immunol.,* 114, 385, 1988.
161. **Goodwin, J. S., Bankhurst, A. D., and Messner, R. P.,** Suppression of human T cell mitogenesis by prostaglandin, *J. Exp. Med.,* 146, 1719, 1977.
162. **Stobo, J. D., Kennedy, M. S., and Goldyne, M. E.,** Prostaglandin E modulation of the mitogenic response of human T cells. Differential response of T cell subpopulations, *J. Clin. Invest.,* 64, 1188, 1979.
163. **Rappaport, R. S. and Dodge, G. R.,** Prostaglandin E_2 inhibits the production of human interleukin-2, *J. Exp. Med.,* 155, 943, 1982.
164. **Walker, C., Kristensen, F., Bettens, F., and De Weck, A. L.,** Lymphokine regulation of activated (G_1) lymphocytes. I. Prostaglandin E_2 induced inhibition of interleukin-2 production, *J. Immunol.,* 130, 1770, 1983.
165. **Betz, M. and Fox, B. S.,** Prostaglandin E_2 inhibits production of Th1 lymphokines but not Th2 lymphokines, *J. Immunol.,* 146, 108, 1991.
166. **Rincon, M., Tugores, A., Lopes-Rivas, A., Silva, A., Alonso, M., De Landazuri, M. O., and Lopez-Botet, M.,** Prostaglandin E_2 and the increase in intracellular cAMP inhibit the expression of interleukin-2 receptors in human T cells, *Eur. J. Immunol.,* 18, 1791, 1988.
167. **Chouaib, S., Welte, K., Mertelsmann, R., and Dupont, B.,** Prostaglandin E_2 acts at two distinct pathways of T lymphocyte activation: inhibition of interleukin-2 production and down regulation of transferrin receptor expression, *J. Immunol.,* 135, 1172, 1985.
168. **Pollak, R., Dumble, L. J., Wiederkehr, J. C., Maddux, M. S., and Moran, M.,** The immunosuppressive properties of new oral prostaglandin E_1 analogues, *Transplantation,* 50, 834, 1990.

169. **Quill, H., Gaur, A., and Phipps, R. P.,** Prostaglandin E_2 dependent induction of granulocyte-macrophage colony-stimulating factor secretion by cloned murine helper T cells, *J. Immunol.,* 142, 813, 1989.
170. **Li, T.-K. and Fox, B. S.,** Effect of prostaglandin E_2 (PGE_2) on IL-3/granulocyte-macrophage colony-stimulating factor production by T helper cells. Mode of stimulation and presence of costimulation can determine response to PGE_2, *J. Immunol.,* 150, 1680, 1993.
171. **Minakuchi, R., Wacholtz, M. C., Davis, L. S., and Lipsky, P. E.,** Delineation of the mechanism of inhibition of human T cell activation by PGE_2, *J. Immunol.,* 145, 2616, 1990.
172. **Mary, D., Aussel, C., Ferrua, B., and Fehlmann, M.,** Regulation of interleukin-2 synthesis by cAMP in human T cells, *J. Immunol.,* 139, 1179, 1987.
173. **Imboden, J. B., Shoback, D. M., Pattison, G., and Stobo, J. D.,** Cholera toxin inhibits the T cell antigen receptor-mediated increases in inositol triphosphate and cytoplasmic free calcium, *Proc. Natl. Acad. Sci. U.S.A.,* 83, 567, 1986.
174. **Papadogiannakis, N., Nordstrom, T. E., Anderson, L. C., and Wolff, C. H.,** cAMP inhibits the OKT3 induced increase in cytoplasmic free calcium, in the Jurkat T cell line. The degree of inhibition correlates inversely with the amount of CD3 binding ligand used, *Eur. J. Immunol.,* 19, 1953, 1989.
175. **Longley, R. E., Caddigan, D., Harmody, D., Gunasekera, M., and Gunasekera, S. P.,** Discodermolide — a new, marine derived immunosuppressive compound. II. *In vivo* studies, *Transplantation,* 52, 656, 1991.
176. **Longley, R. E., Caddigan, D., Harmody, D., Gunasekera, M., and Gunasekera, S. P.,** Discodermolide — a new, marine derived immunosuppressive compound. I. *In vitro* studies, *Transplantation,* 52, 650, 1991.
177. **Manfro, R. C., Pohanka, E., Tomlanovich, S., Benet, L. Z., and Garovoy, M. R.,** Cyclosporin A and prednisolone, an additive inhibitory effect of cell proliferation and interleukin-2 production, *Transplant. Proc.,* 21, 1457, 1989.
178. **Zeevi, A., Duquesnoy, R., Eiras, G., Rabinowich, S., Todo, S., Makowa, L., and Starzl, T.,** Immunosuppressive effect of FK 506 on *in vitro* lymphocyte activation: synergism with cyclosporine A, *Transplant. Proc.,* 19 (Suppl. 6), 40, 1987.
179. **Johansson, A. and Moller, E.,** Evidence that the immunosuppressive effects of FK 506 and cyclosporine are identical, *Transplantation,* 50, 1001, 1990.
180. **Starzl, T. E., Todo, S., Fung, J., Demetris, A. J., Venkataramman, R., and Jain, A.,** FK 506 for liver, kidney and pancreas transplantation, *Lancet,* 2, 1000, 1989.
181. **Burk, M. D., Omar, G., Thomson, A. W., and Whitting, P. H.,** Inhibition of the metabolism of cyclosporine by human liver microsomes by FK 506, *Transplantation,* 50, 901, 1990.
182. **Vincent, S. H., Wang, R. W., Karanam, B. V., Klimko, M., Alvaro, R., and Chiu, S. H.,** Effects of the immunosuppressant FK 506 and its analog FK 520 on hepatic and renal cytochrome P450 mixed function oxidase, *Biochem. Pharmacol.,* 41, 1325, 1991.
183. **Chung, S. W., Kim, P. C. W., Koh, I. H. J., Fung, L. S., Cole, E. H., Cohen, Z., and Levy, G. A.,** The comparative effects of cyclosporin A and 16,16-dimethyl prostaglandin E_2 on the allogeneic induction of monocyte/macrophage procoagulant activity and the cytokines macrophage procoagulant inducing factor and interleukin-2, *Immunology,* 74, 670, 1991.
184. **Gepner, P., Amor, B., and Fournier, C.,**1,25-Dihydroxyvitamin D_3 potentiates the *in vitro* inhibitory effects of cyclosporin A on T cells from Rheumatoid arthritis patients, *Arthritis Rheum.,* 32, 31, 1989.
185. **Romagnani, S.,** Th1 and Th2 subsets: doubt no more, *Immunol. Today,* 12, 256, 1991.
186. **Wierenga, E. A., Snoek, M., Bos, J. D., Jansen, H. M., and Kapsenberg, M. L.,** Comparison of diversity and function of house dust mite-specific T lymphocyte clones from atopic and non-atopic donors, *Eur. J. Immunol.,* 20, 1519, 1990.
187. **Del Prete, G. F., De Carli, M., Mastromauro, C., Biagiotti, R., Macchia, D., Falagiani, P., Ricci, M., and Romagnani, S.,** Purified protein derivative of *Mycobacterium tuberculosis* and excretory-secretory antigen(s) of *Toxocara canis* expand *in vitro* human T cells with stable and opposite (type 1 T helper or type 2 T helper) profile of cytokine production, *J. Clin. Invest.,* 88, 356, 1991.

188. **Wierenga, E. A., Snoek, M., Jansen, H. M., Bos, J. D., van Lier, R. A., and Kapsenberg, M. L.,** Human atopen specific types 1 and 2 T helper cell clones, *J. Immunol.,* 147, 2942, 1991.
189. **Maggi, E., Parronchi, P., Manetti, R., Simonelli, C., Piccinni, M.-P., Santoni Rugiu, F., De Carli, M., Ricci, M., and Romagnani, S.,** Reciprocal regulatory effects of IFNγ- and IL-4 on the *in vitro* development of human Th1 and Th2 clones, *J. Immunol.,* 148, 2142, 1992.
190. **Ricci, M.,** T cells, cytokines, IgE and allergic inflammation, *Int. Arch. Allergy Immunol.,* 99, 165, 1992.
191. **Clerici, M. and Shearer, G. M.,** A Th1-Th2 switch is a critical step in the etiology of HIV infection, *Immunol. Today,* 14, 107, 1993.
192. **Parronchi, P., De Carli, M., Manetti, R., Simonelli, C., Sampognaro, S., Piccinni, M.-P., Macchia, D., Maggi, E., Del Prete, G., and Romagnani, S.,** IL-4 and IFN (α and γ) exert opposite regulatory effects on the development of cytolytic potential by Th1 or Th2 human T cell clones, *J. Immunol.,* 149, 2977, 1992.
193. **Manetti, R., Parronchi, P., Giudizi, M. G., Piccinni, M.-P., Maggi, E., Trinchieri, G., and Romagnani, S.,** Natural killer cell stimulatory factor (Interleukin-12 [IL-12]) induces T helper type 1 (Th1)-specific immune responses and inhibits the development of IL-4-producing Th cells, *J. Exp. Med.,* 177, 1199, 1993.
194. **Swain, S. L., Huston, G., Tonkonogy, S., and Weinberg, A.,** Transforming growth factor and IL-4 cause helper T cell precursors to develop into distinct effector helper cells that differ in lymphokine secretion pattern and cell surface phenotype, *J. Immunol.,* 147, 2991, 1991.
195. **Fiorentino, D. F., Bond, M. W., and Mossman, T.,** Two types of mouse T helper cell. IV. Th2 clones secrete a factor that inhibits cytokine production by Th1 clones, *J. Exp. Med.,* 170, 2081, 1989.
196. **Del Prete, G., De Carli, M., Almerigogna, F., Giudizi, M. G., Biagiotti, R., and Romagnani, S.,** Human IL-10 is produced by both type 1 helper (Th1) and type 2 helper (Th2) T cell clones and inhibits their antigen specific proliferation and cytokine production, *J. Immunol.,* 150, 353, 1993.
197. **Yssel, H., Malefyt, R., Roncarolo, M. -G., Abrams, J. S., Lahesmaa, R., Spits, H., and De Vries, J.,** IL-10 is produced by subsets of human $CD4^+$ T cell clones and peripheral blood T cells, *J. Immunol.,* 149, 2378, 1992.
198. **Gajewski, T. F., Schell, S. R., and Fitch, F. W.,** Evidence implicating utilisation of different T cell receptor-associated signalling pathways by Th1 and Th2 clones, *J. Immunol.,* 144, 4110, 1990.
199. **Andersson, J., Nagy, S., Groth, C.-G., and Andersson, U.,** Effects of FK 506 and cyclosporin A on cytokine production studied at the single-cell level, *Immunology,* 75, 136, 1992.
200. **Wang, S. C., Zeevi, A., Jordan, M. L., Simmons, R. L., and Tweardy, D. J.,** FK 506, rapamycin and cyclosporine. Effects on IL-4 and IL-10 mRNA levels in a T-helper 2 cell line, *Transplant. Proc.,* 23, 2920, 1991.
201. **Durez, P., Abramowicz, D., Gerard, C., van Mechelen, M., Amraoui, Z., Dubois, C., Leo, O., Velu, T., and Goldman, M.,** *In vivo* induction of interleukin-10 by anti-CD3 monoclonal antibody or bacterial lipopolysaccharide. Differential modulation by cyclosporin A, *J. Exp. Med.,* 177, 551, 1993.
202. **Snijdewint, F. G., Kalinski, P., Wierenga, E. A., Bos, J. D., and Kapsenberg, M. L.,** Prostaglandin E_2 differentially modulates cytokine secretion profiles of human T helper lymphocytes, *J. Immunol.,* 150, 5321, 1993.
203. **Lawrence, R. A., Jones, R. L., and Wilson, N. H.,** Characterisation of the receptors involved in the direct and indirect actions of prostaglandins E and I on the guinea-pig ileum, *Br. J. Pharmacol.,* 105, 271, 1992.

Chapter 7

CYTOKINE CONVERTASE INHIBITORS

Douglas K. Miller

CONTENTS

0-8493-8381-1/96/$0.00+$.50

I. INTRODUCTION

Two cytokines are most prominently associated with inflammatory events: interleukin-1 (IL-1) and tumor necrosis factor α (TNFα), both released largely from monocytic cells (see References 1–3 for reviews). IL-1 is composed of two similarly sized forms: IL-1α and IL-1β.[4] Monocytes produce more IL-1β than IL-1α, and it is this former form of IL-1 that is found most prominently in inflammatory tissues.[5,6] The relative importance of IL-1β vs. IL-1α is emphasized by the observation that neutralization of IL-1β rather than IL-1α is much more effective in the treatment of animal models of inflammation.[7,8] Indeed, the importance of IL-1β relative to IL-1α is emphasized by the evolution of pox viral proteins similar in structure to soluble forms of the Type II IL-1 receptors that neutralize IL-1β, but not IL-1α.[9,10] IL-1β and TNFα, despite their different cellular locations, both need to be proteolytically processed prior to their extracellular release. The discovery and characterization of these activating proteases and the development of specific inhibitors to their activities represents a major new approach to anti-inflammatory therapy.

II. IL-1

A. RECOGNITION OF A pIL-1β CONVERTASE ACTIVITY

1. Nature of Active IL-1β

The biologically active extracellular form of IL-1β has a molecular weight of 17 to 18 kDa.[11–17] When IL-1β was cloned, it was discovered that it was synthesized in monocytes and THP.1 monocytic cells as an inactive 31-kDa precursor protein (pIL-1β) and that the active mature form (mIL-1β) is derived from the carboxyl terminus of the precursor.[18–25] The active mIL-1β has an N-terminal amino acid sequence beginning at Ala117 indicating specific cleavage following Asp116 of the pIL-1β.[16–18,21] pIL-1β has no hydrophobic leader sequence[18,19] and was found only in the cytoplasm and not in the endoplasmic reticulum and Golgi body, as might be expected if it were a normally secreted protein.[26–29] Hence, the extracellular presence of mIL-1β requires both novel proteolytic processing and secretory events.

2. Release of IL-1β From Cells

Cellular IL-1β is composed entirely of the precursor with none of the mIL-1β normally being present. Only when proteolysis occurred during cellular breakage could IL-1β fragments smaller than 31 kDa be isolated from cellular extracts.[29–31] In contrast, the culture medium of monocytic cells contains principally mIL-1β with variable amounts of pIL-1β detected.[20,22,26,29,32,33] Differing amounts of pIL-1β observed in the culture supernatants were probably a reflection of the viability of the cultures, with less viable cells releasing more pIL-1β.[31]

The amount and form of the IL-1β released to culture supernatants is also a function of the type of monocytic cell and the culture conditions. In some cases relatively large amounts of pIL-1β were detected in supernatants from elicited macrophages.[32,34,35] Despite comparable levels of IL-1β synthesis, macrophages release less IL-1β than do monocytes,[35–38] and purified monocytes release eight-fold less IL-1β than do monocytes in whole blood.[39] The decreased LPS-induced secretion of IL-1β by macrophages relative to monocytes occurs at the same time that the macrophages have an elevated TNFα secretion, indicating that they have no loss of sensitivity to stimulation.[37]

Pulse-chase kinetics indicated that the mIL-1β was released rapidly and quantitatively from the endogenous pool of newly synthesized pIL-1β, with a half-life in purified human peripheral blood monocytes of 2.5 h.[20,29,40–42] In contrast, the IL-1α remained largely cell associated. Secretion was directional, with depletion of the intracellular pIL-1β occurring simultaneously with extracellular accumulation of mIL-1β.[20,29] The secretion was unaffected by the presence of cycloheximide, indicating that new protein synthesis was not needed for cleavage and secretion.[20,33] In contrast, newly synthesized IL-1α, which is produced at less than 10% the amount as IL-1β,[35] had a half-life of 15 h. The small amount of pIL-1β that was also released had the same half-life as mIL-1β and was present in the same ratio to mIL-1β during the chase, indicating that the processing of pIL-1β to form mIL-1β occurred prior to its release from cells rather than in the culture supernatant.[29] This was confirmed by the observation that mIL-1β was not produced by incubation of pIL-1β with monocytic cells.[33,35,43,44]

3. Cleavage of pIL-1β by Proteases

During the release of mIL-1β from cells, no intermediate-sized IL-1 forms were found, suggesting an endoproteolytic processing to the mIL-1β form and the absence of aminopeptidase activity.[29,45] A number of exogenously added endoproteases could convert the inactive pIL-1β to an active form yielding different sized fragments with increases in activity corresponding directly to the closeness in size to the 17.5-kDa mIL-1β naturally produced in monocytic cells. Trypsin or plasmin, for example, yielded a 22-kDa form with 40 additional residues N-terminal to the Ala117 start site of the native mIL-1β. Elastase, *Staphylococcus aureus* protease, chymotrypsin, cathepsin G, and mast cell chymase yielded 18-kDa fragments with 12, 5, 3, 3, and 3 additional residues, respectively.[41,46–50] The specific activity of these latter, smaller IL-1β forms approached that of native mIL-1β itself.[48,49] That these extracellular proteases are not necessary, however, for the formation of extracellular mIL-1β in monocytic cultures was confirmed not only by the pulse-chase kinetics, where there was no evidence for a precursor-product relationship between the secreted pIL-1β and mIL-1β,[29] but also by the observation that serine protease inhibitors had no effect on the formation of mIL-1β.[34] Thus, the evidence is strong that pIL-1β cleavage to mIL-1β must occur intracellularly prior to or concomitant with secretion. It is possible, however, that in inflammatory exudates *in vivo* some IL-1 activity could result from cleavage of pIL-1β by bystander proteases such as elastase and cathepsins, since such cleavage could be detected.[30,43,48]

B. INTERLEUKIN-1 CONVERTING ENZYME (ICE)

1. Discovery of ICE

The endoproteolytic activity that produced the correctly processed mIL-1β beginning at Ala[117] could be found when pIL-1β was incubated *in vitro* with extracts of the monocytes themselves.[29,51,52] This pIL-1β converting enzyme (ICE) activity was found only in the cytoplasm; membranes or granules from these cells could not correctly process the pIL-1β.[29,51] Besides monocytes, however, few cells synthesizing IL-1 were found to contain a pIL-1β cleavage activity. Keratinocytes, for example, could synthesize and release pIL-1β in culture, but normally did not proteolytically process the pIL-1β to the 17.5-kDa form unless monocytic cell extracts were first added.[53] Similarly, fibroblasts[31,54] or promonocytic U-937 cells[55] transfected with pIL-1β cDNA could synthesize pIL-1β, but they could neither release or cleave it to the mIL-1β form.

2. Substrate Specificity of ICE

a. Importance of Aspartate

The pIL-1β converting enzyme (ICE; EC 3.4.22.36) observed in the monocytic cells was characterized most clearly by its ability to cleave peptide bonds following an aspartate residue. This is an unusual cleavage site for a mammalian protease; only granzyme B, the serine protease characterized in T-cell cytolytic granules, was found to have a similar P1 specificity.[56,57] The native mIL-1β is formed by a cleavage following Asp116 in the pIL-1β, a residue that is totally conserved in all species of pIL-1β.[58,59] When this Asp in pIL-1β was mutated to an Ala, Asn, or to a D-Asp, the precursor could no longer be cleaved at that position.[51,60,61] Even replacement of that Asp with a similarly charged Glu resulted in a greatly diminished cleavage activity (Figure 1).[60,61] *In vitro* cleavage of pIL-1β by ICE also revealed that it contained a second cleavage site following Asp^{27}, generating a 28-kDa piece.[45,51] In the Asp116Ala mutation, only the 28-kDa species could be formed, and if that second Asp were mutated, very little cleavage of the pIL-1β could be found at all (see Figure 1).[61] Inhibition of cleavage by mutagenesis at either site does not influence cleavage at the other site; this independence of activity indicated that cleavage of pIL-1β is not processive.[61] Denaturation of the pIL-1β substrate had little effect on the specificity of the cleavage by ICE, indicating that the tertiary structure of the substrate has little effect on activity.[60]

b. *Amino Acid Specificity for ICE Activity*

The requirement for Asp in P1 was not the only peptide requirement for ICE cleavage since the enzyme does not cleave following other Asp residues in pIL-1β, in pIL-1α, or in a number of other proteins that were tested.[61] ICE substrate specificity studies performed on 10- or 14-mer spanning peptides (Asn-Glu-Ala-Tyr-Val-His-Asp-Ala-Pro-Val-Arg-Ser-Leu-Asn) indicated that the key residue conferring activity was a P4 hydrophobic group; truncations of the N-terminal to this residue had little effect on activity, whereas removal of the P4 residue led to loss of >99% of activity.[44,60] Substitutions of various amino acids at P4 indicated that Tyr, and to a lesser extent Phe, produced optimal recognition by ICE. In P3 Val was preferred, although a variety of acidic and hydrophobic amino acids could be substituted. While Ala was preferred in P2, this position permitted the greatest flexibility for substitution, so that substitutions such as Lys, Asp, and Phe still retained considerable activity.[62] In the P1′ position, small amino acids such as Gly and, to a lesser extent, Ala were preferred,[60,61] but truncation studies indicated that methylamine was sufficient and that no P1′ amino acid was even required.[44] Truncation studies C-terminal to the scissile bond confirmed that P′ amino acids had little effect on activity,[44] although substitution of the Pro at P2′ with Ala or Gly reduced the activity by 50%.[60,61] The best peptide substrate identified for ICE was Ac-Tyr-Val-Ala-Asp-NH-CH_3 with a V_{max}/K_m of 12 ($k_{cat}/K_m = 4 \times 10^5\ M^{-1}s^{-1}$) which was comparable with that of pIL-1β itself.[44]

c. *Assay of ICE*

Historically, ICE activity was measured by the cleavage of either recombinant pIL-1β[46,47] or ^{35}S-Met labeled *in vitro*-translated pIL-1β[51,61] followed by SDS-PAGE and autoradiography (Figure 1). While the labeled pIL-1β permitted great sensitivity of detection, quantitation of activity could be determined with only limited precision by comparison with a standard curve of increasing dilution of enzyme (Figure 2). Based upon the results of the peptide substrate studies (see above), a variety of peptides could be used as substrates with different ease and sensitivity of detection. The cleaved products could be separated by HPLC and quantitated by OD_{280} (utilizing the absorption of the product Tyr). Alternatively, because of the absence of an amino acid requirement at P1′, photometric leaving groups could be attached to the tetrapeptide and measurement could be made by visible wavelength absorbance, such as with a *p*-nitroanilide (pNA), or more

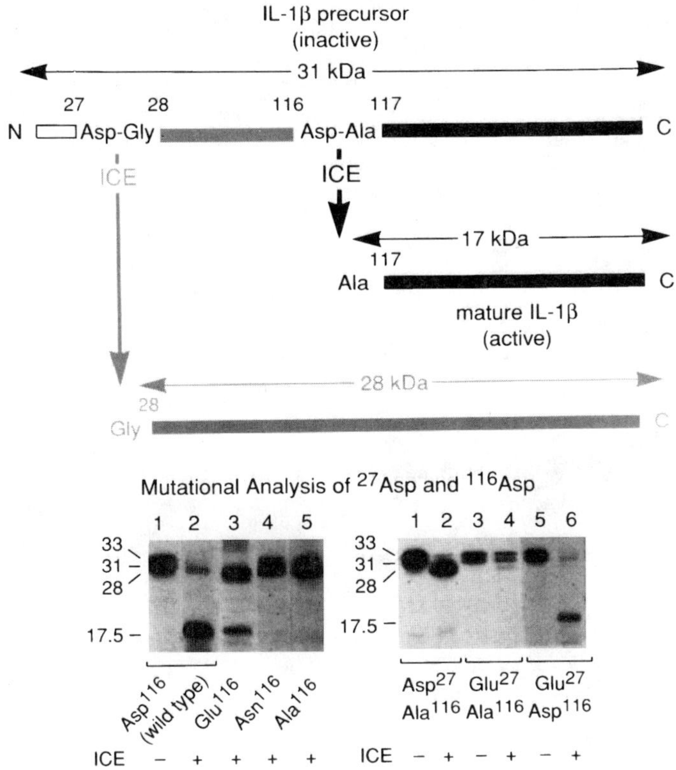

FIGURE 1. Cleavage of pIL-1β by ICE. The stick diagram of pIL-1β (top) indicates the formation of 17-kDa mIL-1β (shown in black) by ICE cleavage at Asp116. ICE cleaves at an alternative site in pIL-1β at Asp27 to generate the 28-kDa form (shown in gray). The effects on ICE cleavage of pIL-1β mutations are shown at the bottom. Asp116 and Asp27 were substituted with the indicated amino acids, and ^{35}S-Met labeled *in vitro*-translated pIL-1β was prepared as a substrate. The pIL-1β was incubated in the presence or absence of added ICE followed by SDS-PAGE and autoradiography. (Taken, in part, from Howard, A.D., Kostura, M.J., Thornberry, N., Ding, G.J.F., Limjuco, G., Weidner, J., Salley, J.P., Hogquist, K.A., Chaplin, D.D., Mumford, R.A., Schmidt, J.A., and Tocci, M.J., *J. Immunol.*, 147, 2964, 1991. With permission.)

sensitively by fluorescence such as with amino-4-methylcoumarin (AMC) of the released groups[44,63,64] (see Figure 2). While these relatively bulky photometric leaving groups have low k_{cat} values, they also have compensatory lower K_m values.[63] The tetrapeptide-pNA and AMC substrates are suitable for use in continuous spectrophotometric and fluorometric assays, respectively. A second continuous fluorometric assay has also been developed measuring resonance energy transfer with the substrate DABCYL-Tyr-Val-Ala-Asp-Ala-Pro-Val-EDANS.[65]

The different assays show different sensitivities to inhibition by salt with the natural pIL-1β substrate being the most salt sensitive (IC_{50} of about 40 m*M* NaCl, see Figure 2), an observation that early on enabled a distinction between ICE and elastase cleavage of pIL-1β.[51] In contrast, ICE cleavage of the tetrapeptide substrates is relatively insensitive to salt inhibition while the 14-mer peptide shows intermediate sensitivity (Miller, D.K. and Thornberry, N.A., unpublished observations).

MEASUREMENT OF ICE ACTIVITY

Substrate	Measurement		Sensitivity U/Assay	Salt Effect (IC_{50})	V_{max}/K_m
	Product	Technique			
^{35}S-Met proIL-1β (31K)	17.5K	SDS gels	0.05U	40mM	12
NEAYVHD–APVRSLN (14 mer)	NEAYVHD	HPLC Absorbance	25U	100mM	1.0
AcYVAD-AMC	AMC	HPLC/Plate Reader Fluorescence	1U	0% at 200mM	1.9
AcYVAD-pNA	pNA	HPLC/Plate Reader Absorbance	10U	ND*	4.4

* Not Determined

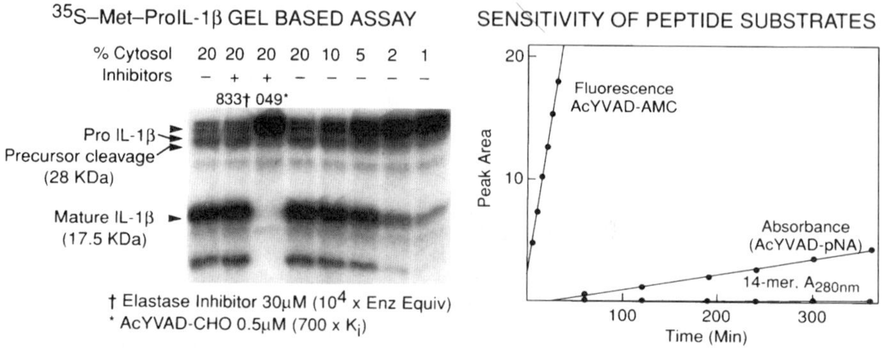

FIGURE 2. Characteristics of four different methods of measurement of ICE activity using ^{35}S-Met labeled *in vitro*-translated pIL-1β (see bottom left, gel based assay) or three separate peptide substrates Asn-Glu-Ala-Tyr-Val-His-Asp-Ala-Pro-Val-Arg-Ser-Leu-Asn (NEAYVHDAPVRSLN), Ac-Tyr-Val-Ala-Asp-amino 4-methylcoumarin (AcYVAD-AMC), and Ac-Tyr-Val-Ala-Asp-*p*-nitroanilide (AcYVAD-pNA). Indicated for each assay is the product that is measured, the technique used for measurement, the limits of sensitivity of ICE in units (1 U = 1 picomol/min at 25°C), the effect of of added NaCl on activity, and the relative activity (V_{max}/K_m). The relative sensitivity to detection of the various peptide products is shown graphically in the lower right. The gel-based assay in the lower left shows a titration of serial dilutions of ICE (% cytosol in the final assay) and the inhibition by L-709,049 (049, Ac-Tyr-Val-Ala-Asp-CHO) and the lack of inhibition by the elastase inhibitor L-680,833 (833, see Knight et al.[135]).

3. ICE Is a Cysteine Protease

ICE activity was found to be competitively inhibited by thiol alkylating agents such as *N*-ethyl maleimide (NEM) and iodoacetate but not by E-64, suggesting that it was an atypical cysteine protease.[44,60,62] Furthermore, ICE is resistant to inhibition by such classic serine protease inhibitors as phenyl methyl sulfonyl fluoride (PMSF), diisopropyl fluorophosphate (DFP), leupeptin, and the aspartyl protease inhibitor pepstatin.[44,60] Further evidence for its atypical nature was the observation that ICE was sensitive to the serine protease inhibitors *N*-tosyl-L-phenylalanine chloromethyl ketone (TPCK) and 3,4-dichloroisocoumarin.[66] Early inhibition studies indicated that while ICE activity was resistant to EDTA (indicating that ICE was not a metalloprotease), there was a puzzling observation of a slow onset of inhibition by 1,10-*(ortho)*-phenanthroline (OPA). The observation that this OPA-induced inhibition was enhanced by Cu^{2+} and inhibited by DTT indicated that the cause was really phenanthroline-metal-catalyzed oxidation of the catalytic cysteine[44,62] (see Reference 67).

The cysteine protease nature of ICE was confirmed by the reaction of ICE with reversible and irreversible inhibitors classically used for the identification and confirmation of cysteine proteases.[68] The peptide aldehyde inhibitor (Ac-Tyr-Val-Ala-Asp-CHO) has a K_i of about 0.8 n*M* and forms a slowly reversible ($t_{1/2}$ = 40 min) thiohemiacetal with the active-site cysteine.[44,66] In contrast, the analog (Ac-Tyr-D-Ala-Ala-Asp-CHO) was inactive (K_i = 1.5 μ*M*).[44] A specific diazomethylketone (Ac-Tyr-Val-Ala-Asp-$COCHN_2$ with a second order rate constant for inactivation of 1.6×10^4 $M^{-1}s^{-1}$) irreversibly inactivated ICE 10,000 times better than the truncated inhibitor carbobenzyloxy-Asp-$COCHN_2$). ICE is also covalently inhibited by cystamine (2-aminoethanethiol disulfide) and glutathione disulfide, an observation useful for the affinity purification of ICE (see below).[44]

The catalytic Cys of ICE reacts tenfold more rapidly with iodoacetate than do other Cys residues or DTT, an observation also supportive of the conclusion that ICE is a cysteine protease. Because of this, ^{14}C-iodoacetate was used to specifically label the catalytic Cys of partially purified ICE. Following purification of the labeled protein (see below), tryptic hydrolysis, and subsequent HPLC purification and sequencing, the catalytic Cys was determined to be Cys^{285}.[44,69] (See below.)

4. Types of ICE Inhibitors

In addition to the peptide aldehyde discussed above, two other classes of reversible ICE inhibitors were identified with lesser potency: peptide ketones and nitriles. For example, the Ac-Tyr-Val-Ala-Asp-CO-$(CH_2)_5$-Ph has a K_i of about 19 n*M*,[70] while the corresponding tetrapeptide nitrile has a K_i of 57 n*M*.[63] A fourth class of ICE inhibitors, which is irreversible in nature, is the β-substituted ketones such as the diazomethylketone described above. An important subclass is the peptide (acyloxy)methyl ketone inhibitors prepared following the lead of Krantz and colleagues for inhibition of cathepsin B and papain.[71,72] These are selective and effective *in vivo* inhibitors and are even more potent than the diazomethyl ketones; the Ac-Tyr-Val-Ala-Asp-$CH_2OC(O)$-[2,6$(CF_3)_2$]Ph, for example, has a second order rate constant for inactivation of 1×10^6 M^{-1} s^{-1}[73] (see also References 74 and 75). The rate of inactivation observed with this class of compound was so rapid that the rate-determining step in inactivation is binding, thus making the inactivation rate dependent upon the rate of diffusion of the compounds.[73] The specificity of these compounds for inhibition of ICE was confirmed by addition of a biotinylated version to crude THP.1 monocytic cell extracts in which the only protein labeled was ICE itself.[73]

An extremely useful protein inhibitor of ICE was the serpin crmA, a member of the family of proteins best known for inhibition of serine proteases.[76] CrmA (cytokine response modifier A), another of a series of IL-1 modifying proteins produced by pox viruses, is produced cytoplasmically and inhibits ICE activity by forming a tight complex with an equilibrium constant of inhibition of $<4 \times 10^{-12}$ *M*.[77,78] The serpin family classically displays a bait peptide region which is cleaved by the serine protease forming a covalent, SDS-stable complex. In the case of crmA and ICE, cleavage probably follows an Asp found in the bait region, but the resultant complex is not stable to SDS-PAGE, unlike most serpin/serine proteinase complexes. Since the crmA utilizes characteristics of a serine protease inhibitor for the inhibition of a cysteine protease, it is an example of a cross-class inhibitor.[78] It would be expected to inhibit serine proteases with the right specificity such as granzyme B.[57]

Other inhibitors of mIL-1β production in whole cell assays include compounds such as the bicyclic imidazoles[79] or a series of antioxidants.[80] These agents, however, are not

ICE inhibitors and are not unique inhibitors of IL-1β, but also inhibit the release of TNFα, IL-6, IL-8, and prostaglandins.

5. Purification of ICE

ICE is present as a relatively rare cytoplasmic protein in monocytic cells; only about 5000 copies of active ICE could be obtained per cell.[69] ICE was purified by two separate approaches: classical protein purification by sequential column fractionations using standard chromatographic techniques including ion exchange and reverse phase high-performance liquid chromatography (RP-HPLC) columns,[69,81,82] and secondly by affinity purification using an immobilized ICE substrate peptide aldehyde.[44,63] Three independent studies identified a protein that migrated at 22 kDa on SDS-PAGE and tracked with activity. In addition, the Merck group observed that a 10-kDa protein comigrated with the 22-kDa protein in an equimolar ratio in the active ICE fractions.[44] Mass spectral analyses of the purified proteins gave molecular weights of 19,866 and 10,248, resulting in their designation as p20 and p10, respectively.[44] In addition, active ICE samples often contained variable amounts of a 16-amino-acid N-terminally extended form of p20 (designated p22) that copurified with p20 on RP-HPLC.[69] The p20 protein contained the active-site Cys that was labeled with ^{14}C-iodoacetate,[44,69,81] and the biotinylated (acyloxy)methyl ketone.[73] Its importance was further emphasized by the observation that antibodies raised against the N-terminal region of p20 would immunoprecipitate ICE activity.[81,82]

Even though p10 does not contain the catalytic Cys or the His necessary for activity (see References 66 and 83), this chain is necessary for ICE activity. DEAE and sulfopropyl ion exchange chromatography of active THP.1 monocytic cell extracts indicated that, in addition to a peak of highly active ICE containing both p20 and p10, there were also two other fractions containing seemingly intact p20 but with differing activities and amounts of p10. One peak contained a 7-kDa (p7) C-terminally truncated fragment of p10 with substantially reduced activity while a second fraction contained no p10 and no measurable ICE activity.[69] (This cleaved form of p10 was also observed subsequently in a recombinant *E. coli* p45 expression system where p20/p7 was purified with a reduced activity[84] [Howard, A.D. et al., manuscript in preparation]). Further analysis indicated that the p20 in the fully active ICE peak migrated as a higher molecular weight complex with p10 on native gels, whereas the p20 fraction with no p10 did not, further emphasizing the importance of p10 in the formation of an active complex (Miller, D.K. and Raju, S.M., unpublished observations).

Active ICE could be purified over 100,000-fold directly from THP.1 cytosolic extracts by the use of a specific peptide aldehyde affinity column.[44,63] Because the P2 position of the peptide substrate allowed a great latitude in substitutions, an Ac-Tyr-Val-Lys-Asp-CHO peptide was prepared and coupled via the ε-amino residue of Lys to Sepharose. Active ICE was bound to the column and slowly released with the Ac-Tyr-Val-Ala-Asp-CHO, followed by destruction of the aldehyde with hydroxylamine, trapping of ICE as the inactive, stable glutathione conjugate and, finally, regeneration of the active ICE by incubation with DTT. This purification produces ICE with a specific activity of 1.8×10^6 units per milligram of enzyme (1 unit produces 1 pmol of AMC/min at 25°C at saturating Ac-Tyr-Val-Ala-Asp-AMC substrate concentrations[63]).

6. Cloning of ICE

Human ICE was cloned either by using probes derived from the subcloning of separate extensive PCR products from both p20 and p10[44] or from the N-terminal sequence of p20.[85] All clones cross hybridized with all of the p20 and p10 probes, indicating that both proteins are encoded on the same mRNA. The longest cDNA insert corresponded to a mRNA transcript of 1.5 kb which contained an open reading frame predicting a protein of 404 amino acids (M_r 45,158; termed p45; see Figures 3a and 4). N-terminal sequencing of the purified proteins indicated the sequences marking the beginning of the p20 and of the p10 within the p45, while fast-atom bombardment mass spectrometry identified the C-terminal sequence of the p20. These observations indicated that the p20 encompassed amino acids Asn120 to Asp297 of the p45 ICE precursor while the p10 encompassed amino acids Ala317 to His404[44] (see Figures 3a and 4). The active site Cys in the previously sequenced peptide was identified as Cys285 13 residues from the C-terminus of p20.

Thus, based upon the results of protein purification and cloning, the p45 ICE precursor is composed of four major domains: (1) the 13-kDa N-terminal pro domain through the Asp119 cleavage site, (2) the p20 domain through the Asp297 cleavage site, (3) the 2-kDa linker peptide through Asp316, and (4) the p10 (see Figures 3a and 4). Asp103 represents an alternative cleavage site within the precursor domain to generate the p22 protein; p22 was produced in roughly equal amounts to p20 during initial cleavage of the p45 precursor, but during purification of ICE it was reduced by continued autocatalytic cleavage to p20.[69,86] The p22 form is catalytically active, as based upon its ability to be labeled by a biotinylated (acyloxy)methyl ketone[87] (Yamin, T.T. et al., manuscript in preparation). All four cleavage sites in the human p45 ICE that generated the p22, p20, and p10 fragments that were found *in vitro* (marked by the N-terminus of p22, the N-terminus of p20, the C-terminus of p20, and the N-terminus of the p10) were distinguished by Asp-X cleavage sites, leading to the suggestion that these cleavages might occur autocatalytically.[44]

Using the human ICE as a probe, the mouse[88,89] and the rat[90] ICE cDNAs were isolated and sequenced. Both encode precursor ICE proteins (402 and 401 amino acids, respectively) of approximately 45 kDa containing the conserved active-site Cys and neighboring amino acids (Figure 3a). The sequence similarity between the murine and human ICE is highest in the p10 and p20 regions (81 and 62% identity, respectively) with fewer identities in the precursor region and in the p20/p10 spanning peptide (53 and 50%, respectively).[88,89]

It is unclear which particular molecular forms of natural murine and rat active ICE correspond to human p20 and p22. Of the four Asp-X cleavage sites defining the human ICE fragments found *in vitro,* only the sites 1, 3, and 4 are conserved in all three cloned species of ICE, and only the tetrapeptide cleavage site at the beginning of p10 (site 4) is totally conserved among these ICE forms. The murine and rat p45 forms contain no second site clearly comparable to the beginning of the human p20. The Asp103 cleavage site that marks the beginning of the p22 protein within the human ICE is conserved in murine and rat ICE, but the tetrapeptide recognition sequence is quite divergent (Figure 3a). Active murine ICE with a defined p20 can be generated by expression in *E. coli* of an artificially truncated version of murine p45 beginning with Asn119,[89] but no data were presented that indicated that such a cleavage could occur spontaneously from murine p45 analogous to that found with human ICE.[91]

a. *ICE RNA Expression*

ICE mRNA is constitutively expressed in many normal leukocytes and leukocyte cell lines. While monocytes contain the largest amount of mRNA, human ICE mRNA was also detected in peripheral blood lymphocytes, neutrophils, resting and activated peripheral blood T cells, the B cell lymphoblastoid line CB23 and, to a smaller extent, Raji cells, but not in HepG2 hepatoma cells.[44,85] Murine ICE mRNA was found in WEHI-3, in IC21, in RLM-11 and EL-4 T cells, and in J774, RAW8.1, and RAW24.7 macrophage cell lines,

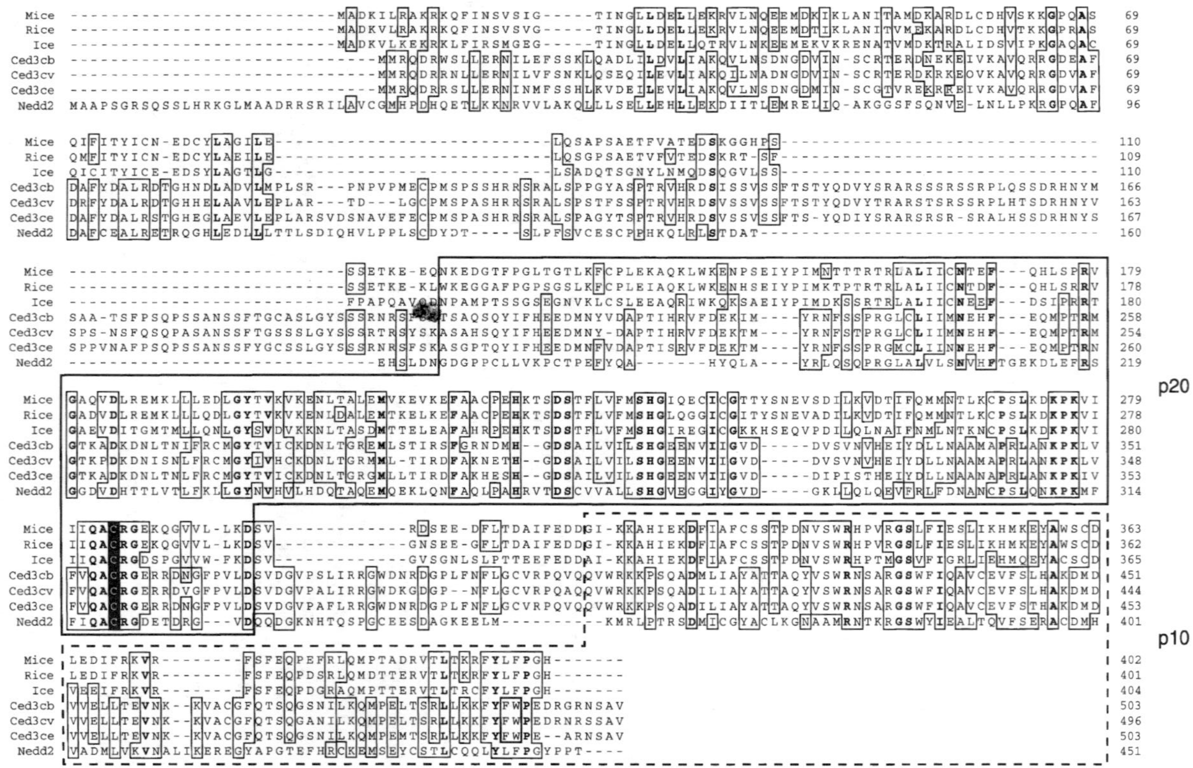

a

FIGURE 3. Alignment **a** and dendrogram **b** of the open reading frames of the three known ICE sequences, human,[44,85] mouse,[88,89] and rat,[90] with the three *Caenorhabditis* species *ced-3* sequences[98] and the murine fetal brain developmental gene *nedd2*[99] using the GCG analysis package.[136] In **a,** the p20 region and p10 region are indicated by solid and dotted line boxes, respectively. Identities between four or more amino acids are individually boxed. Identities between all seven sequences are shown in bold type and the active site Cys is shown in black with white type. The dendrogram in **b** shows graphically the overall similarity of the sequences to each other: the closer the relationship between two sequences, the farther to the right is placed the line connecting them. The results indicate that *nedd2,* while still quite dissimilar from the three species of *ced-3,* is still more closely related to the family of *ced-3* proteins than it is to the family of ICE proteins.

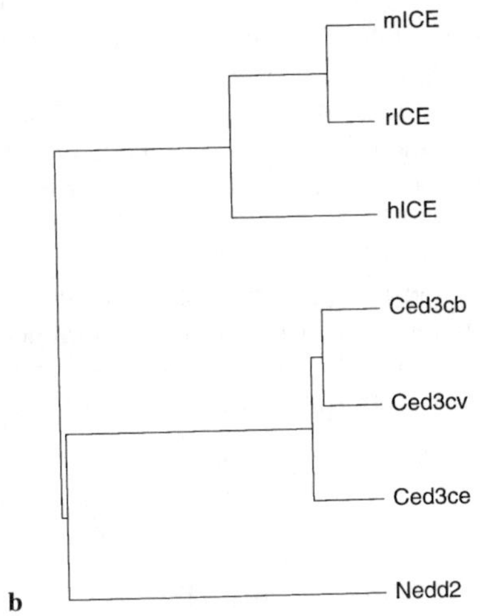

FIGURE 3. (continued)

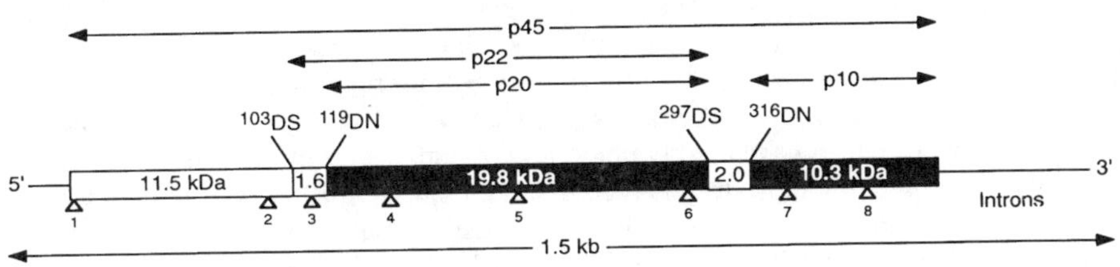

FIGURE 4. Stick diagram of the human ICE protein and gene showing the location of the p20 and p10 subunits (in black), the four Asp-X cleavage sites giving rise to p22, p20, and p10, and the locations of the eight introns found within the open reading frame.

but not in P388D1 cells or 3T3 fibroblasts.[88,89,91] Small amounts of mRNA hybridizing to human ICE were found in L cell fibroblasts and in spleen, heart, brain, placenta, and adrenal glands. Since no detectable IL-1β mRNA was found in the RLM-11 or L cells or in adrenal glands, either there were other ICE substrates present in those cells[88] or the hybridization techniques recognized mRNA of other genes closely related to ICE that might be present in those tissues (see below).

b. Expression of ICE

Proof that the cDNA identified by cloning encoded a protein that could correctly process pIL-1β was determined by transfection of the human cDNA into COS-7 cells. In cells simultaneously cotransfected with both human pIL-1β and human p45 ICE, mIL-1β was formed with the natural sequence (cleavage following Asp116 as determined by radiosequencing of the mIL-1β product), while no cleavage was found in cells lacking transfection by the ICE cDNA.[85] Alternatively, it was shown that extracts of cells transfected by the ICE cDNA could cleave ^{35}S-Met wt pIL-1β but not the ^{35}S-Met Asp116Ala pIL-1β mutant. The expressed ICE protein could also, like the wtICE, generate the 28-kDa fragment by a cleavage following Asp27. This recombinant activity was inhibited by the Ac-Tyr-Val-Ala-Asp-CHO inhibitor, but not by the inactive Ac-Tyr-

D-Ala-Ala-Asp-CHO analog.[44] An ICE active site mutant (Cys285Ala or Cys285Ser) had no pIL-1β cleavage activity.[92,93] ICE activity could be demonstrated when the 32-kDa wt ICE form (lacking the precursor domain, p32) of either the human or the mouse ICE was similarly transfected.[85,89] Thus, the presence of the precursor domain is not necessary for proper folding of ICE or its subsequent cleavage to the p20/p10 form.

Large amounts of active ICE (1:1 ratio of p20/p10) were successfully grown and purified by expression of p45 in both *E. coli* and *Baculovirus*/sf9 cell systems.[84,93] While active enzyme could be prepared from the expression of p32,[66,89] the expression of p20 and p10 individually or simultaneously in *E. coli* did not produce active enzyme, indicating that the larger ICE form was necessary for proper folding.[84] Indeed, refolding of the subunits was necessary to achieve proper ICE activity (see Reference 83).

c. *Location and Structure of the ICE Gene*

The human ICE gene *(IL-1bc)* was initially localized to chromosome 11q23 by *in situ* hybridization to normal human metaphase preparations and Southern blot analysis of rodent-human hybrids.[85] This location was later refined by fluorescence *in situ* hybridization to band location 11q22.2-q22.3.[94] This location lies near the closely linked genes of the D2 dopamine receptor *(DRD2)* and the neural cell adhesion molecule *(NCAM)* and to the matrix metalloproteinase gene cluster of stromelysin 1 and 2 (*STMY1* and *STMY2*) and fibroblast collagenase *(CLG)*.[94] This general location has also been found to be the site of rearrangement in a number of lymphomas and leukemias.[85] Murine *IL-1bc* was localized to the proximal region of chromosome 9 by interspecific backcross analysis and is closely linked to the *Mme* marker and somewhat more distantly to the *ETS-1* and *Thy-1* markers, both of which have their counterparts of human chromosome 11q23. The only known murine mutation close to this location is the curly whisker *(cw)* mutation which is associated in homozygous mice with an abnormal immune response.[88] It is unknown whether defects of IL-1β formation are involved in these diseases.

Both the human and the murine *IL-1bc* genes are very similar.[91,94] Both have a total of 10 exons and all the introns are in identical positions except for intron 9 in the 3′ flanking region. There is a 43% DNA sequence identity in the 5′-flanking region but a 72% sequence identity in the coding region. There is no close correlation between the locations of the introns and the four different regions of the p45 precursor (Figure 4). A portion of the human mRNA in which the 69-bp exon 3 has been spliced out has been found, but it is not certain if this is translated within the cell. Western blot analyses of human monocyte and THP.1 cell extracts to indicate, however, the presence of a small amount of a faster-migrating form of the ICE precursor (M_r of about 43 kDa) that behaves similarly to the p45. This protein may represent the alternatively spliced form of ICE (Yamin, T. T. et al., manuscript in preparation). No such alternative slicing has been found in murine cells.[91,94] Both the human and murine genes are single copy genes with a single initiation site found in the same location relative to the initiator methionine. Neither the human nor murine gene have a TATA box within the promoter region, but they share an identical 14-bp region encompassing the transcription start site.[91,94]

7. Similarly of ICE to Other Proteins

Recently, other proteins have been discovered with structures similar to ICE. Horvitz has found in *Caenorhabditis elegans* a number of genes that are essential for apoptosis, or programed cell death (see Reference 95 for a general review), of a large number of cells, chiefly neural cells, during development.[96] The most essential of these *C. elegans* genes for the induction of the killing is the *ced-3* gene.[97] The recent sequencing of the *ced-3* gene[98] has shown it to be highly homologous to ICE.[98] A sequence comparison of the three species of ICE with three *Caenorhabditis* species of *ced-3* indicates a relatively high sequence identity (Figure 3a), particularly in the p20 (25% identity) and p10 (24% identity) regions, whereas the precursor domain and the p20/p10 spanning peptide have much lower sequence identities (13 and 11%, respectively). The identities within the p20 polypeptide include the active site pentapeptide region as well as the catalytic His237 and other residues comprising the active site of ICE (see below).

A third member of this cysteine protease family has also been identified: *nedd2* is expressed primarily in fetal murine brain but is largely absent in adult brain suggesting that, like *ced-3,* it may also have a role in controlling neural apoptotic cell death.[99] The primary sequence of *nedd2* (Figure 3a) is closer to that of *ced-3* than is ICE (see Figure 3b). *Nedd2* is found on chromosome 6 near the T-cell β-chain, a different location from that of ICE (see above).[99] Cell lines such as neuroblastoma N18 and pheochromocytoma PC12 were found to contain *nedd2,* but there was no obvious effect on the ability of the cells to thrive, indicating that the protein was not normally active. In neural tissues *nedd2* was found closely associated with the expression of Bcl-2,[99] a protein that normally prevents the onset of apoptosis.[100–102] A similar antiapoptotic role is served in *Caenorhabditis* by *ced-9,* a protein with high sequence homology to Bcl-2.[103,104]

The ability for ICE, *ced-3,* and *nedd2* to induce apoptosis has been shown by a series of transfection experiments. In transfected rat-1 cells, p45 and p32 ICE and *ced-3* all induced apoptosis, while the ICE mutants Cys285Gly and Gly287Ser and the *ced-3* mutant Gly360Ser were all inactive.[105] Likewise, in NIH-3T3 cells wtNedd2 induced apoptosis while the active site mutant Cys319Gly was inactive.[99] Individual transfection of p20 or p10 of ICE or the *nedd2* counterparts produced no apoptotic activity[99,105] indicating that, as was seen for pIL-1β cleavage activity by ICE, both polypeptides were required. A number of other *ced-3* mutations also caused a loss of *in vivo* activity including Leu27Phe, Gly65Arg, Ala449Val, Glu483Lys, and Ser486Phe.[98] The first three of these residues involve residues conserved amongst all ICE forms, and the latter two involve residues conserved in both *ced-3* and *nedd2.* The first of these mutations comprised the first two of the seven conserved residues in the precursor domain (see Figure 3a). This suggests that there must be an essential role for the precursor domain in the functioning of this family of proteases independent of the catalytic p20 and p10 domains.

The apoptosis induced by transfection with any of this family of cysteine proteases could be prevented by the simultaneous expression of Bcl-2,[99,105] thus confirming the regulatory role of bcl-2 on activity. Similarly, the pox serpin *crmA* also could prevent either spontaneous apoptosis [106] or that induced by transfection of ICE or *ced-3.*[105] Cotransfection of both *Bcl-2* and *crmA* was not additive, indicating that both were probably acting on the same pathway of regulation, namely, prevention of activity of the cysteine protease.[106]

8. Structure of ICE

Recently two groups have solved the X-ray structure of the active p20/p10 human ICE complexed to either Ac-Tyr-Val-Ala-Asp-CHO[66] or to Ac-Tyr-Val-Ala-Asp-CMK (chloromethylketone)[83] from protein prepared by autoprocessing of the prodomain-free ICE (p32) or from refolded p20 and p10 subunits. They describe a novel protein structure topology. The active form of the enzyme is composed of two sets of p20/p10 heterodimers to form an overall homodimer. The p10 and p20 subunits are folded almost independently with interior regions of the β sheet and exterior, peripheral regions of the α helix (four of each in p20 and two of each in p10). The interface of p10 with each p20 is principally at the central β sheet region to form in the heterodimer an extended region of six strands of β sheet. The homodimer interface is along 2 opposing β sheets of p10 to form a continuous central region of 12 β strands with two-fold symmetry. Thus, p10 is largely hidden from view in the homodimer. In addition, the C-terminus of p20 and the N-terminus of p10 extend away from the heterodimer at opposite sides of the molecule to form additional complexes of β sheets with the respective N-terminus of p10 and C-terminus of p20 in other heterodimers.[66,83]

The homodimer has two active sites at opposite ends of the homodimer forming deep narrow grooves accessible to the solvent. The P1-Asp side chain of the inhibitor lies in a deep pocket in this groove, which is formed by the side chains of Arg179 and Gln283 from p20 and Arg341 and Ser347 from p10 — all of which act to provide hydrogen bonding to the carboxylate of the Asp. His237 lies above the catalytic Cys285 and interacts with the oxyanion of the substate, and the backbone nitrogen of Gly238 hydrogen bonds to the inhibitor ketone.[66,83] All seven of these residues are absolutely conserved in all members of the ICE family (see Figure 3a) suggesting that they all work by a common mechanism.

The remaining substrate binding sites of ICE are formed by the p10 subunit utilizing residues Val338, Trp340, His342, Pro343, and R383 — the latter three of which form the binding site for the critical P4 Tyr.[66] These residues are not all conserved in the *ced-3* and *nedd2* proteins, particularly those recognizing P4 (Figure 3a). Thus, it might be expected that while all the forms of ICE might be cysteine proteases that utilize a P1 Asp, they might recognize a different tetrapeptide sequence and hence act on different protein substrates.

Another key difference between ICE and the *ced-3* and *nedd2* family members is the divergence in structure along the p10 dimer-dimer interface. The X-ray structure indicates that the two heterodimers interact at residues 318–322 and 386–396 of the p10 polypeptide.[66] Whereas these residues are conserved in the three different species of ICE, this is not the case with *ced-3* and *nedd2.* In fact, despite the overall closer similarity of *ced-3* and *nedd2,* there is considerable divergence between them in this region. The region is obviously important to *ced-3* and *nedd2* activity since two of the missense mutations producing inactivity of the *ced-3* lie in this region. It is unclear, however, whether or not *ced-3* and *nedd2* would even form an active complex like that of ICE. Only the site 3 Asp is conserved amongst all the ICE forms and this occurs in a region of divergent sequence (Figure 3a). Furthermore, there is no evidence in *nedd2* of the autocatalytic activation that would even generate ICE-like active fragments.[99]

The observation that the active ICE structure is a homodimer confirms previous enzymatic evidence that ICE behaves as an oligomer. Dilution of active ICE caused a time-dependent loss of activity that was reversed upon reconcentration.[44] The mechanism of action for the noncompetitive inhibitors of ICE, gold thiomalate and auranofin, which bind to cysteine residues adjacent to the two-fold axis of the crystal along the interface in the p10 subunits, may be to force the homodimers apart and reduce activity.[66] Earlier observations that both substrates and inhibitors stabilize the active enzyme and prevent loss of activity through dilution[44] can now be interpreted as holding the p20 and p10 subunits together and presumably permitting the resultant heterodimer to secondarily associate into an active homodimer.

9. Location and Form of ICE in Cells

Initial experiments had shown that the ICE activity in monocytic cells was cytoplasmic and not membrane bound.[51] Analysis of whole monocytic cells and subcellular fractions for the presence of ICE by protein immunoblot indicated that only the p45 precursor form of ICE could be found, and it was found almost exclusively in the cytosolic fraction[86] in the same location as pIL-1β in monocytic cells.[27] *In vitro* assays

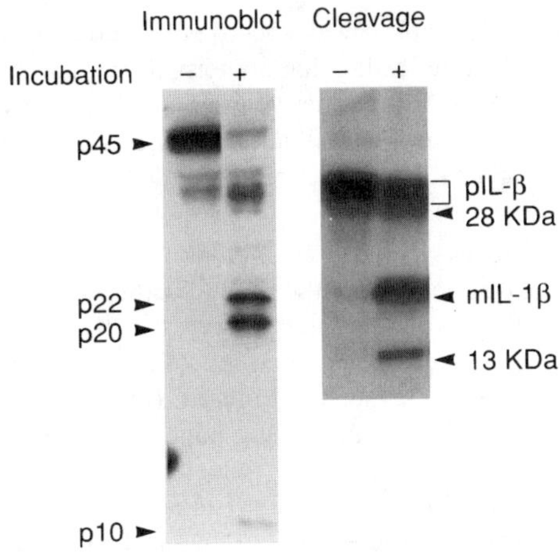

FIGURE 5. Form and activity of ICE present in THP.1 monocytic cell extracts immediately following cell breakage and fractionation (– incubation) or following a 60-min incubation at 30°C (+ incubation). The immunoblot on the left utilized an antibody raised against affinity purified p20/p10 ICE and shows that the p45 form is the only ICE form present in freshly broken cells until the p22, p20, and p10 forms are formed during the incubation. The pIL-1β cleavage assay on the right indicates that the p45-containing sample has little observable activity, and only when the p20/p10 form of the enzyme is formed is activity observed (From Ayala, J.M., Yamin, T.T., Egger, L.A., Chin, J., Kostura, M.J., and Miller, D.K., *J. Immunol.,* 153, 2592, 1994. With permission.)

have shown that this cytosolic form of p45 has no pIL-1β cleavage activity (Figure 5).[86] Only when p45 is broken down to the p20/p10 form of the enzyme can pIL-1β cleavage activity be seen, and no detectable amounts of this form are found in native or LPS-stimulated monocytes or THP.1 monocytic cells by immunoblot and ICE cleavage assays.[86] In contrast to these negative experiments, Singer et al.[107] found immunohistochemical evidence of active ICE in cytosol and on the microvilli of monocyte plasma membranes. Active ICE was detected both by antibodies to the p20/p10 active form as well as a biotin/strepavidin enhancement system for detection of ICE irreversibly labeled with a specific biotinylated (acyloxy)methyl ketone inhibitor (see above). The apparent discrepancy between the undetectable levels of active ICE as seen by biochemical analysis and the observable levels of active ICE as seen by electron microscopy may be a function of the large amplification of the EM system.

10. Autocatalytic Activation of p45 and the Cellular Origin of p20/p10 ICE

Precursor protein p45 is rapidly cleaved *in vitro* to generate p20/p10 ICE.[69,86,87] Active p20/p10 ICE itself can cleave p45, indicating that p45 is an ICE substrate[44] unlike many other proteins.[61] This appears, however, to be an unnatural, inefficient process since the concentration of ICE required for cleavage is relatively high and the pattern of cleavage products is considerably different from the naturally cleaved forms (Figure 5 and Reference 44). Indeed, more recent analysis has shown that p45 can cleave itself *in vitro* and that this autocatalysis is specifically inhibited by high concentrations of ICE inhibitors[87] (Yamin, T.T., Miller, D.K. et al., manuscript in preparation).

Since p45 is very stable in intact cells, either there must be inhibitors of cleavage which are rendered ineffective by cell breakage, or there are cellular activators of p45 breakdown which are brought into contact with p45 by cell breakage. No cellular modulators of p20/p10 pIL-1β cleavage activity could be observed when fresh monocytic cell extracts were added to affinity-purified active ICE, indicating that no normal cellular inhibitors or activators of this form of ICE were present.[86] Preliminary observations using p45 as a substrate did indicate, however, that undetermined cellular factors could accelerate the rate of p45 breakdown (Miller, D.K. and Molineaux, S.M., unpublished observations).

11. Effect of ICE Inhibitors on mIL-1β Release

To confirm that ICE is responsible for the normal cellular secretion of mIL-1β, human peripheral blood monocytes were stimulated to secrete mIL-1β with heat-killed *Staphylococcus aureus* in the presence of varying amounts of the peptide aldehyde inhibitor Ac-Tyr-Val-Ala-Asp-CHO.[44] As can be seen in Figure 6 (top), the release of mIL-1β to the medium was gradually prevented by increasing concentrations of the inhibitor with an IC_{50} of about 1 μ*M*. The amount of pIL-1β released to the medium correspondingly increased, producing no overall inhibition on the total amount of IL-1β secreted, only a shift in the relative proportions of precursor and mature forms. The precursor domain of IL-1β has been found extracellularly in pulse-chase experiments, suggesting that it too was probably secreted after cleavage.[108] A similar tetrapeptide (acyloxy)methyl ketone ICE inhibitor also prevented mIL-1β release (IC_{50} = 0.1 μ*M*) without affecting IL-6, TNFα, prostaglandin E_2, or lactate dehydrogenase release.[74] There was no effect by the inhibitor on either the time course of release of mIL-1β, or any inhibition of the amount of TNF, IL-6, or IL-8 released from the cells.[44,62] The inactive Ac-Tyr-D-Ala-Ala-Asp-CHO inhibitor and other cellular inhibitors such as E-64, pepstatin, phosphoramidon, or an elastase inhibitor had no effect on the mIL-1β release up to 100 μ*M*.[44] These results indicate that (1) processing of pIL-1β is independent of its secretion, (2) ICE is the protease responsible for the pIL-1β cleavage, and (3) the inhibition of mIL-1β production is specific for IL-1β independent of other cytokines.

12. Effect of ICE Inhibitors in Animal Models of Inflammation

The first results of the effectiveness of ICE inhibitors *in vivo* have been recently described. In a murine model of endotoxic shock where LPS was injected intraperitoneally (i.p.), the Ac-Tyr-Val-Ala-Asp-CHO inhibitor administered i.p. inhibited plasma IL-1β levels with an ED_{50} of 2 mg/kg and 100% inhibition at 100 mg/kg. In contrast, administration of an inactive analog, Ac-Tyr-Val-Ala-Asp-CH_2OH, had no effect. Circulating levels of IL-6 or IL-1α were unaffected by the inhibitor, indicating that the ICE inhibitor was specific for inhibition of mIL-1β.[109]

Similar compounds were also effective in a murine LPS-induced fever and in a rat carrageenan foot paw edema assay. A peptide (acyloxy)methyl ketone inhibitor of the type shown to be an irreversible ICE inhibitor[73] was found to be a potent inhibitor of the fever and the paw swelling, with potencies of 0.1 and 5 mg/kg, respectively.[74]

C. CONCLUSIONS

ICE has been shown to be an essential protease enabling the cleavage and, hence, activation of pIL-1β to mIL-1β, the major circulating form of IL-1. Inhibitors of this enzyme specifically inhibit cellular mIL-1β release without effects on other cytokines. While it is still early in the development of potent, orally active ICE inhibitors, even the current peptide-based inhibitors have been shown to be biochemically effective *in vivo*, and in certain animal models of inflammation they may also be therapeutically effective.

A key issue puzzling a number of laboratories studying inhibition of ICE is the observation that while many compounds have been found to be nanomolar inhibitors *in*

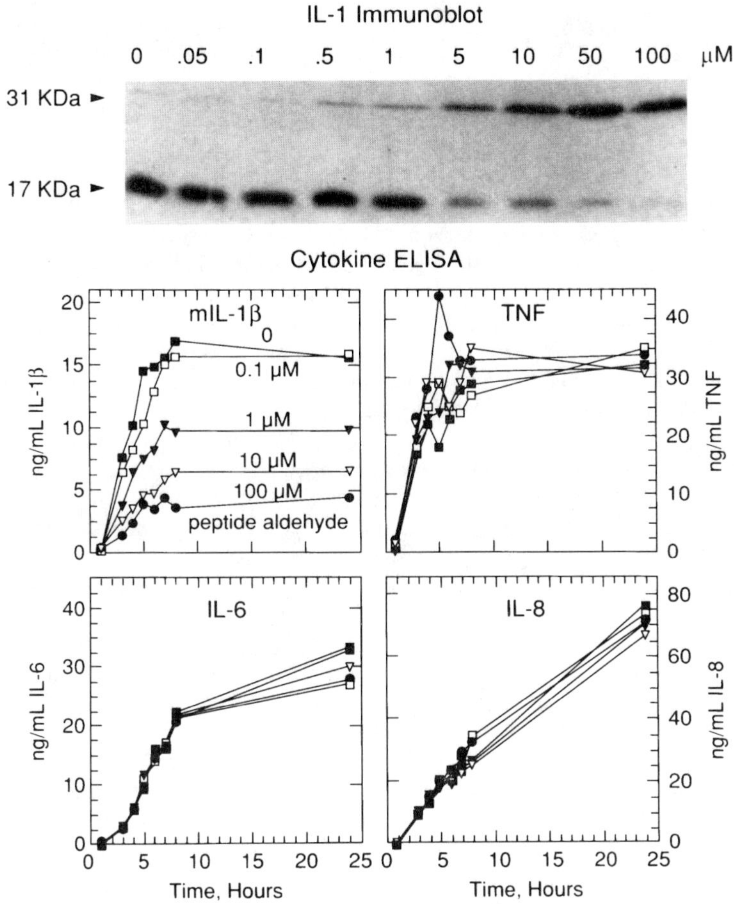

FIGURE 6. Inhibition of mIL-1β production by the Ac-Tyr-Val-Ala-Asp-CHO peptide inhibitor. Top, autoradiogram of the SDS-PAGE separation of IL-1β forms immunoprecipitated from media of ^{35}S-Met labeled human peripheral blood monocytes stimulated with *Staphylococcus aureus* for 5 h in the presence of the indicated concentration of the inhibitor. The two bands shown are the pIL-1β (31 kDa) and mIL-1β (17 kDa). Bottom, ELISA results indicating the time course of release of mIL-1β, TNFα, IL-6, and IL-8 from peripheral blood monocytes during incubation with the indicated concentration of the peptide aldehyde inhibitor. (IL-1 immunoblot from Thornberry, N.A., Bull, H.G., Calaycay, J.R. et al., *Nature,* 356, 768, 1992. With permission.) (ELISA results are from Miller, D.K., Calaycay, J.R., Chapman, K.T. et al., in *Immunosuppressive and Antiinflammatory Drugs,* Allison, A.C., Lafferty, K.J., and Fliri, H., Eds., New York Academy of Science, New York, 1993. With permission.)

vitro, all are little better than micromolar inhibitors in whole cell assays. It is not clear if this reduced potency is a function of cellular penetration of the peptide-based inhibitor, subsequent cellular inactivation of the inhibitor, or action against an unknown form of ICE present in the cells. Thus, the critical issues remaining in ICE research before effective inhibitors can be developed include such questions as what is the true active form of ICE in stimulated monocytic cells? How much of it is present? How long does it remain active? And in which region of the cell is it found? Furthermore, with the discovery that ICE is a member of a more broadly distributed cysteine protease family, it must be determined if ICE has more cellular functions than just activation of pIL-1β.

III. TNFα

A. ORIGIN OF ACTIVE TNF

TNFα (TNF) is a 17-kDa soluble protein originally identified independently as a tumor necrosis and regression factor and as an inducer of tissue wasting associated with chronic infection (cachexia). Like IL-1β, TNF is absent in uninduced monocytes, but after LPS/PMA induction large concentrations are released to the media.[110] The cloning of the cDNA for TNF revealed that it is synthesized as a 26-kDa precursor containing a 76-amino acid N-terminal extension.[111–113] Fractionation of TNF-producing cells indicated that essentially all of the TNF is found on membranes as the precursor, and that essentially no mature TNF (mTNF) is found inside the cells.[114]

Originally the precursor domain was thought to be an extralong signal sequence for secretion, but microsomes contained no protease which cleaved off this domain, indicating that mTNF is not produced as a normal secretory protein.[115] This was confirmed by the observation that the Met[1] and Met[6] residues of the pTNF remained associated with the TNF after synthesis, showing that no intracellular processing had occurred. Because a 20-residue hydrophobic domain flanked by hydrophilic domains was identified within the precursor domain, the 26-kDa precursor appeared to be an integral transmembrane protein. The orientation of the membrane protein was such that the N-terminal domain was cytoplasmic, making the pTNF a Type II membrane protein.[114] Much of that TNF could be found on the cell surface as measured both by flow cytometry[116] and cell surface iodination.[114] Like the mTNF found in cell supernatants, the cell surface pTNF form was active and could kill cells by contact.[114,117]

Pulse-chase studies indicated that the mTNF appeared in the cell supernatants coincident with the loss of the pTNF from cell membranes, each effect with a half-time of 20 to 30 min.[114,118–120] No intermediate forms of the cytokine were found in cells or in the media, indicating that TNF processing occurred in a single step.[114,121] If secretion of the mTNF was prevented by inhibitors (see below), the amounts of total cellular pTNF increased only about twofold, if at all (see Reference 122), with about a tenfold increase at the plasma membrane.[120,121] Larger increases in pTNF probably did not occur because of rapid recycling and downregulation.[119] This was confirmed by immunolocalization studies on inhibited cells showing that the pTNF was sequestered in the Golgi body.[120]

B. OBSERVATIONS OF A pTNF CONVERTASE ACTIVITY

mTNF was formed by cleavage of the Ala76 Val77 bond of the pTNF. Cleavage at this residue was originally observed by the incubation of pTNF with extracts of monocytes, neutrophils, and monocytic cells such as HL-60, THP.1, or Mono Mac 6 cells[120,121,123] or the macrophage line RAW 264.7.[124] pTNF substrates included labeled pTNF itself,[114,123,124] pTNF N-terminally linked with a peptide epitope,[121] or a glutathione-S-transferase linked pTNF.[122] The convertase activity was usually found associated with the particulate fractions of leukocyte membranes.[118,121,123]

1. Cleavage of pTNF by Serine Proteases

Experiments by the Kriegler group[123] indicated that the pTNF processing enzyme was a serine protease largely found in the leukocyte membranes but with some activity also found in granules, particularly neutrophil granules. This activity produced cleavage primarily at the Ala[76] site but also minor cleavages were found at the Val77-Arg78 and the Pro88-Val[89] sites. At least one of the proteases that could produce the correct pTNF cleavage was identified as PR-3,[123] a serine protease associated with elastase and cathepsin G in neutrophil granules and known to prefer cleavage of small aliphatic amino acids at P1.[125,126] This TNF cleavage activity could be inhibited by the nonapeptide Leu-Ala-Gln-Ala-Val-Arg-Ser-Ser-Ser or by a series of peptide diphenyl phosphonates that were irreversible inhibitors of serine proteases.[127] Trypsin was found to cleave the pTNF to 16-kDa and 13.5-kDa forms, but these cleavage products were tenfold less potent than the natural mTNF.[122]

Kim et al. also found that membrane extracts of RAW 264.7 cells contained a membrane-associated pTNF cleavage activity that could be partially inhibited by a variety of serine protease inhibitors — results that suggested that at least some pTNF converting activity could be generated by serine proteases.[124] Furthermore, production of LPS in mice could be inhibited by pretreatment with the serine protease inhibitor α_1-antitrypsin.[128] Consistent with the belief that a serine protease mediated release of mTNF was the observation that the serine protease inhibitor *p*-toluenesulfonyl-L-arginine methyl ester (TAME) inhibited mTNF release without any effect on the amount of TNF mRNA levels.[129] It was later argued, however, that some serine protease inhibitors, such as dichlorisocoumarin, that blocked TNF production inhibited the NF-κβ-induced transcription of TNF and other cytokines, thus indirectly reducing the amount of mTNF produced in these inhibited cell cultures[120] (see also Reference 130).

2. Matrix Metalloproteinase Cleavage of pTNF

More recently, three groups have identified a matrix metalloproteinase (MMP)-like activity responsible for release of cell surface TNF.[120–122] Partial purification of this activity by DEAE ion exchange chromatography and wheat germ agglutinin-agarose chromatography indicated that it was resistant to inhibition by the serine protease inhibitors PMSF and α_1-proteinase inhibitor, the cysteine protease inhibitor iodoacetate, the aspartyl proteinase inhibitor pepstatin, and the metalloprotease inhibitors phosphoramidon, captopril, and tissue inhibitor of metalloproteases (TIMP-2).[121,122] Whole cell assays indicated further that the proteolytic activity was also resistant to the serine protease inhibitors aprotinin, elastinal, and leupeptin; to the cysteine protease inhibitor E64; and to the aminopeptidase inhibitor Bestatin.[122] When assayed in the presence of several serine protease inhibitors, THP.1 cell membranes were found to have an EDTA-inhibitable pTNF cleavage activity which could be restored by Zn^{2+}, but not Ca^{2+}.[121] A chimeric form of mTNF fused to the IFNγ signal sequence product via a Gly-Val cleavage site was not cleaved by the metalloproteinase,[120] again showing its difference from the activity of the signal peptidase.[115]

Incubation of pTNF with purified MMPs such as PUMP (matrilysin, MMP-7), stromelysin-1 (MMP-3), and collagenase (MMP-1) produced mTNF with the identical N-terminal sequence and specific activity as the authentic mTNF standard. The 72- and 92-kDA gelatinases (MMP-9 and MMP-2) gave similar results but were less potent in their cleavage. On the other hand, the bacterial metalloprotease, thermolysin, cleaved pTNF to 17- and 14-kDa bands, but they were ten-fold less potent than the TNF produced by the mammalian MMPs.[122] While no published structure of a TNF-specific membrane MMP is available, a recent report has indicated the presence of at least one membrane-bound form.[131,132]

C. SPECIFIC INHIBITION OF mTNF RELEASE FROM CELLS BY MATRIX METALLOPROTEINASE INHIBITORS

A variety of hydroxamates known to inhibit matrix metalloproteinases were found to inhibit the pTNF cleavage by the MMPs and to specifically inhibit cellular release of mTNF. BB-2284 [3R-(3-*tert*-butoxycarbonyl-1S-methylcarbamoyl-propylcarbamoyl)-2S-(4-hydroxy-phenylsulphanylmethyl)-5-methyl-hexanohydroxamic acid] completely prevented the purified MMPs from *in vitro* cleavage of the pTNF (see above). At the same time BB-2284 inhibited mTNF release from LPS-stimulated human whole blood with an IC_{50} of 400 n*M*, whereas at 100 μ*M* it had no effect on the amount or time course of mIL-1β or IL-8 release.[122] The compound *N*-{D,L-[2-(hydroxyaminocarbonyl)methyl]-4-methylpentanoyl}-L-3-(2′-naphthyl)-alanyl-L-alanine,2-aminoethyl amide (Compound II) blocked TNF production in THP.1 and peripheral blood T cells with an IC_{50} of about 50 to 100 μ*M*, but had little effect on the release of IL-8 or lymphotoxin-α from the cells. TNF message levels were unaffected by the compound. In the presence of 200 μ*M* of Compound II, the amount of cell surface pTNF as measured by FACS analysis increased over 30-fold, consistent with its ability to block cell surface processing of pTNF.[121] A third hydroxamate, GI-129471, inhibited the release of mTNF from LPS-stimulated human peripheral blood monocytes, THP.1 and Mono Mac 6 monocytic cells, OVCA 432 ovarian tumor cells, EHM B-cells, and Jurkat T-cells — all at IC_{50}s of 100 to 200 n*M*. At the same time, there was essentially no inhibition of mIL-1β, IL-6, or IL-2 protein release from, or TNF or IL-1β message accumulation within, the various lines. Furthermore, the processing of other membrane-anchored proteins such as CSF-1, TGFα, and transferrin receptor was unaffected by GI-129471.[120]

D. EFFECTS OF TNF PROCESSING INHIBITORS *IN VIVO*

The hydroxamic acid inhibitors were effective inhibitors of TNF production *in vivo*. BB-2116,an analog of BB-2284 with an IC_{50} on cells of 230 n*M*, gave >95% inhibition with a single i.v. dose of 1 mg/kg in rats,[122] while GI-129471 injected s.c. gave a maximal 80% inhibition at 20 mg/kg in mice.[120] The time of onset for inhibition of mTNF release was very rapid: BB-2116 at 1 mg/kg gave 80% inhibition of TNF production when administered 50 min after the initial injection of LPS and only 10 min before measurement.[122] The *in vivo* inhibition of TNF production enabled mice to survive LPS-induced lethality. Injection of 0.5 mg of Compound II s.c. into mice inhibited TNF production by 84% and enabled a 13% survival. At a 1.5 mg injection dose, 100% survival was obtained.[121]

A similar therapeutic efficacy related to the decrease in circulating TNF may be operating in humans with leprosy. The acute inflammation reaction associated with erythema nodosum leprosum is characterized by elevated serum TNF and IL-1β levels.[133] Treatment of the reaction with thalidomide results in a reduction in TNF levels and an amelioration of symptoms *In vitro,* thalidomide inhibited mTNF formation without affecting secreted levels of IL-1β, IL-6, or GM-CSF or general protein synthesis.[134] Whether or not the thalidomide is inhibiting the MMP associated with TNF convertase activity, as described above, is not known.

E. CONCLUSIONS

These observations suggest that inhibition by certain inhibitors can both substantially reduce circulating mTNF levels and prevent inflammatory states. It is not clear if *in vivo* a single enzyme, such as a specific matrix metalloproteinase, releases the mTNF from its cell surface pTNF form or whether a number of enzymes do so. The compounds recently described and discussed above inhibit a number of MMPs and may be efficacious because they inhibit a broad range of MMP activities. Whether or not inhibition of a specific single protease will effectively reduce *in vivo* TNF will not be known until the specific MMP is identified and unique inhibitors to it are developed. It is possible that in certain inflammatory conditions other proteases may exhibit TNF convertase activity. Such serine proteases as PR-3 or elastase may, for example, release mTNF locally from the pTNF exposed on the cell surface.

REFERENCES

1. **Dinarello, C. A.,** Interleukin-1 and tumor necrosis factor: effector cytokines in autoimmune diseases, *Sem. Immunol.,* 4, 133, 1992.
2. **Aggarwal, B. B. and Vilcek, J.,** Tumor necrosis factors. Structure, function, and mechanism of action, in *Immunology Series,* 56, Rose, N. R. and Trnka, Z., Eds., Marcel Dekker, New York, 1992.
3. **Dayer, J. M. and Fenner, H.,** The role of cytokines and their inhibitors in arthritis, *Bailliere Clin. Rheum.,* 6, 485, 1992.
4. **Oppenheim, J. J., Kovacs, E. J., Matsushima, K., and Durum, S. K.,** There is more than one interleukin-1, *Immunol. Today,* 7, 45, 1986.
5. **Grassi, J., Roberge, C. J., Frobert, Y., Pradelles, P., and Poubelle, P. E.,** Determination of IL-1α, IL-1β, and IL-2 in biological media using specific enzyme immunometric assays, *Immunol. Rev.,* 119, 125, 1991.
6. **Aotsuka, S., Kakamura, K., Nakano, T., Kawakami, M., Goto, M., Okawa-Takatsuji, M., Kinoshita, M., and Yokohari, R.,** Production of intracellular and extracellular interleukin-1α and interleukin-1β by peripheral blood monocytes from patients with connective tissue diseases, *Ann. Rheum. Dis.,* 50, 27, 1991.
7. **Geiger, T., Towbin, H., Cosenti-Vargas, A. O. Z., Arnold, J., Rordorf, C., Glatt, M., and Vosbeck, K.,** Neutralization of interleukin-1β activity *in vivo* with a monoclonal antibody alleviates collagen-induced arthritis in DBA/1 mice and prevents the associated acute-phase response, *Clin. Exp. Rheumatol.,* 11, 515, 1993.
8. **Van den Berg, W. B., Joosten, L. A. B., Helsen, M., and van de Loo, F. A. J.,** Amelioration of established murine collagen-induced arthritis with anti-IL-1 treatment, *Clin. Exp. Immunol.,* 95, 237, 1994.
9. **Spriggs, M. K., Hruby, D. E., Maliszewski, C. R., Pickup, D. J., Sims, J. E., Buller, R. M. L., and VanSlyke, J.,** Vaccinia and cowpox viruses encode a novel secreted interleukin-1-binding protein, *Cell,* 71, 145, 1992.
10. **Alcami, A. and Smith, G. L.,** A soluble receptor for interleukin-1β encoded by vaccinia virus: a novel mechanism of virus modulation of the host response to infection, *Cell,* 71, 153, 1992.
11. **Blyden, G. and Handschumacher, R. E.,** Purification and properties of human lymphocyte activating factor (LAF), *J. Immunol.,* 118, 1631, 1977.
12. **Mizel, S. B.,** Physicochemical characterization of lymphocyte-activating factor (LAF), *J. Immunol.,* 122, 2167, 1979.
13. **Simon, P. L. and Willoughby, W. F.,** The role of subcellular factors in pulmonary immune functions: physicochemical characterization of two distinct species of lymphocyte-activating factor produced by rabbit alveolar macrophages, *J. Immunol.,* 126, 1534, 1981.
14. **Schmidt, J. A., Mizel, S. B., Cohen, D., and Green, I.,** Interleukin-1, a potential regulator of fibroblast proliferation, *J. Immunol.,* 128, 2177, 1982.
15. **Durum, S. K., Schmidt, J. A., and Oppenheim, J. J.,** Interleukin-1: an immunological perspective, *Annu. Rev. Immunol.,* 3, 263, 1985.
16. **Van Damme, J., De Ley, M., Opdenakker, G., Billiau, A., and De Somer, P.,** Homogeneous interferon-inducing 22K factor is related to endogenous pyrogen and interleukin-1, *Nature,* 314, 266, 1985.

17. **Cameron, P., Limjuco, G., Rodkey, J., Bennett, C., and Schmidt, J. A.,** Amino acid sequence analysis of human interleukin-1 (IL-1). Evidence for biochemically distinct forms of IL-1, *J. Exp. Med.,* 162, 790, 1985.
18. **March, C. J., Mosley, B., Larsen, A., Cerretti, D. P., Braedt, G., Price, V., Gillis, S., Henney, C. S., Kronheim, S. R., Grabstein, K., Conlon, P. J., Hopp, T. P., and Cosman, D.,** Cloning, sequence and expression of two distinct human interleukin-1 complementary DNAs, *Nature,* 315, 641, 1985.
19. **Auron, P. E., Webb, A. C., Rosenwasser, L. J., Mucci, S. F., Rich, A., Wolff, S. M., and Dinarello, C. A.,** Nucleotide sequence of human monocyte interleukin-1 precursor cDNA, *Proc. Natl. Acad. Sci. U.S.A.,* 81, 7907, 1984.
20. **Giri, J. G., Lomedico, P. T., and Mizel, S. B.,** Studies on the synthesis and secretion of interleukin-1. I. A 33,000 molecular weight precursor for interleukin-1, *J. Immunol.,* 140, 343, 1985.
21. **Matsushima, K., Copeland, T. D., Onozaki, K., and Oppenheim, J. J.,** Purification and biochemical characterization of two distinct human interleukins-1 from the myeloblastic THP.1 cell line, *Biochemistry,* 25, 3424, 1986.
22. **Limjuco, G., Galuska, S., Chin, J., Cameron, P., Boger, J., and Schmidt, J. A.,** Antibodies of predetermined specificity to the major charged species of human interleukin-1, *Proc. Natl. Acad. Sci. U.S.A.,* 83, 3972, 1986.
23. **DeChiara, T. M., Young, D., Semionow, R., Stern, A. S., Batula-Bernardo, C., Fiedler-Nagy, C., Kaffka, K. L., Kilian, P. L., Yamazaki, S., Mizel, S. B., and Lomedico, P. T.,** Structure-function analysis of murine interleukin-1: biologically active polypeptides are at least 127 amino acids long and are derived from the carboxyl terminus of a 270-amino acid precursor, *Proc. Natl. Acad. Sci. U.S.A.,* 83, 8303, 1986.
24. **Mosley, B., Urdal, D. L., Prickett, K. S., Larsen, A., Cosman, D., Conlon, P. J., Gillis, S., and Dower, S. K.,** The interleukin-1 receptor binds the human interleukin-1α precursor but not the interleukin-1β precursor, *J. Biol. Chem.,* 262, 2941, 1987.
25. **Fenton, M. J., Vermeulen, M. W., Clark, B. D., Webb, A. C., and Auron, P. E.,** Human pro-IL-1β gene expression in monocytic cells is regulated by two distinct pathways, *J. Immunol.,* 140, 2267, 1988.
26. **Bayne, E. K., Rupp, E. A., Limjuco, G., Chin, J., and Schmidt, J. A.,** Immunocytochemical detection of interleukin-1 within stimulated human monocytes, *J. Exp. Med.,* 163, 1267, 1986.
27. **Singer, I. I., Scott, S., Hall, G. L., Limjuco, G., Chin, J., and Schmidt, J. A.,** Interleukin-1β is localized in the cytoplasmic ground substance but is largely absent from the Golgi apparatus and plasma membranes of stimulated human monocytes, *J. Exp. Med.,* 167, 389, 1988.
28. **Bakouche, O., Brown, D. C., and Lachman, L. B.,** Subcellular localization of human monocyte interleukin-1: evidence for an inactive precursor molecule and a possible mechanism for IL-1 release, *J. Immunol.,* 138, 4249, 1987.
29. **Hazuda, D. J., Lee, J. C., and Young, P. R.,** The kinetics of interleukin-1 secretion from activated monocytes. Differences between interleukin-1α and interleukin-1β, *J. Biol. Chem.,* 263, 8473, 1988.
30. **Auron, P. E., Warner, S. J., Webb, A. C., Cannon, J. G., Bernheim, H. A., McAdam, K. J., Rosenwasser, L. J., LoPreste, G., Mucci, S. F., and Dinarello, C. A.,** Studies on the molecular nature of human interleukin-1, *J. Immunol.,* 138, 1447, 1987.
31. **Young, P. R., Hazuda, D. J., and Simon, P. L.,** Human interleukin-1β is not secreted from hamster fibroblasts when expressed constitutively from a transfected cDNA, *J. Cell Biol.,* 107, 447, 1988.
32. **Hogquist, K. A., Unanue, E. R., and Chaplin, D. D.,** Interleukin-1 is processed and released during apoptosis, *Proc. Natl. Acad. Sci. U.S.A.,* 88, 8485, 1991.
33. **Chin, J. and Kostura, M. J.,** Dissociation of IL-1β synthesis and secretion in human blood monocytes stimulated with bacterial cell-wall products, *J. Immunol.,* 151, 5574, 1993.
34. **Beuscher, H. U., Gunther, C., and Rollinghoff, M.,** IL-1β is secreted by activated murine macrophages as biologically inactive precursor, *J. Immunol.,* 144, 2179, 1990.
35. **Hogquist, K. A., Unanue, E. R., and Chaplin, D. D.,** Release of IL-1 from mononuclear phagocytes, *J. Immunol.,* 147, 2181, 1991.

36. **Lepe-Zuniga, J. L. and Gery, I.,** Production of intra- and extracellular interleukin-1 (IL-1) by human monocytes, *Clin. Immunol. Immunopathol.,* 31, 222, 1984.
37. **Wewers, M. D. and Herzyk, D. J.,** Alveolar macrophages differ from blood monocytes in human IL-1β release, *J. Immunol.,* 143, 1635, 1989.
38. **Herzyk, D. J., Berger, A. E., Allen, J. N., and Wewers, M. D.,** Sandwich ELISA formats designed to detect 17 kDa IL-1β significantly underestimate 35 kDa IL-1β, *J. Immunol., Methods,* 148, 243, 1992.
39. **Allen, J. N., Herzyk, D. J., Allen, E. D., and Wewers, M. D.,** Human whole blood interleukin-1-β production. Kinetics, cell source, and comparison with TNF-α, *J. Lab. Clin. Med.,* 119, 538, 1992.
40. **Lonnemann, G., Endres, S., Van der Meer, J. W. M., Cannon, J. G., Koch, K. M., and Dinarello, C. A.,** Differences in the synthesis and kinetics of release of interleukin-1α, interleukin-1β, and tumor necrosis factor from human mononuclear cells, *Eur. J. Immunol.,* 19, 1531, 1989.
41. **Matsushima, K., Taguchi, M., Kovacs, E. J., Young, H. A., and Oppenheim, J. J.,** Intracellular localization of human monocyte associated interleukin-1 (IL-1) activity and release of biologically active IL-1 from monocytes by trypsin and plasmin, *J. Immunol.,* 136, 2883, 1986.
42. **Endres, S., Cannon, J. G., Ghorbani, R., Dempsey, R. A., Sisson, S. D., Lonnemann, G., Van der Meer, J. W. M., Wolff, S. M., and Dinarello, C. A.,** *In vitro* production of IL-1β, IL-1α, TNF and IL-2 in healthy subjects: distribution, effect of cyclooxygenase inhibition and evidence of independent gene regulation, *Eur. J. Immunol.,* 19, 2327, 1989.
43. **Black, R., Kronheim, S., Sleath, P., Greenstreet, T., Duke, Virca, G., March, C., and Kupper, T.,** The proteolytic activation of interleukin-1β, in *Progress in Inflammation Research and Therapy,* Birkhauser Verlag, Basel, 1991, 85.
44. **Thornberry, N. A., Bull, H. G., Calaycay, J. R. et al.,** A novel heterodimeric cysteine protease is required for interleukin-1β processing in monocytes, *Nature,* 356, 768, 1992.
45. **Black, R. A., Kronheim, S. R., Sleath, P. R., Greenstreet, T. A., Van Ness, K. P., Glackin, P. J., Griffin, A., Hendrickson, R. C., Merriam, J. A., and March, C. J.,** Identification of a protease that processes interleukin-1β, in *Molecular and Cellular Biology of Cytokines,* Oppenheim, J. J., Powanda, M. C., Kluger, M. J., and Dinarello, C. A., Eds., Wiley-Liss, New York, 1990, 69.
46. **Black, R. A., Kronheim, S. R., Cantrell, M., Deeley, M. C., March, C. J., Prickett, K. S., Wignall, J., Conlon, P. J., Cosman, D., Hopp, T. P., and Mochizuki, D. Y.,** Generation of biologically active interleukin-1β by proteolytic cleavage of the inactive precursor, *J. Biol. Chem.,* 263, 9437, 1988.
47. **Hazuda, D., Webb, R. L., Simon, P., and Young, P.,** Purification and characterization of human recombinant precursor interleukin-1β, *J. Biol. Chem.,* 264, 1689, 1989.
48. **Hazuda, D. J., Strickler, J., Kueppers, F., Simon, P. L., and Young, P. R.,** Processing of precursor interleukin-1β and inflammatory disease, *J. Biol. Chem.,* 265, 6318, 1990.
49. **Mizutani, H., Schechter, N., Lazarus, G., Black, R. A., and Kupper, T. S.,** Rapid and specific conversion of precursor interleukin-1β (IL-1β) to an active IL-1 species by human mast cell chymase, *J. Exp. Med.,* 174, 821, 1991.
50. **Kobayashi, Y., Oppenheim, J. J., and Matsushima, K.,** Human pre-interleukin-1α and β: structural features revealed by limited proteolysis, *Chem. Pharm. Bull.,* 39, 1513, 1991.
51. **Kostura, M. J., Tocci, M. J., Limjuco, G., Chin, J., Cameron, P., Hillman, A. G., Chartrain, N. A., and Schmidt, J. A.,** Identification of a monocyte specific pre-interleukin-1β convertase activity, *Proc. Natl. Acad. Sci. U.S.A.,* 86, 5227, 1989.
52. **Black, R. A., Kronheim, S. R., and Sleath, P. R.,** Activation of interleukin-1β by a co-induced protease, *FEBS Lett.,* 247, 386, 1989.
53. **Mizutani, H., Black, R., and Kupper, T. S.,** Human keratinocytes produce but do not process pro-interleukin-1 (IL-1) beta. Different strategies of IL-1 production and processing in monocytes and keratinocytes, *J. Clin. Invest.,* 87, 1066, 1991.
54. **Fuhlbrigge, R. C., Fine, S. M., Unanue, E. R., and Chaplin, D. D.,** Expression of membrane interleukin-1 by fibroblasts transfected with murine pro-interleukin-1α cDNA, *Proc. Natl. Acad. Sci. U.S.A.,* 85, 5649, 1988.
55. **Howard, A. D., Chartrain, N., Ding, G. F., Kostura, M. J., Limjuco, G., Schmidt, J. A., and Tocci, M. J.,** Probing the role of interleukin-1β convertase in interleukin-1β secretion, *Agents Actions,* Suppl. 35, 77, 1991.

56. **Poe, M., Blake, J. T., Boulton, D. A., Gammon, M., Sigal, N. H., Wu, J. K., and Zweerink, H. J.,** Human cytotoxic lymphocyte granzyme B, *J. Biol. Chem.,* 266, 98, 1991.
57. **Odake, S., Kam, C., Narasimhan, L., Poe, M., Blake, J. T., Krahenbuhl, O., Tschopp, J., and Powers, J. C.,** *Biochemistry,* 30, 2217, 1991.
58. **Cannon, J. G., Clark, B. D., Wingfield, P., Schmeissner, U., Losberger, C., Dinarello, C. A., and Shaw, A. R.,** Cloning, expression, biologic properties, and transcription during endotoxemia, *J. Immunol.,* 142, 2299, 1989.
59. **Schmidt, J. A. and Tocci, M. J.,** Interleukin-1, in *The Handbook of Experimental Pharmacology. I. Peptide Growth Factors and their Receptors,* Sporn, M. and Roberts, A., Eds., Springer-Verlag, Berlin, 1990, 473.
60. **Sleath, P. R., Hendrickson, R. C., Kronheim, S. R., March, C. J., and Black, R. A.,** Substrate specificity of the protease that processes human interleukin-1 beta, *J. Biol. Chem.,* 265, 14526, 1990.
61. **Howard, A. D., Kostura, M. J., Thornberry, N., Ding, G. J. F., Limjuco, G., Weidner, J., Salley, J. P., Hogquist, K. A., Chaplin, D. D., Mumford, R. A., Schmidt, J. A., and Tocci, M. J.,** IL-1 converting enzyme requires aspartic acid residues for processing of the IL-1β precursor at two distinct sites and does not cleave 31 kDa IL-1α, *J. Immunol.,* 147, 2964, 1991.
62. **Miller, D. K., Calaycay, J. R., Chapman, K. T., Howard, A. D., Kostura, M. J., Molineaux, S. M., and Thornberry, N. A.,** The IL-1β converting enzyme as a therapeutic target, in *Immunosuppressive and Antiinflammatory Drugs,* Allison, A. C., Lafferty, K. J., and Fliri, H., Eds., New York Academy of Sciences, New York, 1993, 133.
63. **Thornberry, N.,** Interleukin-1β converting enzyme, *Methods Enzymol.,* 244, 615, 1994.
64. **Reiter, L. A.,** Peptidic *p*-nitroanilide substrates of interleukin-1β-converting enzyme, *Int. J. Peptide Protein Res.,* 43, 87, 1994.
65. **Pennington, M. W. and Thornberry, N. A.,** Synthesis of a fluorogenic interleukin-1β converting enzyme substrate based on resonance energy transfer, *Peptide Res.,* 7, 72, 1994.
66. **Wilson, K. P., Black, J. F., Thomson, J. A., Kim, E. E., Griffith, J. P., Navia, M. A., Murcko, M. A., Chambers, S. P., Aldape, R. A., Raybuck, S. A., and Livingston, D. J.,** Structure and mechanism of interleukin-1β converting enzyme, *Nature,* 370, 270, 1994.
67. **Kobashi, K. and Horecker, B. L.,** *Arch. Biochem. Biophys.,* 121, 178, 1967.
68. **Shaw, I.,** Cysteinyl proteases and their selective inactivation, *Adv. Enzymol. Rel. Areas Mol. Biol.,* 63, 271, 1990.
69. **Miller, D. K., Ayala, J. M., Egger, L. A. et al.,** Purification and characterization of active human interleukin-1β converting enzyme from THP.1 monocytic cells, *J. Biol. Chem.,* 268, 18062, 1993.
70. **Mjalli, A. M. M., Chapman, K. T., MacCoss, M., and Thornberry, N. A.,** Phenyl alkyl ketones as potent reversible inhibitors of interleukin-1β converting enzyme, *Med. Chem. Lett.,* 3, 2689, 1993.
71. **Krantz, A., Copp, L. J., Coles, P. J., Smith, R. A., and Heard, S. B.,** Peptidyl (acyloxy)methyl ketones and the quiescent affinity label concept: the departing group as a variable structural element in the design of inactivators of cysteine proteinases, *Biochemistry,* 30, 1991.
72. **Robinson, V. J., Pauls, H. W., Coles, P. J., Smith, R. A., and Krantz, A.,** Carbon-13 NMR characterization of the papain aduct formed by peptidyl acyloxy-, aryloxy-, and chloromethyl ketone irreversible inhibitors, *Bioorg. Chem.,* 20, 42, 1992.
73. **Thornberry, N. A., Peterson, E. P., Zhao, J. J., Howard, A. D., Griffin, P. R., and Chapman, K. T.,** Inactivation of interleukin-1β converting enzyme by peptide (acyloxy)methyl ketones, *Biochemistry,* 33, 3934, 1994.
74. **Heng, R., Payne, T. G., Revesz, L., and Weidmann, B.,** New peptide derivatives inhibit IL-1β release — for treating rheumatoid arthritis, osteoarthritis, inflammatory bowel disease, septic shock, asthma etc., European Patent Office, WO9309135-A1, 1993.
75. **Dolle, R. E., Hoyer, D., Prasad, C. V. C., Schmidt, S. J., Helaszek, C. T., Miller, R. E., and Ator, M. A.,** P_1 aspartate-based peptide α-((2,6-Dichlorobenzoyl)oxy)methyl ketones as potent time-dependent inhibitors of interleukin-1β-converting enzyme, *J. Med. Chem.,* 37, 563, 1994.
76. **Carrell, R. W. and Travis, J.,** *Trends Biochem. Sci.,* 10, 20, 1985.
77. **Ray, C. A., Black, R. A., Kronheim, S. R., Greenstreet, T. A., Sleath, P. R., Salvesen, G. S., and Pickup, D. J.,** Viral inhibition of inflammation: cowpox virus encodes an inhibitor of the interleukin-1β converting enzyme, *Cell,* 69, 597, 1992.

78. **Komiyama, T. K., Ray, C. A., Pickup, D. J., Howard, A. D., Thornberry, N. A., Peterson, E. P., and Salvesen, G.,** Inhibition of interleukin-1β converting enzyme by the cowpox virus serpin CrmA, *J. Biol. Chem.,* 19331, 1994.
79. **Lee, J. C., Badger, A. M., Griswold, D. E., Dunnington, D., Truneh, A., Votta, B., White, J. R., Young, P. R., and Bender, P. E.,** Bicyclic imidazoles as a novel class of cytokine biosynthesis inhibitors, *Ann. N.Y. Acad. Sci.,* 696, 149, 1993.
80. **Eugue, E. M., Delustro, B., Rouhafza, S., Wilhelm, R., and Allison, A. C.,** Coordinate inhibition by some antioxidants of TNFα, IL-1β, and IL-6 production by human peripheral blood mononuclear cells, *Ann. N.Y. Acad. Sci.,* 696, 171, 1993.
81. **Kronheim, S. R., Mumma, A., Greenstreet, T., Glackin, P. J., Van Ness, K., March, C. J., and Black, R. A.,** Purification of interleukin-1β converting enzyme, the protease that cleaves the interleukin-1β precursor, *Arch. Biochem. Biophys.,* 296, 698, 1992.
82. **Ammirati, M. J., Mansour, M., LaLiberte, R., Lanzetti, A. J., and Danley, D. E.,** Partial purification of prointer leukin-1β converting enzyme (ICE) allows N-terminal sequence and anti-ICE antibody production, in The Protein Society, Proc. 6th Symposium, p. 77, San Diego, CA, July 25–29, 1992.
83. **Walker, N. P. C., Talanian, R. V., Brady, K. D., Dang, L. C., Bump, N. J., Ferenz, C. R., Franklin, S., Ghayur, T., Hackett, M. C., Hammill, L. D., Herzog, L., Hugunin, M., Houy, W., Mankovich, J. A., McGuiness, L., Orlewica, E., Paskind, M., Pratt, C. A., Reis, P., Summani, A., Terranova, M., Welch, J. P., Xiong, L., Moller, A., Tracey, D. E., Kamen, R., and Wong, W. W.,** Crystal structure of the cysteine protease interleukin-1β-converting enzyme: a $(p20/p10)_2$ homodimer, *Cell,* 78, 343, 1994.
84. **Howard, A. D., Palyha, O. C., Ding, G. J.-F. et al.,** High-level production and characterization of functional human interleukin-1β converting enzyme in *Baculovirus* and *E. coli* expression systems, *J. Cell. Biochem.,* 17B, 146, 1993.
85. **Cerretti, D. P., Kozlosky, C. J., Mosley, B., Nelson, N., Van Ness, K., Greenstreet, T. A., March, C. J., Kronheim, S. R., Druck, T., Cannizzaro, L. A., Huebner, K., and Black, R. A.,** Molecular cloning of the interleukin-1β converting enzyme, *Science,* 256, 97, 1992.
86. **Ayala, J. M., Yamin, T. T., Egger, L. A., Chin, J., Kostura, M. J., and Miller, D. K.,** Interleukin-1 beta converting enzyme is present in monocytic cells as an inactive 45-kDa precursor, *J. Immunol.,* 153, 2592, 1994.
87. **Yamin, T. T., Ayala, J., and Miller, D. K.,** *In vitro* activation of interleukin-1β converting enzyme (ICE), *FASEB J.,* 8, A644, 1994.
88. **Nett, M., Cerretti, D., Black, R., and Chaplin, D.,** Molecular cloning of the murine interleukin-1β (IL-1β) processing enzyme cDNA, *FASEB J.,* 6, A2056, 1992.
89. **Molineaux, S. M., Casano, F. J., Rolando, A. M., Peterson, E. P., Limjuco, G., Chin, J., Griffin, P. R., Calaycay, J. R., Ding, G. J.-F., Yamin, T.-T., Palyha, O. C., Luell, S., Fletcher, D., Miller, D. K., Howard, A. D., Thornberry, N. A., and Kostura, M. J.,** Interleukin-1β (IL-1β) processing in murine macrophages requires a structurally conserved homologue of human IL-1β converting enzyme, *Proc. Natl. Acad. Sci. U.S.A.,* 90, 1809, 1993.
90. **Shivers, B. D., Giegel, D. A., and Keane, K. M.,** Molecular cloning of rat interleukin-1β-converting enzyme: distribution and regulation, *J. Cell. Biochem.,* 17B, 119, 1993.
91. **Casano, F. J., Rolando, A. M., Mudgett, J. S., and Molineaux, S. M.,** The structure and complete nucleotide sequence of the murine gene encoding interleukin-1β converting enzyme (ICE), *Genomics,* 20, 474, 1994.
92. **Howard, A. D., Ding, G. J.-F., Rolando, A. M., Palyha, O. C., Peterson, E. P., Casano, F. J., Bayne, E. K., Donatelli, S., Ayala, J. M., Egger, L. A., Miller, D. K., Raju, S. M., Yamin, T. T., Jackson, J., Chapman, K. T., Thornberry, N. A., Schmidt, J. A., Tocci, M. J., and Molineaux, S. M.,** Human interleukin-1β converting enzyme: a mutational analysis of proenzyme activation, *J. Cell. Biochem.,* 17B, 113, 1993.
93. **Wang, X.-M., Helaszek, C. T., Winter, L. A., Lirette, R. P., Dixon, D. C., Ciccarelli, R. B., Kelley, M. M., Malinowski, J. J., Simmons, S. J., Huston, E. E., Koehn, J. A., Kratz, D., Bruckner, R. C., Graybill, T., Ator, M. A., Lehr, R. V., and Stevis, P. E.,** Production of active human interleukin-1β-converting enzyme in a baculovirus expression system, *Gene,* 145, 273, 1994.

94. **Cerretti, D. P., Hollingsworth, L. T., Kozlosky, C. J., Valentine, M. B., Shapiro, D. N., Morris, S. W., and Nelson, N.,** Molecular characterization of the gene for human interleukin-1β converting enzyme *(IL1BC), Genomics,* 20, 468, 1994.
95. **Arends, J. J. and Wyllie, A. H.,** Apoptosis: mechanisms and roles in pathology, in *Int. Rev. Exp. Path.*, Richter, G. W. and Epstein, M. A., Ed., Academic Press, New York, 1991, 223.
96. **Ellis, R. E., Yuan, J., and Horvitz, H. R.,** Mechanisms and functions of cell death, *Annu. Rev. Cell Biol.,* 7, 663, 1991.
97. **Yuan, J. and Horvitz, H. R.,** The *Caenorhabditis elegans* genes *ced-3* and *ced-4* act autonomously to cause programmed cell death, *Dev. Biol.,* 138, 33, 1990.
98. **Yuan, J., Shaham, S., Ledoux, S., Ellis, H. M., and Horvitz, H. R.,** The *C. elegans* cell death gene *ced-3* encodes a protein similar to mammalian interleukin-1β-converting enzyme, *Cell,* 75, 641, 1993.
99. **Kumar, S., Kinoshita, M., Noda, M., Copeland, N. G., and Jenkins, N. A.,** Induction of apoptosis by the mouse *Nedd2* gene, which encodes a protein similar to the product of the *Caenorhabditis elegans* cell death gene *ced-3* and the mammalian IL-1β-converting enzyme, *Genes Dev.,* 8, 1613, 1994.
100. **Hockenbery, D. M., Zutter, M., Hickey, W., Nahm, M., and Korsmeyer, S. J.,** Bcl2 protein is topographically restricted in tissues characterized by apoptotic cell death, *Proc. Natl. Acad. Sci. U.S.A.,* 88, 6961, 1991.
101. **Korsmeyer, S. J.,** Bcl-2: an antidote to programmed cell death, *Cancer Surv.,* 15, 105, 1992.
102. **Williams, G. T. and Smith, C. A.,** Molecular regulation of apoptosis: genetic controls on cell death, *Cell,* 74, 777, 1993.
103. **Hengartner, M. O., Ellis, R. E., and Horvitz, H. R.,** *Caenorhabditis elegans* gene *ced-9* protects cells from programmed cell death, *Nature,* 356, 494, 1992.
104. **Hengartner, M. O. and Horvitz, H. R.,** *C. elegans* cell survival gene *ced-9* encodes a functional homolog of the mammalian proto-oncogene *bcl-2, Cell,* 76, 665, 1994.
105. **Miura, M., Zhu, H., Rotello, R., Hartweig, E. A., and Yuan, J.,** Induction of apoptosis in fibroblasts by IL-1β-converting enzyme, a mammalian homolog of the *C. elegans* cell death gene *ced-3, Cell,* 75, 653, 1993.
106. **Gagliardini, V., Fernandez, P. A., Lee, R. K., Drexler, H. C., Rotello, R. J., Fishman, M. C., and Yuan, J.,** Prevention of vertebrate neuronal death by the *crmA* gene, *Science,* 263, 826, 1994.
107. **Singer, I. I., Scott, S., Chin, J., Kostura, M. J., Miller, D. K., Chapman, K., and Bayne, E. K.,** Interleukin-1β converting enzyme (ICE) is localized on the external cell-surface membranes and in the cytoplasmic ground substance of activated human monocytes by immuno-EM, *J. Exp. Med.,* in press, 1995.
108. **Higgins, G. C., Foster, J. L., and Postlethwaite, A. E.,** Interleukin-1β propeptide is detected intracellularly and extracellularly when human monocytes are stimulated with LPS *in vitro, J. Exp. Med.,* 180, 607, 1994.
109. **Fletcher, D., Agarwal, L., Chapman, K., Chin, J., Egger, L., Limjuco, G., Luell, S., MacIntyre, D., Peterson, E., Thornberry, N., and Kostura, M.,** A synthetic inhibitor of interleukin-1 converting enzyme (ICE) prevents endotoxin-induced IL-1β production *in vitro* and *in vivo, J. Interferm. Cytokine Res.,* 15, 243, 1995.
110. **Vassalli, P.,** The pathophysiology of tumor necrosis factors, *Annu. Rev. Immunol.,* 10, 411, 1992.
111. **Pennica, D., Nedwin, G. E., Hayflick, J. S., Seeburg, P. H., Derynck, R., Palladino, M. A., Kohr, W. J., Aggarwal, B. B., and Goeddel, D. V.,** Human tumor necrosis factor: precursor structure, expression and homology to lymphotoxin, *Nature,* 312, 724, 1984.
112. **Shirai, T., Yamaguchi, H., Ito, H., Todd, C. W., and Wallace, R. B.,** Cloning and expression in *Escherichia coli* of the gene for human tumour necrosis factor, *Nature,* 313, 803, 1985.
113. **Wang, A. M., Creasey, A. A., Ladner, M. B., Lin, L. S., Strickler, J., Van Arsdell, J. N., Yamamoto, R., and Mark, D. F.,** Molecular cloning of the complementary DNA for human tumor necrosis factor, *Science,* 228, 149, 1985.
114. **Kriegler, M., Perez, C., DeFay, K., Albert, I., and Lu, S. D.,** A novel form of TNF/cachectin is a cell surface cytotoxic transmembrane protein: ramifications for the complex physiology of TNF, *Cell,* 53, 45, 1988.

115. **Muller, R., Marmenout, A., and Fiers, W.,** Synthesis and maturation of recombinant human tumor necrosis factor in eukaryotic systems, *FEBS Lett.,* 197, 99, 1986.
116. **Hofsli, E., Bakke, O., Nonstad, U., and Espevik, T.,** A flow cytometric and immunofluorescence microscopic study of tumor necrosis factor production and localization in human monocytes, *Cell. Immunol.,* 122, 405, 1989.
117. **Perez, C., Alber, I., DeFay, K., Zachariades, N., Gooding, L., and Kriegler, M.,** A nonsecretable cell surface mutant of tumor necrosis factor (TNF) kills by cell-to-cell contact, *Cell,* 63, 251, 1990.
118. **Jue, D.-M., Sherry, B., Luedke, C., Manogue, K. R., and Cerami, A.,** Processing of newly synthesized cachectin/tumor necrosis factor in endotoxin-stimulated macrophages, *Biochemistry,* 29, 8371, 1990.
119. **Pradines-Figueres, A. and Raetz, C. R. A.,** Processing and secretion of tumor necrosis factor α in endotoxin-treated Mono Mac 6 cells are dependent on phorbol myristate acetate, *J. Biol. Chem.,* 267, 23261, 1992.
120. **McGeehan, G. M., Becherer, J. D., Bast, R. C., Jr., Boyer, C. M., Champion, B., Connolly, K. M., Conway, J. G., Furdon, P., Karp, S., Kidao, S., McElroy, A. B., Nichols, J., Pryzwansky, K. M., Schoenen, F., Sekut, L., Truesdale, A., Verghese, M., Warner, J., and Ways, J. P.,** Regulation of tumour necrosis factor-α processing by a metalloproteinase inhibitor, *Nature,* 370, 558, 1994.
121. **Mohler, K. M., Sleath, P. R., Fitzner, J. N., Cerretti, D. P., Alderson, M., Kerwar, S. S., Torrance, D. S., Otten-Evans, C., Greenstreet, T., Weerawarna, K., Kronheim, S. R., Peterseen, M., Gerhart, M., Kozlosky, C. J., March, C. J., and Black, R. A.,** Protection against a lethal dose of tumour necrosis factor processing, *Nature,* 370, 218, 1994.
122. **Gearing, A. J. H., Beckett, P., Christodoulou, M., Churchill, M., Clements, J., Davidson, A. H., Drummond, A. H., Galloway, W. A., Gilbert, R., Gordon, J. L., Leber, T. M., Mangan, M., Miller, K., Nayee, P., Patel, S., Thomas, W., Wells, G., Wood, L. M., and Woolley, K.,** Processing of tumour necrosis factor-α precursor by metalloproteinases, *Nature,* 370, 555, 1994.
123. **Kriegler, M., Perez, C., Halenbeck, R. F., Jewell, D. A., and Koths, K. E.,** Compositions for the Inhibition of Protein Hormone Formation and Uses Thereof, Patent WO 94/00555, Cetus Oncology Corp., 1994.
124. **Kim, K.-U., Kwon, O.-J., and Jue, D.-M.,** Pro-tumour necrosis factor cleavage enzyme in macrophage membrane/particulate, *Immunology,* 80, 134, 1993.
125. **Campanelli, D., Melchior, M., Fu, Y., Nakata, M., Shuman, H., Nathan, C., and Gabay, J. E.,** Cloning of cDNA for proteinase 3: a serine protease, antibiotic, and autoantigen from human neutrophils, *J. Exp. Med.,* 172, 1709, 1990.
126. **Rao, N. V., Wehner, N. G., Marshall, B. C., Gray, W. R., Gray, B. H., and Hoidal, J. R.,** Characterization of proteinase-3 (PR-3), a neutrophil serine proteinase, *J. Biol. Chem.,* 266, 9540, 1991.
127. **Oleksyszyn, J. and Powers, J. C.,** Irreversible inhibition of serine proteases by peptide derivatives of (α-aminoalkyl)phosphonate diphenyl esters, *Biochemistry,* 30, 485, 1991.
128. **Niehorster, M., Tiegs, G., Schade, U. F., and Wendel, A.,** *In vivo* evidence for protease-catalysed mechanism providing bioactive tumor necrosis factor α, *Biochem. Pharmacol.,* 40, 1601, 1990.
129. **Scuderi, P.,** Suppression of human leukocyte tumor necrosis factor secretion by the serine protease inhibitor *p*-toluenesulfonyl-L-arginine methyl ester (TAME), *J. Immunol.,* 143, 168, 1989.
130. **Henkel, T., Machleidt, T., Alkalay, I., Kronke, M., Ber-Neriah, Y., and Baeuerle, P. A.,** Rapid proteolysis of Iκβ-α is necessary for activation of transcription factor NF-κβ, *Nature,* 365, 182, 1993.
131. **Sato, H., Takino, T., Okada, Y., Cao, J., Shinagawa, A., Yamamoto, E., and Seiki, M.,** A matrix metalloproteinase expressed on the surface of invasive tumour cells, *Nature,* 370, 61, 1994.
132. **Vassalli, J.-D. and Pepper, M. S.,** Membrane proteases in focus, *Nature,* 370, 14, 1994.
133. **Sarno, E. N., Grau, G. E., Vieira, L. M., and Nery, J. A.,** Serum levels of tumour necrosis factor-alpha and interleukin-1 beta during leprosy reactional states, *Clin. Exp. Immunol.,* 84, 103, 1991.
134. **Sampaio, E. P., Sarno, E. N., Galilly, R., Cohn, Z. A., and Kaplan, G.,** Thalidomide selectively inhibits tumor necrosis factor α production by stimulated human monocytes, *J. Exp. Med.,* 173, 699, 1991.

135. **Knight, W. B., Green, B. G., Chabin, R. M., Gale, P., Maycock, A., Weston, H., Kuo, D. W., Westler, W. M., Dorn, C. P., Finke, P. E., Hagmann, W. K., Hale, J. J., Liesch, J., MacCoss, M., Navia, M. A., Shah, S. K., Underwood, D., and Doherty, J. B.,** Specificity, stability, and potency of monocyclic β-lactam inhibitors of human leucocyte elastase, *Biochemistry,* 31, 8160, 1992.
136. **Devereux, J., Haeberli, P., and Smithies, O.,** *Nucl. Acids Res.,* 12, 387, 1984.

Chapter 8

OLIGONUCLEOTIDE-BASED INHIBITORS OF CYTOKINE EXPRESSION AND FUNCTION

C. Frank Bennett and Stanley T. Crooke

CONTENTS

I. INTRODUCTION

Cytokines represent a diverse family of immunoregulatory proteins. During the past several years, the number of known cytokines and the roles of cytokines have increased at an extraordinary pace, as witnessed by this volume (see Chapters 1 to 3). Although molecular biological tools have supported the rapid advance in our knowledge of cytokines, the enormous complexity of these proteins, receptors, and autoregulatory processes in normal and diseased states impedes the development of a definitive understanding of these systems. Furthermore, although monoclonal antibodies and soluble receptors have proven to be useful, the discovery and development of pharmacological agents of sufficient specificity to support dissection of these complex immunoregulatory pathways are a daunting challenge.

0-8493-8381-1/96/$0.00+$.50

Because of the need for quantum advances in selectivity if truly isotypic specificity is to be achieved with pharmacological agents, we have chosen to explore new chemical and pharmacological paradigms. Specifically, we have attempted to systematically understand the pharmacological properties of oligonucleotides and analogs as they interact with principal classes of receptors, RNA, and proteins.

In this chapter, we provide a general introduction to oligonucleotide pharmacology and a review of the effects of various oligonucleotides on cytokines and cytokine receptors. The data suggest that properly designed and carefully characterized oligonucleotides are valuable tools in the dissection of the roles and pathways of cytokines and cytokine receptors. Furthermore, the data support optimism that these drugs may prove to be therapeutically valuable.

II. OLIGONUCLEOTIDES AS THERAPEUTIC AGENTS

One of the earliest uses of nucleic acids in the clinic was the administration of copolymers of polyriboinosinic:polyribocytidylic acid, poly (rI):poly (rC), to stimulate the immune system for the treatment of viral infections and cancer.[1–3] Induction of cytokines such as interferon and interleukins appear to be the primary mechanism for the pharmacological activity of poly (rI):poly (rC). In addition to inducing interferon, poly (rI):poly (rC) induces (2′–5′)-oligoadenylate synthetase and double-stranded RNA-activated kinase, both of which may play a role in the antiviral activity described for this class of agents. In the clinic, the use of poly (rI):poly (rC) has met with disappointing results primarily due to toxicity and weak induction of interferons.[4,5] Ampligen, an analog of poly (rI):poly (rC) in which uracil is inserted in some positions in the polyribocytidylate strand to destabilize the duplex, has been reported to be less toxic than poly (rI):poly (rC), yet still retaining antiviral and interferon-inducing activity.[6] Ampligen is currently being evaluated for therapeutic effects in HIV patients, where it has shown some activity, and in cancer patients.[7–9]

More recently, oligonucleotides have been exploited for their sequence-specific interactions with cellular nucleic acid and nonnucleic acid targets.[10–15] Due to the high degree of selectivity and acceptable toxicity profile, this class of chemicals hold great promise as a new generation of therapeutic agents. Different strategies in which investigators design oligonucleotides to inhibit expression of a particular gene product include transcription factor decoys, formation of triple-strand interactions with DNA, and antisense oligonucleotides. The latter application of oligonucleotides has been extensively used to inhibit the expression of a large number of viral and host cell gene products.[14,16,17]

A. ANTISENSE OLIGONUCLEOTIDES

Antisense oligonucleotides oligonucleotides employ the chemical interactions defined by Watson-Crick base pairing to bind to their receptor, i.e., the target RNA. Watson-Crick recognition has been extensively characterized. Thus, if the linear sequence of a target RNA is known, in theory one could design an oligonucleotide which will hybridize to the target RNA and inhibit the expression of the encoded protein. Although the concept of antisense oligonucleotides is rather simple, there are a number of factors which should be considered during the course of designing experiments utilizing this class of compounds and interpreting data obtained from such experiments. As discussed in more detail below, there are multiple chemical classes of oligonucleotides which could be used for antisense experiments. Unfortunately, no chemical class of oligonucleotides has proven to have all the properties desired of an antisense oligonucleotide, each having certain advantages and disadvantages. Another consideration is the length of the oligonucleotide used for experiments. In general, the longer the oligonucleotide the greater affinity it will have for a target RNA; however, the trade-off is that longer oligonucleotides have a greater probability of interacting with nontarget RNAs in a productive manner. Therefore, there is an optimum length for an oligonucleotide. We have found the optimum length for phosphorothioate oligonucleotides to be between 16 and 21 bases.[18–20] Sequences that are enriched for guanine and cytidine (which form three hydrogen bonds) may allow use of shorter oligonucleotides than those enriched in either adenine or thymine residues (which form two hydrogen bonds) to obtain similar affinities for their target RNAs. Modified oligonucleotides which exhibit higher affinity would also allow use of shorter oligonucleotides.

Selection of the appropriate target sites on a mRNA to which the oligonucleotide will hybridize may be the difference between a successful experiment and an unsuccessful one. A mRNA which is 1000 bases in length would have 980 potential target sites for a series of 20-base oligonucleotides. Obviously, this number is far too large to synthesize a complete set of oligonucleotides. Therefore, narrowing down the selection process would be greatly desired. One strategy would be to target those RNA sequences which are known to play an important role in the regulation of expression. The most common mRNA target site reported in the literature is the AUG translation initiation codon[14] (Table 1). Oligonucleotides which hybridize to sequences surrounding the AUG codon would be predicted to either block assembly of the ribosomal subunits or prevent movement of the ribosome, resulting in translation arrest. As is evident from the numerous examples in the literature, oligonucleotides which target the AUG translation initiation codon are often effective. However, we have found in many cases that oligonucleotides which target other regions in the mRNA, such as the 5′- and 3′-untranslated regions were more effective.[18,21–23] It is not clear why oligonucleotides which hybridize to 3′-untranslated region sequences would be more effective than other regions of the mRNA; possibly, this region of the mRNA is more accessible to the oligonucleotide. Other positions on the mRNA to which oligonucleotides exhibit activity include the 5′-terminus of a transcript, 5′-untranslated regions, and regions of the mRNA predicted to form stable stem loop structures.[18,24] It is becoming increasingly clear that regulation of gene expression not only occurs at the DNA level but also at the posttranscriptional level, primarily by the interaction of proteins with the mRNA.[25,26] These proteins may recognize either linear sequence motifs or secondary and tertiary structure elements in the RNA, both of which are obvious sites to direct oligonucleotides. As many cytokine gene products contain regulatory elements in the 3′-untranslated region of their mRNA, this area of the mRNA would be of interest to target with antisense oligonucleotides.[27]

There are multiple mechanisms by which oligonucleotides may terminate expression of the target protein which could also influence selection of an RNA target site.[16,18,28,29] Several mechanisms require the oligonucleotide to occupy a site on the target RNA and by doing so either block ribosome assembly on the mRNA or movement of the ribosome down the mRNA strand, resulting in translational arrest. Oligonucleotide binding to the 5′-terminus of a nascent transcript could block synthesis of a 5′-cap or function of the 5′-cap which has been implicated in nuclear transport and binding translation initiation factors.[30–32] Attachment of reactive groups to oligonucleotides resulting in covalent binding to the RNA has improved their activity in several biological systems.[33–36]

A second general mechanism by which oligonucleotides may terminate protein expression is cleavage of target RNA. This may be accomplished by enzymatic cleavage or chemical cleavage. Several studies have suggested that RNase H-mediated hydrolysis of the target RNA may be a primary mechanism by which deoxyoligonucleotides terminate gene expression.[18,23,37–39] RNase H is a ubiquitous enzyme which selectively cleaves the

TABLE 1
Antisense Oligonucleotides Which Inhibit Expression of Cytokine Genes

Target	Chemistry	Sequence	Target Site	Assay	Cell Type	Ref.
Prothymosin	P=O	GTCTACGGCTGCGTCTGACAT	AUG	↓ Proliferation	Myeloma	190
	P=O	GGTGTCTACGGCTGCG[a]	AUG	↓ Protein		
	P=O	CATCATCTTCAGCTGC	Coding	No effect on mRNA		
	P=O	CCAACTACTTATAGTACAG	3′-UTR			
Interleukin-1β	P=S	TCCGACGAGACCCTA[a]	AUG	↓ IL-1β, α, and TNF	PMN	191
	P=S	CCGGAACCCGGAGTT	Coding			
Interleukin-1β	P=O	CTCAGGTACTTCTGCCAT	AUG	↓ LAK activity ↓ Protein	PBL	148
Interleukin-1α	P=O	TTTGGCCATCTTGACTTC	AUG	↑ Population doublings ↓ mRNA, protein	Endothelial	149
Interleukin-1α	P=O	AGGAACTTTGGCCAT	AUG	↓ Proliferation ↓ mRNA	T lymphocyte	143
IL-1 receptor	P=S, P=O	CACTTTCATATTCTCCAT[a]	Murine AUG	↓ PGE_2 production	Fibroblasts	153
	P=S, P=O	TCTGAGTAACACTTTCAT	Human AUG	↓ Receptors		
	P=S	GAGACAAATGAGCCCCAG	Murine 5′-UTR	↓ Leukocyte recruitment		
	P=S	GCCGAGTGGTAAGTGTGT	Murine Coding			
Interleukin-2	P=S	ACAGGATGCGAGCTGCAT	AUG	↓ Proliferation	T lymphocytes	192
Interleukin-2	P=O	CTGCATGCTGTACAT	AUG	↓ Proliferation ↓ mRNA	T lymphocytes	142
Interleukin-2	P=O	CTGCATGCTGTACAT	AUG	↑ IgE, IgG, IgM	B lymphocytes	151
Interleukin-4	P=O	GGGGTTGAGACCCAT	AUG	↓ Proliferation ↓ mRNA	T lymphocytes	142
Interleukin-4	P=O	GGGGTTGTGTCGCAT[a]	AUG	↓ IgE, ↑ IgG, ↑ IgM	B lymphocytes	151
	P=O	CATCAATAGCTCTGTGCCGT	AUG			

Interleukin-4	P=O	GGGGTTGTGTCCCAT[a]	AUG	↓ IgE, ↓ IgG_{2a}	B lymphocytes	152
	P=O	GGGGTTGTGTCCCATCAATA	AUG			
	P=O	CATCAATAGCTCTGTGCCGT	AUG			
	P=S	GGGGTTGTGTCCAT[a]	AUG			
Interleukin-6	P=O	TCCTGGGGGTACTGG	CODING	↓ Proliferation	Kaposi's sarcoma	144
Interleukin-6	P=O	TCCTGGGGGTACTGG	CODING	↓ Proliferation	Myeloma	145
Interleukin-8	P=O	TCTCAGCCCTCTTCAAAAACTTCTC		↓ Proliferation	Melanoma	146
	P=O	CAAGAGAGCCACGGCCAGCTTGGA[a]				
GM-CSF	P=O	GCTCTGCAGCCACAT	AUG	↓ Protein	Endothelial	154
	P=O	CAGGCTCTGCAGCCACAT	AUG			
G-CSF	P=O	GGTGGCAGGTCCAGCCAT	AUG	↓ Protein	Endothelial	154
CSF-1	P=O	GCCCCCGCGCGGTCAT	AUG	↓ Proliferation	Monocyte line	147

[a] Most active oligonucleotide out of a series tested.

RNA strand of an RNA-DNA heteroduplex.[40] Mammalian cells appear to express several distinct forms of RNase H activity which may be regulated by the proliferation state and transcriptional activity of the cells.[41–45] Although mammalian RNase H enzymes are poorly characterized, bacterial and viral RNase H enzymes are well characterized and could provide insights into the design of oligonucleotides which would serve as better substrates for RNase H.

RNase L is a cellular ribonuclease activated by (2′–5′)-oligoadenylate. Torrence et al. covalently attached (2′–5′)-oligoadenylate to an oligonucleotide to direct RNase L degradation to a targeted RNA. They were able to demonstrate hydrolysis of the targeted RNA in a cell-free system.[46] It remains to be determined if such an approach will work in intact cells. Ribozymes are RNA molecules that catalyze biochemical reactions such as hydrolysis of RNA.[47–50] There are several examples in which vectors expressing ribozymes have been transfected into cells resulting in inhibition of viral replication or cellular oncogene expression.[51–55] A major limitation of ribozymes is that RNA is rapidly degraded in biological systems. To overcome this problem, Sioud et al. used cationic liposomes to deliver preformed ribozymes targeting tumor necrosis factor mRNA to cells.[56] Delivery of ribozyme to the cell with cationic liposomes reduced tumor necrosis factor mRNA by 90%.

Most eukaryotic mRNAs undergo a series of complex processing events within the nucleus of the cell. Interruption of these processing events with oligonucleotides would be predicted to be an efficient mechanism to block expression of a target protein.[57] Alternatively, oligonucleotides may be used to correct a genetic defect, as recently demonstrated by Dominski and Kole, in which they used antisense oligonucleotides to redirect splicing of a mutated β-globin pre-mRNA to restore correct splicing.[58] There are many additional mechanisms by which oligonucleotides may be utilized to inhibit expression of a target protein which have yet to be explored.

The mechanisms by which oligonucleotides enter cells are not well understood.[59–62] Most cells in culture internalize negatively charged oligonucleotides such as phosphorothioate oligonucleotides. However, there appears to be some variability in the extent of oligonucleotide internalization with different cell types.[60–64] Internalization of the oligonucleotide appears to be an active process which is markedly reduced by low temperatures or ATP depletion. There is some suggestion that specific proteins on the surface of cells bind oligonucleotides, although it has not been demonstrated that these proteins are responsible for their internalization.[65,66] Particularly relevant to this discussion are studies of internalization of oligonucleotides in cells of myeloid lineage. Cells of B cell lineage appear to bind and internalize more oligonucleotide than T cells.[67,68] Stimulation of cells with either a B cell or T cell mitogen further increased the amount of oligonucleotide which associated with the different cell population, suggesting that proliferating lymphocytes internalize more oligonucleotide than quiescent lymphocytes.

Even though most cells are capable of internalizing oligonucleotide, we have had difficulty in demonstrating decreased expression of targeted gene products by adding oligonucleotides to the culture medium of several cell types — keratinocytes being a notable exception.[69,70] In many instances, cationic liposomes markedly enhanced the biological activity of phosphorothioate oligonucleotides in a variety of cell types.[18,69,71,72] The mechanism by which cationic liposomes enhance activity of phosphorothioate oligonucleotides appears to be due in part to increasing the amount of oligonucleotides that associate with cells, and also to change the subcellular distribution of the oligonucleotide so that more oligonucleotide is available for interactions with the target mRNA. It remains to be determined how useful cationic liposomes will be for *in vivo* applications of oligonucleotides.

Although the concept of antisense oligonucleotides is easily appreciated, it is not a straightforward process to go from the DNA sequence of the gene of interest to a drug. A number of factors can contribute to successful application of antisense technology, or lack thereof, including stability of the oligonucleotide in the biological system tested, interaction with molecules other than the target nucleic acid, productive interactions with the target RNA, and both cellular and animal pharmacokinetics. Finally, it should be noted that the pharmacokinetics behavior of oligonucleotides in cell culture has not predicted their behavior *in vivo*. Therefore, studies addressing the pharmacokinetics of oligonucleotides in cell culture, at present are not useful for determining their value *in vivo*.

B. TRIPLE HELIX FORMATION

Triple helical structures for synthetic DNA polymers were first described in 1957.[73] However, only recently has their potential for inhibition of gene expression been studied.[74–79] The mechanisms by which oligonucleotides bind to double-stranded nucleic acids, forming a triple-strand structure, are less well-characterized than formation of a duplex structure. Three classes of triple helixes have been characterized for natural nucleotides.[80] The best-characterized structure occurs between an oligonucleotide containing cytosine and thymine which binds parallel to the polypurine strand by Hoogsteen base pairing. Cytosine has to be protonated to form the two hydrogen bonds with the C-G base pair, therefore, the stability of the triplex structure is pH dependent. A second triple helical structure can occur with oligonucleotides containing guanine and adenine. In this case, the oligonucleotide binds in an antiparallel manner to the target polypurine strand by reverse Hoogsteen hydrogen bonding.[81] The third type of structure occurs with oligonucleotides containing guanine and thymine. Both TA-T and CG-G base triplets can form Hoogsteen and reverse Hoogsteen hydrogen bonds. In general, Hoogsteen bonds are preferred for homopolymeric sequences and reverse Hoogsteen bonds preferred for more mixed sequences.[76,82,83] Because only two hydrogen bonds are used to form the third strand, Hoogsteen base pairing is less stable than Watson-Crick base pairing.

Current strategies to form triple helix structures require polypurine stretches in the duplex DNA, limiting the number of potential target sites. The interleukin-2 receptor α-subunit promoter contains a 15-base homopurine-homopyrimidine stretch which overlaps with the NF-kB site. Grigoriev et al.[84,85] designed a homopyrimidine oligonucleotide which hybridized to this sequence *in vitro*. Oligonucleotides preincubated with plasmid prior to transfection into cells inhibited expression of the interleukin-2 receptor α construct but not the HIV long-terminal repeat construct, thus demonstrating selectivity.[84] The effect of triple-strand oligonucleotides on interleukin-2 receptor expression on T cells has not been reported, so it is not known if polypyrimidine oligonucleotides are capable of interacting with the interleukin-2 receptor promoter region in its natural context. With recent oligonucleotides modifications, including synthesis of base modifications which extend the recognition of the third strand, inhibition of gene expression by formation of triple helical structures may become more widely applicable if other technical challenges can be overcome.[86–90]

C. OLIGONUCLEOTIDE INTERACTIONS WITH NONNUCLEIC ACID TARGETS

Like other chemical classes, oligonucleotides are capable of interacting with other macromolecules through hydrogen bonds, ionic interactions and hydrophobic interactions. They are also capable of forming complex secondary and tertiary structures. Thus, oligonucleotides may interact in a sequence-specific manner or in a structure-specific manner with nonnucleic acid targets. Perhaps the best example of this in biological systems is the binding in a sequence-specific manner of transcription factors and restriction endonucleases to DNA. One strategy which exploits this interaction is to utilize double-stranded oligonucleotide analogs which contain a binding site for a specific transcription factor. These oligonucleotides would compete with the genomic DNA for binding to the transcription factor, resulting in a decrease in expression of those genes regulated by the transcription factor. This strategy has been utilized for octamer and NF-kB regulated genes.[91,92]

Another strategy for identification of oligonucleotide-based therapeutic agents screens pools of randomized oligonucleotides, selecting those oligonucleotides which interact with a specific ligand or interfere with a biological process. There are numerous strategies for selection of oligonucleotides from a large pool; some involve enzymatic synthesis and PCR amplification, while others involve chemical synthesis and fractionation of complex oligonucleotide pools.[11,13,93] An example of the former was provided by Bock et al., who used a direct binding assay followed by PCR amplification of those binding oligonucleotides to identify a DNA oligonucleotide which bound to and inhibited thrombin.[13,94] This DNA oligonucleotide was found to adopt a highly compact, symmetrical structure consisting of two tetrads of guanosine base pairs and three loops.[95] This G-quartet-like structure is similar to the structure described for telomeric DNA sequences.[96,97] Similar G-quartet-like structures have been shown to interact with vimentin, macrophage scavenger receptors, type II phospholipase A_2 (Bennett, C. F. et al., manuscript in preparation), and HIV envelope protein gp120.[54,98,99] In each of these examples the oligonucleotides contained a G-quartet structure as the minimum pharmacophore. However, not all G-quartet-containing oligonucleotides bind to the same set of proteins mentioned above,suggesting that the bases surrounding the G-quartet structure contribute to the selectivity of such oligonucleotides. For example, it was found that the oligonucleotide which inhibited thrombin did not inhibit type II phospholipase A_2 and the oligonucleotide which bound to gp120 was a relatively weak inhibitor of type II phospholipase A_2 (Bennett, C. F., unpublished data).

It should also be noted that direct interaction with cellular proteins could account for the sequence-specific, nonantisense-mediated effects observed for some oligonucleotides originally designed as antisense reagents. Thus it is important to do a number of controls to demonstrate that the oligonucleotides are working by the mechanism proposed.

D. MEDICINAL CHEMISTRY OF OLIGONUCLEOTIDES

There have been several recent publications which have reviewed chemical modification of oligonucleotides, therefore this topic will not be discussed in depth.[15,100–103] The purpose of discussing the medicinal chemistry of oligonucleotides in this section will be to highlight some advantages which certain modifications offer and to emphasize that although much progress has been achieved using first-generation oligonucleotides, application of newer oligonucleotide analogs to antisense research and development opens up greater potential for this class of compounds.

The first oligonucleotides used as antisense reagents were natural deoxyoligonucleotides.[10] They still are the most widely used class of oligonucleotides for antisense research.[17] The reagent needed to synthesize this class of oligonucleotides are readily available, synthesis is well characterized, they support RNase H enzyme activity, and metabolites are not predicted to be toxic at low concentrations. However, natural deoxyoligonucleotides are readily metabolized by serum and cellular nucleases, limiting their application to cell culture-based assays. In the last several years a large number of chemical modifications have been made either on the phosphate backbone of the oligonucleotide, on sugar and base of the nucleoside, or by attachment of pendant groups to the oligonucleotides to enhance their biological activity.

Oligonucleotides modified on the phosphate backbone in which one of the nonbridging oxygens has been substituted with a methyl group, methylphosphonate, or a sulfur phosphorothioate, show markedly enhanced stability of the oligonucleotide to serum and cellular nuclease.[104–108] Phosphorothioate deoxyoligonucleotides retain a negative charge on the phosphate and will support RNase H activity. They are the second most widely used class of oligonucleotides in cell-based assay systems and the most widely used class of oligonucleotides in animal experiments. Methylphosphonate oligonucleotides lack a negative charge on the backbone, resulting in more hydrophobic molecules. Methylphosphonate oligonucleotides do not support RNase H-mediated hydrolysis of the hybridized RNA. There are only a limited number of reports which demonstrate activity in cells for methylphosphonate oligonucleotides.[109–113] Examples of additional modifications which retain a phosphorus in the backbone include phosphorodithioate, phosphotriester, and boranophosphate oligonucleotides.[114–116]

More recently, oligonucleosides have been made in which the phosphorus has been replaced in the backbone.[117–124] Although there are limited data demonstrating that such backbone modified oligonucleotides are active in biological systems, the biophysical behavior of these compounds is especially encouraging. In particular,one class of compounds termed peptide nucleic acid (PNA) has the potentially interesting ability to strand displace double-stranded DNA.[118,125] The backbone of PNA consists of aminoglycine units in which the base is attached to the glycine nitrogen of the backbone, thus eliminating both the phosphorus and the ribose sugar.[120] PNA will hybridize to both RNA and DNA by Watson-Crick hydrogen bonding and has a higher affinity for both DNA and RNA compared to natural nucleic acids.[126]

Several modifications of the pentofuranose ring such as 2′-*O*-methyl, *O*-allyl, and *O*-propyl exhibit increased affinity for RNA compared to deoxyoligonucleotides and may also exhibit increased nucleases resistance.[127–131] Oligonucleotides which contain a 2′-fluoro markedly enhance the affinity of the oligonucleotides towards RNA, but do not exhibit enhanced nuclease resistance compared to the deoxyoligonucleotides.[130,132] The trade-off for the increased affinity conferred by 2′-modifications is that they do not support RNase H-mediated hydrolysis of the target RNA. One mechanism which has been employed to take advantage of the greater affinity that 2′-substituted oligonucleotides offer, yet maintain RNase H activity, is to utilize chimeric oligonucleotides containing both 2′-modified nucleosides and deoxynucleosides.[39,127,133] α-Oligonucleotides are a novel class of compounds in which the C′-1 anomeric carbon atom is inverted compared to natural β-oligonucleotides.[134,135] α-Oligonucleotides hybridize in a parallel orientation to complementary β-DNA or RNA, with similar stability as β-oligonucleotides.[135] Data suggest that α-oligonucleotides are more stable to serum nucleases but they do not support RNase H activity. At this time, biological data for α-oligonucleotides are limited due in part to the difficulty in obtaining α-nucleoside synthons.

The purine and pyrimidine bases of nucleosides have been extensively modified due, in part, to the interest in nucleosides as anticancer or antiviral agents.[136] To be effective as antisense oligonucleotides, modifications of the bases should not interfere with normal Watson-Crick base pairing, therefore only a limited number of nucleoside analogs would be suitable for use as antisense agents. Several pyrimidine and purine modifications have been described which enhance hybridization such as 5-methyl-2′-deoxycytidine, 5-propynyl-2′-deoxycytidine, and deoxyuridine, 2-amino-2′-deoxyadenosine.[72,103,136–138] A potential advantage of base modifications which enhance hybridization affinity is that they are less likely to affect RNase H recognition of the heteroduplex than sugar or backbone modifications.

Finally, a number of molecules have been conjugated to oligonucleotides to change their physical behavior or chemical reactivity, or interaction with biological processes.

For example cholesterol has been added to oligonucleotides to enhance their association with cells, or intercalating agents have been conjugated to enhance binding to the target nucleic acid.[34,139–141]

As can be seen from the above discussion many oligonucleotide modifications offer certain advantages over natural DNA or RNA, however, they also have some limitations. The optimal use of these modifications may be to combine different modifications in the same molecule to offset limitations of any single modification. As an example, oligonucleotides may be modified on the backbone to enhance nuclease resistance, on the base or sugar to increase affinity for the target RNA, and conjugated to a ligand to enhance pharmacokinetic properties.

III. INHIBITION OF CYTOKINE ACTION WITH OLIGONUCLEOTIDES

A. CYTOKINES

Antisense oligonucleotides have been used to inhibit the expression of a large number of cytokines (Table 1). In many cases the oligonucleotides were used as a tool to demonstrate that the cytokine serves as an autocrine growth factor for the targeted cell population (Table 1). For example, antisense oligonucleotides to interleukin-1α and interleukin-4 were used to demonstrate that the respective interleukins were autocrine growth factors for Th type 2 cells,[142,143] while interleukin-2 antisense oligonucleotides were used to demonstrate that interleukin-2 is an autocrine growth factor for Th type 1 cells.[142] The growth inhibitory effects of the oligonucleotides were reversed by the addition of the respective cytokines back to the culture. Similarly, interleukin-6 was shown to be an autocrine growth factor for Kaposi's sarcoma-derived cells and human myeloma cell lines, interleukin-8 was shown to be an autocrine growth factor for human melanoma cells, and colony stimulating factor-1 (CSF-1) was shown to be an autocrine growth factor for murine monocytes.[144–147] In each case, phosphodiester deoxyoligonucleotides were used to inhibit the expression of the target cytokine.

An antisense, but not sense, oligonucleotide to interleukin-1β inhibited lymphokine-activated killer cell induction.[148] The effect of the antisense oligonucleotide was attenuated by inclusion of interleukin-1β protein in the culture medium. Oligonucleotides to interleukin-1α have been used to prevent senescence in human umbilical vein endothelial cells and block the growth inhibitory effects of poly (rI):poly (rC) in endothelial cells.[149,150] These studies led to the conclusion that interleukin-1α expression inhibited endothelial cell growth, ultimately leading to cellular senescence.

Benbernou et al. used antisense oligonucleotides designed to hybridize to murine interleukin-4 to inhibit interleukin-4-mediated IgE synthesis in *Nippostrongylus brasilienis*-infected rat spleen cells.[151,152] Two out of the three antisense oligonucleotides used inhibited IgE and IgG2a synthesis. Comparison of the sequences between mouse and rat interleukin-4 mRNA in the regions where the oligonucleotides were designed to hybridize revealed several sequence differences between rat and mouse interleukin-4 mRNA. The most active oligonucleotide contained mismatches in 3 out of 15 bases, while an oligonucleotide which did not exhibit activity had 3 out of 20 bases which should not hybridize to rat interleukin-4 mRNA. Although the effects of the active oligonucleotides were reversed by the addition of interleukin-4 to the cells, the authors were unable to demonstrate an effect of the oligonucleotides on interleukin-4 production due to the low levels produced in the cells. Therefore, mechanistically it was not proven that the oligonucleotides were in fact acting by inhibiting interleukin-4 synthesis.

B. CYTOKINE RECEPTORS

Another approach to blocking the activity of cytokine function would be to design antisense oligonucleotides which target cytokine receptors. A potential advantage of targeting cytokine receptors is that in many cases local application of the oligonucleotide would be feasible. Burch and Mahan used phosphodiester and phosphorothioate oligonucleotides to inhibit the expression of murine and human type 1 interleukin receptors.[153] They demonstrated a decrease in interleukin-1 binding sites by 55% and a corresponding reduction in interleukin-1-mediated prostaglandin E2 synthesis. The effects of the oligonucleotides were enhanced by encapsulating them in liposome formulations. These studies suggest that, at least in the case of fibroblast type 1 interleukin-1 receptors, the reduction in receptor number correlates with biological effects. Perhaps most noteworthy was the fact that subcutaneous injection of the oligonucleotides in mice blocked the recruitment of neutrophils to skin following injection of interleukin-1, demonstrating *in vivo* activity for the oligonucleotides.

C. CYTOKINE-INDUCED PRODUCTS

In addition to directly inhibiting the synthesis of cytokines, antisense oligonucleotides have been used to partially inhibit the effect of cytokines by target gene products which are induced by the cytokines, or targeting a protein in the signal transduction pathway for the activation of a target cells. We have used phosphorothioate oligonucleotides to block the induction of intercellular adhesion molecule-1 (ICAM-1), vascular cell adhesion molecule-1 (VCAM-1), and E-selectin by tumor necrosis factor, interleukin-1, interleukin-4, and interferon-γ.[18,22] In each case, a series of phosphorothioate oligonucleotides of 18 to 21 bases in length were designed to hybridize to different regions on the respective mRNA. The most active oligonucleotides for each target were found to hybridize to sequences in the 3′-untranslated regions of the mRNA. The oligonucleotides were shown to selectively inhibit the targeted protein and block adherence of HL-60 cells to cytokine-activated endothelial cells. Oligonucleotides were also described which selectively inhibited synthesis of granulocyte/monocyte-colony stimulating factor (GM-CSF) and granulocyte-colony stimulating factor (G-CSF) in interleukin-1-activated endothelial cells.[154] Thomae et al. inhibited the synthesis of an inducible form of nitric oxide synthase and nitric oxide production with antisense oligonucleotides in vascular smooth muscle cells.[155] Oligonucleotides which target protein kinase C-α mRNA were used to inhibit interleukin-1α induction of cyclooxygenase expression in human endothelial cells, suggesting that interleukin-1 utilizes this particular protein kinase C isoenzyme as part of the signal transduction pathway.[156] In addition, Dean et al. used phosphorothioate deoxyoligonucleotides designed to hybridize to human protein kinase C-α to block phorbol ester induction of ICAM-1 mRNA in human carcinoma cells.[23]

D. TRANSCRIPTION FACTORS

The NF-kB transcription factor complex is a pleiotropic activator that is involved in the induction of a number of cytokine products, cytokine receptors, and cytokine-induced gene products. The active complex is composed of two subunits designated p50 and p65, both of which exhibit sequence similarity to the oncogene *rel.* The p50 and p65 subunits can form both homodimers and heterodimers. Narayanan et al. have identified and characterized phosphorothioate antisense oligonucleotides to both the p50 and p65 subunits.[157,158] Although the researchers did not examine the effects of the oligonucleotides on a cytokine-regulated process, they did demonstrate that antisense oligonucleotides to p65 inhibited adhesion of a wide variety of cells to extracellular matrix and blocked

CD11b induction on HL-60 cells.[157,158] Similar findings were observed with double-stranded phosphorothioate oligonucleotides containing three repeats of the NF-kB binding site consensus sequence.[94] In addition, the NF-kB transcription factor decoy blocked tumor necrosis factor-induction of ICAM-1 mRNA.[94]

IV. *IN VIVO* ACTIVITY OF OLIGONUCLEOTIDES

A. PHARMACOKINETICS

The pharmacokinetics of natural phosphodiester deoxyoligonucleotides, methylphosphonate deoxyoligonucleotides, and phosphorothioate deoxyoligonucleotides have been studied in rodents following parenteral administration[159–166] and phosphorothioates have been studied in monkeys and humans following intradermal and intravitreal injections.[167] Several radiolabeling methods that provide useful results have been employed including uniform labeling of the backbone with ^{32}P or ^{35}S (or ^{3}H for methylphosphonates). More recently, we have studied phosphorothioate deoxyoligonucleotides labeled in the 2-position of thymines and have developed a postsynthesis labeling method that tritiates a slowly exchanging proton at position 8 in purines.[168]

As expected, phosphodiester deoxyoligonucleotides are cleared rapidly from plasma after intravenous administration, principally by rapid degradation by nucleases.[169] However, it has been reported that association with lipoproteins can enhance the stability and change the distribution of a phosphodiester deoxyoligonucleotide.[167] Furthermore, prolonged infusion of a phosphodiester deoxyoligonucleotide complementary to a region in N-myc reduced N-myc expression, suggesting that chronic infusions may extend the duration of activity of unmodified oligonucleotides.[170]

Methylphosphonate oligonucleotides are nuclease stable, but have also been reported to be rapidly cleared from plasma after intravenous administration. However, in contrast to phosphodiester oligonucleotides, the rapid disappearance from plasma is due primarily to excretion in the urine and secondarily to tissue distribution.[159] The major organs that accumulated the drug were kidney, liver, lung, muscle, and spleen, but more than 70% of the administered dose was found in the urine within 60 min. As methylphosphonate oligonucleotides are insoluble in water, they are typically synthesized with a 5-terminal phosphate to enhance solubility. This may explain the high level of the drug in urine, but one might expect phosphatases to remove the phosphate relatively rapidly, resulting in a water-insoluble molecule that would be unlikely to be excreted in the urine unless conjugated. Perhaps future studies will resolve these issues.

Phosphorothioate deoxyoligonucleotides have been most extensively studied. The pharmacokinetics of a 20-mer phosphorothioate oligodeoxynucleotide were determined after single i.v. or i.p. doses of 30 mg/kg.[161] The oligonucleotide was labeled at each internucleotide linkage with ^{35}S. The compound was shown to be bioavailable after i.p. administration, to have broad peripheral tissue distribution, and to be cleared primarily by renal excretion. Gel electrophoretic analysis demonstrated significant, albeit slow, metabolism in the liver, kidney, and intestines.[161] A 27-mer phosphorothioate deoxyoligonucleotide labeled with ^{35}S was reported to display biexponential elimination in plasma with an elimination half-life in excess of 40 h after a single i.v. dose of 4.8 mg/kg in the rat.[163]

In our laboratories, we have studied the pharmacokinetics of several phosphorothioate deoxyoligonucleotides administered by several routes in several species. We have shown that these compounds bind to serum albumin and α_2 macroglobulin with relatively low affinity. The estimated Kd for human albumin is approximately 10 μm and binding of a representative phosphorothioate deoxyoligonucleotide, ISIS 2105, was not inhibited by aspirin, penicillin, or warfarin (Leeds et al., manuscript in preparation).

After a dose of 3.6 mg/kg of ^{14}C-ISIS 2105 given i.v. to rats, the radioactivity disappeared rapidly from plasma. Distribution was rapid ($t_{1/2}$ 0.4 h) and extensive. The postdistribution volume of distribution was 107.6 ml and the drug was found in all peripheral organs. The liver, kidneys, bone marrow, and skin accumulated the most drug. Clearance was due to slow metabolism and the elimination half-life was 51 h.[165]

When ^{14}C-ISIS 2105 was administered intradermally at 3.6 mg/kg to rats, it was rapidly and extensively absorbed and residual drug left at the injection site was metabolized slowly. Postabsorption pharmacokinetic behavior was essentially equivalent to that observed after an i.v. dose.[166]

^{14}C-ISIS 2105 has also been administered to humans with genital warts. The drug was administered intradermally at the base of penile genital warts and shown to infiltrate the warts and to be rapidly absorbed. Metabolism of the residual drug at the injection site and pharmacokinetics in humans appeared to be very similar to those found in rats.[167]

When ^{3}H-ISIS 3082, another phosphorothioate deoxynucleotide, was administered with ^{14}C-cationic lipid to mice, the only major change in pharmacokinetics was a shift in distribution such that a higher fraction of the i.v. dose was found in lung and liver compared to drug formulated in the absence of cationic lipid.[171]

Other studies that are not yet published confirm phosphorothioate deoxyoligonucleotides are well absorbed after administration to a variety of parenteral sites and that pharmacokinetics support once-a-day or every-other-day dosing.

B. PHARMACOLOGY

In the past two years several studies have been published describing *in vivo* pharmacological activities of oligonucleotides.[21,153,170,172–184] The majority of studies either treated tumor xenografts systemically or applied the oligonucleotide locally. Several studies have demonstrated effects of antisense phosphodiester and phosphorothioate oligonucleotides when administered into the brain.[172,178,182,184] The level of nuclease activity in cerebrospinal fluid is low, which may explain the activity of phosphodiester oligonucleotides in these models.[185] Mice injected s.c. with a phosphorothioate antisense oligonucleotide to murine type I interleukin-1 receptor exhibited decreased neutrophil migration to the site following injection of interleukin-1.[153] However, it was not demonstrated that the oligonucleotides reduced interleukin-1 receptors in the skin. Antisense oligonucleotides designed to inhibit the expression of the p65 subunit of NF-kB have demonstrated antitumor activity in several systems.[173,183,186] Antisense oligonucleotides that targeted the p50 subunit did not exhibit antitumor activity. Considering the central role that NF-kB plays in initiation and propagation of an immune response, it was somewhat surprising that there was no detectable change in leukocyte populations in mice treated with a murine-specific p65 NF-kB antisense oligonucleotide. Recently, we have demonstrated *in vivo* activity of a murine specific ICAM-1 antisense oligonucleotide in several murine inflammatory models, including allograft rejection[187] and inflammatory bowel disease Bennett, C. F. et al., manuscript submitted). In addition we have directly demonstrated that phosphorothioate oligonucleotides can specifically reduce the level of a targeted gene product in normal tissue[188] and that human protein kinase α antisense oligonucleotides inhibit human tumor xenograft growth in nude mice (Dean, N. M. et al., manuscript submitted).

C. TOXICOLOGY

Although studies describing the *in vivo* toxicology of antisense oligonucleotides have not been reported, we have found that phosphorothioate antisense oligonucleotides appear to be well tolerated at doses which produce pharmacological activity. There are two reports of splenomegaly in mice treated with phosphorothioate oligonucleotides directed against the *rev* gene of HIV-1 and mink cell-focus forming envelope gene product.[180,189] These effects were proposed to be due to inhibition of an endogenous retroviral-like gene product which attenuates the immune response. However, similar effects have been observed with other phosphorothioate oligonucleotides which do not target retrovirus sequences, suggesting that splenomegaly may be a common effect of phosphorothioate oligonucleotides (Kornbrust, D., unpublished data).

V. CONCLUSIONS

The substantial progress in the development of oligonucleotide therapeutics has supported the use of oligonucleotide-based drugs to explore the roles of cytokines and to understand cytokine pathways. Given the current data in animals, there is growing cause for optimism that oligonucleotides designed to inhibit production of cytokines or their receptors may be of therapeutic value as well.

REFERENCES

1. **Catalano, L. W. and Baron, S.,** Protection against herpes virus and encephalomyocarditis virus encephalitis with a double-stranded RNA inducer of interferon, *Proc. Soc. Exp. Biol. Med.,* 133, 684, 1970.
2. **Stringfellow, D. A., Overall, J. C., and Glasgow, L. A.,** Interferon inducers in therapy of infection with encephalomyocarditis virus in mice. I. Effect of single doses of polyriboinosinic-polyribocytidylic acid and tilorone hydrochloride on viral pathogenesis, *J. Infect. Dis.,* 130, 470, 1974.
3. **Champney, K. J., Levine, D. P., Levy, H. B., and Martin Lerner, A.,** Modified polyriboinosinic-polyribocytidylic acid complex. Sustained interferonemia and its physiological associates in humans, *Infect. Immunol.,* 25, 831, 1979.
4. **Hill, D. A., Baron, S., and Levy, H. B.,** Clinical studies of induction of interferon by polyinosinic-polycytidylic acid. In: *Perspectives in Virology,* Pollard, M., Ed., Academic Press, New York, 1971, p. 198–222.
5. **Droller, M. J.,** Immunotherapy of metastatic renal cell carcinoma with polyinosinic-polycytidylic acid, *J. Urol.,* 137, 202, 1986.
6. **Ts'o, P. O. P., Alderfer, J. L., and Levy, J.,** An integrated and comparative study of the antiviral effects and other biological properties of polyinosinic acid-polycytidylic acid and its mismatched analogues, *Mol. Pharmacol.,* 12, 299, 1976.
7. **Carter, W. A., Brodsky, D. R., and Pellegrino, M. G.,** Clinical, immunological and virological effects of Ampligen, a mismatched double stranded RNA, in patients with AIDS or AIDS-related complex, *Lancet,* 1, 1286, 1987.
8. **Armstrong, J. A., McMahon, D., Huang, X.-L., Pazin, G. J., Gupta, P., Rinaldo, C. R., Jr., Schoenfeld, D. A., Gaccione, P., Tripoli, C. A., Bensasi, S., and Ho, M.,** A phase I study of ampligen in human immunodeficiency virus-infected subjects, *J. Infect. Dis.,* 166, 717, 1992.
9. **McMahon, D., Winkelstein, A., Huang, X.-L., Armstrong, J., Pazin, G., Rinaldo, C., Tripoli, C., and Ho, M.,** Acute reactions associated with the infusion of ampligen, *AIDS,* 6, 235, 1992.
10. **Zamecnik, P. C. and Stephenson, M. L.,** Inhibition of Rous sarcoma virus replication and cell transformation by a specific oligodeoxyribonucleotide, *Proc. Natl. Acad. Sci. U.S.A.,* 75, 289, 1978.
11. **Tuerk, C. and Gold, L.,** Systematic evolution of ligands by exponential enrichment: RNA ligands to bacteriophage T4 DNA polymerase, *Science,* 249, 505, 1990.
12. **Irvine, D., Tuerk, C., and Gold, L.,** SELEXION — systematic evolution of ligands by exponential enrichment with integrated optimization by nonlinear analysis, *J. Mol Biol.,* 222, 739, 1991.

13. **Bock, L. C., Griffin, L. C., Latham, J. A., Vermaas, E. H., and Toole, J. J.,** Selection of single-stranded DNA molecules that bind and inhibit human thrombin, *Nature,* 355, 564, 1992.
14. **Crooke, S. T.,** Therapeutic applications of oligonucleotides, *Annu. Rev. Pharmacol. Toxicol.,* 32, 329, 1992.
15. **Crooke, S. T. and Lebleu, B.,** *Antisense Research and Applications,* CRC Press, Boca Raton, FL, 1993.
16. **Crooke, S. T.,** Progress toward oligonucleotide therapeutics: pharmacodynamic properties, *FASEB J.,* 7, 533, 1993.
17. **Stein, C. A. and Cheng, Y.-C.,** Antisense oligonucleotides as therapeutic agents. Is the bullet really magical?, *Science,* 261, 1004, 1993.
18. **Chiang, M.-Y., Chan, H., Zounes, M. A., Freier, S. M., Lima, W. F., and Bennett, C. F.,** Antisense oligonucleotides inhibit intercellular adhesion molecule 1 expression by two distinct mechanisms, *J. Biol. Chem.,* 266, 18162, 1991.
19. **Monia, B. P., Johnston, J. F., Ecker, D. J., Zounes, M., Lima, W. F., and Freier, S. M.,** Selective inhibition of mutant Ha-ras mRNA expression by antisense oligonucleotides, *J. Biol. Chem.,* 267, 19954, 1992.
20. **Cowsert, L. M., Fox, M. C., Zon, G., and Mirabelli, C. K.,** *In vitro* evaluation of phosphorothioate oligonucleotides targeted to the E2 mRNA of papillomavirus. Potential treatment for genital warts, *Antimicrob. Agents Chemother.,* 37, 171, 1993.
21. **Perlaky, L., Saijo, Y., Busch, R. K., Bennett, C. F., Mirabelli, C. K., Crooke, S. T., and Busch, H.,** Growth inhibition of human tumor cell lines by antisense oligonucleotides designed to inhibit p120 expression, *Anti-Cancer Drug Design,* 8, 3, 1993.
22. **Bennett, C. F., Condon, T., Grimm, S., Chan, H., and Chiang, M.-Y.,** Inhibition of endothelial cell-leukocyte adhesion molecule expression with antisense oligonucleotides, *J. Immunol.,* 152, 3530, 1994.
23. **Dean, N. M., McKay, R., Condon, T. P., and Bennett, C. F.,** Inhibition of protein kinase C-alpha expression in human A549 cells by antisense oligonucleotides inhibits induction of ICAM-1 mRNA, *J. Biol. Chem.,* 269, 16416, 1994.
24. **Lima, W. F., Monia, B. P., Ecker, D. J., and Freier, S. M.,** Implication of RNA structure on antisense oligonucleotide hybridization kinetics, *Biochemistry,* 31, 12055, 1992.
25. **Jackson, R. J.,** Cytoplasmic regulation of mRNA function: the importance of the 3′ untranslated region, *Cell,* 74, 9, 1993.
26. **Sachs, A. B.,** Messenger RNA degradation in eukaryotes, *Cell,* 74, 413, 1993.
27. **Caput, D., Beutler, B., Hartog, K., Thayer, R., Brown-Shimer, S., and Cerami, A.,** Identification of a common nucleotide sequence in the 3′-untranslated region of mRNA molecules specifying inflammatory mediators, *Proc. Natl. Acad. Sci. U.S.A.,* 83, 1670, 1986.
28. **Helene, C. and Toulme, J.-J.,** Specific regulation of gene expression by antisense, sense and antigene nucleic acids, *Biochim. Biophys. Acta,* 1049, 99, 1990.
29. **Bennett, C. F. and Crooke, S. T.,** Regulation of endothelial cell adhesion molecule expression with antisense oligonucleotides, *Adv. Pharmacol.,* 28, 1, 1994.
30. **Baker, B. F., Miraglia, L., and Hagedorn, C. H.,** Modulation of eucaryotic initiation factor-4E binding to 5′-capped oligoribonucleotides by modified anti-sense oligonucleotides, *J. Biol. Chem.,* 267, 11495, 1992.
31. **Baker, B. F.,** 'Decapitation' of a 5′-capped oligoribonucleotide by *o*-phenanthroline:Cu(II), *J. Am. Chem. Soc.,* 115, 3378, 1993.
32. **Baker, B. F.,** The 5′ Cap of mRNA. Biosynthesis, function, and structure as related to antisense drugs. In: *Antisense Research and Applications,* Crooke, S. T. and Lebleu, B. Eds., CRC Press, Boca Raton, FL, 1993, pp. 37–53.
33. **Miller, P. S., Bhan, P., Cushman, C. D., Kean, J. M., and Levis, J. T.,** Antisense oligonucleoside methylphosphonates and their derivatives, *Nucleosides Nucleotides,* 10, 37, 1991.
34. **Saison-Behmoaras, T., Tocque, B., Rey, I., Chassignol, M., Thuong, N. T., and Helene, C.,** Short modified antisense oligonucleotides directed against Ha-Ras point mutation induce selective cleavage of the mRNA and inhibit T24 cells proliferation, *EMBO J.,* 10, 1111, 1991.
35. **Helene, C. and Toulme, J.-J.,** Control of gene expression by oligonucleotides covalently linked to intercalating agents and nucleic acid-cleaving reagents. In: *Oligonucleotides: Antisense Inhibitors of Gene Expression,* Cohen, J. S., Ed., CRC Press, Boca Raton, FL, 1989, pp. 137–172.

36. **Miller, P. S.,** Antisense oligodeoxyribonucleoside methylphosphonates and their psoralen conjugates, *Clin. Chem.,* 39, 713, 1993.
37. **Walder, R. Y. and Walder, J. A.,** Role of RNase H in hybrid-arrested translation by antisense oligonucleotides, *Proc. Natl. Acad. Sci. U.S.A.,* 85, 5011, 1988.
38. **Giles, R. V. and Tidd, D. M.,** Enhanced RNase H activity with methylphosphonodiester/phosphodiester chimeric antisense oligodeoxynucleotides, *Anti-Cancer Drug Design,* 7, 37, 1992.
39. **Monia, B. P., Lesnik, E. A., Gonzalez, C., Lima, W. F., McGee, D., Guinosso, C., Kawasaki, A. M., Cook, P. D., and Freier, S. M.,** Evaluation of 2′ modified oligonucleotides containing deoxy gaps as antisense inhibitors of gene expression, *J. Biol. Chem.,* 268, 14514, 1993.
40. **Crouch, R. J.,** Ribonuclease H. From discovery to 3D structure, *New Biologist,* 2, 771, 1990.
41. **Cooper, R. J., Duff, P. M., Olivier, A., Craig, R. K., and Keir, H. M.,** Multiple ribonuclease H activities from BHK-21/C13 cells, *FEBS Lett.,* 45, 38, 1974.
42. **Cathala, G., Rech, J., Huet, J., and Jeanteur, P.,** Isolation and characterization of two types of ribonucleases H in Krebs II ascites cells, *J. Biol. Chem.,* 254, 7353, 1979.
43. **Cooper, R. J., Duff, P. M., Olivier, A., Craig, R. K., and Keir, H. M.,** Purification, subunit structure, and serological analysis of calf thymus ribonuclease H I, *J. Biol. Chem.,* 255, 9434, 1980.
44. **Muller, W. E. G., Geurtsen, W., Zahn, R. K., and Arendes, J.,** Cell cycle-dependent alterations of the two types of ribonucleases H in L5178y cells, *FEBS Lett.,* 110, 119, 1980.
45. **Eder, P. S. and Walder, J. A.,** Ribonuclease H from K562 human erythroleukemia cells, *J. Biol. Chem.,* 266, 6472, 1991.
46. **Torrence, P. F., Maitra, R. K., Leisak, K., Khamnei, S., Zhou, A., and Silverman, R. H.,** Targeting RNA for degradation with a (2′–5′) oligoadenylate-antisense chimera, *Proc. Natl. Acad. Sci. U.S.A.,* 90, 1300, 1993.
47. **Been, M. D. and Cech, T. R.,** RNA as an RNA polymerase: net elongation of an RNA primer the Tetrahymena ribozyme, *Science,* 239, 1412, 1988.
48. **Uhlenbeck, O. C.,** A small catalytic oligoribonucleotide, *Nature,* 328, 596, 1987.
49. **Cech, T. R.,** RNA as an enzyme, *Biochem. Int.,* 18, 7, 1989.
50. **Cech, T. R.,** Ribozyme engineering, *Curr. Opin. Struct. Biol.,* 2, 605, 1992.
51. **Sarver, N., Cantin, E. M., Chang, P. S., Zaia, J. A., Ladne, P. A., Stephens, D. A., and Rossi, J. J.,** Ribozymes as potential anti-HIV-1 therapeutic agents, *Science,* 247, 1222, 1990.
52. **Lisziewicz, J., Sun, D., Smythe, J., Lusso, P., Lori, F., Louie, A., Markham, P., Rossi, J., Reitz, M., and Gallo, R. C.,** Inhibition of human immunodeficiency virus type 1 replication by regulated expression of a polymeric Tat activation response RNA decoy as a strategy for gene therapy in AIDS, *Proc. Natl. Acad. Sci. U.S.A.,* 90, 8000, 1993.
53. **Scanlon, K. J., Jiao, L., Funato, T., Wang, W., Tone, T., Rossi, J. J., and Kashani-Sabert, M.,** Ribozyme-mediated cleavage of c-fos mRNA reduces gene expression of DNA synthesis enzymes and metallothionein, *Proc. Natl. Acad. Sci. U.S.A.,* 88, 10, 1991.
54. **Ojwang, J. O., Hampel, A., Looney, D. J., Wong-Staal, F., and Rapapport, J.,** Inhibition of human immunodeficiency virus type 1 expression by a hairpin ribozyme, *Proc. Natl. Acad. Sci. U.S.A.,* 89, 10802, 1992.
55. **Yu, M., Ojwang, J., Yamada, O., Hampel, A., Rapapport, J., Looney, D., and Wong-Staal, F.,** A hairpin ribozyme inhibits expression of diverse strains of human immunodeficiency virus type 1, *Proc. Natl. Acad. Sci. U.S.A.,* 90, 6340, 1993.
56. **Sioud, M., Natvig, J. B., and Forre, O.,** Preformed ribozyme destroys tumour necrosis factor mRNA in human cells, *J. Mol. Biol.,* 224, 831, 1992.
57. **Mayeda, A., Hayase, Y., Inoue, H., Ohtsuka, E., and Ohshima, Y.,** Surveying *cis*-acting sequences of pre-mRNA by adding antisense 2′-*O*-methyl oligoribonucleotides to a splicing reaction, *J. Biochem. (Tokyo),* 108, 399, 1990.
58. **Dominski, Z. and Kole, R.,** Restoration of correct splicing in thalassemic pre-mRNA by antisense oligonucleotides, *Proc. Natl. Acad. Sci. U.S.A.,* 90, 8673, 1993.
59. **Jaroszewski, J. W. and Cohen, J. S.,** Cellular uptake of antisense oligodeoxynucleotides, *Adv. Drug Deliv. Rev.,* 6, 235, 1991.
60. **Crooke, R. M.,** *In vitro* toxicology and pharmacokinetics of antisense oligonucleotides, *Anti-Cancer Drug Design,* 6, 609, 1991.

61. **Crooke, R. M.,** Cellular uptake, distribution and metabolism of phosphorothioate, phosphodiester, and methylphosphonate oligonucleotides. In: *Antisense Research and Applications,* Crooke, S. T. and Lebleu, B., Eds., CRC Press, Boca Raton, FL, 1993, pp. 427–449.
62. **Gao, W.-Y., Storm, C., Egan, W., and Cheng, Y.-C.,** Cellular pharmacology of phosphorothioate homooligodeoxynucleotides in human cells, *Mol. Pharmacol.,* 43, 45, 1993.
63. **Wu-Pong, S., Weiss, T. L., and Hunt, C. A.,** Antisense c-myc oligonucleotide cellular uptake, *Pharm. Res.,* 9, 1010, 1992.
64. **Iversen, P. L., Zhu, S., Meyer, A., and Zon, G.,** Cellular uptake and subcellular distribution of phosphorothioate oligonucleotides into cultured cells, *Antisense Res. Dev.,* 2, 211, 1992.
65. **Loke, S. L., Stein, C. A., Zhang, X. H., Mori, K., Nakanishi, M., Subasinghe, C., Cohen, J. S., and Neckers, L. M.,** Characterization of oligonucleotide transport into living cells, *Proc. Natl. Acad. Sci. U.S.A.,* 86, 3474, 1989.
66. **Yakubov, L. A., Deeva, E. A., Zarytova, V. F., Ivanova, E. M., Ryte, A. S., Yurchenko, L. V., and Vlassov, V. V.,** Mechanism of oligonucleotide uptake by cells. Involvement of specific receptors?, *Proc. Natl. Acad. Sci. U.S.A.,* 86, 6454, 1989.
67. **Iversen, P. L., Crouse, D., Zon, G., and Perry, G.,** Binding of antisense phosphorothioate oligonucleotides to murine lymphocytes is lineage specific and inducible, *Antisense Res. Dev.,* 2, 223, 1992.
68. **Zhao, Q., Matson, S., Herrera, C. J., Fisher, E., Yu, H., and Krieg, A. M.,** Comparison of cellular binding and uptake of antisense phosphodiester, phosphorothioate, and mixed phosphorothioate and methylphosphonate oligonucleotides, *Antisense Res. Dev.,* 3, 53, 1993.
69. **Bennett, C. F., Chiang, M.-Y., Chan, H., Shoemaker, J. E. E., and Mirabelli, C. K.,** Cationic lipids enhance cellular uptake and activity of phosphorothioate antisense oligonucleotides, *Mol. Pharmacol.,* 41, 1023, 1992.
70. **Nestle, F. O., Mitra, R. S., Bennett, C. F., Chan, H., and Nickoloff, B. J.,** Cationic lipid is not required for uptake and inhibitory activity of ICAM-1 phosphorothioate antisense oligonucleotides in keratinocytes, *J. Invest. Dermatol.,* 103, 569, 1994.
71. **Bennett, C. F., Chiang, M.-Y., Chan, H., and Grimm, S.,** Use of cationic lipids to enhance the biological activity of antisense oligonucleotides, *J. Liposome Res.,* 3, 85, 1993.
72. **Wagner, R. W., Matteucci, M. D., Lewis, J. G., Gutierrez, A. J., Moulds, C., and Froehler, B. C.,** Antisense gene inhibition by oligonucleotides containing C-5 propyne pyrimidines, *Science,* 260, 1510, 1993.
73. **Felsenfeld, G., Davies, D. R., and Rich, A.,** Studies on the formation of two- and three-stranded polyribonucleotides, *J. Am. Chem. Soc.,* 79, 2023, 1957.
74. **Moser, H. E. and Dervan, P. B.,** Sequence-specific cleavage of double helical DNA by triple helix formation, *Science,* 238, 645, 1987.
75. **Francois, J. C., Saison-Behmoaras, T., and Helene, C.,** Sequence-specific recognition of the major groove of DNA by oligodeoxynucleotides via triple helix formation. Footprinting studies, *Nucl. Acids Res.,* 16, 11431, 1988.
76. **Cooney, M., Czernuszewicz, G., Postel, E. H., Flint, S. J., and Hogan, M. E.,** Site-specific oligonucleotide binding represses transcription of the human c-myc gene in vitro, *Science,* 241, 456, 1988.
77. **Helene, C.,** The anti-gene strategy: control of gene expression by triplex-forming-oligonucleotides, *Anticancer Drug Des.,* 6, 569, 1991.
78. **Hogan, M. E.,** Inhibition of transcription by triple-helix-forming oligonucleotides, *Clin. Chem.,* 39, 712, 1993.
79. **Mohan, V., Cheng, Y.-K., Marlow, G. E., and Montgomery Pettitt, B.,** Molecular recognition of Watson-Crick base-pair reversals in triple-helix formation. Use of nonnatural oligonucleotide bases, *Biopolymers,* 33, 1317, 1993.
80. **Sun, J.-S. and Helene, C.,** Oligonucleotide-directed triple-helix formation, *Curr. Opin. Struct. Biol.,* 3, 345, 1993.
81. **Beal, P. A. and Dervan, P. B.,** Second structural motif for recognition of DNA by oligonucleotide-directed triple-helix formation, *Science,* 251, 1360, 1991.
82. **Durland, R. H., Kessler, D. J., Gunnell, S., Duvic, M., Pettitt, B. M., and Hogan, M. E.,** Binding of triple helix forming oligonucleotides to sites in gene promoters, *Biochemistry,* 30, 9246, 1991.

83. **Sun, J. S., De Bizemont, T., Duval-Valentin, G., Montenay-Garestier, T., and Helene, C.,** Extension of the range of recognition sequences for triple-helix formation by oligonucleotides containing guanines and thymines, *C.R. Acad. Sci. III,* 313, 585, 1991.
84. **Grigoriev, M., Praseuth, D., Robin, P., Hemar, A., Saison-Behmoaras, T., Dautry-Varsat, A., Thuong, N. T., Helène, C., and Harel-Bellan, A.,** A triple helix-forming oligonucleotide-intercalator conjugate acts as a transcriptional repressor via inhibition of NF κB binding to interleukin-2 receptor α-regulatory sequence, *J. Biol. Chem.,* 267, 3389, 1992.
85. **Grigoriev, M., Praseuth, D., Guieysse, A. L., Robin, P., Thuong, N. T., Helène, C., and Harel-Bellan, A.,** Inhibition of gene expression by triple helix-directed DNA cross-linking at specific sites, *Proc. Natl. Acad. Sci. U.S.A.,* 90, 3501, 1993.
86. **Koh, J. S. and Dervan, P. B.,** Design of a nonnatural deoxyribonucleoside for recognition of GC base pairs by oligonucleotide-directed triple helix formation, *J. Am. Chem. Soc.,* 114, 1470, 1992.
87. **Griffin, L. C., Kiessling, L. L., Beal, P. A., Gillespie, P., and Dervan, P. B.,** Recognition of all four base pairs of double-helical DNA by triple-helix formation. Design of nonnatural deoxyribonucleosides for pyrimidine-purine base pair binding, *J. Am. Chem. Soc.,* 114, 7976, 1992.
88. **Matteucci, M., Lin, K.-Y., Butcher, S., and Moulds, C.,** Deoxyoligonucleotides bearing neutral analogues of phosphodiester linkages recognize duplex DNA via triple-helix formation, *J. Am. Chem. Soc.,* 113, 7767, 1991.
89. **Stilz, H. U. and Dervan, P. B.,** Specific recognition of CG base pairs by 2-deoxynebularine within the purine.purine.pyrimidine triple-helix motif, *Biochemistry,* 32, 2177, 1993.
90. **Kessler, D. J., Pettitt, B. M., Cheng, Y.-K., Smith, S. R., Jayaraman, K., Vu, H. M., and Hogan, M. E.,** Triple helix formation at distant sites: hybrid oligonucleotides containing a polymeric linker, *Nucl. Acids Res.,* 21, 4810, 1993.
91. **Bielinska, A., Shivdasani, R. A., Zhang, L., and Nabel, G. J.,** Regulation of gene expression with double-stranded phosphorothioate oligonucleotides, *Science,* 250, 997, 1990.
92. **Eck, S. L., Perkins, N. D., Carr, D. P., and Nabel, G. J.,** Inhibition of phorbol ester-induced cellular adhesion by competitive binding of NF-kappaB *in vivo, Mol. Cell. Biol.,* 13, 6530, 1993.
93. **Ecker, D. J., Vickers, T. A., Hanecak, R., Driver, V., and Anderson, K.,** Rational screening of oligonucleotide combinatorial libraries for drug discovery, *Nucl. Acids Res.,* 21, 1853, 1993.
94. **Griffin, L. C., Tidmarsh, G. F., Bock, L. C., Toole, J. J., and Leung, L. L. K.,** *In vivo* anticoagulant properties of a novel nucleotide-based thrombin inhibitor and demonstration of regional anticoagulation in extracorporeal circuits, *Blood,* 81, 3271, 1993.
95. **Wang, K. Y., McCurdy, S., Shea, R. G., Swaminathan, S., and Bolton, P. H.,** A DNA aptamer which binds to and inhibits thrombin exhibits a new structural motif for DNA, *Biochemistry,* 32, 1899, 1993.
96. **Williamson, J. R., Raghuraman, M. K., and Cech, T. R.,** Monovalent cation-induced structure of telomeric DNA: the G-quartet model, *Cell,* 59, 871, 1989.
97. **Scaria, P. V., Shire, S. J., and Shafer, R. H.,** Quadruplex structure of $d(G_3T_4G_3)$ stabilized by K^+ or Na^+ is an asymmetric hairpin dimer, *Proc. Natl. Acad. Sci. U.S.A.,* 89, 10336, 1992.
98. **Pearson, A. M., Rich, A., and Krieger, M.,** Polynucleotide binding to macrophage scavenger receptors depends on the formation of base-quartet-stabilized four-stranded helices, *J. Biol. Chem.,* 268, 3546, 1993.
99. **Wyatt, J. R., Vickers, T. A., Roberson, J. L., Buckheit, R. W., Klimkait, T., DeBaets, E., Davis, P. W., Rayner, B., Imbach, J. L., and Ecker, D. J.,** Combinatorially selected G-quartet structure is a potent inhibitor of HIV envelope-mediated cell fusion, *Proc. Natl. Acad. Sci. U.S.A.,* 91, 1356, 1994.
100. **Uhlmann, E. and Peyman, A.,** Antisense oligonucleotides: a new therapeutic principle, *Chem. Rev.,* 90, 544, 1990.
101. **Sanghvi, Y. S., Hoke, G. D., Zounes, M. C., Freier, S. M., Martin, J. F., Chan, H., Acevedo, O. L., Ecker, D. J., Mirabelli, C. K., Crooke, S. T., and Cook, P. D.,** Synthesis and biological evaluation of antisense oligonucleotides containing modified pyrimidines, *Nucleosides Nucleotides,* 10, 345, 1991.
102. **Cook, P. D.,** Medicinal chemistry strategies for antisense research. In: *Antisense Research and Applications,* Crooke, S. T. and Lebleu, B. Eds., CRC Press, Boca Raton, FL, 1993, pp. 149–187.

103. **Milligan, J. F., Matteucci, M. D., and Martin, J. C.,** Current concepts in antisense drug design, *J. Med. Chem.,* 36, 1923, 1993.
104. **Wickstrom, E.,** Oligodeoxynucleotide stability in subcellular extracts and culture media, *J. Biochem. Biophys. Methods,* 13, 97, 1986.
105. **Tidd, D. M. and Warenius, H. M.,** Partial protection of oncogene, anti-sense oligodeoxynucleotides against serum nuclease degradation using terminal methylphosphonate groups, *Br. J. Cancer,* 60, 343, 1989.
106. **Akhtar, S., Kole, R., and Juliano, R. L.,** Stability of antisense DNA oligodeoxynucleotide analogs in cellular extracts and sera, *Life Sci.,* 49, 1793, 1991.
107. **Williams, D. M., Benseler, F., and Eckstein, F.,** Properties of 2′-fluorothymidine-containing oligonucleotides. Interaction with restriction endonuclease *Eco*RV, *Biochemistry,* 30, 4001, 1991.
108. **Hoke, G. D., Draper, K., Freier, S. M., Gonzalez, C., Driver, V. B., Zounes, M. C., and Ecker, D. J.,** Effects of phosphorothioate capping on antisense oligonucleotide stability, hybridization and antiviral efficacy versus herpes simplex virus infection, *Nucl. Acids Res.,* 19, 5743, 1991.
109. **Agris, C. H., Blake, K. R., Miller, P. S., Reddy, M. P., and Ts'o, P. O. P.,** Inhibition of vesicular stomatitis virus protein synthesis and infection by sequence-specific oligodeoxyribonucleoside methylphosphonates, *Biochemistry,* 25, 6268, 1986.
110. **Smith, C. C., Aurelian, L., Reddy, M. P., Miller, P. S., and Ts'o, P. O. P.,** Antiviral effect of an oligo(nucleoside methylphosphonate) complementary to the splice junction of herpes simplex virus type 1 immediate early pre-mRNAs 4 and 5, *Proc. Natl. Acad. Sci. U.S.A.,* 83, 2787, 1986.
111. **Kulka, M., Smith, C. C., Aurelian, L., Fishelevich, R., Meade, K., Miller, P., and Ts'o, P. O. P.,** Site specificity of the inhibitory effects of oligo(nucleoside methylphosphonate)s complementary to the acceptor splice junction of herpes simplex virus type 1 immediate early mRNA 4, *Proc. Natl. Acad. Sci. U.S.A.,* 86, 6868, 1989.
112. **Laurence, J., Sikder, S. K., Kulkosky, J., Miller, P., and Ts'o, P. O. P.,** Induction of chronic human immunodeficiency virus infection is blocked *in vitro* by a methylphosphonate oligodeoxynucleoside targeted to a U3 enhancer element, *J. Virol.,* 65, 213, 1991.
113. **Chang, E. H., Miller, P. S., Cushman, C., Devadas, K., Pirollo, K. F., Ts'o, P. O. P., and Yu, Z. P.,** Antisense inhibition of *ras* p21 expression that is sensitive to a point mutation, *Biochemistry,* 30, 8283, 1991.
114. **Marshall, W. S. and Caruthers, M. H.,** Phosphorodithioate DNA as a potential therapeutic drug, *Science,* 259, 1564, 1993.
115. **Marcus-Sekura, C. J., Woerner, A. M., Shinozuka, K., Zon, G., and Quinnan, G. V., Jr.,** Comparative inhibition of chloramphenicol acetyltransferase gene expression by antisense oligonucleotide analogues having alkyl phosphotriester, methylphosphonate and phosphorothioate linkages, *Nucl. Acids Res.,* 15, 5749, 1987.
116. **Sood, A., Shaw, R. B., and Spielvogel, B. F.,** Boron-containing nucleic acids. Synthesis of oligonucleoside boranophosphates, *J. Am. Chem. Soc.,* 112, 9000, 1991.
117. **Matteucci, M.,** Deoxyoligonucleotide analogs based on formacetal linkages, *Tetrahedron Lett.,* 31, 2385, 1990.
118. **Nielsen, P. E., Egholm, M., Berg, R. H., and Buchardt, O.,** Sequence-selective recognition of DNA by strand displacement with a thymine-substituted polyamide, *Science,* 254, 1497, 1991.
119. **Debart, F., Vasseur, J.-J., Sanghvi, Y. S., and Cook, P. D.,** Intermolecular radical C-C bond formation. Synthesis of a novel dinucleoside linker for non-anionic antisense oligonucleosides, *Tetrahedron Lett.,* 33, 2645, 1992.
120. **Egholm, M., Buchardt, O., Nielsen, P. E., and Berg, R. H.,** Peptides nucleic acids (PNA). Oligonucleotide analogues with an achiral peptide backbone, *J. Am. Chem. Soc.,* 114, 1895, 1992.
121. **Sanghvi, Y. S. and Cook, P. D.,** Towards second-generation synthetic backbones for antisense oligonucleosides. In: *Nucleosides, Nucleotides as Antiviral and Antitumor Agents,* Chu, C. K. and Baker, D. C., Eds., Plenum Press, New York, 1993, pp. 309–322.
122. **Debart, F., Vasseur, J.-J., Sanghvi, Y. S., and Cook, P. D.,** Synthesis and incorporation of methyleneoxyl (methylimino) linked thymidine dimer into antisense oligonucleotides, *Bioorg. Med. Chem. Lett.,* 2, 1479, 1992.

123. **Vasseur, J.-J., Debart, F., Sanghvi, Y. S., and Cook, P. D.,** Oligonucleosides: synthesis of a novel methylhydroxylamine-linked nucleoside dimer and its incorporation into antisense sequences, *J. Am. Chem. Soc.,* 114, 4006, 1992.
124. **De Mesmaeker, A., Lebreton, J., Waldner, A., Fritsch, V., Wolf, R. M., and Freier, S. M.,** Amides as substitute for the phosphodiester linkage in antisense oligonucleotides, *Synlett,* 10, 733, 1993.
125. **Hanvey, J. C., Peffer, N. C., Bisi, J. E., Thomson, S. A., Cadilla, R., Josey, J. A., Ricca, D. J., Hassman, C. F., Bonham, M. A., Au, K. G., Carter, S. G., Bruckenstein, D. A., Boyd, A. L., Noble, S. A., and Babiss, L. E.,** Antisense and antigene properties of peptide nucleic acids, *Science,* 258, 1481, 1992.
126. **Egholm, M., Buchardt, O., Christensen, L., Behrens, C., Freier, S. M., Driver, D. A., Berg, R. H., Kim, S. K., Norden, B., and Nielsen, P. E.,** PNA hybridizes to complementary oligonucleotides obeying the Watson-Crick hydrogen bonding rules, *Nature,* 365, 490, 1993.
127. **Inoue, H., Hayase, Y., Iwai, S., and Ohtsuke, E.,** Sequence dependent hydrolysis of RNA using modified oligonucleotide splints and RNase H, *FEBS Lett.,* 215, 327, 1987.
128. **Iribarren, A. M., Sproat, B. S., Neuner, P., Sulston, I., Ryder, U., and Lamond, A. I.,** 2′-*O*-alkyl oligoribonucleotides as antisense probes, *Proc. Natl. Acad. Sci. U.S.A.,* 87, 7747, 1990.
129. **Guinosso, C. J., Hoke, G. D., Freier, S. M., Martin, J. F., Ecker, D. J., Mirabelli, C. K., Crooke, S. T., and Cook, P. D.,** Synthesis and biophysical and biological evaluation of 2′-modified antisense oligonucleotides, *Nucleosides Nucleotides,* 10, 259, 1991.
130. **Lesnik, E. A., Guinosso, C. J., Kawasaki, A. M., Sasmor, H., Zounes, M., Cummins, L. L., Ecker, D. J., Cook, P. D., and Freier, S. M.,** Oligodeoxynucleotides containing 2′-*O*-modified adenosine. Synthesis and effects on stability of DNA:RNA duplexes, *Biochemistry,* 32, 7832, 1993.
131. **Lamond, A. I. and Sproat, B. S.,** Antisense oligonucleotides made of 2′-*O*-alkylRNA. Their properties and applications in RNA biochemistry, *FEBS Lett.,* 325, 123, 1993.
132. **Kawasaki, A. M., Casper, M. D., Freier, S. M., Lesnik, E. A., Zounes, M. C., Cummins, L. L., Gonzalez, C., and Cook, P. D.,** Uniformly modified 2′-deoxy-2′-fluoro phosphorothioate oligonucleotides as nuclease-resistant antisense compounds with high affinity and specificity for RNA targets, *J. Med. Chem.,* 36, 831, 1993.
133. **Agrawal, S., Mayrand, S. H., Zamecnik, P. C., and Pederson, T.,** Site-specific excision from RNA by RNase H and mixed-phosphate-backbone oligodeoxynucleotides, *Proc. Natl. Acad. Sci. U.S.A.,* 87, 1401, 1990.
134. **Bertrand, J.-R., Imbach, J.-L., Paoletti, C., and Malvy, C.,** Comparative activity of α- and β-anomeric oligonucleotides on rabbit β globin synthesis. Inhibitory effect of cap targeted α-oligonucleotides, *Biochem. Biophys. Res. Commun.,* 164, 311, 1989.
135. **Imbach, J.-L.,** Alpha-oligonucleotides. In: *Antisense Research and Applications,* Crooke, S. T. and Lebleu, B., Eds., CRC Press, Boca Raton, FL, 1993.
136. **Sanghvi, Y. S.,** Heterocyclic base modifications in nucleic acids and their applications in antisense oligonucleotides. In: *Antisense Research and Applications,* Crooke, S. T. and Lebleu, B., Eds., CRC Press, Boca Raton, FL, 1993, pp. 273–288.
137. **Muraoka, M., Miles, H. T., and Howard, F. B.,** Copolymers of adenylic and 2-aminoadenylic acids. Effect of progressive changes in hydrogen bonding and stacking on interaction with poly(uridylic acid), *Biochemistry,* 19, 2429, 1980.
138. **Lamm, G. M., Blencowe, B. J., Sproat, B. S., Iribarren,, A. M., Ryder, U., and Lamond, A. I.,** Antisense probes containing 2-aminoadenosine allow efficient depletion of U5 snRNP from HeLa splicing extracts, *Nucl. Acids Res.,* 19, 3193, 1991.
139. **Letsinger, R. L., Zhang, G. R., Sun, D. K., Ikeuchi, T., and Sarin, P. S.,** Cholesteryl-conjugated oligonucleotides: synthesis, properties, and activity as inhibitors of replication of human immunodeficiency virus in cell culture, *Proc. Natl. Acad. Sci. U.S.A.,* 86, 6553, 1989.
140. **Krieg, A. M., Tonkinson, J., Matson, S., Zhao, Q., Saxon, M., Zhang, L.-M., Bhanja, U., Yakubov, L., and Stein, C. A.,** Modification of antisense phosphodiester oligodeoxynucleotides by a 5′ cholesteryl moiety increases cellular association and improves efficacy, *Proc. Natl. Acad. Sci. U.S.A.,* 90, 1048, 1993.
141. **Manoharan, M.,** Designer antisense oligonucleotides: conjugation chemistry and functionality placement. In: *Antisense Research and Applications,* Crooke, S. T. and Lebleu, B., Eds., CRC Press, Boca Raton, FL, 1993, pp. 303–349.

142. **Harel-Bellan, A., Durum, S., Muegge, K., Abbas, A. K., and Farrar, W. L.,** Specific inhibition of lymphokine biosynthesis and autocrine growth using antisense oligonucleotides in TH1 and Th2 helper T cell clones, *J. Exp. Med.,* 168, 2309, 1988.
143. **Zubiaga, A. M., Munoz, E., and Huber, B. T.,** Production of IL-1 alpha by activated Th type 2 cells. Its role as an autocrine growth factor, *J. Immunol.,* 146, 3849, 1991.
144. **Miles, S. A., Rezai, A. R., Salazar-Gonz;aaalez, J. F., Meyden, M. V., Stevens, R. H., Logan, D. M., Mitsuyasu, R. T., Taga, T., Hirano, T., Kishimoto, T., and Martínez-Maza, O.,** AIDS Kaposi sarcoma-derived cells produce and respond to interleukin-6, *Proc. Natl. Acad. Sci. U.S.A.,* 87, 4068, 1990.
145. **Levy, Y., Tsapis, A., and Brouet, J.-C.,** Interleukin-6 antisense oligonucleotides inhibit the growth of human myeloma cell lines, *J. Clin. Invest.,* 88, 696, 1991.
146. **Schadendorf, D., Moller, A., Algermissen, B., Worm, M., Sticherling, M., and Czarnetzki, B. M.,** IL-8 produced by human malignant melanoma cells *in vitro* is an essential autocrine growth factor, *J. Immunol.,* 151, 2667, 1993.
147. **Birchenall-Roberts, M. C., Ferrer, C., Ferris, D., Falk, L. A., Kasper, J., Gretchen, W., and Ruscetti, F. W.,** Inhibition of murine monocyte proliferation by a colony-stimulating factor-1 antisense oligodeoxynucleotide. Evidence for autocrine regulation, *J. Immunol.,* 145, 3290, 1990.
148. **Fujiwara, T. and Grimm, E. A.,** Specific inhibition of interleukin-1β gene expression by an antisense oligonucleotide. Obligatory role of interleukin-1 in the generation of lymphokine-activated killer cells, *Cancer Res.,* 52, 4954, 1992.
149. **Maier, J. A. M., Voulalas, P., Roeder, D., and Maciag, T.,** Extension of the life-span of human endothelial cells by an interleukin-1α antisense oligomer, *Science,* 249, 1570, 1990.
150. **Garfinkel, S., Haines, D. S., Brown, S., Wessendorf, J., Gillespie, D. H., and Maciag, T.,** Interleukin-1α mediates an alternative pathway for the antiproliferative action of poly(I-C) on human endothelial cells, *J. Biol. Chem.,* 267, 24375, 1992.
151. **Benbernou, N., Matsiota-Bernard, P., and Guenounou, M.,** Effect of cytokine-specific antisense oligonucleotides on the immunoglobulin production by rat spleen cells *in vitro*, *Biochimie,* 75, 55, 1993.
152. **Benbernou, N., Matsiota-Bernard, P., and Guenounou, M.,** Antisense oligonucleotides to interleukin-4 regulate IgE and IgG2a production by spleen cells from *Nippostrongylus brasiliensis*-infected rats, *Eur. J. Immunol.,* 23, 659, 1993.
153. **Burch, R. M. and Mahan, L. C.,** Oligonucleotides antisense to the interleukin-1 receptor mRNA block the effects of interleukin-1 in cultured murine and human fibroblasts and in mice, *J. Clin. Invest.,* 88, 1190, 1991.
154. **Segal, G. M., Smith, T. D., Heinrich, M. C., Ey, F. S., and Bagby, G. C.,** Specific repression of granulocyte-macrophage and granulocyte colony-stimulating factor gene expression in interleukin-1-stimulated endothelial cells with antisense oligodeoxynucleotides, *Blood,* 80, 609, 1992.
155. **Thomae, K. R., Geller, D. A., Billiar, T. R., Davies, P., Pitt, B. R., Simmons, R. L., and Nakayama, D. K.,** Antisense oligodeoxynucleotide to inducible nitric oxide synthase inhibits nitric oxide synthesis in rat pulmonary artery smooth muscle cells in culture, *Surgery,* 114, 272, 1993.
156. **Maier, J. A. M. and Ragnotti, G.,** An oligomer targeted against protein kinase C prevents interleukin-1 induction of cyclooxygenase expression in human endothelial cells, *Exp. Cell Res.,* 205, 52, 1993.
157. **Narayanan, R., Higgins, K. A., Perez, J. R., Coleman, T. A., and Rosen, C. A.,** Evidence for differential functions of the p50 and p65 subunits of NF-kB with a cell adhesion model, *Mol. Cell Biol.,* 13, 3802, 1993.
158. **Sokoloski, J. A., Sartorelli, A. C., Rosen, C. A., and Narayanan, R.,** Antisense oligonucleotides to the p65 subunit of NF-κB block CD11b expression and alter adhesion properties of differentiated HL-60 granulocytes, *Blood,* 82, 625, 1993.
159. **Chem, T.-L., Miller, P. S., Ts'o, P. O., and Colvin, O. M.,** Disposition and metabolism of oligonucleoside methylphosphonate following a single i.v. injection in mice, *Drug Metab. Dispos.,* 185, 815, 1990.
160. **Goodarzi, G., Watabe, M., and Watabe, K.,** Organ distribution and stability of phosphorothioated oligodeoxyribonucleotides in mice, *Biopharm. Drug Dispos.,* 13, 221, 1992.

161. **Agrawal, S., Temsamani, J., and Tang, J. Y.,** Pharmacokinetics, biodistribution, and stability of oligodeoxynucleotide phosphorothioates in mice, *Proc. Natl. Acad. Sci. U.S.A.,* 88, 7595, 1991.
162. **Inagaki, M., Togawa, K., Carr, B. I., Ghosh, K., and Cohen, J. S.,** Antisense oligonucleotides. Inhibition of liver cell proliferation and *in vivo* disposition, *Transplant. Proc.,* 24, 2971, 1992.
163. **Iverson, P.,** *In vivo* studies with phosphorothioate oligonucleotides: pharmacokinetics prologue, *Anti-Cancer Drug Design,* 6, 531, 1992.
164. **Zendegui, J. G., Vasquez, K. M., Tinsley, J. H., Kessler, D. J., and Hogan, M. E.,** *In vivo* stability and kinetics of absorption and disposition of 3′-phosphopropyl amine oligonucleotides, *Nucl. Acids Res.,* 20, 307, 1992.
165. **Cossum, P. A., Sasmor, H., Dellinger, D., Truong, L., Cummins, L., Owens, S., Markham, P. A., Shea, J. P., and Crooke, S. T.,** Disposition of the 14C-labeled phosphorothioate oligonucleotide ISIS 2105 after intravenous administration to rats, *J. Pharmacol. Exp. Ther.,* 1994, in press.
166. **Cossum, P. A., Truong, L., Owens, S. R., Markham, P. M., Shea, J. P., and Crooke, S. T.,** Pharmacokinetics of a 14C-labeled phosphorothioate oligonucleotide, ISIS 2105, after intradermal administration in rats, *J. Pharmacol. Exp. Ther.,* in press, 1994.
167. **Crooke, S. T., Grillone, L. R., Tendolkar, A., Garret, A., Fratkin, M. J., Leeds, J., and Barr, W. H.,** A pharmacokinetic evaluation of ^{14}C-labeled afovirsen sodium in a patient with genital warts, *Clin. Pharm. Therap.,* 56, 641, 1994.
168. **Graham, M. J., Freier, S. M., Crooke, R. M., Ecker, D. J., Maslova, R. N., and Lesnik, E. A.,** Tritium labeling of antisense oligonucleotides by exchange with tritiated water, *Nucl. Acids Res.,* 21, 3737, 1993.
169. **De Smidt, P. C., Le Doan, T., De Falco, S., and Van Berkel, T. J. C.,** Association of antisense oligonucleotides with lipoproteins prolongs the plasma half-life and modifies the tissue distribution, *Nucl. Acids Res.,* 19, 4695, 1991.
170. **Whitsell, L., Rosolen, A., and Neckers, L. M.,** *In vivo* modulation of N-myc expression by continuous perfusion with an antisense oligonucleotide, *Antisense Res. Dev.,* 1, 343, 1991.
171. **Bennett, C. F., Zuckerman, J. E., Kornbrust, D., Sasmor, H., Leeds, J. M., and Crooke, S. T.,** Pharmacokinetics in mice of a [^{3}H]-labeled phosphorothioate eligonucleotide formulated in the presence and absence of a cationic lipid, *J. Controlled Release,* in press.
172. **Chiasson, B. J., Hooper, M. L., Murphy, P. R., and Robertson, H. A.,** Antisense oligonucleotide eliminates *in vivo* expression of c-fos in mammalian brain, *Eur. J. Pharmacol.,* 227, 451, 1992.
173. **Kitajima, I., Shinohara, T., Minor, T., Bibbs, L., Bilakovics, J., and Nerenberg, M.,** Human T-cell leukemia virus type I tax transformation is associated with increased uptake of oligodeoxynucleotides *in vitro* and *in vivo*, *J. Biol. Chem.,* 267, 25881, 1992.
174. **Ratajczak, M. Z., Kant, J. A., Luger, S. M., Huiya, N., Zhang, J., Zon, G., and Gewirtz, A. M.,** *In vivo* treatment of human leukemia in a scid mouse model with c-myb antisense oligodeoxynucleotides, *Proc. Natl. Acad. Sci. U.S.A.,* 89, 11823, 1992.
175. **Simons, M., Edelman, E. R., DeKeyser, J.-L., Langer, R., and Rosenberg, R. D.,** Antisense c-myb oligonucleotides inhibit arterial smooth muscle cell accumulation *in vivo*, *Nature,* 359, 67, 1992.
176. **Offensperger, W.-B., Offensperger, S., Walter, E., Teubner, K., Igloi, G., Blum, H. E., and Gerok, W.,** *In vivo* inhibition of duck hepatitis B virus replication and gene expression by phosphorothioate modified antisense oligodeoxynucleotides, *EMBO J.,* 12, 1257, 1993.
177. **Amaratunga, A., Morin, P. J., Kosik, K. S., and Fine, R. E.,** Inhibition of kinesin synthesis and rapid anterograde axonal transport in vivo by an antisense oligonucleotide, *J. Biol. Chem.,* 268, 17427, 1993.
178. **Heilig, M., Engel, J. A., and Söderpalm, B.,** *C-fos* antisense in the nucleus accumbens blocks the locomotor stimulant action of cocaine, *Eur. J. Pharmacol.,* 236, 339, 1993.
179. **Morishita, R., Gibbons, G. H., Ellison, K. E., Nakajima, M., Zhang, L., Kaneda, Y., Ogihara, T., and Dzau, V. J.,** Single intraluminal delivery of antisense cdc2 kinase and proliferating-cell nuclear antigen oligonucleotides results in chronic inhibition of neointimal hyperplasia, *Proc. Natl. Acad. Sci. U.S.A.,* 90, 8474, 1993.

180. **Mojcik, C. F., Gourley, M. F., Klinman, D. M., Krieg, A. M., Gmelig-Meyling, F., and Steinberg, A. D.,** Administration of a phosphorothioate oligonucleotide antisense to murine endogenous retroviral MCF *env* causes immune effects *in vivo* in a sequence-specific manner, *Clin. Immunol. Immunopathol.,* 67, 130, 1993.
181. **Osen-Sand, A., Catsicas, M., Staple, J. K., Jones, K. A., Ayala, G., Knowles, J., Grenningloh, G., and Catsicas, S.,** Inhibition of axonal growth by SNAP-25 antisense oligonucleotides *in vitro* and *in vivo, Nature,* 364, 445, 1993.
182. **Pollio, G., Xue, P., Zanisi, M., Nicolin, A., and Maggi, A.,** Antisense oligonucleotide blocks progesterone-induced lordosis behavior in ovariectomized rats, *Mol. Brain Res.,* 19, 135, 1993.
183. **Higgins, K. A., Perez, J. R., Coleman, T. A., Dorshkind, K., McComas, W. A., Sariento, U. M., Rosen, C. A., and Narayanan, R.,** Antisense inhibition of the p65 subunit of NF-kB blocks tumorigenicity and causes tumor regression, *Proc. Natl. Acad. Sci. U.S.A.,* 90, 9901, 1993.
184. **Wahlestedt, C., Pich, E. M., Koob, G. F., Yee, F., and Heilig, M.,** Modulation of anxiety and neuropeptide Y-Y1 receptors by antisense oligodeoxynucleotides, *Science,* 259, 528, 1993.
185. **Whitesell, L., Geselowitz, D., Chavany, C., Fahmy, B., Walbridge, S., Alger, J. R., and Neckers, L. M.,** Stability, clearance, and disposition of intraventricularly administered oligodeoxynucleotides. Implications for therapeutic application within the central nervous system, *Proc. Natl. Acad. Sci. U.S.A.,* 90, 4665, 1993.
186. **Kitajima, I., Shinohara, T., Bilakovics, J., Brown, D. A., Xiao, X., and Nerenberg, M.,** Ablation of transplanted HTLV-1 tax-transformed tumors in mice by antisense inhibition of NF-κB, *Science,* 258, 1792, 1992.
187. **Stepkowski, S.M., Tu, Y., Condon, T. P., and Bennett, C. F.,** Blocking of heart allograft rejection by intercellular adhesion molecule-1 antisense oligonucleotides alone or in combination with other immunosuppresive modalities, *J. Immunol.,* 153, 5336, 1994.
188. **Dean, N. M. and McKay, R.,** Inhibition of protein kinase c-alpha expression mice after systemic administration of phosphorothioate antisense oligodeoxynucleotides, *Proc. Natl. Acad. Sci. U.S.A.,* 91, 11762, 1994.
189. **Branda, R. F., Moore, A. L., Mathews, L., McCormack, J. J., and Zon, G.,** Immune stimulation by an antisense oligomer complementary to the *rev* gene of HIV-1, *Biochem. Pharmacol.,* 45, 2037, 1993.
190. **Sburlati, A. R., Manrow, R. E., and Berger, S. L.,** Prothymosin alpha antisense oligomers inhibit myeloma cell division, *Proc. Natl. Acad. Sci. U.S.A.,* 88, 253, 1991.
191. **Manson, J., Brown, T., and Duff, G.,** Modulation of interleukin-1β gene expression using antisense phosphorothioate oligonucleotides, *Lymphokine Res.,* 9, 35, 1990.
192. **Stepkowski, S. M., Tian, L., and Kloc, M.,** Interleukin-2 antisense oligonucleotides inhibit T-cell function, *Trans. Proc.,* 25, 125, 1993.

Naturally Occurring Inhibitors and Neutralizing Antibodies

Chapter 9

INTERLEUKIN-1 RECEPTOR ANTAGONIST

Steven M. Opal

CONTENTS

0-8493-8381-1/96/$0.00+$.50

I. INTRODUCTION

Interleukin-1 receptor antagonist (IL-1ra) is a naturally occurring, regulatory protein unique in cytokine biology in that its inhibitory action occurs at the cell receptor level. Interleukin-1 is a potent pro-inflammatory cytokine which has the capacity to induce a multitude of immunologic and metabolic effects in virtually every organ system in the human body.[1,2] While most interleukin-1 actions occur locally and are essential to normal homeostasis and cellular maintenance (see Chapter 1), unregulated release of interleukin-1 activates cytokine networks and precipitates systemic inflammatory reactions which may be disadvantageous to the host.[3,4] Compelling evidence now exists that excess levels of interleukin-1 induce systemic hypotension and create a metabolic and hemodynamic state which resembles septic shock.[3–5] Interleukin-1 activity, therefore, must be tightly regulated to ensure that the beneficial effects of low levels of IL-1 are preserved while the harmful effects of excess quantities of IL-1 are minimized.

The control of interleukin-1 action is one of the most complex and sophisticated regulatory systems known to exist within the human immune system and occurs at various levels including synthesis, processing, and release through the final binding of the functional ligand to its cellular receptor.[1,2,6,7] Posttranscriptional regulation of IL-1 synthesis occurs by alterations in messenger RNA stability.[7,8] There is evidence that IL-1 may downregulate its own activity in target tissues by decreasing IL-1 receptor type-1 mRNA stability.[9] Following translation into the 31-kDa precursor proteins of IL-1 alpha and IL-1 beta, the rate of release of IL-1 may differ according to the cell type and level of activation of producer cell lines[10] as well as the degree of apoptosis in the cell population. The interleukin-1 beta precursor molecule is inactive until cleaved by cysteine proteases into the active 17.5-kDa mature protein (see Chapter 7).[12] This proteolytic cleavage is carried out at the cell surface by a specific enzyme known as the interleukin-1 beta converting enzyme.[13,14] The activity of this enzyme may vary in different cell lines and affect the amount of functional IL-1 beta activity released into the systemic circulation. IL-1 alpha, on the other hand, is active as either the 31-kDa form or the 17-kDa cleavage product form.[15] The final mature protein of either IL-1 alpha or beta may bind to either of two separate cellular receptors at the tissue level. The density of receptors available for IL-1 binding may be downregulated by IL-1 itself or by other mediators.[16]

In addition to the aforementioned control mechanisms, interleukin-1 has another regulatory mechanism which has yet to be described for other cytokines. The presence of an IL-1 inhibitory protein which induces a cellular blockade at the receptor level is apparently a novel cytokine control system. This regulatory protein, now known as interleukin-1 receptor antagonist (IL-1ra), has become the focus of extensive research in an effort to understand its functional significance as a control mechanism in the normal host inflammatory response. IL-1ra may have a potential therapeutic role as an immunomodulator in a wide variety of inflammatory, infectious, and neoplastic diseases. This review will summarize the present knowledge of the physiology of IL-1ra and describe the current experience with IL-1ra as a novel therapeutic agent.

II. BIOLOGY OF IL-1ra

A. DISCOVERY AND INITIAL CHARACTERIZATION

A heterogeneous group of naturally occurring proteins and other molecules has been described which interfere with interleukin-1 activity.[17] Substances have been found in culture supernatants, synovial fluid, and other body fluids which bind to interleukin-1 in solution and may interfere nonspecifically with bioassays for IL-1. There are also natural inhibitors which interfere with the activity of other cytokines including interleukin-2 and tumor necrosis factor-alpha (TNF). Barak et al.[18] have described a specific interleukin-1 inhibitor in cultured supernatants from a myelomonocytic cell line. This IL-1 inhibitor has an apparent molecular mass of 52 kDa and interferes with specific IL-1 activity *in vitro*.[18] Autoantibodies to interleukin-1-alpha have been isolated from normal human serum. The functional activity of these autoantibodies as a physiologic carrier of IL-1 or a regulator of pro-inflammatory cytokine activity remains speculative at present.[19]

A specific interleukin-1 receptor antagonist was independently isolated and characterized by Arend and colleagues in 1985[20] and Seckinger and co-workers in 1987.[21,22] Arend's group described a 22 to 25-kDa inhibitory protein which was found in culture supernatants of human monocytes cultured on adherent IgG.[20] This protein was found to specifically inhibit IL-1 activity. Dayer's group isolated a natural IL-1 inhibitor from the urine of febrile patients with acute monocytic leukemia.[21] This urine-derived IL-1 inhibitor was found to be approximately 25 to 35 kDa in size and to inhibit (specifically) IL-1 activity in bioassays of prostaglandin synthesis and thymocyte proliferation. This group subsequently demonstrated that the activity of this IL-1 inhibitor was functional at the tissue receptor level.[22,23] The inhibitor blocked the binding of ^{125}I-IL-1 alpha to IL-1 receptors on a murine thymoma cell line.[22] It was subsequently shown that the IL-1 inhibitor originally isolated from cultured human monocytes also functioned as competitive inhibitor of IL-1 ligand-receptor binding.[24] Hannum and colleagues[24] demonstrated that the monocyte-derived IL-1 inhibitor consisted of three derivatives of a common native protein inhibitor. These molecules were purified and found to consist of a species of approximately 18 kDa as well two larger derivatives of approximately 22 kDa. The two larger forms were found to be glycosylated versions of the same 18-kDa molecule. All three forms of this native IL-1 inhibitor possessed equally potent IL-1 receptor binding activity, suggesting that the N-linked glycosylation site was not essential to its inhibitory properties.

Eisenberg and colleagues[25] have successfully cloned and sequenced the IL-1 inhibitor derived from human monocytes. This protein was shown to possess interleukin-1 receptor binding capacities and yet lacked agonist activity. This IL-1 inhibitor is now known as interleukin-1 receptor antagonist based upon its mechanism of action. The urine-derived IL-1 receptor antagonist is probably identical to the monocyte-derived molecule.[26,27]

B. ORGANIZATION OF THE HUMAN IL-1ra GENE

The gene mediating IL-1ra synthesis is located on the long arm of human chromosome 2. It has been mapped by Lennard et al.[28] to the 2q13–14.1 region of the human genome in close proximity to the synthetic genes for both IL-1 alpha and IL-1 beta, as well as type I and type II IL-1 receptor genes.[29,30] The close genetic linkage of all the known members of the IL-1 family supports the hypothesis that IL-1 ligands and their antagonist originated by divergent evolution from a single ancestral IL-1 gene.[31]

The structural components of the IL-1ra gene are depicted in Figure 1. The open reading frame for the secretory form of monocyte-derived interleukin-1 receptor antagonist encodes a single 177-amino acid protein and is derived from four exons. The short 14-nucleotide 5′-untranslated sequence, the 25-amino acid leader sequence, and the first 14 amino acids of the amino terminus of the mature protein are located on the first exon. The last three exons of IL-1ra share considerable sequence homology to the final three exons of the human IL-1 alpha and IL-1 beta genes.[28] Eisenberg and colleagues[31] have proposed that the human IL-1ra gene was derived from a partial duplication event of an ancestral IL-1 gene followed by transposition of a nonhomologous DNA sequence accounting for the first exon, first intron, and first two codons of exon 2 of the current

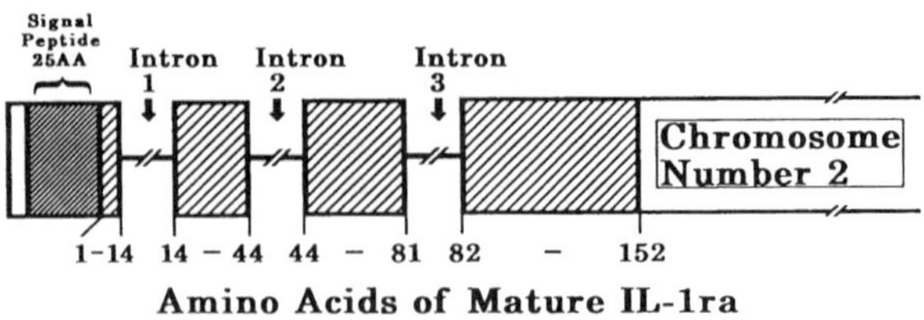

FIGURE 1 Schematic representation of the organization of the Interleukin-1 receptor antagonist gene. Exons are depicted by the large bars and intron locations are indicated by the thin lines. The shaded region indicates the translated IL-1ra protein produced by human monocytes. The first 25 amino acids at the amino terminus serve as signal peptides which are cleaved during secretion to generate the final 152-amino acid extracellular form of IL-1ra.

IL-1ra gene. By sequence analysis and mutation rate calculations, it has been estimated that the gene for IL-1ra diverged from IL-1 approximately 350 million years ago. This places the evolutionary event in the Carboniferous period (the age of amphibians in vertebrate evolution). The IL-1ra gene segregated about 50 million years before the subsequent divergence of the ancestral IL-1 gene into separate IL-1 alpha and IL-1 beta genes.[31,32] The human IL-1ra gene shares up to 77% sequence homology with rat and mouse IL-1ra gene products. A 26% amino acid sequence homology is found between human IL-1ra and human IL-1 beta; 19% homology exists between human IL-1ra and IL-1 alpha.[31]

Analysis of the 5′-untranslated sequences upstream from the IL-1ra gene reveals a consensus TATA box of 38 base pairs before the start of the cDNA sequence and 52 base pairs before the ATG start codon. Several possible transcription factor binding sites are found in the 5′ upstream region of the gene.[28] An LPS-sensitive promoter site of IL-1ra synthesis has been described within the 1700-base pair upstream region of the IL-1ra gene.[33] The 1,133 nucleotide, 3′-untranslated region of the IL-1ra mRNA lacks the AUUUA sequence found in many cytokine mRNA sequences.[25] This sequence of five base pairs may affect the stability of mRNA of a number of inflammatory cytokines including TNF-alpha and IL-1.[34]

Careful analysis of intron 2 of the human IL-1ra gene reveals a series of 86-base pair, tandem, direct repeat sequences.[35] The number of tandem repeats in this sequence varies and results in differences in the size of this intron. The functional significance of this polymorphism is unclear, but Tarlow et al.[36] have described potential protein-binding sites within these repeat regions. It is speculated that this allelic polymorphism may result in differential regulation of the level of IL-1ra synthesis in individuals with certain inflammatory disease states. The clinical significance of these findings awaits further study.

A structural variant of IL-1ra is produced by human keratinocytes and epithelial cell lines.[37] This molecule is known as intracellular IL-1ra (ic IL-1ra) and is derived from an alternative first exon which is spliced with the remaining exons via an internal acceptor site after base pair 87 of the cDNA. This variant lacks the 25-amino acid leader sequence of the secretory IL-1ra (s IL-1ra) product and thus remains an intracellular protein within epithelial cells.[37] The location of the alternative transcriptional start site of icIL-1ra is not known, but it does not appear to be within the 6 kilobase pairs of DNA upstream from the start site of the secretory IL-1ra gene in human monocytes.[28] IL-1ra extracted from keratinocytes possesses activity similar to the recombinant monocyte secretory form of IL-1ra.[38] Additionally, up to 50% of secretory IL-1ra produced by mononuclear cells remains cell associated. The functional significance of the intracellular form of IL-1ra is not clearly defined at the present time. It appears that the intracellular form of IL-1ra may be important in physiologic homeostasis within the fibroblast following IL-1 beta synthesis and release.[39] Intracellular IL-1ra may regulate growth and differentiation of IL-1 responsive neoplastic cells.[40] Since considerable amounts of IL-1 alpha and IL-1 beta remain in an intracellular location, ample opportunity is available for intracellular forms of interleukin-1 receptor antagonist to play an important role in the regulation of IL-1 actions within IL-1-producing cell lines.[41]

C. STRUCTURE OF IL-1ra

The monocyte-derived translated form of secretory IL-1ra is synthesized as a precursor protein which is 177 amino acids in length. The first 25 amino acids function as a signal peptide. The signal peptide is cleaved by peptidases at the cell membrane during the release of the final extracellular form of the molecule. The mature secretory IL-1ra molecule is a 152-amino acid protein which exists in nonglycosylated and glycosylated forms. The nonglycosylated form has a molecular weight of 17,115 Da. The glycosylated forms vary in weight up to 26 kDa depending on the degree of glycosylation.[24,25,42] The N-glycosylation site is located at Asn-84 of the mature IL-1ra molecule. Intracellular IL-1ra of keratinocytes lacks the 25-amino acid signal peptide of sIL-1ra. It consists of the identical 152-amino acid sequence of the secretory form of IL-1ra with 7 additional amino acids at its amino terminus.[38]

The tertiary structure of sIL-1ra has recently been deduced by examination of the hydrophilicity plots,[25] homologous amino acid locations,[31] and crystallographic analysis.[43] The three-dimensional structure of human IL-1ra resembles that of IL-1 alpha and IL-1 beta and consists of a pyramidal structure with a central beta-stranded core.[44] A total of 12 beta strands are present with 6 strands arranged in an antiparallel fashion to form a beta barrel. The hydrophobic core region of IL-1ra contains conserved amino acids with other IL-1 ligands, as does the vertex portion of the beta-barrel structure. The open end of the three-dimensional structure contains the binding site(s) for IL-1 receptors. Interleukin-1 beta has two discontinuous binding sites (termed site A and site B) on the exterior surface of the molecule. By site-specific mutagenesis and mutein analysis, the IL-1ra region homologous to the A site of IL-1 beta appears to be critical for binding to the type 1 interleukin-1 receptor.[43]

An amino acid substitution at position 145 of IL-1ra from lysine to aspartic acid results in a molecule with partial agonist activity. Conversely, aspartic acid to lysine substitution in the homologous region of IL-1 beta results in partial loss of agonist activity.[45] Apparently, the charge of the amino acid at this critical point in the amino acid sequence is essential in determining the functional activity of IL-1 ligands. Other amino acid substitutions of IL-1 beta result in muteins with wild-type binding to IL-1 receptors but reduced agonist activity.[46] These mutated molecules behave as functional receptor antagonists to IL-1 action. It is likely that multiple amino acids in the IL-1ra molecule participate as contact points for receptor binding and inhibition of signal transduction.

D. MECHANISM OF ACTION OF IL-1ra

IL-1ra induces IL-1 blockade by competitive inhibition at the cellular receptor level. IL-1ra binds with similar avidity to IL-1 receptors as the IL-1 ligands but lacks any intrinsic agonist activity. If sufficient quantities of IL-1ra are available, the receptor antagonist will occupy all the IL-1 receptors and sterically inhibit IL-1 ligand-receptor interactions. This results in a functional blockade of IL-1 activity at the cellular receptor level.[24,42]

Two structurally similar yet functionally and immunologically distinct IL-1 receptors have been described in human tissues. Both receptors have extracellular binding sites which consists of three immunoglobulin super family domains arranged in linear array. The type I IL-1 receptor is 80 kDa in size and is the dominant receptor type on T cells, endothelial cells, and hepatocytes.[47] The type II interleukin-1 receptor is 68 kDa in size and predominates on the cell surface of B cells, neutrophils, and certain bone marrow elements.[48–50] Both receptors have high-affinity extracellular binding sites for IL-1 (apparent dissociation constant ranges from 0.1 to 20 n*M*) and are expressed on cell surfaces with several hundred to several thousand receptor sites per cell.

Both type I and type II receptors may coexist on some cell lines.[48] The intracytoplasmic domain of the type II receptor is short and the degree to which type II receptor contributes to signal transduction and activation of IL-1 activity is unclear. A monoclonal antibody which specifically binds to the type I IL-1 receptor (35F5) induces a blockade of IL-1ra activity similar to that seen with IL-1ra.[51] This monoclonal antibody inhibited IL-1-mediated neutrophil egress from the bone marrow, neutrophilia, release of acute phase proteins, IL-6 production, and corticosterone production. These results indicate that the type I IL-1 receptor may be the principal mediator of IL-1 effects *in vivo.*

It has been speculated that the type II receptor may function as an inactive receptacle for circulating forms of IL-1 at the cell surface. The binding of IL-1 to type II receptors would decrease the amount of IL-1 available for binding of type I receptors. This would attenuate the amount of signal transduction and intracellular message from IL-1 and provide an additional downregulatory mechanism for IL-1 activity in target tissues. Alternatively, it is possible that type II receptors may participate in intracellular signaling pathways in some tissues or may facilitate the activity of type 1 receptors.[52] The intracellular tail of the type II receptor may activate other cytoplasmic elements which then translate the signal to the nucleus. The physiologic role of the type II IL-1 receptor needs to be characterized.

Interleukin-1 receptor antagonist binds with essentially equal avidity as IL-1 alpha and IL-1 beta to type I IL-1 receptors.[53] The dissociation constant (Kd) for IL-1ra to type I IL-1 receptors on T cells is 0.1 n*M,* which is similar to the Kd for IL-1 in this cell line. However, IL-1ra binds to type II receptors with lower affinity than IL-1 alpha or IL-1 beta. In human neutrophils, the affinity of IL-1ra to type II receptors is 25-fold lower than IL-1 beta and 200-fold lower than IL-1 alpha.[54]

Binding of IL-1 alpha or IL-1 beta to type I receptors results in internalization of the receptor-ligand complex, phosphorylation of the intracytoplasmic segment of the IL-1 receptor, and signal transduction to the nucleus.[55] The precise nature of the intranuclear transcriptional activator(s) of IL-1 action is currently unknown. However, the IL-1ra:IL-1 receptor complex remains on the cell surface and fails to stimulate any intracellular message.[53] There is no evidence of upregulation or downregulation of IL-1 receptor density on the cell surface. IL-1ra binding to the receptor is reversible and receptors previously occupied by IL-1ra may serve as subsequent binding sites for IL-1 ligands.[53]

Since as few as 5% of IL-1 receptors need to be occupied by natural IL-1 ligands in order to trigger IL-1 action, a large molar excess of IL-1ra must be present in the system to generate a complete blockade of IL-1 activity.[56] A tenfold molar access of IL-1ra is necessary to inhibit IL-1-induced synthesis of TNF, IL-1, or IL-6 from human monocytes *in vitro.*[57] *In vivo,* even greater concentrations of IL-1ra are necessary to block IL-1 activity. As much as a 1000-fold excess of IL-1ra was necessary to inhibit IL-induced fever[58] or hypotension[59] in rabbits. A 10,000-fold excess of IL-1ra was necessary to block IL-1-induced neutrophilia in the systemic circulation in an experimental mouse model.[51]

E. REGULATION OF IL-1ra SYNTHESIS

The synthesis of IL-1ra is mediated independent of the promoters of IL-1 alpha and IL-1 beta synthesis.[28] While activation of the receptor antagonist and IL-1 ligand synthesis is induced by a similar set of stimuli, the relative concentrations of IL-1ra vs. IL-1 alpha or IL-1 beta vary according to the stimulus, the cell line, and the degree of cellular differentiation.[27] The genes for IL-1ra and IL-1 alpha and beta are closely linked but exist on different genetic loci with separate promoters.[28,31] Numerous studies have demonstrated that substantial disparities exist in the rate of synthesis of IL-1 and the receptor antagonist, depending upon the reaction conditions.

LPS-stimulated monocytes *in vitro* produce similar quantities of IL-1 beta and IL-1ra and have similar transcriptional rates and mRNA stability.[60] In human volunteers, however, infusion of LPS resulted in a >100-fold excess of IL-1ra in the systemic circulation when compared to circulating IL-1 beta levels.[61] Monocytes cultured on adherent IgG exhibit low levels of IL-1 beta transcription and protein synthesis while high levels of IL-1ra protein are expressed.[60] Nonadherent monocytes produced 75% less IL-1ra than stationary mononuclear cells while IL-1 beta synthesis remained unchanged.[62] Human peripheral blood monocytes produce minute quantities of IL-1ra and IL-1 beta in the absence of human serum. However, the addition of as little as 1% AB serum enhanced IL-1ra production seven-fold whereas IL-1 beta synthesis remained unchanged. Soluble IgG markedly stimulated IL-1ra synthesis by peripheral blood mononuclear cells as did granulocyte-macrophage colony-stimulating factor (GM-CSF), yet IL-1 beta levels remained unchanged.[62] The results with IgG are of interest in that intravenous IgG has been used therapeutically in systemic inflammatory states.[63,64] It has been speculated that the therapeutic administration of IgG may promote the production of IL-1ra in these patients, which might modulate the systemic effects of IL-1.[62]

Adherent mononuclear cells which have differentiated *in vitro* into macrophages produce large amounts of IL-1ra yet appear to lose their ability to respond further to LPS or adherent IgG with enhanced production of IL-1ra.[65] Transforming growth factor-beta (TGF-beta),[66] GM-CSF,[62] and IL-3[65] are also known to induce IL-1ra production in isolated human monocytes. IL-4 has been shown to downregulate IL-1 synthesis yet upregulated IL-1ra synthesis.[68] The recently characterized inhibitory cytokine IL-13 is structurally and functionally related to IL-4.[69] Interleukin-13 also appears to diminish IL-1 production and to stimulate IL-1ra activity.[70] Similarly, the anti-inflammatory cytokine, IL-10, strongly inhibits IL-1 alpha synthesis by human monocytes while it potentiates the expression of IL-1ra.[71]

The synthesis of the intracellular form of IL-1ra by keratinocytes also appears to be differentially regulated when compared to IL-1 alpha synthesis. During the differentiation of keratinocytes, IL-1ra production increased greater than twofold while IL-1 alpha production remained unchanged.[38]

Differences in the synthetic ratio between IL-1ra and IL-1 may be determined by the genetic background of the host and this may contribute to variations in the host response to infectious and inflammatory stimuli. Recent evidence suggests that the ultimate outcome in patients with Lyme arthritis is predicted by the synovial fluid ratio of IL-1ra to IL-1 beta. Patients with high IL-1ra:IL-1 beta ratios had a limited inflammatory arthritis which rapidly resolved following infection with *Borrelia burgdorferi.* Patients with low IL-1ra:IL-1 beta ratios, however, had prolonged and often persistent synovitis.[72] Other examples of imbalances between cytokines and natural anticytokine molecules have been described. Imbalances between TNF alpha and soluble TNF receptor concentrations similarly may determine the outcome in patients with severe meningococcemia.[73] The

TABLE 1
Evidence that IL-1ra Blocks IL-1 Activity *In Vitro*

Experimental Model	Ref.
Proliferation and CSF production myelolastic leukemia cells	75
Nitric oxide-dependent vasodilator in blood vessels	76
Synthesis of IL-1, TNF, and IL-6 in GM-CSF from monocytes	57
Adhesiveness of endothelial cells for neutrophils	52
Lymphocyte proliferation	21
Collagenase production	56
PGE_2 synthesis in fibroblasts in synovial cells	21
Bone resorption	77
Cartilage matrix degradation	78

cytokine profile in patients with lethal infections was characterized by high TNF levels with low levels of soluble TNF receptors. Whether a similar relationship between IL-1 receptor antagonist and IL-1 will be found in patients with sepsis or other inflammatory conditions remains to be seen.

III. THERAPEUTIC POTENTIAL OF IL-1ra

A. INHIBITION OF IL-1 ACTIVITY *IN VITRO*

Provided that sufficient quantities of IL-1ra are used, it has been shown to be able to inhibit the actions of IL-1 in virtually every experimental system tested to date. A summary of *in vitro* data on the inhibition of IL-1 effects by IL-1ra is provided in Table 1.

Results of these experiments demonstrate the ability of IL-1ra to attenuate a multitude of IL-1 induced physiologic responses. IL-1ra might prove to be useful as a therapeutic agent in disease states in which IL-1 has a pathophysiologic role. Considerable clinical experience is accumulating with IL-1ra as an anticytokine therapy in a variety of disease states. Experimental work is now in progress in septic shock, inflammatory bowel disease, inflammatory arthritis, graft vs. host disease, diabetes mellitus, lymphoreticular malignancies, and a variety of other inflammatory conditions.[7,79] Current experimental and clinical experience with IL-1ra in each of these disease states is reviewed in the following sections.

B. EXPERIMENTAL BASIS FOR THE USE OF IL-1ra IN SEPSIS

Intravenous administration of recombinant human IL-1-beta to rabbits at 5 μg/kg resulted in a shock-like state with systemic hypotension, decreased systemic vascular resistance, increased cardiac output, leukopenia, and thrombocytopenia.[80] Moreover, addition of IL-1 and TNF at doses of each cytokine that were insufficient to induce hypotension alone, resulted in a synergistic hemodynamic effect with the induction of severe hypotension. Waage and Espevik[81] have also demonstrated that IL-1 injections potentiate the lethal effect of TNF in mice. Elevated levels of the pro-inflammatory cytokines, TNF and IL-1, are frequently found simultaneously in the systemic circulation in human sepsis.[3,82] Inhibition of either cytokine has proven to be of benefit in a number of experimental sepsis models.[83,84]

IL-1ra has been shown to inhibit endotoxin-induced neutrophil accumulation in the peritoneal cavity in an experimental mouse model.[51] IL-1ra has also been shown to reduce the acute inflammatory response with neutrophil infiltration in the rat lung following intratracheal administration of bacterial endotoxin.[85] IL-1ra provides 50% protection in an otherwise lethal challenge of *E. coli* endotoxin following intraperitoneal injection in mice.[86] IL-1ra protected rabbits given a dose of *E. coli* endotoxin which resulted in lethality in nine out of ten placebo-treated control animals. A dose-response effect was observed in this model where 100 mg/kg of IL-1ra produced an 90% survival, whereas 30 mg/kg resulted in an intermediate level of mortality and 10 mg/kg produced no improvement in survival over the placebo control group.[87]

Injection of heat-killed *E. coli* at 2.0×10^9 colony forming units (CFU)/kg results in profound hypotension, tachycardia, leukopenia, thrombocytopenia, and high mortality in anesthetized rabbits. Injection of IL-1ra at 10 mg/kg prior to the *E. coli* challenge greatly reduced hypotension, leukopenia, and prevented death in this experimental model.[88] Intravenous inoculation of approximately 10^{11} CFU of live *E. coli* in baboons resulted in severe hypotension and death in the majority of animals within 24 h. Seven animals challenged with this otherwise lethal inoculum of intravenous *E. coli* were protected from lethality and had less severe hypotension following IL-1ra administration.[89] Analysis of the cytokine profiles observed in this experimental model revealed that IL-1ra had no effect on circulating levels of TNF-alpha, which increased and then fell precipitously following the *E. coli* injection. Circulating levels of interleukin-1, on the other hand, were significantly reduced by the administration of interleukin-1 receptor antagonist. IL-1ra has been shown to block the autocrine activation of IL-1 synthesis by IL-1 itself. Monocytes express IL-1 receptors on their cell surface and these receptors are susceptible to IL-1ra blockade.[57] This may account for the attenuation of IL-1 levels observed in these animals. While no consistent effect was observed on circulating IL-8 levels, IL-6 levels were significantly reduced by the administration of IL-1ra in this primate model of *E. coli* sepsis.[89] In other experimental systems, IL-1ra has been shown to inhibit IL-8 production by human mononuclear cells.[90] The mechanism by which IL-1ra attenuates the chemoattractant activity of IL-8 *in vivo* is currently unknown.

Experimental evidence exists that IL-1ra may also be beneficial in Gram-positive bacterial sepsis. The administration of heat-killed *Staphylococcus epidermidis* to anesthetized rabbits resulted in profound hypotension. These animals had elevated serum levels of both IL-1 beta and TNF following the infusion of this Gram-positive bacterium despite the absence of measurable endotoxin in the systemic circulation.[91] Pretreatment with human recombinant interleukin-1 receptor antagonist prevented hypotensive shock in this animal model of Gram-positive sepsis.[92]

Interleukin-1 receptor antagonist has also proven to be efficacious in infection models of Gram-negative sepsis. A single intraperitoneal injection of IL-1ra decreased lethality in a bacteremic model of *Klebsiella pneumoniae* infection in neonatal rats.[93] Interestingly, multiple doses of IL-1ra actually increased mortality in this animal model. The effect of a 7-day continuous infusion of IL-1ra (maintaining a systemic blood level of approximately 12.5 μg/ml) was studied in a neutropenic rat model of *Pseudomonas aeruginosa* infection. IL-1ra provided 37% protective efficacy against lethality from septic shock in these immunocompromised animals.[84]

C. CLINICAL EXPERIENCE WITH IL-1ra IN THE TREATMENT OF SEPSIS

1. Phase I Trial

The safety and pharmacokinetics of the intravenous administration of recombinant human interleukin-1 receptor antagonist in human volunteers was first analyzed in a phase 1 trial prior to the use of IL-1ra in clinical trials of human sepsis.[94] The recombinant form of IL-1ra is identical to the naturally occurring, monocyte-derived, secretory form of IL-1ra except that it is nonglycosylated and has a single additional amino acid (methionine) at its N-terminus.[95] A total of 25 normal volunteers received an i.v. dose of interleukin-1 receptor antagonist as a continuous infusion in an escalating dose trial. Subjects received IL-1ra at doses from 1 to 10 mg/kg over 3 h without any hemodynamic sequelae. It was anticipated that IL-1ra would be well tolerated in that this is a naturally occurring human protein known to be in the systemic circulation in a variety of infectious and inflammatory states in humans.

The pharmacokinetics of IL-1ra in normal human volunteers demonstrate an apparent volume of distribution of the molecule of approximately 0.16 L/kg. The alpha phase serum half-life was 21 min and the beta elimination phase revealed a terminal serum half-life of 108 min in subjects with normal renal function. The pharmacokinetics were similar over a large range of i.v. doses of IL-1ra. The mean peak serum levels reached 29 μg/ml in the 10 mg/kg dose group of this study. The infusion was well tolerated without evidence of systemic toxicity, hematologic or biochemical abnormalities, or immunologic reactions. No anti-IL-1ra antibodies were detected in these subjects following IL-1ra infusion. IL-1ra administration resulted in an attenuation of IL-6 levels when the subject's peripheral blood mononuclear cells were isolated and stimulated with endotoxin *ex vivo*.[94] It is remarkable that no toxicity was observed in this clinical study. Previous human trials with recombinant IL-1 as a bone marrow stimulant suggested that the highest tolerable i.v. dose would be approximately 10 ng/kg.[96] At this dose, subjects experienced severe myalgia, headache, and malaise. Interleukin-1 receptor antagonist, however, can be given at up to one million times this dose (10 mg/kg) without any apparent systemic toxicity. This would indicate that human recombinant IL-1ra is essentially devoid of any intrinsic agonist activity and should prove safe to administer to septic patients.[94]

2. Phase II Trial

A phase II trial in patients with sepsis syndrome was conducted in the U.S. in 1991. This was a open-label, randomized, placebo-controlled, multicenter trial designed to assess the safety, pharmacokinetics, efficacy, and cost effectiveness of interleukin-1 receptor antagonist in patients who met the sepsis syndrome criteria.[95] The study consisted of 99 patients who were randomized into a placebo group or one of three treatment groups with IL-1ra given at the following doses: 17 mg/h (low-dose); 67 mg/h (mid-dose); 133 mg/h (high-dose). Each patient randomized to the IL-1ra treatment group received an initial 100 mg bolus over 60 s followed by a continuous infusion of one of the three doses of IL-1ra over the next 72 h. Patients were then monitored for 28 days for evidence of drug toxicity, organ dysfunction, and ultimate outcome in this septic population. The information which has been made available from this study thus far would indicate that these were severely ill patients with a mean APACHE II score at study entry of 22.3. Of the 99 patients enrolled in this trial, two-thirds were hypotensive at study entry and 30% were bacteremic. By an intention-to-treat analysis, a dose-dependent survival advantage was observed with IL-1ra therapy ($p = 0.015$. The 28-day all-cause mortality rate in the placebo group was 44%, while the low-dose IL-1 group had a 32% mortality. The mid-dose IL-1ra treatment resulted in a 25% mortality rate and the high-dose IL-1ra treatment resulted in a 16% mortality rate.[97]

Elevated plasma IL-6 levels at study entry predicted a fatal outcome with nonsurvivors having a mean plasma IL-6 level of 8.5 ± 23.1 ng/ml while survivors had a mean IL-6 level at study entry of 0.6 ± 1.4 ng/ml ($p = 0.006$). Treatment with IL-1ra resulted in a more rapid decrement of IL-6 levels over the first 24 h when compared to the placebo group.[98] These results are similar to data obtained in experimental models of sepsis where IL-1ra was shown to attenuate the IL-6 response in septic primates.[89]

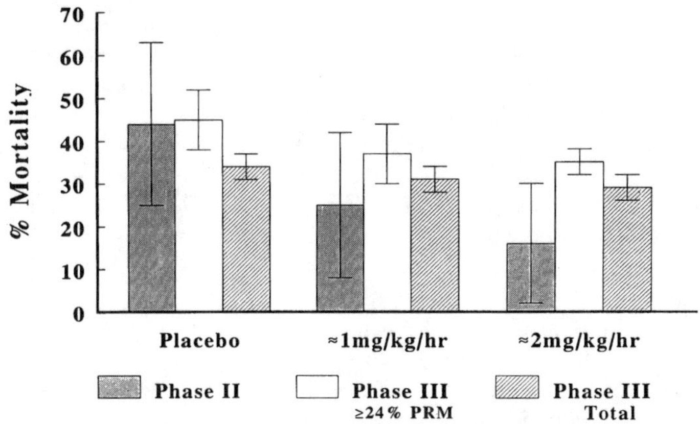

FIGURE 2 Comparison of the 28-day all-cause mortality rates with 95% confidence intervals for the phase II and phase III IL-1ra sepsis trials. The 1 mg/kg/h and 2 mg/kg/h infusion dose utilized in the phase III trial approximates the mid-dose (67 mg/h), and the 2 mg/kg/h dose approximates the high-dose (133 mg/h) employed in the phase II trial. Results of the phase III trial include both the total study population (n = 893) and those patients in the phase III trial with a predicted risk of mortality of ≥24% at study entry (n = 595).

3. Phase III Trials

Based on the favorable results of the phase I and phase II clinical trials, a multicenter phase III IL-1ra sepsis trial was conducted in 1992 in North America and Europe. This trial was designed as a 900-patient double-blind, placebo-controlled, prospective trial in patients with sepsis syndrome. Patients were randomized into one of three treatment arms: (1) placebo; (2) 1 mg/kg/h of IL-1ra; (3) 2 mg/kg/h of IL-1ra. The 1 mg/kg/h dose approximated the mid-dose (67 mg/h) of the phase II trial and the 2 mg/kg/h dose of IL-1ra approximated the high-dose (133 mg/h) of the phase II trial. The study drug was preceded by a 100 mg i.v. bolus of IL-1ra in the two IL-1ra treatment groups or an equal volume of buffered vehicle in the placebo group. The patients received the test article for a 72-h continuous infusion and then were followed for a total of 28 days. A total of 893 patients actually received the test article following randomization. The 28-day all-cause mortality rate of the three treatment groups did not differ significantly: placebo 34%; 1 mg/kg IL-1ra 31%; and 2 mg/kg IL-1ra 29% ($p = 0.22$, generalized Wilcoxon statistic). A comparison of mortality rates in the phase II and phase III trials are presented in Figure 2. A total of 713 patients randomized in this trail were characterized as being in shock at study entry. Analysis of this subgroup of patients also failed to show a statistically significant survival advantage of IL-1ra therapy ($p = 0.23$).[99]

A retrospective analysis of outcome in this trial as a function of a predicted risk of mortality[100] suggested that a potential survival benefit may be confined to those patients with severe sepsis. In patients who had a predicted risk of mortality at study entry of 24% or greater (n = 595), a statistically significant increase in survival time was detected ($p = 0.03$; Wilcoxon rank-sum test). A 22% reduction in point mortality was observed with IL-1ra therapy in this group of patients; 45% — placebo; 37% — 1 mg/kg/h IL-1ra; and 35% — 2 mg/kg/h IL-1-ra (Figure 2). By a retrospective analysis, the beneficial effects of IL-1ra in this trial were most apparent in those patients with confirmed infections (n = 718), in patients with Gram-negative bacterial infections and mixed infections (n = 583), and in patients with high plasma endotoxin levels at study entry (n = 380).[101] Further analysis demonstrated a salutory effect of IL-1ra in septic patients with major organ dysfunction at study entry. Patients with disseminated intravascular coagulation, adult respiratory distress syndrome, acute renal failure, or sepsis-induced hepatobiliary dysfunction derived significant survival benefit by this *post hoc* analysis.[102] A confirmatory multicenter clinical trial is now in progress in patients with severe sepsis and septic shock.

D. EXPERIMENTAL EVIDENCE OF IL-1ra ACTIVITY IN OTHER PATHOLOGIC STATES

1. Inflammatory Arthritis

Interleukin-1 has been implicated as a principal mediator in the pathogenesis of numerous forms of inflammatory arthritis including rheumatoid arthritis.[103] The receptor antagonist may participate in the regulation of joint inflammation. Interleukin-1 receptor antagonist is detectable in the circulation of many patients with juvenile rheumatoid arthritis.[104] IL-1ra is found in the synovial fluid of the majority of patients with rheumatoid arthritis.[105,106] The relative concentration of synovial fluid levels of IL-1ra and IL-1 beta may determine the outcome of the inflammatory arthritis associated with Lyme disease.[72] Interleukin-1 is known to produce leukocyte infiltration into joint tissues and to induce degradation of articular cartilage proteolglycan and activate collagenase activity.[103] Injection of interleukin-1 receptor antagonist decreases the inflammatory response to IL-1 in a rabbit model of inflammatory arthritis.[107] IL-1ra is known to block type II collagen-induced arthritis in mice.[108] IL-1ra significantly reduced joint swelling and inflammation in the streptococcal cell wall-induced arthritis model in Lewis rats.[109] The potential utility of IL-1ra as a disease-modifying agent in rheumatoid arthritis or other inflammatory arthritides will be determined in ongoing clinical trials.

2. Inflammatory Bowel Disease

Interleukin-1 has been shown to mediate inflammatory changes and necrosis of gastrointestinal tissues in an immune complex-induced model of colitis in rabbits.[110] Pretreatment with IL-1ra in this experimental model reduces the degree of inflammation and necrosis. The severity of a streptococcal cell wall-induced model of enterocolitis in rats was also attenuated by IL-1ra.[111] The role of IL-1ra in modulating disease activity in inflammatory bowel disease in humans is currently unknown. Clinical trials with IL-1ra are planned in this group of patients to determine treatment efficacy.

3. Neoplasia

Neoplastic cells from some patients with lymphoreticular malignancies express IL-1 beta mRNA and synthesize detectable levels of IL-1 protein.[112–114] Antibodies to IL-1 beta have been shown to reduce the production of colony stimulating factors and to inhibit proliferative responses in these cell lines.[114,115] Interleukin-1 receptor antagonist has been shown to decrease spontaneous proliferation and colony stimulating factor production from peripheral blood- and bone marrow-derived acute myelogenous leukemia cells.[75] Similar findings have been noted in patients with chronic myelogenous leukemia.[116] Addition of up to 15 ng/ml of recombinant human IL-1ra inhibited spontaneous and IL-1-stimulated cellular responses in acute leukemia cells by up to 85%.[75] A disproportionate amount of IL-1 beta was produced relative to IL-1ra production in some leukemic cells. Only 1 of 11 acute myelogenous leukemia cell lines examined by Rambaldi et al.[75] expressed IL-1ra while all the cell lines produced IL-1. These results suggest that an imbalance may exist in IL-1 responsive tumor cell lines between IL-1 and the receptor antagonist. Whether the therapeutic administration of IL-1ra will prove to be efficacious in certain IL-1 responsive leukemias will require adequately controlled clinical trials.

IL-1 has recently been shown to be ubiquitously expressed in endometrial carcinoma cell lines. The intracellular form of IL-1ra has also been isolated in endometrial carcinoma cells. An imbalance between intracellular IL-1ra and IL-1-mediated events such as growth and differentiation of endometrial neoplastic cells may be of pathogenetic significance in this form of human neoplasia[40] and perhaps other forms of human malignancies. There is evidence that IL-1 may play a role in the regulation of the life span of endothelial cells. An antisense oligodeoxynucleotide to the human IL-1-alpha transcript has been shown to prolong the proliferative potential of cultured human endothelial cells.[117] The precise physiologic role of intracellular interleukin-1 and IL-1ra as regulators of cellular life cycles in normal and neoplastic cells is currently unknown. If intracellular levels of interleukin-1 regulate the life cycle of human cell lines, it may be possible to modulate this effect with anti-IL-1 strategies such as interleukin-1 receptor antagonist.

4. CNS Effects

IL-1 is known to be synthesized by normal brain cells *in vivo* and regulate physiologic functions such as body temperature, pituitary hormone secretion, and slow-wave sleep.[112,119] The administration of IL-1ra into the lateral ventricle of rabbit cerebral tissues reduced spontaneous slow-wave sleep and blocked IL-1-induced fever.[120] Interleukin-1 receptor antagonist has recently been shown to be synthesized within mammalian brain tissue including the hippocampus and hypothalamus.[121] This finding indicates paracrine regulation of IL-1 action may occur at the receptor level within the central nervous system. Interleukin-1 has been shown to promote the synthesis of the amyloid precursor peptide, a major component of amyloid deposits found in patients with Alzheimer's disease.[122] Interestingly, IL-1ra inhibits serum amyloid A protein synthesis in cytokine-treated mice.[123] Whether a similar IL-1ra effect would occur in humans is unknown. Cytokine synthesis is known to occur within the CNS after head injury[124] and interleukin-1 synthesis may occur within the CNS during systemic inflammatory reactions.[125] While the capability of IL-1ra to inhibit IL-1 activation within the human central nervous system has yet to be demonstrated, experimental evidence suggests that modulation of CNS effects of IL-1 may be possible.

IL-1 administered systemically or via intraventricular injections produce loss of social activity, hypersomnia, anorexia, and weight loss in rats. Pretreatment with human recombinant IL-1ra will block the anorexia, weight loss, and loss of social behavior in IL-1-treated rats.[126] Bacterial endotoxin will also produce loss of social activity and hypersomnia in rats. These behavioral changes are attenuated by pretreatment with human recombinant interleukin-1 receptor antagonist.[127]

TNF, IL-1, and perhaps other cytokines play an important role in the pathogenesis of inflammation within the cerebrospinal fluid as seen in bacterial meningitis.[128,129] Use of corticosteroids prior to the administration of antimicrobial agents has been shown to decrease cytokine generation and prevent neurologic sequelae in patients with bacterial meningitis.[130,131] There is evidence that anti-TNF monoclonal antibody administered intrathecally and intravenously will decrease brain edema and improve outcome in experimental bacterial meningitis.[132] Whether similar effects would be found with other anticytokines such as IL-1ra remains to be demonstrated.

Interleukin-1 promotes the synthesis of locally active neurotrophic growth factors which participate in the regeneration of damaged neurons.[133] Interleukin-1 receptor antagonist has been shown to inhibit the regeneration of severed peripheral neurons.[134] These results serve to illustrate the potential advantageous and deleterious effects of interleukin-1 in human physiology. While it might be desirable to inhibit certain deleterious effects of IL-1, this blockade will simultaneously disrupt other advantageous

TABLE 2
Summary of Inhibitory Effects of IL-1RA in Other Experimental Models

Disease Models	Ref.
Delay of onset of spontaneous diabetes in BB rats	135
Graft-vs.-host disease in mice	136
Acute inflammatory response to LPS or IL-1 following intratracheal injection in rats	85
Release of neutrophil precursors from bone marrow in mice	51
Carrageenan-induced pleurisy	137
IL-1 induced inhibition of insulin production in pancreatic islets	138
Pulmonary infiltration with eosinophils in bronchoalveolar lavage fluid following antigen challenge	139

properties of IL-1. This concept must be carefully considered when developing clinical indications for the therapeutic modulation of IL-1.

5. Other Pathologic Conditions

Interleukin-1 receptor antagonist has been shown to attenuate IL-1 effects in a multitude of experimental models of IL-1-induced inflammation. Favorable effects have been observed with IL-1ra in such diverse systems as diabetes mellitus, asthma, graft-vs.-host disease, and other inflammatory states. The therapeutic implications of IL-1ra in these pathologic states await further clinical investigation. A summary of these experimental findings is provided in Table 2.

IV. POTENTIAL HAZARDS OF IL-1ra TREATMENT

Interleukin-1 is critical to the host inflammatory response to a myriad of potentially pathogenic infectious agents.[140] Blockade of IL-1 activity may prove to be disadvantageous to the infected host in that an appropriate inflammatory response may be inhibited at a critical early phase of the host-pathogen interaction. While IL-1 and TNF are clearly detrimental when produced systemically in excess quantities,[1–7] small amounts of these cytokines functioning in the microcirculation may be essential for coordination of the early host response. Inhibition of IL-1 activity has been shown to promote disseminated listeriosis in experimental animals.[141] Furthermore, IL-1ra has been shown to increase mortality when given at high doses in a neonatal rat model of Gram-negative sepsis.[93] In the same rat model, smaller quantities of IL-1ra were shown to be protective. The addition of small quantities of recombinant IL-1 alone or with TNF proved to be of survival benefit in C3H/Hej mice following intraperitoneal injection with a virulent *E. coli* strain.[142]

These studies indicate that a dynamic balance exists between the advantageous and disadvantageous effects of pro-inflammatory cytokines in different disease states. The net effect of IL-1 synthesis may be beneficial to the host in early stages of localized infection. The net effect of excessive IL-1 synthesis in the later stages of systemic infection may be deleterious to the host. The relative merits of IL-1ra therapy in systemic infection may depend upon the timing of administration of IL-1ra, the level of IL-1 blockade, and the severity of illness in individual patients. The results of the phase III IL-1ra sepsis trial would indicate that the most severely ill patients derive benefit from IL-1ra, while those who are less severely ill experience no apparent benefit.[100] These observations, if substantiated in further clinical studies, would indicate the need to carefully select patients for treatment with IL-1ra in the face of an ongoing infectious process. Moreover, the chronic administration of IL-1ra for other indications such as inflammatory arthritis may result in an excess risk of infectious diseases in IL-1ra-treated patients.

Another potential disadvantage with IL-1ra therapy as a therapeutic modality is its short serum half-life. The elimination half-life of IL-1ra human volunteers with normal renal function is 1.8 h.[94] This will necessitate a continuous infusion of IL-1ra if interleukin-1 blockade is indicated for an extended time period. Additionally, the need to generate a large molar excess of IL-1ra over IL-1 concentrations will require the provision of large therapeutic doses of IL-1ra. Despite the fact that human recombinant IL-1ra is generated in a high-yield prokaryotic cloning system by *E. coli* fermentation, the large quantities required place constraints on the amount of material that can be synthesized and the economic feasibility of the use of IL-1ra as a therapeutic agent. Innovations in production techniques for IL-1ra or the generation of genetic constructs of IL-1ra with prolonged serum inhibitory activity may be needed in the future.

Theoretical concerns exist regarding the potential immunogenicity of human recombinant IL-1ra. Conformational epitopes may differ between the prokaryotic recombinant products and the naturally occurring form of IL-1ra. Additionally, the current form of human recombinant IL-1ra has an additional methionine amino acid at its amino terminus. It is conceivable that this additional amino acid may generate an immune response in some individuals.

No significant allergic reactions have been described in either the phase II or phase III sepsis trial with IL-1ra.[98,101] Anti-IL-1ra antibodies were not detected in human volunteer studies.[101] Nonetheless, the potential for generation of anti-human IL-1ra antibodies will need to be carefully monitored in clinical studies.

V. SUMMARY

IL-1ra is a naturally occurring anticytokine protein which effectively blocks the activity of IL-1 alpha and IL-1 beta at the receptor level. The receptor antagonist has been shown to inhibit IL-1-mediated effects in a variety of experimental systems. Preliminary evidence in human clinical trials would indicate that IL-1ra may be safely administered and may prove to be of therapeutic value in the treatment of severe sepsis. The ultimate clinical utility of IL-1ra in sepsis and numerous other inflammatory disease states remains to be demonstrated by carefully controlled clinical trials. The physiologic role of IL-1ra in regulating IL-1 activity both in the systemic circulation and in the intracellular environment are fascinating questions which may have profound effects on our understanding of human biology. The availability of interleukin-1 receptor antagonist provides a molecular mechanism to specifically probe the function of interleukin-1 in normal physiology and in a variety of pathologic states. The dynamic balance between IL-1 and IL-1ra may determine the outcome in a large number of pathophysiologic processes. Whether the therapeutic modulation of IL-1 action with the receptor antagonist will ultimately prove to be of clinical utility remains an exciting yet unproven possibility and will be the focus of considerable attention in the future.

ACKNOWLEDGMENTS

The author would like to thank Gail Armstrong for preparing this manuscript and C. J. Fisher, Jr. (Cleveland, OH), J. P. Pribble, S. P. Eisenberg, and R. C. Thompson (Boulder, CO) for their helpful suggestions.

ADDENDUM

Since the original submission of this manuscript the results of the first phase III trial of IL-1ra for the treatment of sepsis syndrome have been published.[143] A second phase III trial designed to determine the efficacy of a high dose (2 mg/kg/h for 72 h) of IL-1ra vs. placebo in patients with severe sepsis was terminated for lack of efficacy after 696 patients were enrolled. An interim analysis projected that continuation of the trial to the original target of 1200 patients would not have sufficient power to achieve statistical significance in demonstrating a survival advantage with administered IL-1ra. Detailed analyses of the results of this clinical trial are underway, but as of September, 1994 they have not been published.

REFERENCES

1. **Dinarello, C. A.,** Biology of interleukin-1, *FASEB J.,* 2, 108, 1988.
2. **Dinarello, C. A.,** Interleukin-1, *Rev. Infect. Dis.,* 6, 51, 1984.
3. **Dinarello, C. A.,** The pro-inflammatory cytokines interleukin-1 and tumor necrosis factor and treatment of the septic shock syndrome, *J. Infect. Dis.,* 136, 1177, 1991.
4. **Bone, R. C.,** Pathogenesis of sepsis, *Ann. Intern. Med.,* 115, 457, 1991.
5. **Filkins, J. P.,** Monokines and the metabolic pathophysiology of septic shock, *Fed. Proc.,* 44, 300, 1985.
6. **Dinarello, C. A.,** Interleukin-1 and interleukin-1 antagonism, *Blood,* 77, 1627, 1991.
7. **Dinarello, C. A. and Thompson, R. C.,** Blocking interleukin-1: interleukin-1 receptor antagonist *in vivo* and *in vitro, Immunol. Today,* 12, 404, 1991.
8. **Fenton, M. T., Clark, B. D., Collins, K. S., Webb, A. C., Rich, A., and Auron, P. E.,** Transcriptional regulation of the human pro-interleukin-1-beta gene, *J. Immunol.,* 138, 1372, 1987.
9. **Ye, K., Koch, K. C., and Dinarello, C. A.,** Interleukin-1 beta down regulates gene and surface expression of interleukin-1 receptor type I by destabilizing its mRNA whereas interleukin-2 increases its expression, *Clin. Res.,* 39, 469A, 1991.
10. **Hazuda, D. J., Lee, J. C., and Young, P. R.,** Kinetics of interleukin-1 secretion from activated monocytes, *J. Biol. Chem.,* 263, 8473, 1990.
11. **Hogquist, K. A., Nett, M. A., Unanue, E. R., and Chaplin, D. D.,** Interleukin-1 is processed and released during apoptosis, *Proc. Natl. Acad. Sci. U.S.A.,* 88, 8485, 1991.
12. **Hazuda, D. J., Webb, R. L., Simon, P., and Young, P.,** Purification and characterization of human recombinant precursor interleukin-1-beta, *J. Biol. Chem.,* 264, 1689, 1989.
13. **Huzuda, D. J., Stickler, J., Kueppers, F., Simon, P. L., and Young, P. R.,** Processing of precursor interleukin-1-beta and inflammatory disease, *J. Biol. Chem.,* 265, 6318, 1990.
14. **Sleath, P. R., Hendrickson, R. C., Kronheim, S. R., March, C. J., and Black, R. A.,** Substrate specificity of the protease that processes human interleukin-1-beta, *J. Biol. Chem.,* 265, 14526, 1990.
15. **Hazuda, D., J., Stickler, J., Simon, P., and Young, P. R.,** Structure-function mapping of interleukin-1 precursors. Cleavage leads to a confirmational change in the mature protein, *J. Biol. Chem.,* 266, 7081, 1991.
16. **Curtis, B. M., Widmer, M. B., deRoos, P., and Qwarnstrom, E. E.,** IL1 and its receptor are translated to the nucleus, *J. Immunol.,* 144, 1495, 1990.
17. **Larrick, J. W.,** Native interleukin-1 inhibitors, *Immunol. Today,* 10, 61, 1989.
18. **Barak, V., Treves, A. J., Yanai, P., Halperin, M., Wasserman, D., Biran, S., and Braun, S.,** Interleukin-1 inhibitory activity secreted by a human myelomonocytic cell line (M20), *Eur. J. Immunol.,* 16, 1449, 1986.
19. **Bendtzen, K., Svenson, M., Jinsson, V., and Hippe, E.,** Autoantibodies to cytokines — friends or foes?, *Immunol. Today,* 11, 167, 1990.

20. **Arend, W. P., Joslin, F. G., and Massoni, R. J.,** Effects of immune complexes on production by human monocytes of interleukin-1 or an interleukin-1 inhibitor, *J. Immunol.,* 134, 3868, 1985.
21. **Balavoine, J.-F., deRochemonteix, B., Williamson, K., Seckinger, P., Cruchaud, A., and Dayer, J.-M.,** Prostaglandin E_2 and collagenase production by fibroblasts in synovial cells is regulated by urine-derived human interleukin-1 and inhibitor(s), *J. Clin. Invest.,* 78, 1120, 1986.
22. **Seckinger, P., Lowenthal, J. P., Williamson, K., Dayer, J.-M., and MacDonald, H. R.,** A urine inhibitor of interleukin-1 activity that blocks ligand binding, *J. Immunol.,* 139, 1546, 1987.
23. **Seckinger, P., Williamson, K., Balavoine, J.-F., Mach, B., Mazzei, G., Shaw, A., and Dayer, J.-M.,** A urine inhibitor of interleukin-1 activity affects both interleukin-1 alpha and 1 beta but not tumor necrosis factor, *J. Immunol.,* 139, 1541, 1987.
24. **Hannum, C. H., Wilcox, C. J., Arend, W. P., Joslin, F. G., Dripps, D. J., Heimdal, P. L. Armes, L. G., Sommer, A., Eisenberg, S. P., and Thompson, R. C.,** Interleukin-1 receptor antagonist activity of a human interleukin-1 inhibitor, *Nature (London),* 343, 336, 1990.
25. **Eisenberg, S. P., Evans, R. J., Arend, W. P., Verderber, A., Brewer, M. T., Hannum, H. C., and Thompson, R. C.,** Primary structure and functional expression from complementary DNA of a human interleukin-1 receptor antagonist, *Nature (London),* 343, 331, 1990.
26. **Massie, G. J., Seckinger, P. L., Dayer, J.-M., and Shaw, A. R.,** Purification and characterization of a 26-kDa competitive inhibitor of interleukin-1, *Eur. J. Immunol.,* 20, 683, 1990.
27. **Arend, W. P.,** Interleukin-1 receptor antagonist a new member of the interleukin-1 family, *J. Clin. Invest.,* 88, 1445, 1991.
28. **Lennard, A., Gorman, P., Carrier, M., Griffiths, S., Scotney, H., Sheer, D., and Solari, R.,** Cloning and chromosome mapping of the human interleukin-1 receptor antagonist gene, *Cytokine,* 4, 83, 1992.
29. **Webb, A. C., Collins, K. L., Auron, P. E., Eddy, R. L., Nakai, H., Byers, M. G., Haley, L. L., Henry, W. M., and Show, T. B.,** Interleukin-1 gene (IL-1) assigned to long arm of chromosome 2, *Lymphokine Res.,* 5, 77, 1986.
30. **Copeland, N. G., Silan, C. M., Kingsley, D. M., Jenkins, N. A., Cannizzaro, L. M., Croce, C. M., Huebner, K., and Sims, J. E.,** Chromosomal location of murine and human IL-1 receptor genes, *Genomics,* 9, 44, 1991.
31. **Eisenberg, S. P., Brewer, M. T., Verderber, E., Heimdal, P., Brandhuber, B. J., and Thompson, R. C.,** Interleukin-1 receptor antagonist (IL-1ra) is a member of the IL-1 gene family. Evolution of a cytokine control mechanism, *Proc. Natl. Acad. Sci. U.S.A.,* 88, 5232, 1991.
32. **Young, P. R. and Sylvester, D.,** Cloning of rapid IL-1-beta proteins. Differential evolution of IL-1-alpha and IL-1-beta proteins, *Protein Eng.,* 2, 545, 1989.
33. **Smith, M. F., Arend, W. P., and Gutierrez-Hartmann, A.,** Isolation of characterization of the IL-1ra promotor, *FASEB J.,* 5, A970, 1991.
34. **Malter, J. S.,** Identification of an AUUUA-specific messenger RNA binding protein, *Science,* 246, 644, 1989.
35. **Steinkasserer, A., Koelble, K., and Sim, R. B.,** Length variation within intron 2 of the human interleukin-1 receptor antagonist protein gene (IL-1rn), *Nucl. Acid Res.,* 19, 5095, 1991.
36. **Tarlow, J. K., Blakemore, A. I. F., Lennard, A., Solari, R., Hughes, H. N., Steinkasserer, A., and Duff, G. W.,** Polymorphism in human interleukin-1 receptor antagonist gene intron 2 is caused by variable numbers of an 86-bp tandem repeat, *Hum. Genet.,* 91, 403, 1991.
37. **Haskill, S., Martin, G., VanLee, L., Morris, J. P., Peace, A., Bigler, C. F., Jaffe, G. J., Hammerberg, C., Sporn, S. N., Fong, S., Arend, W. P., and Ralph, P.,** cDNA cloning of an intracellular form of interleukin-1 receptor antagonist associated with epithelium, *Proc. Natl. Acad. Sci. U.S.A.,* 88, 3681, 1991.
38. **Bigler, C. F., Norris, D. A., Weston, W. L., and Arend, W. P.,** Interleukin-1 receptor antagonist production by keratinocytes, *J. Invest. Dermatol.,* 98, 38, 1992.
39. **Chan, S. L., Hammerberg, C., Kang, K., Sabb, P., Tavakkol, A., and Cooper, K. D.,** Human dermal fibroblast interleukin-1 receptor antagonist (IL-1ra) and interleukin-1-beta (IL-1-beta) mRNA and protein are co-stimulated by phorbol ester: implications for a homeostatic mechanism. *J. Invest. Dermatol.,* 99, 315, 1992.
40. **VanLee, L., Haskill, S., Jaffe, G. J., and Fowler, W. C. E., Jr.,** Expression of interleukin-1 and interleukin-1 receptor antagonists in endometrial cancer, *Gynecol. Oncol.,* 42, 161, 1991.

41. **Stevenson, F. L., Locksley, R. M., and Lovett, T. H.,** The extracellular cytokine-interleukin-1, is synthesized on free polyribosomes and distributed intracellularly as a cytoskeletal-associated protein, *Clin. Res.,* 39, 290A, 1991.
42. **Arend, W. P., Joslin, F. G., Thompson, R. C., and Hannum, C. H.,** An IL-1 inhibitor from human monocytes. Production and characterization of biological properties, *J. Immunol.,* 143, 1851, 1989.
43. **Eisenberg, S. P., Evans, R. J., Vigers, G. P. A., Bray, J., Caffes, T., Childs, J. D., Dripps, D. J., Thompson, R. C., and Brandhuber, B. J.,** Interleukin-1 receptor antagonist (IL-1ra) lacks one of the two receptor binding sites for IL-1-beta based on structure/function analysis, *Lymphokine Cytokine Res.,* 12, 376, 1993.
44. **Labriola-Thompkins, E., Chandrand, C., Kaffka, K. L., Biondi, D., Graves, B. J., Hatada, M., Madison, V. S., Karas, J., Kilian, P. O., and Ju, G.,** Identification of the discontinuous binding site in human interleukin-1-beta for the type I interleukin-1 receptor, *Proc. Natl. Acad. Sci. U.S.A.,* 88, 11182, 1991.
45. **Ju, G., Labriola-Thompkins, E., Campen, C. A., Benjamin, W. R., Karas, J., Polcinski, J., Biondi, D., Kaffka, K. L., Kilin, P. L., Eisenberg, S. P., and Evans, R. J.,** Conversion of the interleukin-1 receptor antagonist into an agonist by site-specific mutagenesis, *Proc. Natl. Acad. Sci. U.S.A.,* 88, 2658, 1991.
46. **Gehrke, L., Jobling, S. O., Paik, S. L. K., McDonald, B., Rosenwasser, L. J., and Auron, P. E.,** A point mutation uncouples human interleukin-1 beta biological activity in receptor binding, *J. Biol. Chem.,* 265, 2922, 1990.
47. **Urdal, D. L., Call, S. M., Jackson, J. L., and Dower, S. K.,** Affinity purification and chemical analysis of the interleukin-1 receptor, *J. Biol. Chem.,* 263, 2870, 1988.
48. **Munoz, E., Zubiaga, A. L., Sims, J. E., and Huber, B. T.,** Interleukin-1 signal transduction pathways. I. Two function IL-1 receptors are expressed in T cells, *J. Immunol.,* 146, 136, 1991.
49. **Granowitz, E. V., Mancilla, J., Clark, B. W., and Dinarello, C. A.,** Interleukin-1 receptor antagonist competitively inhibits the binding of interleukin-1 to the type II interleukin-1 receptor, *Clin. Res.,* 39, 462A, 1991.
50. **Chizzonite, R., Truitt, T., Kilian, P. L., Stern, A. S., Nunes, P., Parker, K. P., Kaffka, K. L., Chua, A. O., Lugg, D. K., and Gubler, U.,** Two high-affinity interleukin receptors represent separate gene products, *Proc. Natl. Acad. Sci. U.S.A.,* 86, 8029, 1989.
51. **McIntyre, K. W., Stepan, G. J., Kolinsky, K. D., Benjamin, W. R., Plocinski, J. M., Kaffka, K. L., Kampen, C. A., Chizzonite, R. A., and Kilian, P. L.,** Inhibition of interleukin-1 (IL-1) binding and bioactivity *in vitro* and modulation of acute inflammation *in vivo* by IL-1 receptor antagonist and anti-IL-1 receptor monoclonal antibody, *J. Exp. Med.,* 173, 931, 1991.
52. **Carter, D. B., Deibel, M. R., Dunn, C. J. et al.,** Purification, cloning, expression and biological characterization of an interleukin-1 receptor antagonist protein, *Nature (London),* 344, 633, 1990.
53. **Dripps, D. J., Brandhuber, B. J., Thompson, R. C., and Eisenberg, S. P.,** Interleukin-1 receptor antagonist binds to the 80 kDa IL-1 receptor but does not initiate IL-1 signal transduction, *J. Biol. Chem.,* 266, 10331, 1991.
54. **Granowitz, E. V., Clark, B. D., Mancilla, J., and Dinarello, C. A.,** Interleukin-1 receptor antagonist competitively inhibits the binding of interleukin I to type II interleukin-1 receptor, *J. Biol. Chem.,* 266, 14147, 1991.
55. **Mizzel, S. B., Kilian, P. L., Lewis, J. C., Paganelli, K. A., and Chizzonite, R. A.,** The interleukin-1 receptor. Dynamics of IL-1 binding and internalization in T cells and fibroblasts, *J. Immunol.,* 138, 2906, 1987.
56. **Arend, J. P., Welgus, H. G., Thompson, R. C., and Eisenberg, S. P.,** Biological properties of recombinant human monocyte-derived interleukin-1 receptor antagonist, *J. Clin. Invest.,* 85, 1694, 1990.
57. **Granowitz, E. V., Clark, B. D., Vannier, E., Callahan, M. V., and Dinarello, C. A.,** Interleukin-1 receptor antagonist inhibits interleukin-1 induced cytokine synthesis by human monocytes, *Clin. Res.,* 39, 462A, 1991.
58. **Ikejima, T., Zhang, X., and Dinarello, C. A.,** Interleukin-1 receptor antagonist (IL-1ra) reduces IL-1 mediated fever in the rabbit but not LPS or *Staphylococcus epidermidis*-induced fever, *Clin. Res.,* 39, 462A, 1991.

59. **Ohlsson, K., Bjork, P., Bergenfeldt, M., Hageman, R., and Thompson, R. C.,** Interleukin-1 receptor antagonist reduces mortality from endotoxin shock, *Nature (London),* 348, 550, 1990.
60. **Arend, W. P., Smith, M. F., Jr., Janson, R. W., and Joslin, F. G.,** Interleukin-1 receptor antagonist and interleukin-1 beta production in human monocytes are regulated differently, *J. Immunol.,* 147, 1530, 1991.
61. **Granowitz, E. V., Santos, A. A., Poutsiaka, D. D., Kannon, J. G., Wilmore, D. W., Wolff, S. M., and Dinarello, C. A.,** Production of interleukin-1 receptor antagonist during experimental endotoxaemia, *Lancet,* 338, 1423, 1991.
62. **Poutsiaka, D. D., Clark, B. D., Vannier, E., and Dinarello, C. A.,** Production of interleukin-1 receptor antagonist and interleukin-1 beta by peripheral blood mononuclear cells is differentially regulated, *Blood,* 78, 1275, 1991.
63. **Sullivan, K. M., Kopecky, K. J., Jocom, J., et al.,** Immunomodulatory and antimicrobial efficacy of intravenous immunoglobulin in bone marrow transplantation, *New Engl. J. Med.,* 323, 705, 1990.
64. **Furusho, K., Kamiya, T., Nakano, H., Kiyosawa, N., Shinomiya, K., Hayashidera, T., Tamura, T., Hirose, O., Manabe, Y., Yokoyama, T., Kawarano, M., Bab, K., Bab, K., and Mori, C.,** High-dose intravenous gamma globulin for Kawasaki disease, *Lancet,* 2, 1055, 1984.
65. **Janson, R. W., Hance, K. R., and Arend, W. P.,** The production of interleukin-1 receptor antagonist by human in vitro derived macrophages: the effects of LPS and GM-CSF, *J. Immunol.,* 147, 4218, 1991.
66. **Turner, M., Chantry, D., Katsikis, P., Berger, A., Brennan, F. M., and Felmann, M.,** Induction of the interleukin-1 receptor antagonist protein by transforming growth factor-beta, *Eur. J. Immunol.,* 21, 1635, 1991.
67. **Chosay, J. G. and Tracey, D. E.,** Induction of interleukin-1 (IL-1) receptor antagonist protein (IRAP-HM) from cultured human monocytes, *Fed. Am. Soc. Exp. Biol. J.,* 4, A1711, 1990.
68. **Vannier, E., Miller, L. C., and Dinarello, C. A.,** Coordinated anti-inflammatory effects of interleukin-4: interleukin-4 suppresses interleukin-1 production but up-regulates gene expression synthesis of interleukin-1 receptor antagonist, *Proc. Natl. Acad. Sci. U.S.A.,* 89, 4076, 1992.
69. **Minty, A., Chalon, P., Derocg, J. M., Dumont, X., Guillemot, J. C., Kaghad, M., Labit, C., Leplatois, P., Liauzun, P., and Miloux, B.,** Interleukin-13: a novel human lymphokine regulating inflammatory and immune responses, *Nature,* 362, 248, 1992.
70. **de Waal Malefyt, R.,** Interleukin-13 is a novel Th2 cytokine with antiinflammatory activities: comparison with IL-4 and IL-10, *Intensive Care Med.,* 20 (Suppl. 1) (Abstr.), 212, 1994.
71. **Howard, M., O'Garra, A., Ishida, H., de Waal Malefyt, R., and de Vries, J.,** Biological properties of interleukin-10, *J. Clin. Immunol.,* 12, 239, 1992.
72. **Miller, L. C., Lynch, E. A., Isa, S., Logan, G. W., Dinarello, C. A., and Steere, A. C.,** Balance of synovial fluid IL-1-beta and IL-1 receptor antagonist and recovery from Lyme arthritis, *Lancet,* 341, 146, 1993.
73. **Girardin, E., Rux-Lombard, P., Grau, G. E., Suter, P., Gallati, H., and Dayer, J.-M.,** The J-5 Study Group. Imbalance between tumor necrosis factor-alpha and soluble TNF receptor concentrations in severe meningococcaemia, *Immunology,* 76, 20, 1992.
74. **Dinarello, C. A., Gelfand, J. A., and Wolff, S. M.,** Anticytokine strategies in the treatment of the systemic inflammatory response syndrome, *J. Am. Med. Assoc.,* 269, 1829, 1993.
75. **Rambaldi, A., Torcia, M., Bettoni, S., Vannier, E., Barbui, T., Dinarello, C. A., and Lozzolino, F.,** Modulation of cell proliferation and cytokine production in acute myeloblastic leukemia by recombinant IL-1 receptor antagonist and lack of its expression by leukemic cells, *Blood,* 76, 114A, 1990.
76. **Petitclerc, E., Abel, S., DeBlois, D., Poubelle, P. E., and Marceau, F.,** Effects of interleukin-1 receptor antagonist on three types of responses to interleukin-1 in rabbit isolated blood vessels, *J. Cardiovasc. Pharmacol.,* 19, 821, 1992.
77. **Seckinger, P., Klein-Nulend, J., Alander, C., Thompson, R. C., Dayer, J.-M., and Raisz, L. G.,** Natural and recombinant human interleukin-1 receptor antagonists block the effects of Interleukin-1 on bone resorption and prostaglandin production, *J. Immunol.,* 145, 4181, 1990.
78. **Smith, R. J., Chin, J. E., Sam, L. M., and Justen, J. M.,** Biologic effects of an interleukin-1 receptor antagonist protein on interleukin-1-stimulated cartilage erosion and chondrocyte responsiveness, *Arthritis Rheum.,* 34, 78, 1991.

79. **Dayer, J.-M.,** Immunomodulating functions of tumor necrosis factor and interleukin-1 inhibitors, *J. Rheumatol.,* Suppl. 32, 59, 1992.
80. **Okusawa, S., Gelfand, J. A., Ikejima, T., Connolly, R. J., and Dinarello, C. A.,** Interleukin-1 induces a shock-like state in rabbits; synergism with tumor necrosis factor and effect of cyclooxygenase inhibition, *J. Clin. Invest.,* 81, 1162, 1988.
81. **Waage, A. and Espevik, T.,** Interleukin-1 potentiates the lethal effects of tumor necrosis factor-alpha/cachectin in mice, *J. Exp. Med.,* 167, 1987, 1988.
82. **Casey, L. C., Balk, R. A., and Bone, R. C.,** The plasma, cytokine, and endotoxin levels correlate with survival in patients with sepsis syndrome, *Ann. Intern. Med.,* 119, 1771, 1993.
83. **Opal, S. M., Cross, A. S., Kelley, N. M., Sadoff, J. C., Bodmer, M. W., Palardy, J. E., and Victor, G. H.,** Efficacy of a monoclonal antibody directed against tumor necrosis factor in protecting neutropenic rats from lethal infection with *Pseudomonas aeruginosa, J. Infect. Dis.,* 161, 1181, 1990.
84. **Opal, S. M., Palardy, J. E., Romulo, R. L. C., and Smith, C.,** Interleukin-1 receptor antagonist in an actual infection model of *Pseudomonas* sepsis, *Infect. Dis. Soc. Am. Meet.,* 12 (Abstr.), 1993.
85. **Ulich, T. R., Yin, S. M., Guo, K., del Castillo, J., Eisenberg, S. P., and Thompson, R. C.,** The intratracheal administration of endotoxin and cytokines. III. The interleukin-1 (IL-1) receptor antagonist inhibits endotoxin IL-1-induced acute inflammation, *Am. J. Pathol.,* 138, 521, 1991.
86. **Alexander, H. R., Doherty, G. M., Buresh, C. M., Venzon, D. J., and Norton, J. A.,** A recombinant human receptor antagonist to interleukin-1 improves survival after lethal endotoxemia in mice, *J. Exp. Med.,* 173, 1024, 1991.
87. **Ohlsson, K., Björk, B., Bergenfeldt, M., Hageman, R., and Thompson, R. C.,** Interleukin-1 receptor antagonist reduces mortality from endotoxin shock, *Nature,* 348, 550, 1990.
88. **Wakabayashi, G., Gelfand, J. A., Burke, J. F., Thompson, R. C., and Dinarello, C. A.,** A specific receptor antagonist for interleukin-1 prevents E. coli-induced shock in rabbits, *Fed. Am. Soc. Exp. Biol.,* 5, 338, 1991.
89. **Fischer, E., Marano, M. A., VanZee, K. J., Rouck, C. S., Hawes, A. S., Thompson, W. A., DeForge, L., Kenney, J. S., Remick, D. G., Bloedow, D. C., Thompson, R. C., Lowry, S. F., and Moldawer, L. L.,** Interleukin-1 receptor blockade improves survival and hemodynamic performance in *Escherichia coli* septic shock, but fails to alter host responses to sublethal endotoxemia, *J. Clin. Invest.,* 89, 1551, 1992.
90. **Porat, R., Poutsiaka, D. D., Miller, L. C., Granowitz, E. V., and Dinarello, C. A.,** Interleukin-1 (IL-1) receptor blockade reduces endotoxin in *Borrelia burgdorferi*-stimulated IL-8 synthesis in human mononuclear cells, *Fed. Am. Soc. Exp. Biol.,* 6, 2482, 1992.
91. **Wakabayashi, G., Gelfand, J. A., Jung, W. K., Connolly, R. J., Burke, J. F., and Dinarello, C. A.,** *Staphylococcus epidermidis* induces complement activation, tumor necrosis factor in interleukin-1, a shock-like state and tissue injury in rabbits without endotoxemia, comparison to *Escherichia coli, J. Clin. Invest.,* 87, 1925, 1991.
92. **Aiura, K., Gelfand, J. A., Burke, J. F., Thompson, R. C., and Dinarello, C. A.,** Interleukin-1 (IL-1) receptor antagonist prevents *Staphylococcus epidermidis*-induced hypotension and reduces circulating levels of tumor necrosis factor-alpha and IL-1-beta in rabbits, *Infect. Immunity,* 61, 3342, 1993.
93. **Mancilla, J., Garcia, P., and Dinarello, C. A.,** The interleukin-1 receptor antagonist can either reduce or enhance the lethally of *Klebsiella pneumoniae* sepsis in newborn rats, *Infect. Immunity,* 61, 926, 1993.
94. **Granowitz, E. V., Porat, R., Mier, J. W., Pribble, J. P., Stiles, D. M., Bloedow, D. C., Catalano, M. A., Wolff, S. M., and Dinarello, C. A.,** Pharmacokinetics, safety and immunomodulatory effects of human recombinant interleukin-1 receptor antagonist in healthy humans, *Cytokine,* 4, 353, 1992.
95. **Fisher, C. J., Jr., Slotman, G. J., Opal, S. M., Pribble, J. P., Bone, R. C., Emmanuel, G., Ng, D., Bloedow, D. C., and Catalano, M. A.,** The IL-1 Sepsis Syndrome Study Group. Initial evaluation of human IL-1 receptor antagonist (IL-1ra) in the treatment of sepsis syndrome; a randomized, open-labelled, placebo-controlled trial, *Crit. Care Med.,* 22, 12, 1994.
96. **Walsh, C. E., Liu, J. M., Anderson, S. M., Russio, J. L., Nienhuis, A. S., and Young, M. S.,** A trial of recombinant interleukin-1 in patients with severe refractory aplastic anemia, *Br. J. Haematol.,* 80, 106, 1992.

97. **Fisher, C. J., Jr., Slotman, G. S., Opal, S. M., Pribble, J., and Catalano, M.,** The IL-1ra Sepsis Study Group. Interleukin-1 receptor antagonist reduces mortality in patients with sepsis syndrome, Am. Coll. Chest Phys. Annu. Meet., San Francisco, CA, (Abstr.) 1991.
98. **Opal, S. M., Fisher, C. J., Jr., Slotman, G. J., Bone, R. C., Emmanuel, G., Pribble, J. P., Stiles, D. M., Catalano, M. A., Ng, D., and LaBrecque, J. F.,** The IL-1ra Sepsis Study Group. The phase II interleukin-1 receptor antagonist sepsis syndrome trial: analysis of clinical, cytokine, and microbial features with outcome, Intersci. Conf. Antimicrobial Agents and Chemotherapy, 1570 (Abstr.), 1992.
99. **Fisher, C. J., Slotman, G. J., Opal, S. M., Dhainaut, J.-F., and Pribble, J. P.,** The IL-1ra Sepsis Syndrome Study Group. Human recombinant interleukin-1 receptor antagonist in the treatment of patients with sepsis syndrome, *Circ. Shock,* Suppl. 1, 11.8 (Abstr.), 1993.
100. **Knaus, W. A., Harrell, F. E., Fisher, J. C., Jr., Wagner, D. P., Opal, S. M., Sadoff, J. C., Hein, M., Walawander, C. A., Conboy, K., and Grasela, J. H.,** The clinical evaluation of new drugs for sepsis: a prospective study design based on survival analysis, *J. Am. Med. Assoc.,* 270, 1233, 1993.
101. **Opal, S. M., Sadoff, J. C., Fisher, C. J., Dhainaut, J.-F., Pribble, J. P., Stiles, D. M., and Knaus, W. A.,** The IL-1ra Phase III Sepsis Study Group. Initial findings of the IL-1 receptor antagonist (IL-1ra) phase III trial in patients with severe sepsis, Intersci. Conf. Antimicrobial Agents and Chemotherapy, 20 (Abstr.), 1993.
102. **Pribble, J., Fisher, C., Opal, S., Balk, R., Slotman, G., Reines, D., Shelly, M., Dhainaut, J.-F., Stiles, D., Knaus, W., Sadoff, J., and Labrecque, J.,** The IL-1ra Phase III Sepsis Syndrome Study Group. Human recombinant interleukin-1 receptor antagonist (IL-1ra) increases survival time in patients with sepsis syndrome and end organ dysfunction, *Soc. Crit. Care Med. Meet.,* R-0413 (Abstr.), 1994.
103. **Arend, W. P. and Dayer, J.-M.,** Cytokines and cytokine inhibitors or antagonists in rheumatoid arthritis, *Arthritis Rheum.,* 33, 305, 1990.
104. **Prieur, A.-M., Kaufmann, M.-T., Griscelli, C., and Dayer, J.-M.,** Specific interleukin-1 inhibitor in serum and urine of children with systemic juvenile chronic arthritis, *Lancet,* 2, 1240, 1987.
105. **Malyak, M., Joslin, F. G., and Arend, W. P.,** Synovial fluid in IL-1ra levels as determined by a modified sandwich ELISA, *Arthritis Rheum.,* 33, S149, 1990.
106. **Firestein, G. S., Berger, A. E., Chapman, D. L., Tracey, D. E., Chosay, J. G., and Zvaifler, N. J.,** IL-1 receptor antagonist protein (IRAP) production and gene expression by rheumatoid arthritis (RA) in osteoarthritis (OA) synovium, *Clin. Res.,* 39, 291A, 1991.
107. **Henderson, B., Thompson, R. C., Hardingham, T., and Lewthwaite, J.,** Inhibition of interleukin-1-induced synovitis in articular cartilage proteoglycan loss in the rabbit knee by recombinant human interleukin-1 receptor antagonist, *Cytokine,* 3, 246, 1991.
108. **Wooley, P. H., Whalen, J. D., Chapman, D. L., Berger, A. E., Aspar, D. G., Richard, K. A., and Staite, N. D.,** The effect of an interleukin-1 receptor antagonist protein on Type II collagen and antigen-induced arthritis in mice, *Arthritis Rheum.,* 33, S20, 1990.
109. **Schwab, J. H., Anderle, S. K., Brown, R. R., Dalldorf, F. G., and Thompson, R. C.,** Pro- and anti-inflammatory roles of interleukin-1 and recurrence of bacterial cell wall-induced arthritis in rats, *Infect. Immunity,* 59, 4436, 1991.
110. **Campion, G. V., Lebsack, M. E., Pauls, C. C., Martindale, J. J., and Catalano, M. A.,** A dose and regimen-ranging study of recombinant human IL-1 receptor antagonist in patients with rheumatoid arthritis, *Arthritis Rheum.,* 36, S39, 1993.
111. **Cominelli, F., Nast, C. C., Clark, B. D., Schindler, R., Llerena, R., Eysselein, V. E., Thompson, R. C., and Dinarello, C. A.,** Interleukin-1 (IL-1) gene expression, synthesis, and affect of IL-1 receptor blockade in rabbit immune complex colitis, *J. Clin. Invest.,* 86, 972, 1990.
112. **Sartor, R. B., Holt, L. C., Bender, D. E., Murphy, M. E., McCall, R. D., and Thompson, R. C.,** Interleukin-1 and interleukin-1 receptor antagonist gene expression correlate with activity of inflammation and spontaneously relapsing enterocolitis in rats, *Gastroenterology,* 100, A613, 1991.
113. **Nakamura, M., Kanakura, Y., Furukawa, Y., Ernst, T. J., and Griffin, J. D.,** Demonstration of interleukin-1 beta transcripts in acute myelogenous leukemic cells by in situ hybridization, *Leukemia,* 4, 466, 1990.
114. **Bagby, G. C. J., Dinarello, C. A., Neerhout, R. C., Ridgeway, D., and McCall, E.,** Interleukin-1 dependent paracrine granulopoiesis in chronic granulocytic leukemia of the juvenile type, *J. Clin. Invest.,* 82, 1430, 1988.

115. **Griffin, J. D., Rambaldi, A., Vellenga, E., Young, D. C., Octaporicz, D., and Kannistra, S. A.,** Secretion of interleukin-1 by acute myeloblastic leukemia cells *in vitro* induces endothelial cells to secrete colony stimulating factors, *Blood,* 70, 1218, 1979.
116. **Cozzolino, F., Rubartelli, A., Aldinucci, D., Sitia, R., Torcia, M., Shaw, A., and DiGuglielmo, R.,** Interleukin-1 as an autocrine growth factor for acute myeloid leukemia cells, *Proc. Natl. Acad. Sci. U.S.A.,* 86, 2369, 1989.
117. **Estrov, Z., Kurzrouck, R., Wetzler, M., Kantarjian, H., Blake, M., Harris, D., Gutterman, J. U., and Talpaz, M.,** Suppression of CML colony growth by IL-1-receptors: a novel application for inhibitors of IL-1 activity, *Blood,* 78 (Suppl. 6), 1476, 1991.
118. **Maier, J. A. M., Voulalas, P., Roeder, D., and Maciag, T.,** Extension of the life span of human endothelial cells by an interleukin-1 alpha antisense oligomer, *Science,* 249, 1570, 1990.
119. **Bernton, E. W., Beach, J. E., Holaday, J. W., Smallridge, R. C., and Fein, H. G.,** Release of multiple hormones by a direct action of interleukin-1 on pituitary cells, *Science,* 238, 519, 1987.
120. **Dinarello, C. A.,** Interleukin-1 and the pathogenesis of the acute-phase response. *New Engl. J. Med.,* 311, 1413, 1984.
121. **Opp, M. R. and Krueger, J. M.,** Interleukin-1 receptor antagonist blocks interleukin-1-induced sleep and fever, *Am. J. Physiol.,* 260, R453, 1991.
122. **Licinio, J., Wong, M. L., and Gold, P. W.,** Localization of the interleukin-1 receptor antagonist mRNA in rat brain, *Endocrinology,* 129, 562, 1991.
123. **Altstiel, L. D. and Sperber, K.,** Cytokines in Alzheimer's disease, *Prog. Neuropsychopharmacol. Biol. Psychiatry,* 15, 481, 1991.
124. **Numerof, R. P., Sipe, J. D., Zhang, Y., Dinarello, C. A., and Mier, J. W.,** Use of an IL-1 receptor antagonist (IL-1ra) in IL-2-treated mice, *Lymphokine Res.,* 9 (Abstr.), 612, 1990.
125. **McClain, C., Cohen, D., Phillips, R., Ott, L., and Young, B.,** Increased plasma and ventricular fluid interleukin-6 levels in patients with head injury, *J. Lab. Clin. Med.,* 118, 225, 1991.
126. **Dinarello, C. A. and Wolff, S. M.,** The role of interleukin-1 in disease, *New Engl. J. Med.,* 328, 106, 1993.
127. **Bluthe, R. M., Parnet, P., Dantzer, R., and Kelley, K. W.,** Interleukin-1 receptor antagonist blocks effects of IL-1-alpha and IL-1-beta on social behavior and body weight in mice, *Neurosci. Res. Commun.,* 15, 151, 1991.
128. **Bluthe, R. M., Dantzer, R., and Kelley, K. W.,** Effects of interleukin-1 receptor antagonist on the behavioral affects of lipopolysaccharide in rat, *Brain Res.,* 573, 318, 1992.
129. **Saez-Llorens, X., Ramilo, O., Mustafa, M., Mertsola, J., and McCracken, G. H., Jr.,** Molecular pathophysiology of bacterial meningitis: current concepts and therapeutic implications, *J. Pediatr.,* 5116, 671, 1990.
130. **Tunkle, A. R., Wispelwey, B., and Scheld, W. M.,** Bacterial meningitis: recent advances in pathophysiology and treatment, *Ann. Intern. Med.,* 112, 610, 1990.
131. **Odio, C. M., Faingezicht, I., Paris, M., Nassar, M., Baltodano, A., Rogers, J., Saez-Llorens, X., Olsen, K. D., and McCracken, G. H.,** The beneficial effects of early dexamethasone administration in infants and children with bacterial meningitis, *New Engl. J. Med.,* 324, 1525, 1991.
132. **Kim, K. S., Was, C. A., Cross, A. S., and Opal, S. M.,** Modulation of blood-brain barrier permeability by tumor necrosis factor and antibody to tumor necrosis factor in the rat, *Lymphokine Cytokine Res.,* 11, 593, 1992.
133. **Friedman, W. J., Larkfors, L., Ayer-LeLievre, C., Ebendal, T., Olson, L., and Person, H.,** Regulation of beta-nerve growth factor expression by inflammatory mediators in hippocampal cultures, *J. Neurosci. Res.,* 27, 374, 1990.
134. **Guenard, V., Dinarello, C. A., Weston, P. J., and Aebischer, P.,** Peripheral nerve regeneration is impeded by interleukin-1 receptor antagonist released from a polymeric guidance channel, *J. Neurosci. Res.,* 29, 396, 1991.
135. **Dayer-Metroz, M.-D., Duhamel, D., Rufer, N., Izui, S., Carmichael, D., Wollheim, C. B., Thompson, R. C., and Dayer, J.-M.,** Interleukin-1 receptor antagonist delays spontaneous autoimmune diabetes in BB rat, *Eur. J. Clin. Invest.,* (Abstr.), 1991.
136. **McCarthy, P. L., Abhyankar, S., Neben, S., Newman, G., Sieff, C., Thompson, R. C., Burakoff, S. J., and Farrara, J. L. M.,** Inhibition of interleukin-1 by an interleukin-1 receptor antagonist prevents graft vs. host disease, *Blood,* 78, 1915, 1991.

137. **Meyers, K., Czachewski, C., Haggman, R., Thompson, R. C., and Coffey, J.,** Effect of indomethacin and dexamethasone treatment on the response of interleukin-1 receptor antagonist in carrageenan-induced pleurisy in the rat, 5th Int. Conf. Inflammation Res. Assoc., A47 (Abstr.), 1990.
138. **Dayer-Metroz, M.-D., Hollheim, C. B., Seckinger, P., and Dayer, J.-M.,** A natural interleukin-1 (IL-1) inhibitor counteracts the inhibitory effect of IL-1 on insulin production in cultured rat pacreatic islets, *J. Autoimmunity,* 2, 163, 1989.
139. **Selig, W. and Tocker, J.,** Effect of interleukin-1 receptor antagonist on antigen-induced pulmonary responses in guinea pigs, *Eur. J. Pharmacol.,* 213, 331, 1992.
140. **Dinarello, C. A.,** Role of interleukin-1 in infectious diseases, *Immunol. Rev.,* 127, 119, 1992.
141. **Havell, E. A., Moldawer, L. L., Helfgott, D., Kilian, P. L., and Sehgal, P. B.,** Type I interleukin-1 receptor blockade exacerbates murine listeriosis, *J. Immunol.,* 148, 1486, 1992.
142. **Cross, A. S., Sadoff, J. C., Kelly, N., Bernton, E., and Gemski, P.,** Pre-treatment with recombinant murine tumor necrosis alpha/cachetin and murine interleukin-1 alpha protects mice from lethal bacterial infection, *J. Exp. Med.,* 169, 2021, 1989.
143. **Fisher, C. J., Dhainaut, J.-F., Opal, S. M., Pribble, J. P., Balk, R. A., Slotman, G. J., Iberti, T. J., Rackow, A. C., Shapiro, M. J., Greenman, R. L., Reines, H. D., Shelly, M. P., Thompson, R. W., LaBreque, J. F., Catalano, M. A., Knaus, W. A., and Sadoff, J. C.,** The phase III Sepsis Syndrome Study Group. Recombinant human interleukin-1 receptor antagonist in the treatment of patients with sepsis syndrome: results from a randomized, double-blind, placebo-controlled trial, *J. Am. Med. Assoc.,* 271, 1836, 1994.

Chapter 10

TNFα NEUTRALIZATION BY BIOLOGICAL ANTAGONISTS

Mark W. Bodmer and Roland Foulkes

CONTENTS

I. INTRODUCTION

Tumor Necrosis Factor-α (TNF) derives its name from an apparent ability to kill transformed cells *in vitro* and to cause hemorrhagic necrosis of tumors in experimental systems. The earliest recorded observations of antitumor effects of bacterial products were in the 1890s with the work of Coley.[1] The first description of the biological activity of TNF was as a tumor cytotoxic/necrotic activity in the serum of mice challenged with bacteria (for a review see Old[2]). It has since become clear that TNF plays a wide variety of roles in host defense, metabolic status, and the coordination of inflammatory responses and their sequelae.[3]

TNF may play a pathogenic role not only in septic shock (as well as in other infectious conditions such as malaria[4,5]) but also in a wide variety of subacute and chronic diseases. These include rheumatoid arthritis,[6–10] inflammatory bowel disease,[11] allograft rejection,[12–14] HIV infection,[15–19] and some adverse drug reactions such as the OKT3-induced syndrome.[20] A TNF antagonist, therefore, has the potential to address several important clinical indications.

0-8493-8381-1/96/$0.00+$.50

This review will address the extent to which TNF plays a role in pathological responses through both animal models and clinical studies using specific antagonists of TNF biological activity.

Two main types of antagonists have been used in these studies. The first are monoclonal antibodies which are selected to be capable of binding to and neutralizing TNF. Antibodies are bifunctional adaptor molecules with antigen recognition sites in the variable domains at one end, and an isotype-dependent capacity to recruit cell- and complement-mediated effector functions in the constant domains at the other end. Thus, the choice of a reagent with suitable biological activity for therapy requires the selection of an antibody with both high-potency TNF neutralizing activity in the variable region, and the functionally optimal isotype. IgG isotypes fall into two general classes with respect to their ability to elicit effector functions. Human IgG1 and 3 and murine IgG2a have high affinity for all FcγR cellular receptors and are good complement activators. Human IgG2 and 4 and murine IgG1 are low or null in these activities. These known activities do not clearly predict which class of function would be preferred for neutralizing a soluble circulating mediator *in vivo.* This issue is further addressed below with respect to models of septic shock. Most monoclonal antibodies are of murine origin and in humans they elicit anti-immunoglobulin immune responses which limit their clinical use. Recombinant engineering techniques are being used to create "humanized" versions of rodent antibodies which retain the binding activity of the parent molecule in the relatively small antigen-binding region, but otherwise are composed of human sequences.[21,22]

A second type of antagonist is based on the extracellular portions of the cellular receptors for TNF. These polypeptide domains are naturally shed from the cell surface, and appear to act as natural soluble binding proteins and endogenous antagonists of TNF. Recombinant versions have been used either in native form, or in the form of "immunoadhesins" in which they are linked to immunoglobulin heavy-chain constant domains.[23] This latter modification is important both in increasing the neutralizing potency of the receptor by causing it to dimerize on the immunoglobulin structure, and also in increasing the plasma residence time by several orders of magnitude.

A substantial body of experience has built up, principally using monoclonal antibodies, indicating where TNF blockade may be of benefit. The three areas where the most clinical interest is currently focused are infectious disease and shock, inflammatory bowel disease, and rheumatoid arthritis. These are reviewed in detail below, followed by a short section outlining several other pathologies where there is an emerging role for TNF.

II. INFECTIOUS DISEASE AND SHOCK

TNF has emerged as one of a number of host-derived mediators which, either singly or in combination, appear able to mediate some of the systemic inflammatory events which followed a septic challenge. There is a large body of evidence from both experimental animal studies and clinical monitoring implicating inappropriate production of TNF as a key factor in the degeneration of patients with septic shock. The magnitude of the contribution of any given mediator to clinical status remains open to debate. However, for the purposes of the development of new therapies, the primary focus has been to select individual mediators whose antagonism may be of clinical benefit. Prominent recent examples of this include Interleukin-1 (IL-1), Platelet Activating Factor (PAF), and TNF.

In humans, numerous assay studies in septic patients have attempted to correlate the plasma levels of TNF and related cytokines with the clinical course and prognosis.[24–30] Although these studies reveal a general trend to an association, the results are somewhat patchy and may be assay dependent. In a few cases of acute illness a clear clinical picture emerges. In less acute cases, it may be that plasma TNF is either below assay detection limits or absent; is associated with cells, soluble receptors, or nonspecific serum binding proteins; or is produced and/or sequestered into other body microenvironments. Direct administration of low doses of endotoxin to human volunteers elicits a measurable rise in plasma TNF.[31] This is an important observation, as endotoxin is thought to be a primary pathogenic driver in Gram-negative septic shock.

More recently, genetic disruption of the cellular receptors for TNF in transgenic mice[32,33] has been shown to render them both resistant to endotoxin challenge and more sensitive to infection by *Listeria monocytogenes,* resistance to which had been previously shown to be TNF dependent.

A septic challenge causes a very wide variety of effects that can broadly be put into three categories. First, there are physiological changes associated mainly with deterioration in function in most vital organ systems. Second, there are cellular effects, characterized by changes in adhesion and activation states of endothelial and circulating inflammatory cells. Third, a large number of host mediators are elaborated which are likely to be orchestrating the physiological and cellular effects. The use of monoclonal antibodies as specific TNF antagonists in rodent and nonhuman primate models of septic shock demonstrate a role for TNF in this complex cascade of events.

TNF administered to experimental animals[34] and humans elicits many of the physiological and tissue disturbances characteristic of clinical septic shock. The earliest demonstration that *in vivo* enodotoxic shock may be treated by TNF blockade was in a mouse model of endotoxemia using a rabbit polyclonal antiserum to murine TNF.[35] The antiserum was shown to cause an increase in the median lethal dose of endotoxin in Balb/C mice. A similar protective effect was later shown using the TNF receptor, immunoadhesin.[23]

More clinically relevant than endotoxemia is the use of bacterial challenges. Studies in rodents,[36,37] pigs,[38,39] and nonhuman primates[40–43] have shown protection by neutralizing antibodies against the lethal effects of shock induced by sepsis. Silva et al.[36] used a monoclonal antibody to recombinant murine TNF, TN3 19.12,[44] to look at the role of TNF in Gram-negative shock. A protective effect was seen in mice receiving an LD_{90} dose of *Escherichia coli* when given shortly prior to the intravenous bacteria. Less benefit was seen in mice given *Pseudomonas aeruginosa* or in cyclophosphamide-treated mice challenged with *Klebsiella pneumoniae.*

In a further attempt to mirror the genesis of clinical septic shock, Opal et al.[37] developed a rat model in which the gut is colonized with *P. aeruginosa* after which the animals are rendered neutropenic with cyclophosphamide. The neutropenia allows the bacteria to infect systemically, leading to death of all animals within 7 to 10 days. Treatment with TN3 19.12, which cross-neutralizes rat TNF, resulted in a 50% survival rate. When combined with ciprofloxacin, all animals survived. In this system, the protective effects of anti-TNF were seen not only with prophylactic administration, but also when treatment was initiated after the appearance of clinical signs of disease.[45]

A seminal study of anti-TNF in lethal bacteremia in the baboon was reported by Tracey et al.[40] Animals were passively immunized against TNF with an Fab'_2 fragment of a monoclonal antibody to human TNF which was cross reactive. This protected against an LD_{100} dose of i.v. *E. coli* given 2 h later. Strangely, the protective effect was much diminished if the interval between antibody and bacterial infusion was reduced to 1 h. A later study by Hinshaw et al.[41] addressed this question. Using a full immunoglobulin in a model similar to that of Tracey et al., they were able to show protective effects against the LD_{100} challenge when antibody was given 30 min after the start of the bacterial infusion. Although this does not define a treatment window that would be of use clinically, it does remove the apparent need for preconditioning of the animals for the protective effect to be shown. These animal systems use very large bolus challenges which do not accurately mirror the clinical course of the development of sepsis and shock, apart, perhaps, from the neutropenic rat. They do define a clear pathophysiological role for TNF in Gram-negative shock and endotoxemia. Whether this activity is relevant at the point of clinical diagnosis and intervention remains to be determined.

The role of TNF in morbidity and mortality associated with Gram-positive infection and shock is more controversial. These infections are very important, accounting for 40 to 50% of clinical septic shock. Like endotoxin, cell wall components of Gram-positive organisms are capable of inducing TNF secretion both *in vitro* and *in vivo.*[46,47] In a baboon model of lethal *Staphylococcus aureus*-induced shock, anti-TNF monoclonal antibody was highly effective in preventing multiple organ failure and preventing death.[42] These results were achieved in the presence of the antibiotic, ceftriaxone, which on its own had only modest effects at prolonging survival. Results in a study of murine Gram-positive challenge have been less clear cut.[48]

A second area of controversy surrounding the clinical potential of TNF blockade in infectious disease concerns differences in response to anti-TNF depending on the site of infection. All models in which beneficial effects have been seen primarily involve systemic challenge. Bagby et al.[49] compared the efficacy of anti-TNF treatment on survival following intravascular or intraperitoneal instillation of *E. coli.* The i.v. challenge gave high TNF plasma levels in controls and responded to anti-TNF. The treatment was not effective in the i.p. animals. Echtenacher et al.[50] induced peritonitis by cecal ligation and puncture (CLP). In this case anti-TNF was deleterious, suggesting a requirement for endogenous TNF for recovery from CLP. Indeed, Alexander et al.[51] have demonstrated a protective effect of exogenous recombinant TNF in a similar model of CLP in rats.

These studies raise an important issue relating to the balance between the protective and destructive effects of the endogenous TNF response to infection. TNF probably plays a role in primary host defense against infection. In circumstances where the etiology of disease is related to infection per se, removing TNF may be deleterious in the absence of covering antimicrobial therapy. The responses in the intravascular models may be dominated by a systemic inflammatory response which results, at least in part, from an inappropriate overproduction of TNF. The possibility of compromising host defense by TNF blockade has been a key safety issue in clinical studies, although to date there have been no reports of adverse events relating to infection from studies involving nearly 2000 patients treated with anti-TNF antibodies.

As mentioned earlier, one area of interest in the design of antibodies to suppress TNF relates to isotype differences in the antibody constant region. We have investigated the effect of isotype on *in vivo* therapeutic activity in a neutropenic rat model of lethal *Pseudomonas* infection.[52] Antihuman TNF monoclonal antibodies (MAbs) do not cross-react with rodent TNF, so for this study it was necessary to create two reagents based on the hamster MAb, TN3 19.12,[44] which neutralizes both mouse and rat TNF. The hamster constant region was replaced with either the *inactive* murine IgG1 (equivalent to human IgG4) or the *active* murine IgG2a (equivalent to human IgG1). Note the difference in nomenclature between human and mouse IgG with respect to Fc function. These reagents were compared for their ability to protect the lethally infected rats. The inactive murine IgG1 gave better protection (~65% survival) than the active IgG2a (~20% survival). Both antibodies were administered as sole therapy after the first physiological evidence of sepsis became apparent. The biological basis for this difference is not clear at present. It is possible that since TNF is a trimer presenting multiple epitopes for antibody binding, relatively efficient immune complex formation may result. Immune complexes of functionally active antibodies have pro-inflammatory activity in their own right, and this may explain the isotype difference. Whatever the explanation, efficacy is clearly related to isotype and the functionally inactive murine IgG1 is indicated for a mouse monoclonal anti-TNF antibody.

All of the above results provide a reasonable basis for postulating that a TNF antagonist may be of therapeutic benefit in clinical septic shock.[53] Three small clinical studies have been reported in which a murine anti-TNF MAb has been administered to septic patients.[54–56] These give some limited support for the concept, at least in those acutely ill patients in whom cytokine levels are elevated. More recently, two large clinical studies, one centered in North America and one in Europe, have been completed. These studies compare a single administration of a murine monoclonal antibody to TNF with placebo in patients with septic shock, with a prospective stratification of patients into either a shock or no shock stratum. Although publication of the results is still awaited at the time of writing, data on the 994 patients enrolled in the North American study were presented in October, 1993 at the 33rd Interscience Conference on Antimicrobial Agents and Chemotherapy (ICAAC). In patients not in shock at study entry, no effect of the antibody was seen, either positive or negative. In patients in shock at entry, there was a clear trend to a reduction in mortality at both dose levels (7.5 and 15 mg/kg) used in the study. The antibody was well tolerated with very few adverse events that were judged antibody related. This antibody has now been entered into a larger clinical trial with the aim of confirming the activity in septic shock patients.

The fact that efficacy was limited to the 50% of patients with frank shock (hemodynamic instability) at entry compares well with the results of the animal model studies. These are patients, selected from a very heterogeneous group, who are in an acute phase of inflammatory degeneration during which the processes involving mediators such as TNF are likely to be active. This is the cohort likely to be most similar to the experimental animal experiencing an acute septic/inflammatory challenge.

III. INFLAMMATORY BOWEL DISEASE

The inflammatory bowel diseases (IBD), which include Crohn's disease (CD) and ulcerative colitis (UC), are debilitating diseases characterized by chronic inflammation of intestinal mucosa and lamina propria. Although the events which trigger their onset (such as dietary, bacterial, viral, or environmental) remain unknown, they are associated with a marked leukocyte infiltrate and local production of soluble mediators. CD is a condition of chronic cell-mediated immunopathology with an infiltration of activated ($CD25^+$) $CD4^+$ T-cells into the intestinal lamina propria.[57] The antigen against which these T-cells are directed is obscure but is likely to be the intestinal epithelial cell[58] which stains strongly positive for HLA-DR in patients with active but not passive disease. However, although lymphocyte-mediated cytotoxicity against epithelial cells has been demonstrated *in vitro,*[59] it remains unclear whether such a mechanism operates *in vivo.*[58] Like CD, there is also an influx of immune cells into the submucosa of UC patients although there is a lower T-cell ($CD25^+$, $CD3^+$, $CD4^+$) and higher macrophage ($CD25^+$, $CD3^-$, $CD4^+$, HLA-DR^+) presence in the latter.[57] Indeed, the pathology of UC may also be autoimmune in origin as it is often associated with the production of anticolon antibodies by cells within the colonic mucosa.[60] Despite these differences in T-cell infiltrate and in the distribution and localization of the inflammation, both are characterized by the presence of inflammatory cells such as neutrophils and macrophages.[61] As such, the potential array of soluble mediators produced by immune and inflammatory cells is large and may vary depending on the type and severity of the disease.[62] It is not surprising, therefore, that increased TNF immunoreactivity has been demonstrated in the lamina propria in both CD and UC patients.[63] The cells are most likely of macrophage origin and specific TNF immunoreactivity, mRNA levels (assessed by PCR, Andus et al.[64]) or *in situ* hybridization[65] and TNF as well as IFN-γ, IL-1β, and IL-6 secretion by cultured lamina propria MNC,[66] are all increased in active disease compared to normals. Hence, the effect of such high local concentrations of TNF are likely to be adverse.

Further indirect evidence for the involvement of TNF in the pathology of IBD has been obtained by measuring TNF activity in the stools from diseased individuals. In the stools of children with active CD or UC, TNF concentrations were 994 pg/g (CD) or 276–5982 pg/g (UC) compared to around 50 pg/g in normal children, those with nonspecific diarrhea, or those with inactive disease.[11,67] Indeed, it has been suggested that routine assessment of stool TNF levels may represent a useful marker of disease activity in these patients. There are also reports of enhanced circulatory levels of TNF in IBD; increased plasma levels being reported in 58% of CD and UC patients[68] and higher levels of the p55 (but not p75) TNF-R associated with both conditions.[69]

TNF therefore may be a key mediator in the pathogenesis of IBD either by a direct cytotoxic action or as an orchestrator of the inflammatory cascade. Blood-borne TNF itself can reduce intestinal blood flow and limit fuel availability to the gut mucosa,[70] but it is more likely that its effect due to local production is the most important. The release of other inflammatory cytokines thought to be important in IBD, such as IL-1β,[65,71,72] IL-6,[67,73] and IL-8[74] can be stimulated by TNF. The recruitment of inflammatory cells by these soluble mediators, and the expression of the appropriate adhesion molecules that have been reported to be associated with IBD, such as E-selectin, VCAM-1,[75] and ICAM-1[76] can all be influenced by local TNF production. TNF-mediated increases in vascular thrombosis and inducible NOS expression[77] in IBD may represent other effector mechanisms of this cytokine.

Hence, TNF may represent a credible therapeutic target for intervention for the treatment of CD and UC. There are a number of animal models in which the hypothesis has been tested. In an acetic acid/immune complex-mediated model of colonic inflammation in rats the maximum increase in fecal water content (a measure of disease severity) coincided with an influx of inflammatory cells into the colonic mucosa.[78] Continuous treatment with a specific antirodent TNF MAb (chimeric mouse/hamster TN3 19.12) prevented the diarrheic response and abrogated the inflammatory cell infiltrate. In this model there is no TNF detectable in the plasma so the therapeutic effect of the antibody is likely to be through an inhibition of local TNF activity. Similar results were obtained with cTN3 19.12 in a second rat model. Feeding degraded carrageenan solution to rats also induces diarrhea, the extent of which can be prevented by prophylactic antibody treatment or is markedly reduced if treatment is withheld until the onset of symptoms occur (Ward, P., personal communication).

Although neither of these induced models in rats can be considered as representative of CD or UC, they do mimic some of the symptoms of these diseases and allow the investigation of mechanisms involved. Hence, in acetic acid-induced colitis, changes in vascular permeability and granulocyte recruitment as seen in UC can be effectively studied, whereas the carrageenan model induces a more patchy, diffuse inflammation more indicative of CD. In contrast, the cotton-top tamarin (CTT) provides a very good model of human UC. The disease occurs spontaneously in a proportion of these animals and is characterized by an acute colonic inflammation with infiltrating neutrophils seen in the lamina propria and the development of eccentric crypt abscesses. The animals develop diarrhea and an acute wasting disease, resulting in the animals' demise if not treated. Active disease is associated with an increase in fecal TNF concentration although plasma TNF levels are in the normal range. Treatment of colitic animals with a short course of humanized antihuman TNF MAb (CDP571)[21] reverses the weight loss and diarrhea with a rapid (<6 days) onset of action. Moreover, there is a marked reduction in the disease severity as assessed histologically on colonic biopsies, with improvements persisting in some animals 12 months after the last treatment.[79] In addition, some CTT become resistant to treatment with both olsalazine and prednisone. Treatment of these animals with CDP571 also has been shown to induce remission in some animals (Watkins, P., personal communication).

Hence, data from animal models would seem to support the rationale for a therapeutic study in human IBD aimed at reducing the effect of TNF. Indeed, in a recent report[80] a 12-year-old girl with severe unresponsive CD showed complete endoscopic remission following treatment with anti-TNF MAb. These investigators have now extended their open study to more than 10 CD patients with a reported 100% remission rate, as presented at the Fifth International TNF Congress in June, 1994.

IV. RHEUMATOID ARTHRITIS

The mechanisms involved in the immune and inflammatory responses leading to cartilage and bone destruction in rheumatoid arthritis (RA) are poorly understood. The disease is characterized by inflammation and cellular proliferation of the synovial lining (pannus formation) which results in the invasion of the adjacent cartilage matrix and its subsequent erosion. The pannus and cartilage tissue become infiltrated by cells of the immune system and inflammatory leukocytes. Activated T-cells are found within the subintimal layer of the synovium, but not normally at the pannus/cartilage junction. This area is associated with a predominantly PMN, macrophage, fibroblast, and plasma cell infiltrate. Hence, as in many other inflammatory disorders, there is a wealth of soluble mediators that can be detected within the rheumatoid joint which may contribute to progressive erosion.

Although TNF is generally considered as a product of leukocytes it can also be produced by postcapillary endothelial cells, chondrocytes, osteoclasts, and fibroblasts. All of these cells are found in the rheumatoid joint and TNF can be detected both in synovial fluid and synovial tissue from patients.[81] Increased expression of TNF has been identified in cells within the synovial lining and at the cartilage pannus junction, especially around the area of cartilage erosion.[82] The net effect of local TNF secretion would be not only to amplify the inflammatory process by its effect on leukocyte migration and activation but also to increase synoviocyte proliferation, since the latter effect can be reduced by anti-TNF MAb treatment.[83] In addition, TNF may stimulate chondrocytes and osteoclasts to break down cartilage and bone.[84] TNF added to cultured cartilage *in vitro* enhances matrix breakdown and inhibits proteoglycan synthesis.[85] This may be through an effect of TNF on chondrocytes resulting in the production of specific proteinases (such as collagenase) although some of these enzymes may be produced by an effect of TNF on the synovium itself. Indeed, TNF production by normal chondrocytes may be important in autocrine control of cartilage homeostasis.[82] Hence, the balance of cytokine regulation of cartilage and bone turnover may be disturbed in RA.[86] Simulation of TNF overproduction was achieved by injecting TNF into rabbit knee joints, which caused depletion of proteoglycan from the articular cartilage. TNF is also an autocrine factor associated with bone turnover and therefore, potentially, with bone remodeling and joint destruction. TNF stimulates bone resorption by increasing osteoclast formation and activity and by increasing osteoblast-mediated osteoclast function.[84]

The role of TNF in animal models of RA has been studied extensively. In rats with collagen II-induced polyarthritis, infusion of TNF resulted in the development of a more severe disease than in noninfused arthritic rats.[10] This was manifested by an increased arthritic index and radiographic score although there was no upregulation of humoral responses to collagen II. Furthermore, a transgenic mouse which expresses a 3′-modified human TNF transgene develops chronic inflammatory polyarthritis.[87] Joints of these animals show hyperplasia of the synovial membrane and polymorphonuclear and lymphocytic infiltration into the synovium. Pannus formation, articular cartilage destruction, and fibrous tissue deposition were also prevalent. In the collagen II model, treatment of mice with either anti-TNF MAbs or a recombinant soluble TNF receptor prevented the development of polyarthritis.[9] This was seen as a reduction both in footpad swelling and in synovial inflammation. Since no effect of anti-TNF treatment was seen when given at 2 months after disease onset, the authors concluded that TNF was important in the early inflammatory phase of the disease but not during the erosion phase. Similar striking effects of early anti-TNF treatment have also been reported in a model of antigen-induced arthritis in rabbits.[88] In this case, a 3-day course of anti-TNF MAb (rat antirabbit TNF, R6) reduced the extent of knee joint swelling and synovial leukocyte infiltration by about 50%. Evidence of a beneficial effect of neutralizing TNF has also been shown when treatment is withheld until the onset of disease.[89] Mice treated with collagen II were given anti-TNF (TN3 19.12) twice a week once the first signs of disease were apparent. In this study, the progression of disease, in terms of paw swelling and number and severity of joints involved, was arrested. Furthermore, if the human TNF transgenic mouse which develops polyarthritis[87] is treated with antihuman TNF MAb, then all symptoms are ameliorated (Kollias, G., personal communication).

Such preclinical studies have suggested that neutralization of TNF may be a therapeutically useful approach to reduce the inflammation in RA and possibly the erosive process itself. Indeed, a mouse/human chimeric anti-TNF MAb (cA_2) has been studied in an open trial in patients with severe RA.[90] In 20 patients given 20 mg/kg cA_2, significant improvements were seen in the Ritchie score, the swollen joint count, and in serum levels of C-reactive protein. Hence, in this albeit uncontrolled trial, neutralization of TNF does ameliorate symptoms of RA, possibly by reducing the inflammatory component of the disease. Since no radiological studies were performed, it is unclear whether such treatment can affect the erosion process. Since then, there have been two double-blind placebo-controlled studies in RA with anti-TNF MAbs (cA_2[119] and CDP571). Both of these studies were initially reported at the 1994 National Meeting of the American College of Rheumatology, and broadly confirmed the earlier published results. Further clinical studies with these reagents are now needed to determine if they have an important role in the clinical management of RA, particularly with respect to any disease-modifying activity.

V. OTHER TNF-MEDIATED PATHOLOGIES

There is a long list of conditions in which there is some evidence that TNF may be involved. In some cases, such as transplantation and graft-vs.-host disease, and ischemia/reperfusion injury, the potential mechanism is consistent with the idea that TNF is an ubiquitous pro-inflammatory agent induced by many different "challenges" to tissue. Others, such as the suppression of insulin responsiveness in noninsulin-dependent diabetes and potentiation of HIV replication relate to more selective actions of TNF which appear to fall outside the pro-inflammatory model. A selection of these pathologies is discussed below in the light of data on the effect of TNF antagonism *in vitro* or *in vivo*.

A. TRANSPLANTATION AND GRAFT–VS.–HOST DISEASE

Many investigators have reported elevated serum or plasma TNF levels associated with complications following a variety of clinical transplantation procedures. These include allografts of kidney,[91] liver,[92] heart,[93] pancreas,[94] and bone marrow.[95] In each case, the appearance of TNF correlates with the onset of rejection or, in the case of bone marrow transplantation, graft-vs.-host disease (GvHD).

Intervention studies using anti-TNF antibodies in animal models have attempted to discover whether TNF is an effector of these responses or merely a marker. In mouse models of GvHD, anti-TNF was shown to reduce both mortality and tissue damage caused by the reaction.[96,97] A small clinical study has examined the effect of a murine monoclonal antibody to TNF in 19 patients with severe acute GvHD refractory to conventional (steroid) therapy.[98] Partial responses were seen in 14 of the patients during the period of treatment, but in all cases GvHD recurred after discontinuation, suggesting that long-term control cannot be achieved, at least not in this severely ill population.

To date there have been no clinical studies of anti-TNF in organ allografts. In rat cardiac heterotopic grafts, some prolongation has been seen[12] and there are at least two reports of positive interactions between anti-TNF and cyclosporine.[13,99] In an immunologically very severe model of skin grafting in rhesus monkeys, Stevens et al.[100] observed a modest prolongation of graft survival when anti-TNF was combined with an antibody to γ-interferon. In summary, it is not clear to what extent TNF blockade will be of use in transplantation. Although it does seem to be important in the effector arm of the rejection episode, the immunological drive remains and the process restarts when therapy is withdrawn. Further progress here may depend on the availability of low molecular weight TNF suppressors which can be administered over long periods of time.

B. ISCHEMIA/REPERFUSION INJURY

Certain types of injury caused by, for instance, major trauma or surgical procedures, result in a transient ischemia which may be either local or systemic. Reperfusion of ischemic tissue leads to a vascular inflammatory response which can have very severe clinical consequences. A part of this response may be mediated by TNF. Colletti et al.[101] reported the effects of hepatic ischemia/reperfusion injury in the rat, showing induction of TNF in the plasma and remote tissue damage in the lung as well as locally at the site of challenge. Pretreatment with anti-TNF antiserum completely prevented the development of pulmonary edema and reduced the elevation of serum glutamic pyruvic transaminase used as a marker of hepatic damage. Two groups have reported the effect of anti-TNF antibodies in models of systemic ischemia/reperfusion. In rats, anti-TNF was able to reduce mortality to a lethal challenge of blood removal and replacement.[102] In a similar, but sublethal model in rabbits carried out in our own laboratory, treatment resulted in a significant reduction in tissue damage.

C. HIV REPLICATION

Replication of HIV is activated by the inducible transcription factor, NF-κB[103] which, in turn, is induced by TNF.[104] HIV expression can be induced by TNF in macrophage lines and T-cell clones chronically infected with virus.[15] Infusion of recombinant TNF in a small number of patients with AIDS-related Kaposi's sarcoma appeared to cause an increase in the HIV p24 antigen level, a marker of viral replicative activity.[105] Several groups have reported either elevations in serum TNF[19,106] or increased cellular production of TNF,[18,107] broadly correlating with disease severity. These results provide a mechanistic basis for considering the use of a TNF blocker to reduce infectious HIV burden, although there is conflicting evidence in the literature of the ability of HIV to induce TNF expression in mononuclear cells.[108,109]

De novo infection of freshly isolated peripheral blood mononuclear cells with HIV has been shown to induce TNF expression.[16] Blocking the activity of this endogenous TNF with neutralizing antibodies reduces the virus production in these assays. A similar observation has been made using *N*-acetyl-L-cysteine (NAC). Treatment of cells with NAC increases intracellular glutathione. This leads to nonresponsiveness to TNF, including the loss of the HIV-inducing activity.[110] The definition of the potential clinical role of anti-TNF in HIV-infected patients has not developed significantly since the early 1990s, and we must now await results from intervention studies in primate models and the clinic.

D. OBESITY-LINKED INSULIN RESISTANCE

An early description of the biological activity of TNF was as a "cachectin", capable of inducing a systemic metabolic wasting.[111] Production of TNF in late stages of cancer is thought to be involved in the cachexia seen in these patients.[112,113] Paradoxically, recent reports suggest that TNF may be induced in adipocytes of both genetically obese rodents and of obese human subjects.[114,115] This induction of TNF appears to cause a downregulation of insulin sensitivity, which is a major defect in obese noninsulin-dependent diabetics. In experimental animals, systemic administration of either soluble TNF receptor of anti-TNF antibody is able to normalize the glucose response to insulin challenge. It has also been shown that TNF mRNA levels can be at least partially normalized in obese animals treated with the insulin-sensitizing drug, pioglitazone.[116]

This raises the intriguing possibility that TNF blockade may be able to restore insulin-regulated glucose balance and ameliorate some of the clinical effects of compensatory insulin oversecretion. A small clinical study is currently underway to determine the effect of an anti-TNF antibody in obese diabetics.

VI. CONCLUDING REMARKS

We have reviewed here the TNF-mediated pathologies for which the clinical interest in the use of antagonists is greatest. There are many others — such as B-cell malignancies,[117] multiple sclerosis (as modeled by EAE[118]), cerebral malaria,[5] and adverse reactions to the immunosuppressive antibody OKT3.[20] The next two to three years should see the publication of the results of a wide variety of clinical studies which will begin to reveal the true value of TNF blockade in human healthcare.

REFERENCES

1. **Coley Nauts, H., Fowler, G. A., and Bogatko, F. H.,** A review of the influence of bacterial infection and of bacterial products (Coley's toxins) on malignant tumours in man, *Acta Med. Scand.,* Suppl. 274:29–97, 1953.
2. **Old, L. J.,** Tumor necrosis factor, *Sci. Am.*, 258:41–49, 1988.
3. **Beutler, B. and Cerami, A.,** Cachetin: more than a tumor necrosis factor, *N. Engl. J. Med.,* 316:379–385, 1987.
4. **Grau, G. E., Piguet, P.-F., Vassalli, P., and Lambert, P.-H.,** TNF and other cytokines in cerebral malaria: experimental and clinical data, *Immunol. Rev.,* 112, 1989.

5. **Kwiatkowski, D., Molyneux, M. E., Stephens, S., et al.,** Anti-TNF therapy inhibits fever in cerebral malaria, *Q. J. Med.,* 86:91–98, 1993.
6. **Brennan, F. M., Chantry, D., Jackson, A. J., et al.,** Inhibitory effect of TNFα antibodies on synovial cell IL-1 production in rheumatoid arthritis, *Lancet,* 2(8657):244–247, 1989.
7. **Yocum, D. E., Esparza, L., Dubry, S., et al.,** Characteristics of TNF production in rheumatoid arthritis, *Cell. Immunol.,* 122:131–145, 1989.
8. **Wilkinson, L. S. and Edwards, J. C. W.,** Binding of antibodies raised against TNFα to blood vessels and macrophages in inflamed synovial tissue, *Rheumatol. Int.,* 11:19–25, 1991.
9. **Piguet, P. F., Grau, G. E., Vesin, C., et al.,** Evolution of collagen arthritis in mice is arrested by treatment with anti-TNF antibody or a recombinant soluble TNF receptor, *Immunology,* 77:510–514, 1992.
10. **Brahn, E., Peacock, D. J., Banquerigo, M. N., and Lui, D. Y.,** Effects of TNFα on collagen arthritis, *Lymphokine Cytokine Res.,* 11:253–256, 1992.
11. **Braegger, C. P., Nicholls, S., Murch, S. H., Stephens, S., and MacDonald, T. T.,** Tumour necrosis factor alpha in stool as a marker of intestinal inflammation, *Lancet,* 339:89–91, 1992.
12. **Imagawa, D. K., Millis, J. M., Seu, P., Olthoff, K. M., Hart, J., et al.,** The role of tumor necrosis factor in allograft rejection, *Transplantation,* 51:057–062, 1991.
13. **Seu, P., Imagawa, D. K., Wasef, E., Olthoff, K. M., et al.,** Monoclonal anti-tumor necrosis factor-α antibody treatment of rat cardiac allografts: synergism with low-dose cyclosporine and immunohistological studies, *J. Surg. Res.,* 50:520–528, 1991.
14. **Noronha, I. L., Eberlein-Gonska, M., Hartley, B., et al.,** In situ expression of TNF-alpha, interferon-gamma, and IL-2 receptors in renal allograft biopsies, *Transplantation,* 54:1017–1024, 1992.
15. **Folks, T. M., Clouse, K. A., Justement, J., Rabson, A., et al.,** Tumor necrosis factor α induces expression of human immunodeficiency virus in a chronically infected T-cell clone, *Proc. Natl. Acad. Sci. U.S.A.,* 86:2365–2368, 1989.
16. **Vyakarnam, A., McKeating, J., Meager, A., and Beverley, P. C.,** Tumour necrosis factors (α, β) induced by HIV-1 in peripheral blood mononuclear cells potentiate virus replication, *AIDS,* 4:21–27, 1990.
17. **Aboulafia, D., Miles, S. A., Saks, S. R., and Mitsuyasu, R. T.,** Intravenous recombinant TNF in the treatment of AIDS-related Kaposi's Sarcoma, *AIDS,* 2:54–58, 1989.
18. **Roux-Lombard, P., Modoux, C., Cruchard, A., and Dayer, J. M.,** Purified blood monocytes from HIV-1 infected patients produce high levels of TNFα and IL-1, *Clin. Immunol. Immunopathol.,* 50:374–384, 1989.
19. **Mintz, M., Rapaport, R., Oleske, J. M., Connor, E. M., et al.,** Elevated serum levels of TNF are associated with progressive encephalopathy in children with acquired immunodeficiency syndrome, *AJDC,* 143:771, 1989.
20. **Charpentier, B., Hiesse, C., Lantz, O., Ferran, C., et al.,** Evidence that antihuman tumor necrosis factor monoclonal antibody prevents OKT3-induced acute syndrome, *Transplantation,* 54(6):997–1002, 1992.
21. **Stephens, S., Emtage, S., Vetterlein, O., Athwal, D., et al.,** Clinical experience with CDP571, an engineered anti-TNF-alpha antibody, *J. Cell. Biochem.,* Suppl. 18D, 1994.
22. **Knight, D. M., Trinh, H., Le, J., Siegel, S., et al.,** Construction and initial characterization of a mouse-human chimeric anti-TNF antibody, *Mol. Immunol.,* 30(16):1443–53, 1993.
23. **Ashkenazi, A., Marsters, S. A., Capon, D. J., et al.,** Protection against endotoxic shock by a tumor necrosis factor receptor immunoadhesin, *Proc. Natl. Acad. Sci. U.S.A.,* 88:10535–10539, 1991.
24. **Waage, A., Halstensen, A., and Espevik, T.,** Association between tumor necrosis factor in serum and fatal outcome in patients with meningococcal disease, *Lancet,* 1(8529):355–357, 1987.
25. **Waage, A., Brandtzaeg, P., Halstensen, A., et al.,** The complex pattern of cytokines in serum from patients with meningococcal septic shock, *J. Exp. Med.,* 169:333–338, 1989.
26. **Girardin, E., Grau, G. E., Dayer, J.-M., et al.,** TNF and IL-1 in the serum of children with severe infectious purpura, *N. Engl. J. Med.,* 319:397–400, 1988.
27. **Debets, J. M. H., Kampmeijer, R., van der Linden, M. P. M. H., et al.,** Plasma TNF and mortality in critically ill septic patients, *Crit. Care Med.,* 17:489, 1989.

28. **Girardin, E., Berner, M. E., Grau, G. E., et al.,** Serum TNF in newborns at risk for infections, *Eur. J. Pediatr.*, 149:645–647, 1990.
29. **Cannon, J. G., Tompkins, R. G., Gelfand, J. A., et al.,** Circulating IL-1 and TNF in septic shock and experimental endotoxin fever, *J. Infect. Dis.*, 161:79–84, 1990.
30. **Calandra, T., Baumgartner, J.-D., Grau, G. E., et al.,** Prognostic values of TNF/Cachectin, IL-1 Interferon-a and Interferon-g in the serum of patients with septic shock, *J. Infect. Dis.*, 161:982–987, 1990.
31. **Michie, H. R., Manogue, K. R., Spriggs, D. R., et al.,** Detection of circulating TNF after endotoxin administration, *N. Engl. J. Med.*, 318:1, 1988.
32. **Pfeffer, K., Matsuyama, T., Kundig, T. M., Wakeham, A., et al.,** Mice deficient for the 55-kDa tumor necrosis factor receptor are resistant to endotoxic shock, yet succumb to *L. monocytogenes* infection, *Cell,* 73:457–467, 1993.
33. **Rothe, J., Lesslauer, W., Lotscher, H., Lang, Y., et al.,** Mice lacking the tumour necrosis factor receptor 1 are resistant to TNF-mediated toxicity but highly susceptible to infection by *Listeria monocytogenes, Nature,* 364:798–802, 1993.
34. **Tracey, K. J., Beutler, B., Lowry, S. F., Merryweather, J., et al.,** Shock and tissue injury induced by recombinant human cachectin, *Science,* 234:470–474, 1986.
35. **Beutler, B., Milsark, I. W., and Cerami, A. C.,** Passive immunization against cachectin/tumor necrosis factor protects mice from lethal effect of endotoxin, *Science,* 229:869–871, 1985.
36. **Silva, A. T., Bayston, K. F., and Cohen, J.,** Prophylactic and therapeutic effects of a monoclonal antibody to TNF in experimental Gram-negative shock, *J. Infect. Dis.*, 162:421–427, 1990.
37. **Opal, S. M., Cross, A. S., Kelly, N. M., Sadoff, J. C., Bodmer, M. W., Palardy, J. E., and Victor, G. H.,** Efficacy of a monoclonal antibody directed against tumor necrosis factor in protecting neutropenic rats from lethal infection with *Pseudomonoas aeruginosa, J. Infect. Dis.*, 161:1148–1152, 1990.
38. **Walsh, C. J., Sugerman, H. J., Mullen, P. G., Carey, P. D., et al.,** Monoclonal antibody to tumor necrosis factor alpha attenuates cardiopulmonary dysfunction in porcine Gram-negative sepsis, *Arch. Surg.*, 127(2):138–44; discussion 144–5, 1992.
39. **Windsor, A. C., Mullen, P. G., Walsh, C. J., Fisher, B. J., et al.,** Delayed tumor necrosis factor alpha blockade attenuates pulmonary dysfunctions and metabolic acidosis associated with experimental Gram-negative sepsis, *Arch. Surg.*, 129(1):80–9, 1994.
40. **Tracey, K. J., Fong, Y., Hesse, D. G., et al.,** Anti-cachectin/TNF monoclonal antibodies prevent septic shock during lethal bacteraemia, *Nature,* 330:662–664, 1987.
41. **Hinshaw, L. B., Tekamp-Olson, P., Chang, A. C. K., Lee, P. A., et al.,** Survival of primates in LD_{100} septic shock following therapy with antibody to tumor necrosis factor (TNFα), *Circ. Shock,* 30:279–292, 1990.
42. **Hinshaw, L. B., Emerson, T. E., Taylor, F. B., Chang, A. C. K., et al.,** Lethal *Staphylococcus aureus*-induced shock in primates: prevention of death with anti-TNF antibody, *J. Trauma,* 33(4):568–573, 1992.
43. **Fiedler, V. B., Loof, I., Sander, E., Voehringer, V., et al.,** Monoclonal antibody to tumor necrosis factor-alpha prevents lethal endotoxin sepsis in adult rhesus monkeys, *J. Lab. Clin. Med.*, 120(4):574–88, 1992.
44. **Sheehan, K. C. F., Ruddle, N. H., and Schreiber, R. D.,** Generation and characterisation of hamster monoclonal antibodies that neutralize murine tumour necrosis factors, *J. Immunol.*, 142(11):3884–3893, 1989.
45. **Cross, A. S., Opal, S. M., Palardy, J. E., Bodmer, M. W., and Sadoff, J. C.,** The efficacy of combination immunotherapy in experimental *Pseudomonas* sepsis, *J. Infect. Dis.*, 167:112–118, 1993.
46. **Wayte, J., Silva, A. T., Krausz, T., and Cohen, J.,** Observations on the role of tumour necrosis factor-alpha in a murine model of shock due to *Streptococcus pyogenes, Crit. Care Med.*, 21(8):1207, 1993.
47. **Zanetti, G., Heumann, D., Gerain, J., Kohler, J., Abbet, P., et al.,** Cytokine production after intravenous or peritoneal Gram-negative bacterial challenge in mice, *J. Immunol.*, 148(6):1890–1897, 1992.
48. **Martin, R. A., Silva, A. T., and Cohen, J.,** Effect of anti-TNF-alpha treatment in antibiotic treated murine model of shock due to *Streptococcus pyogenes, FEMS Microbiol. Lett.*, 110(2):175, 1994.

49. **Bagby, G. J., Plessala, K. J., Wilson, L. A., Thompson, J. J., and Nelson, S.,** Divergent efficacy of antibody to tumor necrosis factor-α in intravascular and peritonitis models of sepsis, *J. Infect. Dis.,* 163:83–88, 1991.
50. **Echtenacher, B., Falk, W., Mannel, D. N., and Krammer, P. H.,** Requirement of endogenous tumor necrosis factor/cachectin for recovery from experimental peritonitis, *J. Immunol.,* 145(11):3762–3766, 1990.
51. **Alexander, H. R., Sheppard, B. C., Jensen, J. C., Langstein, H. N., Buresh, C. M., et al.,** Treatment with recombinant human tumor necrosis factor-alpha protects rats against the lethality, hypotension and hypothermia of Gram-negative sepsis, *J. Clin. Invest.* 88:34–39, 1991.
52. **Suitters, A., Foulkes, R., Opal, S. M., Palardy, J. E., Emtage, J. S., et al.,** Differential effect of isotype on efficacy of anti-tumor necrosis factor α chimeric antibodies in experimental septic shock, *J. Exp. Med.,* 179:849–856, 1994.
53. **Glauser, M. P., Zanetti, G., Baumgartner, J.-D., and Cohen, J.,** Septic shock: pathogenesis, *Lancet,* 338(8769):732–736, 1991.
54. **Exley, E. R., Cohen, J., Buurman, W., et al.,** Monoclonal antibody to TNF in severe septic shock, *Lancet,* 335:1274–1277, 1990.
55. **Vincent, J. L., Bakker, J., Marecaux, G., Schandene, L., et al.,** Administration of anti-TNF antibody improves left ventricular function in septic shock patients. Results of a pilot study, *Chest,* 101(3):810, 1992.
56. **Fisher, C. J., Opal, S. M., Dhainaut, J.-F., et al.,** Influence of anti-TNF monoclonal antibody on cytokine levels in patients with sepsis, *Crit. Care Med.,* 21:318–327, 1993.
57. **Choy, M. Y., Walker-Smith, J. A., Williams, C. B., and MacDonald, T. T.,** Differential expression of CD25 (interleukin-2 receptor) on lamina propria T cells and macrophages in the intestinal lesions in Crohn's disease and ulcerative colitis, *Gut,* 31:1365–1370, 1990.
58. **Targan, S. R., Deem, R. L., and Shanahan, F.,** Immune-mediated cytotoxicity in inflammatory bowel disease, *Trends Inflam. Bowel Dis. Ther. Falk Symp.,* 29–40, 1990.
59. **Shorter, R. G., Cardoza, M. R., ReMilne, S. G., Spencer, R. J., and Huizinga, K. A.,** Modification of *in vitro* cytotoxicity of lymphocytes from patients with chronic ulcerative colitis or chronic granulomatous colitis for allogenic colon epithelial cells, *Gastroenterology,* 58:692–698, 1970.
60. **Hibi, T., Ohara, M., Toda, K., Ogata, H., Iwao, Y., et al.,** *In vitro* anticolon antibody production by mucosal or peripheral blood lymphocytes from patients with ulcerative colitis, *Gut,* 31:1371–1376, 1990.
61. **Glebmann, C. M., and Barrett, K. E.,** Role of inflammatory cell types, in *Inflammatory Bowel Diseases,* Kluwer Academic, Norwell, MA, 1993, 62.
62. **Plevy, S. E., Targan, S. R., Andus, T., and Toyoda, H.,** Differentiation of Crohn's disease from ulcerative colitis mucosal inflammation by levels of TNF-α mRNA, *Gastroenterology,* 104(4), part 2, A764, 1994.
63. **Murch, S. H., Braegger, C. P., Walker-Smith, J. A., and MacDonald, T. T.,** Location of tumour necrosis factor α by immunohistochemistry in chronic inflammatory bowel disease, *Gut,* 34:1705–1709, 1993.
64. **Andus, T., Targan, S. R., Deem, R., and Toyoda, H.,** Quantitative polymerase chain reaction allows the determination of TNF-α mRNA levels in mononuclear cells from mucosal biopsies, in *Inflammatory Bowel Diseases,* Kluwer Academic, Norwell, MA, 1993, 491.
65. **Radema, S. A., Hommes, D. W., Fockens, P., et al.,** Cytokine induction in experimental and ulcerative colitis, *Dig. Surg.,* 9:177–182, 1992.
66. **Steffen, M., Reinecker, H. C., Witthoft, T., Voss, A., Schreiber, S., and Raedler, A.,** Enhanced secretion of TNF-α, IFN-γ, IL-1β and IL-6 by lamina propria mononuclear cells from patients with inflammatory bowel disease, in *Inflammatory Bowel Diseases,* Kluwer Academic, Norwell, MA, 1993, 490.
67. **Nicholls, S., Stephens, S., Braegger, C. P., Walker-Smith, J. A., and MacDonald, T. T.,** Cytokines in stools of children with inflammatory bowel disease or infective diarrhoea, *J. Clin. Pathol.,* 46:757–760, 1993.
68. **Gardiner, K. R., Halliday, M. I., Lloyd, F., Stephens, S., and Rowlands, B. J.,** Circulating tumour necrosis factor in active inflammatory bowel disease, in *Host Defense Dysfunction in Trauma, Shock and Sepsis,* E. Faist et al., Eds., Elsevier, New York, 1993, 699.

69. **Andus, T., Gross, V., Gallati, H., Gerok, W., and Scholmerich, J.,** Concentrations of soluble TNF-receptor p55 but not p75 are significantly elevated in plasma of patients with inflammatory bowel disease (IBD), in *Inflammatory Bowel Disease,* Kluwer Academic, Norwell, MA, 1993, 492.
70. **van Lanschot, J. J. B., Mealy, K., and Wilmore, D. W.,** The effects of tumor necrosis factor on intestinal structure and metabolism, *Ann. Surg.,* 212(6):663–670, 1990.
71. **Ligumsky, M., Simon, P. L., Karmeli, F., and Rachmilewitz, D.,** Role of interleukin-1 in inflammatory bowel disease — enhanced production during active disease, *Gut,* 31:686–689, 1990.
72. **Pullman, W. E., Elsbury, S., Kobayashi, M., Hapel, A. J., and Doe, W. F.,** Enhanced mucosal cytokine production in inflammatory bowel disease, *Gastroenterology,* 102:529–537, 1992.
73. **Stevens, C., Walz, G., Singaram, C., Lipman, M. L., Zanker, B., et al.,** Tumor necrosis factor-α, interleukin-1β, and interleukin-6 expression in inflammatory bowel disease, *Dig. Dis. Sci.,* 37(6):818–826, 1992.
74. **Yahida, Y. R., Ceska, M., Effenberger, F., Kurlak, L., Lindley, I., and Hawkey, C. J.,** Enhanced synthesis of neutrophil-activating peptide-1/interleukin-8 in active ulcerative colitis, *Clin. Sci.,* 82:273–275, 1992.
75. **Koizumi, M., King, N., Lobb, R., Benjamin, C., and Podolsky, D. K.,** Expression of vascular adhesion molecules in inflammatory bowel disease, *Gastroenterology,* 103:840–847, 1992.
76. **Kvale, D., Krajci, P., and Brandzaeg, P.,** Expression and regulation of adhesion molecules ICAM-1 (CD45) and LFA-3 (CD58) in human intestinal epithelial cell lines, *Scand. J. Immunol.,* 35:669–676, 1992.
77. **Middleton, S. J., Shorthouse, M., and Hunter, J. O.,** Increased nitric oxide synthesis in ulcerative colitis, *Lancet,* 341:465–466, 1993.
78. **Ward, P. S., Woodger, S. R., Bodmer, M., and Foulkes, R.,** Anti-tumour necrosis factor α monoclonal antibodies (anti-TNF MAb) are therapeutically effective in a model of colonic inflammation, *Br. J. Pharmacol.,* 110:77, 1993.
79. **Warren, B. F., Watkins, P. E., Foulkes, R., Ward, P., Stephens, S., and Bodmer, M.,** Anti-TNF alpha treatment of a model of human ulcerative colitis, *Gut,* 35 (Suppl. 2):850, 1994.
80. **Derkx, B., Taminiau, J., Radema, S., Stronkhorst, A., et al.,** Tumour-necrosis-factor antibody treatment in Crohn's disease, *Lancet,* 342:173–174, 1993.
81. **Di Giovine, F. S., Nuki, G., and Duff, G.,** Tumour necrosis factor in synovial exudates, *Ann. Rheum. Dis.,* 47:768–772, 1988.
82. **Chu, C. Q., Field, M., Allard, S., Abney, E., et al.,** Detection of cytokines at the cartilage/pannus junction in patients with rheumatoid arthritis: implications for the role of cytokines in cartilage destruction and repair, *Br. J. Rheumatol.,* 31:653–661, 1992.
83. **Alvaro-Gracia, J. M., Zvaifler, N. J., and Firestein, G.,** Cytokines in chronic inflammatory arthritis. V. Mutual antagonism between interferon-gamma and tumor necrosis factor-alpha on HLA-DR expression, proliferation, collagenase production and granulocyte macrophage colony-stimulating factor production by rheumatoid arthritis synoviocytes, *J. Clin. Invest.,* 86:1790–1798, 1990.
84. **MacDonald, B. R. and Gowen, M.,** Cytokines and bone, *Br. J. Rheumatol.,* 31:149–155, 1992.
85. **Saklatvala, J.,** Tumour necrosis factor α stimulates resorption and inhibits synthesis of proteoglycan in cartilage, *Nature,* 322(6079):547–549, 1986.
86. **Brennan, F. M., Maini, R. N., and Feldmann, M.,** TNFα — A pivotal role in rheumatoid arthritis?, *Br. J. Rheumatol.,* 31:293–298, 1992.
87. **Keffer, J., Probert, L., Cazlaris, H., Georgopoulos, S., et al.,** Transgenic mice expressing human tumour necrosis factor: a predictive genetic model of arthritis, *EMBO J.,* 10(13):4025–4031, 1991.
88. **Henderson, B., Blake, S., Lewthwaite, J., Staines, N., Harper, N., et al.,** The effects of monoclonal antibodies to tumour necrosis factor in animal models of chronic arthritis, *Br. J. Pharmacol.,* 104:442, 1991.
89. **Williams, R. O., Feldmann, M., and Maini, R. N.,** Anti-tumor necrosis factor ameliorates joint disease in murine collagen-induced arthritis, *Proc. Natl. Acad. Sci. U.S.A.,* 89:9784–9788, 1992.
90. **Elliott, M. J., Maini, R. N., Feldmann, M., Long-Fox, A., et al.,** Treatment of rheumatoid arthritis with chimeric monoclonal antibodies to tumor necrosis factor α, *Arthritis Rheum.,* 36(12):1681–1690, 1993.

91. **Maury, C. P. J., and Teppo, A. M.,** Raised serum levels of cachectin/tumor necrosis factor α in renal allograft rejection, *J. Exp. Med.,* 166:1132–1137, 1987.
92. **Tilg, H., Vogel, W., Aulitzky, W. E., Herold, M., Konigsrainer, A., et al.,** Evaluation of cytokines and cytokine-induced secondary messages in sera of patients after liver transplantation, *Transplantation,* 49:1074–1080, 1990.
93. **Chollet-Martin, S., Depoix, J. P., Hvass, U., Pansard, Y., et al.,** Raised plasma levels of tumor necrosis factor in heart allograft rejection, *Transplant. Proc.,* 22(1):283–286, 1990.
94. **Grewel, H. P., Kotb, M., Salem, A., et al.,** Elevated tumor necrosis factor levels are predictive for pancreas allograft transplant rejection, *Transplant. Proc.,* 25(1):132–135, 1993.
95. **Holler, E., Kolb, H. J., Moller, A., Kempeni, J., et al.,** Increased serum levels of tumor necrosis factor α precede major complications of bone marrow transplantation, *Blood,* 75(4):1011–1016, 1990.
96. **Piguet, P.-F., Grau, G. E., Allet, B., and Vassalli, P.,** Tumor necrosis factor/cachectin is an effector of skin and gut lesions of the acute phase of graft-vs.-host disease, *J. Exp. Med.,* 166:1280–1289, 1987.
97. **Shalaby, M. R., Fendly, B., Sheehan, K. C., Schreiber, R. D., and Ammann, A. J.,** Prevention of the graft-versus-host reaction in newborn mice by antibodies to tumor necrosis factor-alpha, *Transplantation,* 47:1057–1061, 1989.
98. **Herve, P., Flesch, M., Tiberghien, P., Wijdenes, J., et al.,** Phase I-II trial of a monoclonal anti-tumor necrosis factor α antibody for the treatment of refractory severe acute graft-versus-host disease, *Blood,* 79(12):3362–3368, 1992.
99. **Bolling, S. F., Kunkel, S. L., and Lin, H.,** Prolongation of cardiac allograft survival in rats by anti-TNF and cyclosporin combination therapy, *Transplantation,* 53:283–286, 1992.
100. **Stevens, H. P. J. D., Van der Kwast, T. H., Van der Meide, P. H., et al.,** Synergistic immunosuppressive effects of monoclonal antibodies specific for interferon-gamma and tumor necrosis factor alpha, *Transplantation,* 50:856–861, 1990.
101. **Colletti, L. M., Remick, D. G., Burtch, G. D., et al.,** Role of tumor necrosis factor-α in the pathophysiologic alterations after hepatic ischemia/reperfusion injury in the rat, *J. Clin. Invest.,* 85:1936–1943, 1990.
102. **Redl, H., Nikolai, A., Kneidinger, R., and Schlag, G.,** Endothelial and leucocyte activation in experimental polytrauma and sepsis, *Behring Inst. Mitt.,* 92:218–28, 1993.
103. **Nabel, G., and Baltimore, D.,** An inducible transcription factor activates expression of human immunodeficiency virus in T cells, *Nature,* 326:711–713, 1987.
104. **Osborn, L., Kunkel, S., and Nabel, G. J.,** Tumor necrosis factor α and interleukin-1 stimulate the human immunodeficiency virus enhancer by activation of the nuclear factor kB, *Proc. Natl. Acad. Sci. U.S.A.,* 86:2336–2340, 1989.
105. **Aboulafia, D., Miles, S. A., Saks, S. R., and Mitsuyasu, R. T.,** Intravenous recombinant tumor necrosis factor in the treatment of AIDS-related Kaposi's Sarcoma, *J. AIDS,* 2:54–58, 1989.
106. **Reddy, M. M., Sorrell, S. J., Lange, M., and Grieco, M. H.,** Tumor necrosis factor and HIV P24 antigen levels in serum of HIV-infected populations, *J. AIDS,* 1:436–440, 1988.
107. **Krishnan, V. L., Meager, A., Mitchell, D. M., and Pinching, A. J.,** Alveolar macrophages in AIDS patients: increased spontaneous tumour necrosis factor-alpha production in *Pneumocystus carinii* pneumonia, *Clin. Exp. Immunol.,* 80:156–160, 1990.
108. **Merrill, J. E., Koyanagi, Y., and Chen, I. S. Y.,** Interleukin-1 and tumor necrosis factor-alpha can be induced from mononuclear phagocytes by human immunodeficiency virus type 1 binding to the CD4 receptor, *J. Virol.,* 63(10):4404–4408, 1989.
109. **Munis, J. R., Richman, D. D., and Kornbluth, R. S.,** Human immunodeficiency virus-1 infection of macrophage *in vitro* neither induces tumor necrosis factor (TNF)/cachectin gene expression nor alters TNF/cachectin induction by lipopolysaccharide, *J. Clin. Invest.,* 85:591–596, 1990.
110. **Roederer, M., Staal, F. J. T., Raju, P. A., Ela, S. W., et al.,** Cytokine-stimulated human immunodeficiency virus replication is inhibited by *N*-acetyl-L-cysteine, *Proc. Natl. Acad. Sci. U.S.A.,* 87:4884–4888, 1990.
111. **Beutler, B., and Cerami, A.,** Cachectin and tumour necrosis factor as two sides of the same biological coin, *Nature,* 320:584–588, 1986.

112. **Gelin, J., Moldawer, L. L., Lonnroth, C., Sherry, B., et al.,** Role of endogenous tumor necrosis factor alpha and interleukin-1 for experimental tumor growth and the development of cancer cachexia, *Cancer Res.,* 51(1):415–21, 1991.
113. **Costelli, P., Carbo, N., Tessitore, L., Bagby, G. J., et al.,** Tumor necrosis factor-alpha mediates changes in tissue protein turnover in a rat cancer cachexia model, *J. Clin. Invest.,* 92(6):2783–9, 1993.
114. **Hotamisligil, G. S., Shargill, N. S., and Spiegelman, B. M.,** Adipose expression of tumor necrosis factor-α: direct role in obesity-linked insulin resistance, *Science,* 259:87–91, 1993.
115. **Spiegelman, B. M., Choy, L., Hotamisligil, G. S., Graves, R. A., et al.,** Regulation of adipocyte gene expression in differentiation and syndromes of obesity/diabetes, minireview, *J. Biol. Chem.,* 268(10):6823–6826, 1993.
116. **Hofmann, C., Lorenz, K., Braithwaite, S. S., Colca, J. R., et al.,** Altered gene expression for tumor necrosis factor-alpha and its receptors during drug and dietary modulation of insulin resistance, *Endocrinology,* 134(1):264–70, 1994.
117. **Huang, D., Reittie, J. E., Stephens, S., Hoffbrand, A. V., and Brenner, M. K.,** Effects of anti-TNF monoclonal antibody infusion in patients with hairy cell leukaemia, *Br. J. Haematol.,* 81(2):231–4, 1992.
118. **Santambrogio, L., Hochwald, G. M., Saxena, B., Leu, C. H., et al.,** Studies on the mechanisms by which transforming growth factor-beta (TGF-beta) protects against allergic encephalomyelitis. Antagonism between TGF-beta and tumor necrosis factor, *J. Immunol.,* 151(2):1116–27, 1986.
119. **Elliott, M. J., Maini, R. N., Feldmann, M., et al.,** Randomised double-blind comparison of chimeric monoclonal antibody to tumour necrosis factorα (cA2) versus placebo in rheumatoid arthritis, *Lancet,* 344:1105–10, 1994.

Chapter 11

ANTI-INTERLEUKIN-5 ANTIBODIES AS THERAPEUTIC AGENTS IN ASTHMA AND OTHER EOSINOPHILIC DISEASES

Robert W. Egan, Francis M. Cuss, Shelby P. Umland, and Richard W. Chapman

CONTENTS

0-8493-8381-1/96/$0.00+$.50

ABBREVIATIONS

IL — interleukin
GM-CSF — granulocyte macrophage colony stimulating factor
m — murine
h — human
EPO — eosinophil peroxidase
MBP — major basic protein
ECP — eosinophil cationic protein
EDN — eosinophil derived neurotoxin
MAB — monoclonal antibody
PMA — phorbol myristate acetate
fMLP — N-formyl-Met-Leu-Phe
BAL — bronchoalveolar lavage
CDR — complementarity determining region

I. INTRODUCTION

Although the most observable feature in asthma is the bronchoconstriction associated with an acute "asthmatic attack", asthma is a chronic inflammatory disorder.[1–3] For example, even between "attacks", most people with asthma have a breathing deficit as indicated by a decreased FEV_1.[3] Furthermore, a cardinal hallmark of asthma is increased airway responsiveness to a variety of stimuli.[3] Considerable attention, therefore, is being focused on the role of inflammation in the pathophysiology of asthma, especially as it relates to chronic airway hyperreactivity. Some features of pulmonary inflammation are edema, decreased mucociliary clearance, epithelial damage accompanied by increased neuronal responsiveness, and bronchoalveolar eosinophilia.[2] In treating asthma, therefore, it is essential to focus on inflammation that leads to airway hyperreactivity, rather than simply attenuating the symptoms.

Eosinophils appear to be a major contributor to the tissue destruction that accompanies pulmonary inflammation. Eosinophils are the major cells infiltrating into the lung during asthma.[4] They are produced from myeloid precursors in the bone marrow in response to a series of cytokines, and are released from the bone marrow following an appropriate stimulus. Once in the circulation, they rapidly accumulate in tissue in both the human and animal models.[1,5] Although there are very few eosinophils in the lungs of normal individuals, they accumulate dramatically in the lungs of people with asthma.[2]

Eosinophils are also capable of lysing cells, activating cells, and destroying tissue.[1,6] When activated, they synthesize and release inflammatory cytokines and lipid mediators that can produce edema, bronchoconstriction, and chemotaxis. Concurrently, they secrete a set of cytotoxic enzymes and proteins that can lyse epithelial tissue and are found in the lungs of people with asthma.[6,7] Furthermore, the role of the eosinophil as a protective agent against parasitic infections is limited[8] and there are no other salutory functions known. It should, therefore, be possible to inhibit the actions of the eosinophil without inducing the severe immunosuppressive consequences that would arise from inhibiting the actions of other inflammatory cells such as neutrophils, macrophages, or lymphocytes. For these reasons, the eosinophil appears to be an ideal target for selectively inhibiting the tissue damage that accompanies pulmonary inflammation during asthma.

There is strong but circumstantial evidence for a role of interleukin-5 (IL-5) in asthma because it is an essential cytokine for the maturation of eosinophils.[9,10] In humans the biology of IL-5 is very specific for eosinophil and basophil maturation, growth, activation, and survival.[11–14] By inhibiting IL-5 it should, therefore, be possible to selectively suppress these two cell types that are associated with allergic disorders. Such specificity is not uniform across species; IL-5 is a B-lymphocyte growth factor in mice.[15] In species other than the human, IL-5 inhibitors could, therefore, induce more extensive physiological changes.

Although the presence of IL-3, IL-4, IL-5, IL-10, and granulocyte macrophage-colony stimulating factor (GM-CSF) protein or mRNA has been found in bronchoalveolar lavage fluid and tissue of patients with asthma,[16–18] animal studies indicate that these cytokines do not function in parallel to enhance eosinophil maturation, such that lack of one could be compensated for by another. Specifically inhibiting the action of IL-5 with a monoclonal antibody to IL-5 can completely inhibit bronchoalveolar eosinophilia caused by allergic challenge in guinea pigs, mice, and monkeys.[5,19–21] Therefore, inhibiting the actions of IL-5 by any of a variety of possible mechanisms can suppress one of the alleged causes of asthma — tissue damage caused by eosinophils during pulmonary inflammation.

One of the pathophysiological consequences of damage to pulmonary tissue would be hyperreactive airways. While there is a correlation between pulmonary eosinophilia and asthma,[22,23] the correlation in animal models using airway hyperreactivity as a physiological surrogate for asthma is less convincing. Some studies indicate a correlation[24,25] but, in others, eosinophil accumulation is inhibited under conditions where hyperreactivity persists.[19,26] It has, therefore, been suggested that release of eosinophil contents into the tissue may correlate better with hyperreactivity. Nevertheless, it is clear that selective inhibition of IL-5 can block hyperreactivity, and it is that response in animals that is expected to correlate with the breathing deficit that characterizes the chronic pathology of asthma in humans. Although asthma is the most common disease associated with eosinophilia, there are less prevalent eosinophilic disorders including atopic dermatitis, idiopathic hypereosinophilic syndrome, conjunctivitis, and others.[1,27,28] Each of these disorders should be ameliorated by blocking IL-5.

This chapter reviews the biochemistry and molecular biology of the IL-5 system, the *in vivo* biology of IL-5 including its critical influence on eosinophil function and the use of antibodies to IL-5 in animal models, the role of IL-5 in human diseases, the potential for host responses to anti-IL-5 antibodies, and the mechanisms by which the action of IL-5 can be inhibited.

II. BIOCHEMISTRY AND MOLECULAR BIOLOGY OF IL-5

A. GENE STRUCTURE

The cDNA encoding murine IL-5 (originally named T cell replacing factor or B cell growth factor II) was cloned in 1986 from a T cell line.[15] The deduced amino acid sequence of the mature mIL-5 polypeptide consists of 113 residues with three potential N-linked glycosylation sites. Minimal amino acid sequence homology was found to exist with other cytokines, except for short stretches in the mIL-3, mGM-CSF, and mIFN-γ proteins.[15] Using a mIL-5 cDNA as a probe, an IL-5 cDNA was subsequently isolated from a human T cell leukemia line. This encoded a precursor of 134 amino acids with a N-terminal signal sequence of 22 amino acids and a molecular weight of approximately 13,000 Da.[29] The factor encoded by this cDNA induced IgM synthesis by stimulated human B cells. By functional analysis, a correlation was observed between B cell activity and eosinophil differentiation activity.[30] Furthermore, a single mouse cDNA clone was isolated by expression cloning and the encoded protein enhanced IgA secretion by mouse

B cells and stimulated eosinophil colony formation from human cord blood cells.[31] Subsequently, based on homology with the mouse cDNA, a human cDNA was isolated encoding a protein with similar activities. This established that a single cDNA clone encoded a protein that acts as a growth and differentiation factor for both B cells and eosinophils.

The nucleotide and amino acid sequence homologies of the coding regions of human and murine IL-5 are 77 and 70%, respectively. Of the three cysteine residues in mIL-5, the two C-terminal residues are conserved in hIL-5.[29] Higher homology (80%) was found in the C-terminal compared to the N-terminal half (66%), and this is compatible with data suggesting that the active site is located in the C-terminal region.[32] This high degree of conservation is reflected in the ability of the two proteins to retain significant biological activity with cells of several mammalian species.[33]

The IL-5 gene consists of four exons and three introns,[34] and resides as a single copy in both the human and mouse genome.[29] The IL-5 gene is located on mouse chromosome 11[35] and human chromosome 5,[35,36] within a cluster of the IL-3, GM-CSF, and IL-4 genes. Examination of the exon-intron organization and the location of cysteine codons in the IL-5 gene and other cytokine genes indicates that IL-5 is phylogenetically related to the genes encoding GM-CSF, IL-4, and IL-2.[34] In addition, the mouse and human IL-5 genes show a high degree of homology (77%) in their 5′ flanking regions at nucleotide positions –150 to +1.[34] Within this region are short stretches of conserved sequence motifs, designated CLE 0, CLE 1, and CLE 2, which are also found in the 5′ flanking regions of the IL-3, IL-4, and GM-CSF genes.[34,37] These sequence similarities, as well as evidence of shared sets of DNA binding proteins[38] recognizing this and other sequences, suggest that there may be coordinate regulation of cytokine gene transcription.

B. PROTEIN STRUCTURE

Both mIL-5 and hIL-5 contain potential N-linked glycosylation sites and cysteine residues. Differential glycosylation accounts for the broad range of apparent molecular weights (45 to 60 kDa) of recombinant mIL-5 and hIL-5 produced in mammalian cells. However, deglycosylated IL-5 retains full biological activity.[39] IL-5 exists as a disulfide-linked homodimer[40] and the cysteine residues crosslink the dimer in an antiparallel arrangement.[41] Evidence from reduction and alkylation[42] as well as site-directed mutagenesis[43] indicates that dimerization is essential for biological activity.

The crystal structure of hIL-5 has been reported as a novel two-domain structure with each domain requiring the participation of two chains.[44,45] Each domain shows a high degree of similarity to the cytokine fold found in GM-CSF, IL-3, and IL-4 that is important to receptor binding.[44]

C. BIOSYNTHESIS

IL-5 is produced primarily by T cells.[46] Upon stimulation of resting T cells by accessory cell presentation of antigen in an MHC restricted manner, T cells undergo a series of early metabolic events that culminate in the transcriptional activation of genes that encode cytokines, including IL-5.[47] IL-5, like IL-4, is produced by T cells belonging to the Th2 but not the Th1 subset. This subset, by virtue of the pattern of cytokines synthesized in response to T cell receptor signaling (IL-4, IL-5, as well as IL-3, GM-CSF, and IL-10), promotes humoral immune responses, particularly IgE synthesis, mast cell growth, and eosinophil differentiation.[48] Thus, Th2 cells are thought to control the growth and effector function of those cell types involved in allergic inflammatory responses.

Although T cell cytokine synthesis occurs in a coordinated manner, differential dependency on specific signal transduction pathways has been demonstrated for IL-5.

Protein kinase C activation is obligatory and sufficient[49–51] for IL-5 mRNA induction, although IL-5 gene expression is enhanced by stimuli that increase intracellular calcium. IL-5 gene expression requires *de novo* protein synthesis and, unlike other cytokines, is only partially inhibited by cyclosporin A.[49,51,52] Glucocorticoids are potent inhibitors of IL-5 gene expression.[53]

Other cell types have been reported to synthesize IL-5. For example, mouse mast cell lines transcribe IL-5 mRNA following specific crosslinking of the FcεRI receptor.[54] Furthermore, eosinophils isolated from patients with hypereosinophilic syndrome secrete IL-5 after stimulation with Ig complexes, and IL-5 immunoreactive protein is found to be localized within the granules.[54] IL-5 mRNA has been identified in eosinophils from patients with celiac disease,[56] heart disease,[57] and asthma.[17,58] By immunohistochemistry, IL-5 is associated with mast cells and eosinophils in nasal mucosal specimens isolated from patients with perennial allergic rhinitis[59] and in bronchial mucosal biopsies from asthmatics.[17,60] These findings indicate that autocrine production of IL-5 by eosinophils may also contribute to the recruitment and activation of these cells. *In vivo* depletion of T cells in a mouse model of pulmonary inflammation reduced pulmonary eosinophilia and IL-5 and other cytokine mRNA levels, indicating strong T cell dependency of these responses.[61] Mast cells are also important for the development of pulmonary eosinophilia in allergic mice, because mast cell-deficient mice did not produce as significant a pulmonary eosinophilia in response to antigen challenge and the eosinophil response could be reconstituted by intravenous injections of bone marrow-derived mast cells from congenic normal littermates.[62]

D. RECEPTORS

IL-5 induces its multiple functions through a specific receptor expressed on a restricted number of target cells. The IL-5 receptor (IL-5R) belongs to the Type I cytokine receptor family, which includes the receptors for IL-2, IL-3, IL-4, IL-6, IL-7, GM-CSF, G-CSF, and erythropoietin.[63] These receptors are integral membrane glycoproteins with N-termini directed extracellularly and a single membrane spanning domain.

The mIL-5R is expressed by pre-B and early B cell lines,[64,65] activated but not resting normal B cells,[65] $CD5^+$ peritoneal B cells,[66] and by eosinophils.[30] There are two classes of IL-5R in the mouse (mIL-5R): one binds IL-5 with low affinity (K_d = 30 n*M*) and the other with high affinity (K_d = 150 p*M*). Crosslinking analysis has revealed that two membrane-associated polypeptides of 60 kDa and 130 kDa are involved in IL-5 interactions.[64] A cDNA encoding the 60-kDa protein of 415 amino acids including an N-terminal hydrophobic region, an external domain of 322 amino acids, a transmembrane region of 22 amino acids, and an intracellular tail (54 amino acids) was isolated from a murine pre-B cell by expression cloning.[67] This 60-kDa polypeptide chain expressed in COS7 cells bound mIL-5 with low affinity (K_d = 2 to 10 n*M*) and is comprised of the typical structure of Type I cytokine receptors[68–70] with a motif of four conserved cysteines in the N-terminal region of the extracellular domain, fibronectin Type III domains in the extracellular region,[70] and a conserved W-S-X-W-S motif in the C-terminal extracellular region near the transmembrane domain.[63,68,69] cDNA clones lacking the transmembrane portion have also been isolated and encode soluble receptors.[67] By examining IL-5 binding to the transfected IL-3-dependent myeloid cell line, FDC-P1, it was determined that the high-affinity mIL-5R is a multiprotein complex.[67] Additional information from reconstruction experiments using transfected L cells[71–73] and antibody inhibition studies[74,75] indicated that the high-affinity complex consists of an α subunit (60 kDa) and a β subunit (130 kDa). The β subunit is identical to the β subunit of the mIL-3 and GM-CSFR[67,71,72,76] and is expressed endogenously by the FDC-P1 cell line.[67]

The β subunit, which alone does not bind ligand, is essential for high-affinity binding of IL-5 , and is crosslinked with IL-5 in the presence of the α subunit.[64] The β subunit is responsible not only for high-affinity IL-5 binding, but also for signal transduction.[72] In addition, this chain may stabilize ligand binding to the α chain.[77] However, transfection of the FDC-PI cells, which constitutively express only the β subunit of the IL-5R, with a mutant mIL-5Rα lacking a complete cytoplasmic domain, and the subsequent observation of expression of high-affinity IL-5R without a biological response, indicates that the α subunit does play a role in signal transduction.[72] In addition, the α subunit may be involved in internalization of IL-5.[78]

The human IL-5R (hIL-5R) is expressed as a single class (K_d = 170 to 330 p*M*) in freshly isolated eosinophils[79] and eosinophilic sublines of the promyelocytic leukemia HL60 cell line.[79,80] Affinity crosslinking showed that cell-bound IL-5 was associated with two membrane proteins of 60 and 130 kDa.[80] cDNAs for hIL-5Rα isolated from peripheral blood eosinophils[78,81] and HL60 cells[80] encoded a protein with 70% amino acid homology with the mIL-5-Rα and with structural characteristics of the cytokine Type I receptor family. The 60-kDa recombinant IL-5Rα chain expressed in COS7 cells bound mouse and human IL-5 with relatively high affinity (K_d = 300 p*M*),[81] similar to that of the endogenous receptor in human peripheral blood eosinophils.[79] This high binding affinity seen with the hIL-5Rα is in contrast to the low binding affinity observed with the 60-kDa mIL-5Rα chain.[79] As in the mouse, interaction of the α subunit with the β subunit, which is identical to the β subunit of the human GM-CSFR and IL-3R,[82] fully reconstituted IL-5 binding[81,82] and functional activity.[72,82,83] However, unlike the mouse receptor, association of the hIL-5Rα and β subunits by cotransfection in COS cells resulted in only a modest increase (twofold) in affinity for IL-5, as compared to cells expressing the α chain only.[83] The cytoplasmic domain of the β chain has been implicated in signal transduction.[84]

As in the mouse, soluble receptor forms of the IL-5Rα chain have been identified.[82,83] The soluble hIL-5Rα molecule demonstrated antagonistic activities in both binding and functional assays.[82] Although its role *in vivo* is unknown, the soluble receptor might protect against excessive eosinophil recruitment and activation.

E. SIGNAL TRANSDUCTION

The β subunit of both the mouse and human IL-5R is critical for high-affinity binding and signal transduction resulting in biological responsiveness.[72,84] The cytoplasmic domain of the signal transducing β subunit lacks a consensus sequence for either tyrosine kinase or phosphatase domains.[29,31,67,71–73] However, like IL-3 and GM-CSF, IL-5 induces rapid tyrosine phosphorylation of at least four different proteins in an IL-5-dependent cell line, indicating that tyrosine kinases are involved in the cellular activation pathways.[85] Protein tyrosine kinases that physically associate with cytokine receptors and become activated following ligand binding have been identified. More specifically, it has been demonstrated that IL-3, GM-CSF, and IL-5 (unpublished observations cited in Reference 86) primarily activate JAK2 in response to ligand binding.[86] JAK2 is a member of the JAK family of protein tyrosine kinases which are involved in the signal transducing pathways utilized by the cytokine/IFN superfamily of receptors.[87] The activation of the JAK proteins is associated with their autophosphorylation. Although much is known regarding IFN-induced signal transduction pathways involving JAK proteins and their subsequent activation of members of the STAT family of transcriptional regulatory proteins,[87] much remains to be elucidated regarding similar pathways induced by IL-5.

III. BIOLOGY OF IL-5

IL-5 was discovered as a murine B-cell growth and differentiation factor that stimulates the release of immunoglobulins from activated B cells. However, the major role of IL-5 in the human, and a significant role in other species where it has been studied, is in the growth and function of eosinophils and basophils. The *in vitro* biology of IL-5 and its effects on various cell types has been reviewed extensively.[88–91] This section focuses on the correlation between the *in vitro* and the *in vivo* biology of IL-5.

A. EOSINOPHILS AND BASOPHILS

IL-5 has multiple effects on the maturation and function of eosinophils both *in vitro* and *in vivo*.[88,90,92] *In vitro* studies have identified an important role for IL-5 in eosinophil growth and differentiation,[9,10,90,93,94] in migration,[11,12,90,95] in activation and effector function,[90,96,97] and in survival.[90,98] Although it was once thought that the effects of IL-5 were restricted to the eosinophil, recently IL-5 has demonstrated activity on human basophil growth, differentiation, and activation.[13,14,99] Eosinophils and basophils are two predominant effector cell types involved in allergic inflammation. As well as IL-5, IL-3 and GM-CSF are maturation and growth factors for human eosinophils.[100] Morphologically, eosinophils stimulated by IL-3 and GM-CSF are immature, whereas IL-5 produces growth and differentiation to mature eosinophils. Furthermore, IL-3 and GM-CSF stimulate the production of other granulocytes such as neutrophils, while IL-5 is selective for the eosinophil/basophil lineages.[90]

Eosinophils preferentially accumulate and persist in tissues during allergic reactions.[5,19,101] IL-5 is a chemoattractant for human eosinophils and primes chemotactic responses of the eosinophil toward several other agents.[12,95] Although human neutrophil migratory responses are also primed by IL-5, this activity is weak relative to the effect of IL-5 on eosinophil migration.[102] IL-5 selectively enhances the chemotactic response of eosinophils from normal subjects, but this effect is not seen in atopic patients with asthma or allergic rhinitis.[103] However, the eosinophil chemoattractant response to IL-5 could have become refractory following prior exposure to this cytokine. IL-5 enhances the adhesion of eosinophils to endothelial cells by upregulating the expression of CD11b/18 antigen.[104] On the other hand, IL-5 does not upregulate the activity of other adhesion molecules on human eosinophils.[105]

Mature human eosinophils respond to several external stimuli by synthesizing and releasing mediators and by secreting preformed granule constituents.[106–110] Secretory IgA is a potent activator of eosinophils.[111] Eosinophils possess the low-affinity IgE receptor, CD23,[112] and it has also been reported that they have the FcεRI high-affinity IgE receptor.[113] Unique to the human eosinophil and basophil is the IL-5 receptor, although they also possess receptors for other cytokines, based on their priming and antiapoptotic actions.[114–116] Importantly, IL-5 primes eosinophils for activation by other mediators, notably PAF.[117,118] Similar effects are seen on human basophils where IL-5 primes the cells for histamine release and leukotriene generation to diverse agonists.[14,99] Nevertheless, the eosinophil responds to a unique set of physiological triggers during allergic reactions, largely resulting from a Th2-type lymphocyte response. For example, eosinophils infiltrate into the lung during asthma and into the skin during atopic dermatitis, but not generally into the lung during pneumonia.[119]

Once they are recruited into a site of inflammation, eosinophils are activated to release their cytotoxic products. Mature eosinophils have many properties in common with neutrophils. Both cells respond to many common stimulators such as C5a, PMA, zymosan, and fMLP[106–110] with a respiratory burst, activation of phospholipases, production of leukotrienes, prostaglandins, and platelet activating factor and secretion of preformed granule contents such as peroxidases and proteinases including lysozyme and collagenases.[110,120,121] By releasing EPO and O_2^- that dismutates to hydrogen peroxide, the eosinophils can oxidatively damage adjacent tissue. The secreted Type I and Type III collagenases can also destroy interstitial tissue.

On the other hand, the ability to store and secrete the cationic low molecular weight proteins such as the major basic protein (MBP), eosinophil cationic protein (ECP), and eosinophil-derived neurotoxin (EDN) is unique to the eosinophil.[101,122] These proteins are stored in secretory granules where major basic protein dominates and comprises a morphologically unique core.[122] The genes for these proteins have been identified on various chromosomes and MBP is produced as a precursor with a long anionic carboxy-terminal sequence that is cleaved prior to sequestration in the granule.[123,124] ECP is as basic as MBP (pI = 11), while EDN (pI = 9) is far less cationic. At present, no enzyme activity has been associated with either MBP or ECP, while EDN is a ribonuclease.[125] MBP and ECP can lyse cells[126] and can cause tissue damage at low concentrations.[127] For example, at 500 p*M*, MBP and ECP cause fluid extravasation in the hamster cheek pouch.[127] MBP, but not ECP, also causes muscle fasciculation in this model, despite their similar pI values, so there is more than cationic change involved in the latter response.[127]

Another unique feature of the eosinophil is the Charcot-Leyden crystal protein that is disseminated throughout the cytoplasm, constitutes 7 to 10% of the total eosinophil protein, possesses lysophospholipase activity, is released upon stimulation, and readily crystallizes in the eosinophil and in the surrounding tissue.[119,128–130] Although all inflammatory cells possess peroxidases that metabolize hydrogen peroxide to generate increasingly reactive oxygen species that oxidize or oxygenate low molecular weight and protein substrates,[120,121] the eosinophil peroxidase (EPO) is molecularly unique from the others and can be distinguished by its substrate specificity.[131] Although EPO is as cationic as MBP and ECP, it is not as toxic to tissue in the absence of hydrogen peroxide, further indicating that there is more to the cytotoxicity of MBP and ECP than simply the positive charge.[127] In the presence of hydrogen peroxide, EPO is significantly more toxic to tissue.[127]

IL-5 is also capable of inhibiting eosinophil apoptosis and prolonging eosinophil survival.[90,98] There is a strong correlation between IL-5-dependent eosinophil survival activity and the severity of lung injury in allergic patients following segmental antigen challenge to the lung.[132] Glucocorticoids inhibit IL-5-mediated eosinophil survival,[133,134] which may be an important component of these drugs in the treatment of inflammatory disorders like asthma.

In response to an allergic stimulus in the mouse, eosinophils are rapidly released from the bone marrow into the circulation.[135] IL-5 released in the periphery primarily by T cells is essential to this process.[61] As they are released into the circulation, the level of eosinophils in the bone marrow decreases.[135] Other cytokines such as GM-CSF are also involved, but eosinophils are not released without IL-5.[21] Circulating eosinophils move rapidly into the tissue where they reside and where, when activated, they release their contents. Although eosinophils are normally less than 5% of the total white cells in the circulation, they are greater than 90% of the cells infiltrating into the airways during asthma.[22]

In the human, there is often a correlation between eosinophil accumulation in the lung, levels of ECP in bronchoalveolar lavage, and a breathing deficit as indicated by a decreased FEV_1.[136] However, this does not necessarily indicate that the eosinophils are responsible for the decreased lung function. To evaluate the possible cause and effect relationship it is necessary to attenuate the pulmonary eosinophilia and see improvement, but the agents needed to do that are not available. Furthermore, there are no animal models that precisely duplicate the chronic nature of asthma. However, the lungs of most asthmatics are hyperresponsive to a variety of stimuli, and this hyperreactivity has been duplicated in several animal models.[19,20,24–26]

B. B LYMPHOCYTES

IL-5 was originally isolated as a murine B cell growth and differentiation factor that stimulates the release of immunoglobulins from activated B cells.[88,89,92,137] Further investigations of the activity of IL-5 on murine B cells have revealed that IL-5 stimulates the release of antibodies when given in combination with other cytokines such as IL-2,[138,139] IL-4,[138,140,141] IL-6,[142] and transforming growth factor-β.[143–145] IL-5 is probably important in the production of IgM and IgA antibodies in mice, because transgenic mice expressing elevated levels of IL-5 show increased amounts of these antibodies in the serum.[146] On the other hand, IL-5 is not important to the increased serum IgE in response to helminth infection, because treatment with a monoclonal antibody to IL-5 did not change circulating levels of that antibody.[8] Furthermore, we have recently found that treatment with a monoclonal antibody to IL-5 did not change serum IgE levels in ovalbumin-sensitized mice (unpublished observations), indicating that IL-5 is not important in the synthesis and production of IgE in allergic reactions. These *in vivo* results question the relevance of *in vitro* studies demonstrating that IL-5 augments the production of IgE from murine B cells that are primed with IL-4.[140,141]

In the human, IL-5 has been reported to stimulate the release of IgA from activated mucosal B cells.[148,149] These results suggest that IL-5 is important in the development of mucosal immunity to airborne or ingested pathogens and increased levels of IL-5 are found in the nasal and bronchial biopsies from atopic individuals that have been topically challenged with antigen.[17,150–152] Further studies are required to define the role of IL-5 on mucosal B cell responses in human airways, because the functional role of IL-5 in controlling antibody production from human B cells has been questioned.[137,139,147,153]

IV. EFFECTS OF ANTIBODIES TO IL-5 IN ANIMAL MODELS OF ALLERGIC INFLAMMATORY DISORDERS AND PARASITE INFECTIONS

In general, increased IL-5 production is observed when there is an expansion of Th2 cell populations. This condition is induced during allergic inflammatory disorders and by infection with parasites. An association of eosinophils and neoplasms has also been found in mice, and this too may involve the effects of IL-5.[154]

A. ALLERGIC INFLAMMATORY DISORDERS

1. Mice

Several models of allergic inflammation have been used in mice to investigate the role of IL-5.[155–157] Intradermal injection of antigen into mice produced a cutaneous late-phase reaction that is characterized by increased numbers of eosinophils into the site of injection,[156] and a model of allergic peritonitis has been used to quantify the number of eosinophils recruited into the peritoneal cavity after antigen provocation.[155] In both these models, the tissue eosinophilia was inhibited by treatment with a monoclonal antibody to IL-5, indicating that IL-5 plays an important role in the local accumulation of eosinophils after antigen challenge. Transgenic mice that constitutively express the IL-5 gene show an infiltration of eosinophils into blood and tissues[146,160] and treatment of these mice with a monoclonal antibody to IL-5, or with a monoclonal antibody to the IL-5 receptor, produced a long-lasting (up to 40 days) reduction in the number of circulating eosinophils.[161] Despite the magnitude and lifelong nature of the eosinophilia in these IL-5 transgenic mice, there was no morbidity associated with these abnormalities.

The eosinophilia associated with allergic pulmonary inflammation has also been used to study the role of IL-5.[21,157–159] Antigen challenge to the airways of sensitized mice produced infiltration of eosinophils into the bronchoalveolar fluid[135,157] and eosinophil invasion into airway tissue.[135,158,159] Pulmonary challenge with antigen also decreased bone marrow eosinophils as they migrated from the bone marrow into the blood and lungs.[21,135] Treatment of sensitized, challenged mice with a monoclonal antibody to IL-5 consistently reduced the number of eosinophils appearing in the BAL fluid and airway tissue.[21,157–159] There was also an inhibition of eosinophil efflux from the bone marrow after treatment with an anti-IL-5 antibody.[21,159] Hence, IL-5 is important for the migration of eosinophils from the bone marrow into the blood and lungs in response to antigen challenge.

An important role of IL-5 is to prolong the survival of eosinophils.[90,98] Therefore, inhibition of IL-5 activity during an established pulmonary eosinophilia could cause tissue destruction due to apoptosis of eosinophils and release of their inflammatory mediators. However, administration of an antibody to IL-5 given after antigen challenge, when lung eosinophilia was already established, did not increase tissue damage in the lungs.[21] These results have important therapeutic implications for the potential use of IL-5 inhibitors in the treatment of inflammatory airway disorders.

Bronchial hyperreactivity is an important feature of asthmatic airways and can be demonstrated in sensitized mice following repeated exposure to antigen.[157] In this paradigm, treatment of sensitized, challenged mice with a relatively low dose (1 μg/mouse, i.v.) of monoclonal antibody IL-5 had no effect on bronchial hyperresponsiveness due to antigen challenge, whereas the pulmonary eosinophilia was significantly reduced. These results indicate that eosinophil accumulation is not necessary for the development of bronchial hyperreactivity in mice.

2. Guinea Pigs

Topical administration of recombinant mIL-5 to guinea pig airways increased the number of inflammatory cells (eosinophils and neutrophils) appearing in the BAL fluid and airway tissue.[162,163] These IL-5-treated animals showed signs of bronchial hyperresponsiveness,[164,165] although the magnitude of this effect was small. Furthermore, intraperitoneal administration of alginate-encapsulated IL-5 produced a two- to threefold increase in bronchial hyperreactivity and a small increase in the number of eosinophils in BAL fluid.[166] However, airway hyperresponsiveness was seen before pulmonary eosinophilia, suggesting that these phenomena do not occur simultaneously.[166] In a recent study, human recombinant IL-5 was given topically to guinea pig airways and, even though there was evidence of an inflammatory condition, the lungs did not appear hyperreactive.[167] On the other hand, intratracheal instillation of recombinant human IL-5 into the lungs of sensitized and antigen-boosted guinea pigs was followed by enhanced bronchoconstrictor responses to PAF.[168,169] This may have been due to a functional interaction on basophils, because combined IL-5 and PAF caused the release of mediators such as thromboxane B_2 and histamine that are found in basophils.[169]

Guinea pigs that were actively sensitized to ovalbumin showed increased numbers of eosinophils in the BAL fluid and lungs and demonstrated evidence of bronchial hyperresponsiveness following antigen bronchoprovocation. IL-5 appeared to be an important mediator of these allergic responses, because treatment of guinea pigs with a monoclonal antibody to IL-5 inhibited the development of pulmonary eosinophilia[5,19,170,171] and the bronchial hyperreactivity.[19,171] However, larger doses of the anti-IL-5 antibody were needed to block the hyperreactivity response to antigen challenge,[19] indicating a dissociation between eosinophil accumulation in the lung and the development of airway hyperresponsiveness. Viral respiratory tract infection also produced pulmonary inflammation and bronchial hyperresponsiveness in guinea pigs. In this model, an anti-IL-5 antibody suppressed the development of bronchial hyperresponsiveness, but had no effect on the eosinophil influx into the lungs.[172] These experiments clearly indicate an important role for IL-5 in the bronchial hyperreactivity caused by allergen challenge and respiratory viral infection in guinea pigs.

A characteristic feature in people with asthma and chronic bronchitis is the augmented vagally mediated bronchoconstrictor response. It was recently found in guinea pigs that the granule proteins released from eosinophils act as an endogenous allosteric antagonist at the inhibitory muscarinic M_2 receptor.[173] Treatment of guinea pigs with an antibody to IL-5 protected against neuronal M_2-receptor dysfunction that occurred following antigen bronchoprovocation, probably by inhibiting eosinophil influx into the lungs in response to antigen challenge.[174] Whether IL-5 has effects on neuronal responses in the airway that are independent from its effect on eosinophils remains to be established.

3. Monkeys

Antigen provocation to *Ascaris suum*-sensitive cynomolgus monkeys produces inflammatory cell influx into the airways, bronchial hyperreactivity, and an early and late phase bronchoconstriction. It was recently found that treatment with a monoclonal antibody to IL-5 inhibited the inflammatory cell influx and bronchial hyperreactivity caused by antigen challenge to monkeys.[20] Furthermore, inhibition of this pulmonary eosinophilia and bronchial hyperresponsiveness by the antibody to IL-5 was seen for up to 3 months after a single treatment. These results demonstrate the importance of IL-5 and the inhibitory effects of an anti-IL-5 antibody on the induction of pulmonary inflammation and airway hyperreactivity in nonhuman primates.

B. PARASITE INFECTIONS

The association of increased eosinophils and elevated IgE during parasite infections has been recognized for many years. These responses appear to be regulated by T lymphocytes, because they do not occur in athymic or in T-cell-depleted mice.[8] Recent studies in parasite-infected mice indicate that Th2-derived cytokines, notably IL-4 and IL-5, are important to the increased IgE and to the development of eosinophilia, respectively.[8,175,176]

The importance of IL-5 and eosinophil production in the development of protective immunity to parasites is uncertain. There was no difference in the protective immunity of infected mice treated with a monoclonal antibody to IL-5, even though there was total suppression of the eosinophil response.[8,175–178] On the other hand, recent studies have found enhanced survival of the nematode parasite, *Angiostrongylus cantonesis,* after treatment of infected mice with an anti-IL-5 antibody,[179] and IL-5 may play a role in the protective immunity of mice to *Strongyloides venezuelensis* infection.[177] These results are supported by *in vitro* studies that demonstrate helminthotoxic activity of granule products from eosinophils such as EPO, ECP, MBP, and EDN.[180,181] Furthermore, treatment of parasite-infected mice with an antibody to IL-4 did not change the protective immunity of these animals even though there was suppression of serum IgE levels.[175] These results question the functional role of both IL-4 and IL-5 in the protective immunity of mice to parasite infections.

Experimentally induced pulmonary eosinophilia is produced in mice following topical administration of the parasite *Ascaris suum.*[182] In this model of pulmonary inflammation, treatment with an antimurine IL-5 antibody before challenge with *Ascaris suum* inhibited the infiltration of eosinophils into the BAL fluid.[182] These results suggest that IL-5 is essential for the development of pulmonary eosinophilia after *Ascaris suum* challenge.

V. CLINICAL IMPLICATIONS OF IL-5 MODULATION

It has been appreciated, since Charcot and Leyden more than a century ago, that eosinophils are associated with asthma.[183,184] They are seen in the sputum of asthmatic patients and their numbers increase during asthma attacks.[185] Eosinophilic inflammation, as measured by bronchial biopsy and lavage, is now recognized as a important part of the asthmatic process, after death from severe asthma,[186] in mild asymptomatic asthma,[4] and following antigen challenge.[22] Treatment with nonspecific anti-inflammatory medication such as corticosteroids, improves the symptoms and signs of asthma and reduces the numbers of eosinophils observed in sputum and the systemic circulation.[187–189] Corticosteroids have many effects, but inhibition of release, migration, chemotaxis, and activation of eosinophils, probably through suppression of IL-5,[190] may be a critical element in their proven efficacy in asthma treatment. Interestingly, cyclosporin, a specific T-cell inhibitor that inhibits IL-5 production *in vitro,* also has been shown to have steroid-sparing activity in patients with severe asthma.[191] As has already been discussed, IL-5 is an important, and maybe essential, part of the inflammatory cascade that results in eosinophilic inflammation. Therefore, any strategy that blocks the effects of IL-5 would be predicted to be about as effective as corticosteroids in treating asthma. There are a number of possible approaches to inhibiting the effects of IL-5, but a neutralizing anti-IL-5 monoclonal antibody (MAB) is likely to be the first therapy into the clinic.

IL-5 is an attractive target because it appears to be very specific for eosinophils and basophils in humans. This suggests that inhibition of IL-5 will not be generally immunosuppressive. Eosinophilia is a prominent element of parasitic infestations, but the role it plays in controlling or suppressing these infections is not fully understood. Prudence dictates that patients with parasitic infestations should be avoided early in the development of an anti-IL-5 therapy, at least until the benefits and potential risks of this therapeutic approach in humans can be better determined. However, corticosteroids, which have a profound effect on eosinophilia, have been used extensively and with confidence for half a century. It is possible, therefore, that IL-5, and hence, eosinophils can be inhibited with considerable therapeutic benefit and without significant detrimental effects.

The use of an anti-IL-5 MAB might bring its own unique set of advantages (and challenges) and produce both improvement in symptoms and reduction in medication and hospitalization in asthma. It is only through studies such as these in the human that important insights into the importance of IL-5 and eosinophils, both in pathological situations and in host defense, will be established.

VI. HOST RESPONSE TO ANTIBODIES

Any protein, if recognized as foreign by the immune system, will induce a humoral response. This can occur even when human recombinant proteins are used therapeutically. Naturally occurring immunoglobulins have a long half-life in the systemic circulation, which is approximately 21 days for IgG_4 in humans. The systemic half-life of recombinant MABs, although somewhat shorter than natural immunoglobulins, is very long compared to other therapeutic proteins such as cytokines and, consequently, they require only infrequent administration. If a host response occurs in a significant proportion of patients, the value of the MAB as a therapeutic modality becomes severely limited, as has been the case with murine MABs.

The process of engineering murine MABs to make then more human, and hence less immunogenic, is reviewed elsewhere in detail.[192] Chimeric MABs retain murine variable regions, but have the murine constant regions replaced by human sequences. CDR grafting (humanization), where the murine binding sites are incorporated into a human variable region, further reduces the amount of "foreign" material in the MAB. After grafting, the antibody binding affinity may be reduced or lost, presumably as a result of changes in the tertiary structure of the MAB, and considerable further refinement of the primary sequence may be required to reconstitute the activity of the antibody.[192]

The degree and type of host response and its consequences may vary considerably. Host response can be measured in a number of different ways. The systemic concentration of anti-MAB antibodies can be determined, their site of binding can be identified, and their neutralizing capacity can be assessed. Strong responses, such as those seen in humans following administration of murine MABs, may lead to serious generalized reactions such as anaphylaxis or serum sickness.[193] An antimurine response is usually anti-isotypic in nature, and occurs in almost all patients after multiple doses of the MAB,[193–200] but can occur after the first dose.[193,197] In contrast, anti-idiotypic responses, which may be seen with murine or humanized MABs,[201–206] are of much lower intensity[204] and are generally not associated with any clinical effects. A significant reduction in the survival time of the MAB strongly suggests a host response and may be evident before anti-MAB antibodies are detected. A neutralizing anti-idiotype response may lead to a reduction in the MAB's therapeutic effect[201] and the need for larger doses.

There are a number of studies reporting host responses following administration of murine, chimeric, and humanized MABs to both animals and humans (see Table 1). In cynomolgus monkeys,[207–209] the intensity and type of host response seen after murine MABs is much improved by humanization, and there is a modest improvement in the therapeutic efficacy of the MAB (prolongation of allograft survival in this case). In humans, chimeric[203,204,206] and humanized[195] MABs are also less immunogenic than murine MABs.[193,194,200,203] However, even among chimeric MABs to different antigens, there are strikingly different degrees of host response. For instance, CO17–1A induced very little response in patients,[204] while administration of B72.3 was associated with anti-idiotypic antibodies in more than half the patients.[203] It appears that the host response can be moderated by the concomitant administration of immunosuppressive drugs such as cyclosporin[198] or another MAB,[210] but this is not a practical solution in the majority of cases. It is also likely that modification of the dose and dosing regimen may reduce immunization, but there are few data available to suggest what an optimal protocol might be.

Studies of the degree and consequences of host responses, comparing MABs with different degrees of humanization, have been conducted in animals and humans (Table 2). Because immune responses to a particular MAB are likely to be different between species, animal studies are of limited predictive value. For example, a number of studies have been performed in mice, but the results show little correlation with human data.[192] Monkey studies[207–209] appear to have results closest to humans. The pharmacokinetics and therapeutic effects of murine and humanized MABs to the same antigen (α[207,208] and β[209] subunits of the IL-2 receptor) were compared in cynomolgus monkeys. Administration of the anti-tac and anti-Mikβ1 MABs cause immunosuppression and enhanced cardiac allograft survival. With the humanized MAB, there was a reduction in host responses,

TABLE 1
Host Responses in Clinical Trials of Murine, Chimeric, and Humanized MABs

Antigen	Type[a]	Monoclonal Antibody	No. of Patients	Dosing[b]	Immunogenic Response[c]	Ref.
17–1A	M	CO17–1A	25	400–1600 mg 1–3 doses	6/15 after 1 dose 4/15 after 2 doses	193
17–1A	C	CO17–1A	4@1 dose 6@3 dose	10–40 mg 2-wk intervals	1/10 anti-id (1 pt after 3 doses)	204
TAG-72	M	B72.3	31	0.16–20 mg	16/31	194
TAG-72	C	B72.3	12	3.4–6.9 mg 4 pts 2nd dose 8 wk	7/12 after 1 dose 2/4 after 2 doses	203
GD-2	M	14G2A	12	10–120 mg 4 doses over 8 days	12/12 HAMA	199
GD-2	C	CH14.18	13	5–100 mg	8/13 anti-id	206
CD-33	M	M195	24	3 mg/M^2 2–4 doubled doses	7/24 HAMA +ve	200
CD-33	H	Hu-M195	13	Multiple	None	211
CDw52	H	CAMPATH-1H	8	60–260 mg daily for 5–10 days	0/8 after 1st 3/4 after 2nd	202
CDw52	H	CAMPATH-1H	2	126 mg over 30 days, 85 mg over 43 days	None	210
IL-2Rα	M H	Anti-tac	Monkeys	1 mpk alt. die 1 mpk alt. die	4/4 after 15 days 3/5 after 33 days	207
IL-2Rα	M H	Anti-tac	Monkeys	0.05–5 mpk daily for 14 days	2/4 0/4	208
IL-2Rα	H	Anti-tac	12	0.5–1.5 mpk	None	220
IL-2Rβ	M H	Anti-Mikβ1	Monkeys	1 mpk 1 mpk	4/5 1/5	209
CD-4	M	T4	4	1 mpk over 5 days	3/4 HAMA +ve	196
CD-4	M	16H5	10	3 mpk over 7 days 105–170 mg	5/10 after 1 dose 3/4 after 2 dose	197
CD-4	M	BL4	7	200–400 mg over 10 days	4/7 HAMA +ve	195
CD-4	C	LM-T412	25	10–700 mg (in 1–7 divided doses)	2/25	205
CEA	M	A5B7	3 7	 + Cyclosporin	All pts HAMA +ve Slower onset	198
TNF	C	cA2	8	20 mpk then 10 mpk × 3	4/8 anti-id	201

[a] M = murine, C = chimeric, H = humanized.
[b] hr = hour, d = day, wk = week, mg = milligram, mpk = milligram per kilogram body weight, alt.die = alternate day dosing, pt = patient.
[c] HAMA = human antimouse antibody.

increase in the half-life, and modest prolongation of allograft survival compared to the murine MAB (Table 2).

A number of human studies have been performed which allow comparison of the pharmacokinetics of murine and chimeric or humanized MABs after single and multiple doses. Murine MABs have much shorter half-lives[193,194,199] and remain in the systemic

TABLE 2
Pharmacokinetics in Clinical Studies of Murine, Chimeric, and Humanized Antibodies

Antigen	Type	Monoclonal Antibody	Mean Terminal Half-Life (hours)	Range (hours)	Ref.
17–1A	M	CO17–1A	16.2		193
17–1A	C	CO17–1A	100.5	65–139	204
TAG-72	M	B72.3	65	32–106	194
TAG-72	C	B72.3	224	102–298	203
GD-2	M	14G2A	42	±6	199
GD-2	C	CH14.18	181	±73	206
CD-33	M	M195	52	35–99	200
CD-33	H	Hu-M195	38	23–54	211
IL-2Rα	M	Ant-tac (monkeys)	38	25–43	207
	H		103	84–133	
IL-2Rα	M	Anti-tac (monkeys)	47.8	±9.04	208
	H		213.6	±58.8	
IL-2Rα	H	Anti-tac	104	41–363	220
IL-2Rβ	M	Anti-Mikβ1 (monkeys)	37	±2	209
	H		104	±10	

circulation for a briefer period. For instance, the murine B72.3 MAB had a mean half-life of 65 h,[194] while the chimeric MAB averaged 224 h.[203] Interestingly, the humanized MAB HuM195[211] had a shorter half-life than its murine predecessor,[200] but this appeared to be related to rapid uptake by its target and not a host response. In multiple dose studies, the pharmacokinetics were very reproducible after subsequent doses, but only in the absence of a host response.[203,204] There was a clear reduction in half-life and removal of the MAB from the circulation in those patients who experienced immune responses.[203]

No clinical studies have been reported with comprehensive data comparing host response, MAB pharmacokinetics, and therapeutic effects. Recently, a phase II clinical trial of a single dose of a chimeric anti-TNF MAB (cA_2) in rheumatoid arthritis has been reported.[212] The MAB induced a response in 19 of 24 patients, compared to 2 of 24 after placebo. In a separate study,[201] cA_2 MAB was given to eight patients on four separate occasions. An anti-idiotypic response was seen in four of the patients and, although the therapeutic response and duration of remission was reduced in two, it did not seem to be affected in the others. Neither the full characterization of the anti-MAB antibodies nor the pharmacokinetics of the cA_2 MAB were recorded in this paper, and they are important factors in assessing the likely value of this MAB after multiple dosing.

In summary, measurement of MAB pharmacokinetics and characterization of the anti-MAB responses are an important part of the clinical development of a monoclonal antibody. Murine MABs induce a strong, predominantly anti-isotype reaction. As the murine content of the MAB is reduced, so the host response diminishes, although even with humanized MABs anti-idiotypic antibodies are sometimes observed. A therapeutic MAB circulates at a far higher concentration than a specific, natural immunoglobulin, which is presumably present in insufficient quantities in the circulation to be recognized as foreign. Therefore, because of the requirement for a high systemic concentration, it is possible that even a human MAB may induce an anti-idiotypic response. Clinical experience to date shows clearly that "humanization" reduces the host response to MABs, but that the immunogenicity of any particular MAB can only be determined in comprehensive clinical studies.

VII. STRATEGIES FOR INHIBITING IL-5

A. BINDING IL-5

As described above, administration of high-affinity antibodies against IL-5 is a specific and effective mechanism for blocking its actions, and this approach can lead to a prolonged biological response in animals. Furthermore, this strategy could be extended to man with humanized[213] or human[214] antibodies that would be expected to survive in the circulation for several weeks. Other similar mechanisms such as Fab fragments from anti-IL-5 antibodies or monomeric anti-IL-5 antibodies would act by binding to IL-5, but these molecules would be expected to have shorter circulating lifetimes. The same should be true of administering soluble fragments from the α-subunit of the IL-5 receptor.[82,83] A potential drawback to each of these approaches would be the host antibody response to foreign proteins. Historically, such a response to injected antibodies has usually led to lack of efficacy on multiple dosing, rather than acute anaphylaxis.[201]

B. INHIBITION OF THE BIOSYNTHESIS OF IL-5

IL-5 is produced in the human mainly by T lymphocytes and mast cells.[54,60] The promoter and enhancer regions of the IL-5 gene possess regulatory elements in common with other proteins including cytokines.[51] To date, an IL-5-specific regulatory element or transcription factor has not been identified, and the presence of such an element would be necessary to obtain specific inhibition of IL-5 biosynthesis. That such inhibition could be achieved with small molecules is indicated by the inhibition of protein biosynthesis resulting from association of glucocorticoids with the glucocorticoid response element on several genes.[215] As with all nonpeptide approaches to blocking IL-5, a heterocyclic transcription inhibitor would not induce a host antibody response.

C. INHIBITION OF THE RESPONSE TO IL-5

In much the same way as inhibiting the biosynthesis of IL-5, it could be possible to inhibit the biosynthesis of the IL-5 receptor by disrupting the binding of transcription factors to the upstream promoter region of the IL-5Rα subunit gene. Because the α subunit is specific for IL-5 binding, while the β subunit is shared with IL-3 and GM-CSF, inhibition of α subunit rather than β subunit biosynthesis would be more selective for treating IL-5-based diseases such as asthma.[81] With other promoters, binding of small molecules such as calicheamicin in the minor groove of the DNA within a response element can block transcription factor protein binding and inhibit transcription of the gene.[216]

At the receptor itself, it would also be possible to inhibit the association of IL-5 to IL-5Rα with a small molecule or a mutated IL-5 protein that binds but does not transduce a signal, as with IL-1[217] or IL-4.[218] Because transduction of the IL-5 signal is dependent upon the β subunit that confers high affinity to the IL-5 receptor, disruption of the association of α and β subunits would also inhibit signaling from the IL-5 receptor. Although there are several steps in the signal transduction pathway for β subunit-based receptors that involve kinases[86,87] and phosphatases,[219] there does not appear to be any aspect that is unique to IL-5, so signal transduction is not a preferred site for selective inhibition. On the other hand, blocking the binding of IL-5 to the IL-5Rα by whatever mechanism would be specific to the eosinophil and to the basophil in the human, because only those cells possess the receptor.[11,14]

VIII. CONCLUSIONS

There is strong evidence for a role of IL-5 in asthma and other diseases that involve eosinophils. IL-5 is produced primarily in Th2-type lymphocytes and also in mast cells and is essential for the maturation of eosinophils in the bone marrow, where it acts through a specific Type 1 cytokine receptor. In the human, it appears that only eosinophils and basophils respond to IL-5, so it is unlikely that severe immunosuppression will result from inhibiting this cytokine. Hence, a specific inhibitor of IL-5 could be as effective as a steroid without the immunosuppressive side effects.

One of several mechanisms for inhibiting the actions of IL-5 is to reduce the circulating levels using an anti-IL-5 antibody. The viability of this approach has been demonstrated in various animal models of eosinophil accumulation. In some but not all of these models that encompass several species, inhibition of tissue and BAL eosinophilia correlates with decreased airway hyperreactivity. On the other hand, even a correlation does not conclusively demonstrate that the eosinophils cause the pathophysiology. However, whether the eosinophil is critical or not, IL-5 is essential for hyper-responsiveness because the pathophysiology is blocked consistently by an anti-IL-5 antibody.

Humanized antibodies to IL-5 should, therefore, be effective therapy against IL-5-based disorders such as asthma. Such antibodies can be produced by CDR grafting technology and should be accessible for clinical evaluation before low molecular weight inhibitors of IL-5 are available. However, these antibodies could cause a host response that cannot be accurately predicted from animal studies and must be established in human clinical trials.

ACKNOWLEDGMENTS

The authors thank Mrs. Ginny Citarella and Mrs. Terri Ruhnke for preparing the manuscript.

REFERENCES

1. **Gleich, G. J., Kita, H., and Adolphson, C. R.,** Eosinophils, in *Samters Immunologic Diseases,* 5th ed., Frank, M. N., Austen, K. F., Cloman, H. N., and Unanue, E. R., Eds., Little, Brown, Boston, 1995, 205.
2. **Kay, A. B.,** Asthma and Inflammation, *J. Allergy Clin. Immunol.,* 87, 893, 1991.
3. **Durham, S. R. and Kay, A. B.,** Eosinophils, bronchial hyperreactivity and late phase asthmatic reactions, *Clin. Allergy,* 15, 411, 1985.
4. **Wardlaw, A. J., Dunnette, S., Gleich, G. J., Collins, J. V., and Kay, A. B.,** Eosinophils and mast cells in bronchoalveolar lavage in subjects with mild asthma: relationship to bronchial hyperreactivity, *Am. Rev. Respir. Dis.,* 137, 62, 1988.
5. **Gulbenkian, A. R., Egan, R. W., Fernandez, X., Jones, H., Kreutner, W., Kung, T., Payvandi, F., Sullivan, L., Zurcher, J. A., and Watnick, A. S.,** Interleukin-5 modulates eosinophil accumulation in allergic guinea pig lung, *Am. Rev. Respir. Dis.,* 146, 263, 1992.
6. **Frigas, E., Loegering, D. A., and Gleich, G. J.,** Cytotoxic effects of the guinea pig eosinophil major basic protein on tracheal epithelium, *J. Lab. Invest.,* 42, 35, 1980.

7. **Weiss, S. J., Test, S. T., Eckmann, C. M., Roos, D., and Regiani, S.,** Brominating oxidants generated by human eosinophils, *Science,* 234, 200, 1986.
8. **Coffman, R. L., Seymour, B. W., Hudak, S., Jackson, J., and Rennick, D.,** Antibody to interleukin-5 inhibits helminth-induced eosinophilia in mice, *Science,* 245, 308, 1989.
9. **Clutterbuck, E. K., Hirst, E. M. A., and Sanderson, C. J.,** Human interleukin-5 (IL-5) regulates the production of eosinophils in human bone marrow cultures: comparison and interaction with IL-1, IL-3, IL-6 and GM-CSF, *Blood,* 73, 1504, 1989.
10. **Yamaguchi, Y., Suda, T., Suda, J., Eguchi, M., Miura, Y., Harad, N., Taminaga, A., and Takatsu, K.,** Purified interleukin-5 supports the terminal differentiation and proliferation of murine eosinophilic precursors, *J. Exp. Med.,* 167, 43, 1988.
11. **Resnick, M. B. and Weller, P. F.,** Mechanisms of eosinophil recruitment, *Am. J. Respir. Cell Mol. Biol.,* 8, 349, 1993.
12. **Wang, J. M., Rambaldi, A., Biondi, A., Chen, Z. G., Sanderson, C. J., and Mantovani, A.,** Recombinant human interleukin-5 is a selective eosinophil chemoattractant, *Eur. J. Immunol.,* 19, 701, 1989.
13. **Denburg, J. A., Silver, J. E., and Abrams, J. S.,** Interleukin-5 is a human basophilopoietin: induction of histamine content and basophilic differentiation of HL-60 cells and of peripheral blood basophil-eosinophil progenitors, *Blood,* 77, 1462, 1991.
14. **Hirai, K., Yamaguchi, M., Misaki, Y., Takaishi, T., Ohfa, K., Morita, Y., Ito, K., and Miyamoto, T.,** Enhancement of human basophil histamine release by interleukin-5, *J. Exp. Med.,* 172, 1525, 1990.
15. **Kinashi, T., Harada, N., Severinson, E., Tanabe, T., Sideras, P., Konishi, M., Azuma, C., Tominaga, A., Bergstedt-Lindqvist, S., Takahashi, M., Matsuda, F., Yaoita, Y., Takatsu, K., and Honjo, T.,** Cloning of complementary DNA encoding T-cell replacing factor and identity with B-cell growth factor II, *Nature,* 324, 70, 1986.
16. **Krishnaswamy, G., Liu, M. C., Su, S.-N., Kumai, M., Ziao, H.-Q., Marsh, D. G., and Huang, S.-K.,** Analysis of cytokine transcripts in the bronchoalveolar lavage cells of patients with asthma, *Am. J. Respir. Cell Mol. Biol.,* 9, 279, 1993.
17. **Hamid, Q., Azzawi, M., Ying, S., Moqbel, R., Wardlaw, A. J., Corrigan, C. J., Bradley, B., Durham, S. R., Collins, J. V., Jeffery, P. K., Quint, D. J., and Kay, A. B.,** Expression of mRNA for interleukin-5 in mucosal bronchial biopsies from asthma, *J. Clin. Invest.,* 87, 1541, 1991.
18. **Robinson, D., Hamid, Q., Bentley, A., Ying, S., Kay, A. B., and Durham, S. R.,** Activation of CD_4^+ T cells, increased Th2-type cytokine mRNA expression and eosinophil recruitment in bronchoalveolar lavage after allergen inhalation challenge in patients with atopic asthma, *J. Allergy Clin. Immunol.,* 92, 313, 1993.
19. **Mauser, P. J., Pitman, A., Witt, A., Fernandez, X., Zurcher, J., Kung, T., Jones, H., Watnick, A. S., Egan, R. W., Kreutner, W., and Adams, G. K., III,** Inhibitory effect of the TRFK-5 anti-IL-5 antibody in a guinea pig model of asthma, *Am. Rev. Respir. Dis.,* 148, 1623, 1993.
20. **Mauser, P. J., Pitman, A. M., Fernandez, X., Foran, S. K., Adams, G. K., III, Kreutner, W., Egan, R. W., and Chapman, R. W.,** Effects of an antibody to interleukin-5 in a monkey model of asthma, *Am. J. Respir. Crit. Care Med.,* 152, 467, 1995.
21. **Kung, T. T., Stelts, D., Zurcher, J. A., Watnick, A. S., Jones, H., Mauser, P. J., Fernandez, X., Umland, S., Kreutner, W., Chapman, R. W., and Egan, R. W.,** Mechanisms of allergic pulmonary eosinophilia in the mouse, *J. Allergy Clin. Immunol.,* 94, 1217, 1994.
22. **DeMonchy, J. G., Kauffman, H. F., Venge, P., Koeter, G. H., Jansen, H. M., Sluiter, H. J., and DeVries, K.,** Bronchoalveolar eosinophilia during antigen-induced late asthmatic reactions, *Am. Rev. Respir. Dis.,* 131, 373, 1985.
23. **Gibson, P. G., Manning, P. J., O'Byrne, P. M., Girgis-Gabardo, A., Dolovich, J., Denburg, J. A., and Hargreave, F. E.,** Allergen-induced asthmatic responses. Relationship between increases in airway responsiveness and increases in circulating eosinophils, basophils and their progenitors, *Am. Rev. Respir. Dis.,* 143, 331, 1991.
24. **Hutson, P. A., Church, M. K., Clay, T. P., Miller, P., and Holgate, S. T.,** Early and late-phase bronchoconstriction after antigen challenge of nonanesthetized guinea pigs. I. The association of disordered airway physiology to leukocyte infiltration, *Am. Rev. Respir. Dis.,* 137, 548, 1988.

25. **Bichot, E., Lagenmte, V., Carre, C., Waltmann, J. M., and Braquet, P.,** Bronchial hyperresponsiveness and cellular infiltration in the lung of guinea pigs sensitized and challenged by aerosol, *Clin. Exp. Allergy,* 21, 67, 1991.
26. **Ishida, K., Thompson, R. J., Beattie, L. L., Wiggs, B., and Schellenberg, R. R.,** Inhibition of antigen-induced airway hyperresponsiveness but not acute hypoxia nor airway eosinophilia, by an antagonist of platelet-activating factor, *J. Immunol.,* 144, 3907, 1990.
27. **Butterfield, J. H., Leiferman, K. M., and Gleich, G. J.,** Eosinophil associated diseases, in *Samters Immunologic Diseases,* 5th ed., Frank, M. N., Austen, K. F., Cloman, H. N., and Unanve, E. R., Eds., Little, Brown, Boston, 1995, 501.
28. **Allen, J. N. and Davis, B.,** Eosinophilic lung diseases, *Am. J. Respir. Crit. Care Med.,* 150, 1423, 1994.
29. **Azuma, C., Tanabe, T., Konishi, M., Kinashi, T., Noma, T., Matsuda, F., Yaoita, Y., Takatsu, K., Hammarström, L., Edvard Smith, C. I., Severinson, E., and Honjo, T.,** Cloning of cDNA for human T-cell replacing factor (interleukin-5) and comparison with the murine homologue, *Nucl. Acids Res.,* 14, 9149, 1986.
30. **Sanderson, C. J., Warren, D. J., and Strath, M.,** Identification of a lymphokine that stimulates eosinophil differentiation *in vitro.* Its relationship to IL-3, and functional properties of eosinophils produced in cultures, *J. Exp. Med.,* 162, 148, 1985.
31. **Yokota, T., Coffman, R. L., Hagiwara, H., Rennick, D., Takebe, Y., Yokota, K., Gemmell, L., Shrader, B., Yang, G., Meyerson, P., Luh, J., Hoy, P., Péne, J., Briére, F., Spits, H., Banchereau, J., De Vries, J., Lee, F. D., Aria, N., and Arai, K-I.,** Isolation and characterization of lymphokine cDNA clones encoding mouse and human IgA-enhancing factor and eosinophil colony-stimulating factor activities. Relationship to interleukin-5, *Proc. Natl. Acad. Sci. U.S.A.,* 84, 7388, 1987.
32. **Kodama, S., Tsuruoka, N., and Tsujimoto, M.,** Role of the C-terminus in the biological activity of human interleukin-5, *Biochem. Biophys. Res. Commun.,* 178, 514, 1991.
33. **Sanderson, C. J.,** Eosinophil differentiation factor (Interleukin-5), in *Colony Stimulating Factors: Molecular and Cellular Biology,* Dexter, T. M., Garland, J. M., and Testa, N. G., Eds., Marcel Dekker, New York, 1990, 231.
34. **Tanabe, T., Konishi, M., Mizuta, T., Noma, T., and Honjo, T.,** Molecular cloning and structure of the human interleukin-5 gene, *J. Biol. Chem.,* 34, 16580, 1987.
35. **Takahashi, M., Yoshida, M. C., Satoh, H., Hilgers, J., Yaoita, Y., and Honjo, T.,** Chromosomal mapping of the mouse IL-4 and human IL-5 genes, *Genomics,* 4, 49, 1988.
36. **van Leewen, B. H., Martinson, M. E., Webb, G. C., and Young, I. C.,** Molecular organization of the cytokine gene cluster, involving the human IL-3, IL-4, IL-5, and GM-CSF genes, on human chromosome 5, *Blood,* 73, 1142, 1989.
37. **Miyatake, S., Shlomai, J., Ken-ichi, A., and Ari, N.,** Characterization of the mouse granulocyte-macrophage colony-stimulating factor (GM-CSF) gene promoter. Nuclear factors that interact with an element shared by three lymphokine genes — those for GM-CSF, interleukin-4 (IL-4), and IL-5, *Mol. Cell. Biol.,* 12, 5894, 1991.
38. **Kaushansky, K., Shoemaker, S. G., O'Rork, C. A., and McCarty, J. M.,** Coordinate regulating of multiple human lymphokine genes by Oct-1 and potentially novel 45 and 43 kDa polypeptides, *J. Immunol.,* 152, 1812, 1994.
39. **Tominaga, A., Takahashi, T., Kikuchi, Y., Mita, S., Naomi, S., Harada, N., Yamaguchi, N., and Takatsu, K.,** Role of carbohydrate moiety of IL-5, *J. Immunol.,* 144, 1345, 1990.
40. **Takahashi, T., Yamaguchi, N., Mita, S., Yamaguchi, Y., Suda, T., Tominaga, A., Kikuchi, Y., Miura, Y., and Takatsu, K.,** Structural comparison of murine T-cell (B151K12)-derived T-cell-replacing factor (IL-5) with rIL-5. Dimer formation is essential for the expression of biological activity, *Mol. Immunol.,* 27, 911, 1990.
41. **Minamitake, Y., Kodama, S., Katayama, T., Adachi, H., Tanaka, H., and Tsujimoto, M.,** Structure of recombinant human interleukin-5 produced by Chinese hamster ovary cells, *J. Biochem.,* 107, 292, 1990.
42. **Tsuruoka, N., Funakowshi, K., Kodama, S., and Tsujimoto, M.,** Interaction of interleukin-5 with its receptors on murine leukaemic BCL1 cells and its implication in biological activity, *Cell. Immunol.,* 125, 354, 1990.

43. **McKenzie, A. N. J., Ely, B., and Sanderson, C. J.,** Mutated interleukin-5 monomers are biologically inactive, *Mol. Immunol.,* 28, 155, 1991.
44. **Milburn, M. V., Hassell, A. M., Lambert, M. H., Jordan, S. R., Proudfoot, A. E. I., Graber, P., and Wells, T. N. C.,** A novel dimer configuration revealed by the crystal structure at 24Å resolution of human interleukin-5, *Nature,* 363, 172, 1993.
45. **Guisez, Y., Oefner, C., Winkler, F. K., Schlaeger, E.-J., Zulauf, M., Van der Heyden, J., Plaetinck, G., Cornelis, S., Tavernier, J., Fiers, W., Devos, R., and D'Arcy, A.,** Expression, purification and crystallization of fully active, glycosylated human interleukin-5, *Eur. Biochem. Soc.,* 331, 49, 1993.
46. **Corrigan, C. J. and Kay, A. B.,** T cells and eosinophils in the pathogenesis of asthma, *Immunol. Today,* 13, 501, 1992.
47. **Altman, A., Coggeshall, K. M., and Mustelin, T.,** Molecular events mediating T cell activation, *Adv. Immunol.,* 48, 227, 1990.
48. **Mosmann, T. R., and Coffman, R. L.,** TH_1 and TH_2 cells. Different patterns of lymphokine secretion lead to different functional properties, *Ann. Rev. Immunol.,* 7, 145, 1989.
49. **Van Straaten, J. F. M., Dokter, W. H. A., Stulp, B. K., and Vellenga, B. K.,** The regulation of interleukin-5 and interleukin-3 gene expression in human T cells, *Cytokine,* 6, 229, 1994.
50. **Staynov, D. Z. and Lee, T. H.,** Expression of interleukin-5 and granulocyte-macrophage colony-stimulating factor in human peripheral blood mononuclear cells after activation with phorbol myristate acetate, *Immunology,* 75, 196, 1992.
51. **Naora, H., Altin, J. G., and Young, I. G.,** TCR-dependent and independent signaling mechanisms differentially regulate lymphokine gene expression in the murine T helper clone D10.G4.1, *J. Immunol.,* 152, 5691, 1994.
52. **Bohjanen, P. R., Okajima, M., and Hodes, R. J.,** Differential regulation of interleukin-4 and interleukin-5 gene expression. A comparison of T-cell gene induction by anti-CD3 antibody or by exogenous lymphokines, *Proc. Natl. Acad. Sci. U.S.A.,* 87, 5283, 1990.
53. **Rolfe, F. G., Hughes, J. M., Armour, C. L., and Sewell, W. A.,** Inhibition of interleukin-5 gene expression by dexamethasone, *Immunology,* 77, 494, 1992.
54. **Plaut, M., Pierce, J. H., Watson, C. J., Hanley, H. J., Nordan, R. P., and Paul, W. E.,** Mast cell lines produce lymphokines in response to cross-linkage of Fc epsilon RI or to calcium ionophores, *Nature,* 339, 64, 1989.
55. **Dubucquoi, S., Desreumaux, P., Janin, A., Klein, O., Goldman, M., Tavernier, J., Capron, M., and Capron, A.,** Interleukin-5 synthesis by eosinophils. Association with granules and immunoglobulin-dependent secretion, *J. Exp. Med.,* 179, 703, 1994.
56. **Desreumaux, P., Janin, A., Colombel, J. F., Prin, L., Plumas, J., Emilie, D., Torpier, G., Capron, A., and Capron, M.,** Interleukin-5 messenger RNA expression by eosinophils in the intestinal mucosa of patients with coeliac disease, *J. Exp. Med.,* 175, 293, 1992.
57. **Desreumaux, P., Janin, A., Dubucquoi, S., Copin, M. C., Torpier, G., Capron, A., Capron, M., and Prin, L.,** Synthesis of interleukin-5 by activated eosinophils in patients with eosinophilic heart diseases, *Blood,* 82, 1553, 1993.
58. **Broide, D. H., Paine, M. M., and Firestein, G. S.,** Eosinophils express interleukin-5 and granulocyte macrophage-colony-stimulating factor mRNA at sites of allergic inflammation in asthmatics, *J. Clin. Invest.,* 90, 1414, 1992.
59. **Bradding, P., Feather, I. H., Wilson, S., Bardin, P. G., Heusser, C. H., Holgate, S. T., and Howarth, P. H.,** Immunolocalization of cytokines in the nasal mucosa of normal and perennial rhinitic subjects, *J. Immunol.,* 151, 3853, 1993.
60. **Bradding, P., Roberts, J. A., Britten, K. M., Montefort, S., Dujukanovic, R., Mueller, R., Heusser, C. H., Howarth, P. H., and Holgate, S. T.,** Interleukin-4,-5, and -6 and tumor necrosis factor-α in normal and asthmatic airways. Evidence for the human mast cell as a source of these cytokines, *Am. J. Respir. Cell Mol. Biol.,* 10, 471, 1994.
61. **Garlisi, C. G., Falcone, A., Kung, T. T., Stelts, D., Pennline, K. J., Beavis, A. J., Smith, S. R., Egan, R. W., and Umland, S. P.,** T cells are necessary for Th_2 cytokine production and eosinophil accumulation in airways of antigen-challenged allergic mice, *Clin. Immunol. Immunopathol.,* 75, 75, 1995.

62. **Kung, T. T., Stelts, D., Zurcher, J. A., Jones, H., Umland, S. P., Kreutner, W., Egan, R. W., and Chapman, R. W.,** Mast cells modulate allergic pulmonary eosinophilia in mice, *Am. J. Respir. Cell Biol.,* 12, 404, 1995.
63. **Miyajima, A., Kitamura, T., Harada, N., Yokota, T., and Ken-ichi, A.,** Cytokine receptors and signal transduction, *Annu. Rev. Immunol.,* 10, 295, 1992.
64. **Mita, S., Tominaga, A., Hitoshi, Y., Sakamoto, K., Honjo, T., Akagi, M., Kikuchi, Y., Yamaguchi, N., and Takatsu, K.,** Characterization of high affinity receptors for interleukin-5 on interleukin-5-dependent cell lines, *Proc. Natl. Acad. Sci. U.S.A.,* 86, 2311, 1989.
65. **Mita, S., Harada, N., Naomi, S., Hitoshi, Y., Sakamoto, K., Akagi, M., Tominaga, A., and Takatsu, K.,** Receptors for T cell replacing factor/interleukin-5, *J. Exp. Med.,* 168, 863, 1988.
66. **Hitoshi, Y., Yamaguchi, N., Mita, S., Tominaga, A., and Takatsu, K.,** Distribution of interleukin-5 receptor-positive B cells. Expression of IL-5 receptor on Ly-1 (CD5)-positive B cells, *J. Immunol.,* 144, 4218, 1990.
67. **Takaki, S., Tominaga, A., Hitoshi, Y., Mita, S., Sonada, E., Yamaguchi, N., and Takatsu, K.,** Molecular cloning and expression of the murine interleukin-5 receptor, *EMBO J.,* 9, 4367, 1990.
68. **Gearing, D. P., King, J. A., Gough, N. H., and Nicola, N. A.,** Expression cloning of a receptor for human granulocyte-macrophage stimulating factor, *EMBO J.,* 8, 3667, 1989.
69. **Bazan, J. F.,** Structural design and molecular evaluation of a cytokine receptor superfamily, *Proc. Natl. Acad. Sci. U.S.A.,* 87, 6934, 1990.
70. **Patthy, L.,** Homology of a domain of the growth hormone/prolactin receptor family with type III modules of fibronectin, *Cell,* 61, 13, 1990.
71. **Takaki, S., Mita, S., Kitamura, T., Yonehara, S., Yamaguchi, N., Tominaga, A., Miyajima, A., and Takatsu, K.,** Identification of the second subunit of the murine interleukin-5 receptor: interleukin-3 receptor-like protein, AIC2B is a component of the high affinity interleukin-5 receptor, *EMBO J.,* 10, 2833, 1991.
72. **Takaki, S., Murata, Y., Kitamura, T., Miyajima, A., Tominaga, A., and Takatsu, K.,** Reconstitution of the functional receptors for murine and human interleukin-5, *J. Exp. Med.,* 177, 1523, 1993.
73. **Devos, R., Plaetinck, G., Van der Heyden, J., Cornelis, S., Vanderkerckhove, J., Fierce, W., and Tavernier, J.,** Molecular basis of a high affinity murine interleukin-5 receptor, *EMBO J.,* 10, 2133, 1991.
74. **Rolink, A. G., Melchers, F., and Palacios, R.,** Monoclonal antibodies reactive with the mouse interleukin-5 receptor, *J. Exp. Med.,* 169, 1693, 1989.
75. **Van der Heyden, J., Devos, R., Plaetinck, G., Fache, I., Fiers, W., and Tavernier, J.,** Characterization of the murine IL-5 receptor complex with the use of a panel of monoclonal antibodies, *J. Immunol.,* 147, 3413, 1991.
76. **Gorman, D. M., Itoh, N., Kitamura, T., Scheurs, J., Yonehara, S., Yahhara, I., Arai, K.-I., and Miyajima, A.,** Cloning and expression of a gene encoding an interleukin-3 receptor-like protein; identification of another member of the cytokine receptor gene family, *Proc. Natl. Acad. Sci. U.S.A.,* 87, 5459, 1990.
77. **Takatsu, K.,** Interleukin-5, *Curr. Opin. Immunol.,* 4, 299, 1992.
78. **Mita, S., Takaki, S., Tominaga, A., and Takatsu, K.,** Comparative analysis of the kinetics of binding and internalization of IL-5 in murine IL-5 receptors of high and low affinity, *J. Immunol.,* 151, 6924, 1993.
79. **Migita, M., Yamaguchi, N., Mita, S., Higuchi, S., Hitoshi, Y., Yosha, Y., Tomonaga, M., Matsuda, I., Tominaga, A., and Takatsu, K.,** Characterization of the human IL-5 receptors on eosinophils, *Cell. Immunol.,* 133, 484, 1991.
80. **Plaetinck, G., Van der Heyden, J., Tavernier, J., Faché, I., Tuypens, T., Fischkoff, S., Fiers, W., and Devos, R.,** Characterization of interleukin-5 receptors on eosinophilic sublines from human promyelocytic leukemia (HL-60) cells, *J. Exp. Med.,* 172, 683, 1990.
81. **Murata, Y., Takaki, S., Migita, M., Kikuchi, Y., Tominaga, A., and Takatsu, K.,** Molecular cloning and expression of the human interleukin-5 receptor, *J. Exp. Med.,* 175, 341, 1992.
82. **Tavernier, J., Devos, R., Cornellia, S., Tuypens, T., Van der Heyden, J., Fiers, W., and Plaetinck, G.,** A human high affinity interleukin-5 receptor (IL-5R) is composed of an IL-5-specific α chain and a β chain shared with the receptor for GM-CSF, *Cell,* 66, 1175, 1991.

83. **Tavernier, J., Tuypens, T., Plaetinck, G., Verhee, A., Fiers, W., and Devos, R.,** Molecular basis of the membrane-anchored and two soluble isoforms of the human interleukin-5 receptor α subunit, *Proc. Natl. Acad. Sci. U.S.A.,* 89, 7041, 1992.
84. **Kazuhiro, S., Miyajima, I., Kitamura, T., and Miyajima, A.,** Critical cytoplasmic domains of the common β subunit of the human GM-CSF, IL-3 and IL-5 receptors for growth signal transduction and tyrosine phosphorylation, *EMBO J.,* 11, 3541, 1992.
85. **Murata, Y., Yamaguchi, N., Hitoshi, Y., Tominaga, A., and Takatsu, K.,** Interleukin-5 and interleukin-3 induce serine and tyrosine phosphorylations of several cellular proteins in an interleukin-5 dependent cell line, *Biochem. Biophys. Res.,* 173, 1102, 1990.
86. **Silvennoinen, O., Witthuhn, B. A., Quelle, F. W., Cleveland, J. L., Taolin, Y., and Ihle, J. N.,** Structure of the murine Jak2 protein-tyrosine kinase and its role in interleukin-3 signal transduction, *Proc. Natl. Acad. Sci. U.S.A.,* 90, 8429, 1993.
87. **Darnell, J. E., Jr., Kerr, A. M., and Stark, G. R.,** Jak-STAT pathways and transcriptional activation in response to IFNs and other extracellular signaling proteins, *Science,* 262, 1415, 1994.
88. **Sanderson, C. J., Campbell, H. D., and Young, I. G.,** Molecular and cellular biology of eosinophil differentiation factor (interleukin-5) and its effects on human and mouse B cells, *Immunol. Rev.,* 102, 29, 1988.
89. **Takatsu, K. T., Tominaga, A., Harada, N., Mita, S., Matsumoto, M., Takahashi, T., Kikuchi, Y., and Yamaguchi, N.,** T cell-replacing factor (TRF)/interleukin-5 (IL-5): molecular and functional properties, *Immunol. Rev.,* 102, 107, 1988.
90. **Lopez, A. F., Shannon, M. F., Chia, M.-M., Park, L., and Vadas, M. A.,** Regulation of human eosinophil production and function by interleukin-5, *Immunol. Ser.,* 57, 549, 1992.
91. **McKenzie, A. N. J. and Sanderson, C. J.,** Interleukin-5, in *Interleukins: Molecular Biology and Immunology,* Kishimoto, T., Ed., S. Karger, New York, 1992, 135.
92. **O'Garra, A., Warren, D. J., Holman, M., Popham, A. M., Sanderson, C. J., and Klaus, G. G. B.,** Interleukin-5 (B cell growth factor II/eosinophil differentiation factor) is a mitogen for preactivated murine B lymphocytes, *Proc. Natl. Acad. Sci. U.S.A.,* 83, 5228, 1986.
93. **Warren, D. J. and Moore, M. S.,** Synergism among interleukin-1, interleukin-3, and interleukin-5 in the production of eosinophils from primitive hemopoietic stem cells, *J. Immunol.,* 140, 94, 1988.
94. **Clutterbuck, E. J. and Sanderson, C. J.,** Regulation of human eosinophil precursor production by cytokines: a comparison of recombinant human interleukin-1 (rhIL-1), rhIL-3, rhIL-5, rhIL-6 and rh granulocyte-macrophage colony-stimulating factor, *Blood,* 75, 1774, 1990.
95. **Warringa, R. A. J., Schweizer, R. C., Maikoe, T., Kuijper, P. H. M., Bruijnzeel, P. L. B., and Koenderman, L.,** Modulation of eosinophil chemotaxis by interleukin-5, *Am. J. Respir. Cell Mol. Biol.,* 7, 631, 1992.
96. **Kita, H., Weiler, D., Abu-Ghazaleh, R., Sanderson, C. J., and Gleich, G. J.,** Release of granule proteins from eosinophils cultured with IL-5, *J. Immunol.,* 149, 629, 1992.
97. **Carlson, M., Peterson, C., and Venge, P.,** The influence of IL-3, IL-5 and GM-CSF on normal human eosinophil and neutrophil C3b-induced degranulation, *Allergy,* 48, 437, 1993.
98. **Yamaguchi, Y., Suda, T., Oha, S., Tominaga, K., Miura, Y., and Kasahara, T.,** Analysis of the survival of mature human eosinophils: interleukin-5 prevents apoptosis in mature human eosinophils, *Blood,* 78, 2542, 1991.
99. **Bischoff, S. C., Brunner, T., DeWeck, A. L., and Dahinden, C. A.,** Interleukin-5 modifies histamine release and leukotriene generation by human basophils in response to diverse agonists, *J. Exp. Med.,* 172, 1577, 1990.
100. **Sonoda, Y., Arai, N., and Ogawa, M.,** Humoral regulation of eosinophilopoiesis *in vitro:* analysis of the targets of interleukin-3, granulocyte/macrophage colony-stimulating factor (GM-CSF), and interleukin-5, *Leukemia,* 3, 14, 1989.
101. **Gleich, G. J. and Adolphson, C. R.,** The eosinophilic leukocyte: structure and function, *Adv. Immunol.,* 39, 177, 1986.
102. **Håkansson, L. and Venge, P.,** Priming of eosinophil and neutrophil migratory responses by interleukin-3 and interleukin-5, *APMIS,* 102, 308, 1994.
103. **Sehmi, R., Wardlaw, A. J., Cromwell, O., Kurihara, K., Waltmann, P., and Kay, A. B.,** Interleukin-5 selectively enhances the chemotactic response of eosinophils obtained from normal but not eosinophilic subjects, *Blood,* 79, 2952, 1992.

104. **Walsh, G. M., Hartnell, A., Wardlaw, A. J., Kurihara, K., Sanderson, C. J., and Kay, A. B.,** IL-5 enhances the *in vitro* adhesion of human eosinophils, but not neutrophils, in a leucocyte integrin (CD11/18)-dependent manner, *Immunology,* 71, 258, 1990.
105. **Hansel, T. T., DeVries, J. M., Carballido, J. M., Braun, R. K., Carballido-Perrig, N., Rihs, S., Blaser, K., and Walker, C.,** Induction and function of eosinophil intercellular adhesion molecule-1 and HLA-DR, *J. Immunol.,* 149, 2130, 1992.
106. **Petreccia, D. C., Nauseef, W. M., and Clark, R. A.,** Respiratory burst of normal human eosinophils, *J. Leukemia Biol.,* 41, 283, 1987.
107. **Spry, C. J. F.,** Eosinophil activation, cytotoxicity, and interactions in eosinophils, in *A Comprehensive Review and Guide to the Scientific and Medical Literature,* Spry, C. J. F., Ed., Oxford University Press, New York, 1988, 103.
108. **Yazdanbakhsh, M., Eckmann, C. M., Koenderman, L., Verhoeven, A. J., and Roos, D.,** Eosinophils do respond to fMLP, *Blood,* 70, 379, 1987.
109. **Minnicozzi, M., Anthes, J. C., Siegel, M. I., Billah, M. M., and Egan, R. W.,** Activation of phospholipase D in normodense human eosinophils, *Biochem. Biophys. Res. Commun.,* 170, 540, 1990.
110. **Lehrer, R. I., Ganz, T., Selsted, M. E., Babior, B. M., and Carnutte, J. T.,** Neutrophils and host defense, *Ann. Int. Med.,* 109, 127, 1988.
111. **Abu-Ghazaleh, R. I., Fujisawa, T., Mestecky, J., Kyle, R. A., and Gleich, G. J.,** IgA-induced eosinophil degranulation, *J. Immunol.,* 142, 2393, 1989.
112. **Capron, M., Capron, A., Dessaint, J. B., Torpier, G., Johanson, S. G. O., and Prin, L.,** F_c receptors for IgE on human and rat eosinophils, *J. Immunol.,* 126, 2087, 1981.
113. **Gounni, A. S., Lamkhioued, B., Ochiai, K., Tanaka, Y., Delaporte, E., Capron, A., Kinet, J. P., and Capron, M.,** High affinity IgE receptor on eosinophils is involved in defense against parasites, *Nature,* 367, 183, 1994.
114. **Rothenberg, M. E., Owen, W. F., Siberstein, D. S., Woods, J., Soberman, R. J., Austen, K. F., and Stevens, R. L.,** Human eosinophils have prolonged survival, enhanced functional properties and become hypodense when exposed to human interleukin-3, *J. Clin. Invest.,* 81, 1986, 1988.
115. **Owen, W. F. and Austen, K. F.,** Eosinophils altered phenotypically and primed by culture with granulocyte macrophage colony-stimulating factor and 3T3 fibroblasts generate leukotriene C_4 in response to fMLP, *J. Clin. Invest.,* 83, 1958, 1991.
116. **Her, E., Frazer, J., Austen, K. F., and Owen, W. F., Jr.,** Eosinophil hematopoietins antagonize the programmed cell death of eosinophils, *J. Clin. Invest.,* 88, 1982, 1991.
117. **Coëffier, E., Joseph, D., and Vargaftig, B. B.,** Activation by human recombinant IL-5 of guinea pig eosinophils. Selective priming to PAF and interference of its antagonists, *J. Immunol.,* 147, 2595, 1991.
118. **Fujisawa, T., Abu-Ghazaleh, R., Kita, H., Sanderson, C. J., and Gleich, G. J.,** Regulatory effect of cytokines on eosinophil degranulation, *J. Immunol.,* 144, 642, 1990.
119. **Ottesen, E. A. and Cohen, S. G.,** The eosinophil, eosinophilia and eosinophil-related disorders, in *Allergy: Principles and Practice,* Vol. 2, Middleton, E., Reed, C. E., and Ellis, E. F., Eds., C. V. Mosby, St. Louis, 1978, 584.
120. **Bruynzeel, P. L. B., Kok, P. T. M., Hamelink, M. L., Kijne, A. M., and Verhagen, J.,** Exclusive leukotriene C_4 synthesis by purified human eosinophils induced by opsonized zymosan, *Fed. Eur. Biochem. Soc.,* 189, 350, 1985.
121. **Cockcroft, S.,** G-protein-regulated phospholipase C, D, and A_2-mediated signalling in neutrophils, *Biochim. Biophys. Acta,* 1113, 135, 1992.
122. **Hamann, K. J., Barker, R. I., Ten, R. M., and Gleich, G. J.,** The molecular biology of eosinophil granule proteins, *Int. Arch. Allergy Appl. Immunol.,* 94, 202, 1991.
123. **Banker, R. L., Loegering, D. A., Ten, R. M., Hamann, K. J., Pease, L. R., and Gleich, G. J.,** Eosinophil cationic protein cDNA, *J. Immunol.,* 143, 952, 1989.
124. **Barker, R. L., Gleich, G. J., and Pease, L. R.,** Acidic precursor revealed in human eosinophil granule major basic protein cDNA, *J. Exp. Med.,* 168, 1493, 1988.
125. **Gleich, G. J., Loegering, D. A., Bell, M. A., Checkel, J. L., Ackerman, S. J., and McKean, D. J.,** Biochemical and functional similarities between human eosinophil derived neurotoxin and eosinophil cationic protein: homotology with ribonuclease, *Proc. Natl. Acad. Sci. U.S.A.,* 83, 3146, 1986.

126. **Motojima, S. E., Frigas, E., Loegering, D. A., and Gleich, G. J.,** Toxicity of eosinophil cationic proteins for guinea pig tracheal epithelium *in vitro*, *Am. Rev. Respir. Dis.,* 139, 801, 1989.
127. **Minnicozzi, M., Durán, W. N., Gleich, G. J., and Egan, R. W.,** Eosinophil granule proteins increase microvascular macromolecular transport in the hamster cheek pouch, *J. Immunol.,* 153, 2664, 1994.
128. **Ackerman, S. J., Loegering, D. A., and Gleich, G. J.,** The human eosinophil Charcot-Leyden crystal protein: biochemical characteristics and measurement by radioimmunoassay, *J. Immunol.,* 125, 2118, 1980.
129. **Weller, P. F., Bach, D. S., and Austen, K. F.,** Biochemical characterization of human eosinophil Charcot-Leyden crystal protein (lysophospholipase), *J. Biol. Chem.,* 259, 15100, 1984.
130. **Dvorak, A. M., Weller, P. F., Monahan-Earley, R. A., Letourneau, L., and Ackerman, S. J.,** Ultrastructural localization of Charcot-Leyden crystal protein (lysophospholipase) and peroxidase in macrophages, eosinophils and extracellular matrix of the skin in the hypereosinophilic syndrome, *Lab. Invest.,* 62, 590, 1990.
131. **Strath, M., Warren, D. J., and Sanderson, C. J.,** Detection of eosinophils using an eosinophil peroxidase assay. Its use as an assay for eosinophil differentiating factors, *J. Immunol. Methods,* 83, 209, 1985.
132. **Ohnishi, T., Sur, S., Collins, D. S., Fish, J., Gleich, G. J., and Peters, S. P.,** Eosinophil survival activity identified as interleukin-5 is associated with eosinophil recruitment and degranulation and lung injury 24 hours after segmental antigen lung challenge, *J. Allergy Clin. Immunol.,* 92, 607, 1993.
133. **Wallen, N., Kita, H., Weiler, D., and Gleich, G. J.,** Glucocorticoids inhibit cytokine-mediated eosinophil survival, *J. Immunol.,* 147, 3490, 1991.
134. **Hallsworth, M. D., Litchfield, T. M., and Lee, T. H.,** Glucocorticoids inhibit granulocyte macrophage colony-stimulating factor-1 and interleukin-5 enhanced *in vitro* survival of human eosinophils, *Immunology,* 75, 382, 1992.
135. **Kung, T. T., Jones, H., Adams, G. K., III, Umland, S. P., Kreutner, W., Egan, R. W., Chapman, R. W., and Watnick, A. S.,** Characterization of a murine model of allergic pulmonary inflammation, *Int. Arch . Allergy Immunol.,* 105, 83, 1994.
136. **Ädelroth, E., Rosenhall, L., Johansson, S., Linden, M., and Venge, P.,** Inflammatory cells and eosinophilic activity in asthmatics investigated by bronchoalveolar lavage, *Am. Rev. Respir. Dis.,* 142, 91, 1990.
137. **Bertolini, J. N., Sanderson, C. J., and Benson, E. M.,** Human interleukin-5 induces staphylococcal A cowan 1 strain-activated human B cells to secrete IgM, *Eur. J. Immunol.,* 23, 398, 1993.
138. **McHeyzer-Williams, M. G.,** Combinations of interleukins 2, 4 and 5 regulate the secretion of murine immunoglobulin isotypes, *Eur. J. Immunol.,* 19, 2025, 1989.
139. **Morikawa, K., Oseko, F., Morikawa, S., Imai, K., and Sawad, M.,** Recombinant human IL-5 augments immunoglobulin generation by human B lymphocytes in the presence of IL-2, *Cell . Immunol.,* 149, 390, 1993.
140. **Purkerson, J. M. and Isakson, P. C.,** Interleukin-5 (IL-5) provides a signal that is required in addition to IL-4 for isotype switching to immunoglobulin (Ig)G_1 and IgE, *J. Exp. Med.,* 175, 973, 1992.
141. **Murray, P. D., McKenzie, D. T., Swain, S. L., and Kagnoff, M. F.,** Interleukin-5 and interleukin-4 produced by Peyer's patch T cells selectively enhance immunoglobulin A expression, *J. Immunol.,* 139, 2669, 1987.
142. **Kunimoto, D. Y., Harriman, G. R., and Stober, W.,** Regulation of IgA differentiation in CH12LX B cells by lymphokines. IL-4 induces membrane IgM-positive CH12LX cells to express membrane IgA and IL-5 induces membrane IgA-positive CH12LX cells to secrete IgA, *J. Immunol.,* 141, 713, 1988.
143. **Sonoda, E., Hitoshi, Y., Yamaguchi, N., Ishii, T., Tominaga, A., Araki, S., and Takatsu, K.,** Differential regulation of IgA production by TGF-β and IL-5. TGF-β induces surface IgA-positive cells bearing IL-5 receptor, whereas IL-5 promotes their survival and maturation into IgA-secreting cells, *Cell. Immunol.,* 140, 158, 1992.
144. **Sonoda, E., Matsumoto, R., Hitoshi, Y., Ishii, T., Sugimoto, M., Araki, S., Tominaga, A., Yamaguchi, N., and Takatsu, K.,** Transforming growth factor β induces IgA production and acts additively with interleukin-5 for IgA production, *J. Exp. Med.,* 170, 1415, 1989.

145. **Kim, P. H. and Kagnoff, M. F.,** Transforming growth factor-β1 is a costimulator for IgA production, *J. Immunol.,* 144, 3411, 1990.
146. **Tominaga, A., Takaki, S., Koyama, N., Katoh, S., Matsumoto, R., Migita, M., Hitoshi, Y., Hosoya, Y., Yamaguchi, S., Kanai, Y., Miyazaki, J.-I., Usuku, G., Yamamura, K.-I., and Takatsu, K.,** Transgenic mice expressing a B cell growth and differentiation factor gene (interleukin-5) develop eosinophilia and acute auto antibody production, *J. Exp. Med.,* 173, 429, 1991.
147. **Pene, J., Rousset, F., Briere, F., Chrétien, I., Wideman, J., Bonnefoy, J. Y., and DeVries, J. E.,** Interleukin-5 enhances interleukin-4-induced IgE production by normal human B cells. The role of soluble CD23 antigen, *Eur. J. Immunol.,* 18, 929, 1988.
148. **Ramsay, A. J. and Kohonen-Corish, M.,** Interleukin-5 expressed by a recombinant virus vector enhances specific mucosal IgA responses *in vivo, Eur. J. Immunol.,* 23, 3141, 1993.
149. **Schoenbeck, S., McKenzie, D. T., and Kagnoff, M. F.,** Interleukin-5 is a differentiation factor for IgA B cells, *Eur. J. Immunol.,* 19, 965, 1989.
150. **Durham, S. R., Ying, S., Varney, V. A., Jacobson, M. R., Sudderick, R. M., Mackay, I. S., Kay, A. B., and Hamid, Q. A.,** Cytokine messenger RNA expression for IL-3, IL-4, IL-5 and granulocyte/macrophage-colony-stimulating factor in the nasal mucosa after local allergen provocation: relationship to tissue eosinophilia, *J. Immunol.,* 148, 2390, 1992.
151. **Ying, S., Durham, S. R., Barkans, J., Masuyama, K., Jacobson, M., Rak, S., Löwhagen, O., Moqbel, R., Kay, A. B., and Hamid, Q. A.,** T cells are the principal source of interleukin-5 mRNA in allergen-induced rhinitis, *Am. J. Respir. Cell Mol. Biol.,* 9, 356, 1993.
152. **Bentley, A. M., Meng, Q., Robinson, D. S., Hamid, Q., Kay, A. B., and Durham, S. R.,** Increases in activated T lymphocytes, eosinophils, and cytokine mRNA expression for interleukin-5 and granulocyte/macrophage colony-stimulating factor in bronchial biopsies after allergen inhalation challenge in atopic asthmatics, *Am. J. Respir. Cell Mol. Biol.,* 8, 35, 1993.
153. **Clutterbuck, E., Shields, J. G., Gorden, J., Smith, S. H., Boyd, A., Callard, R. E., Campbell, H. D., Young, I. G., and Sanderson, C. J.,** Recombinant human interleukin-5 is an eosinophil differentiation factor but has no activity in standard human B cell growth factor assays, *Eur. J. Immunol.,* 17, 1743, 1987.
154. **Mahanty, S., and Nutman, T. B.,** The biology of interleukin-5 and its receptor, *Cancer Invest.,* 11, 624, 1993.
155. **Kaneko, M., Hitoshi, Y., Takatsu, K., and Matsumoto, S.,** Role of interleukin-5 in local accumulation of eosinophils in mouse allergic peritonitis, *Int. Arch. Allergy Appl. Immunol.,* 96, 41, 1991.
156. **Iwamoto, I., Tomoe, S., Tomioka, H., Takatsu, K., and Yoshida, S.,** Role of CD_4^+ T lymphocytes and interleukin-5 in antigen-induced eosinophil recruitment into the site of cutaneous late-phase reaction in mice, *J. Leukemia Biol.,* 52, 572, 1992.
157. **Nagai, H., Yamaguchi, S., Inagaki, N., Tsuruoka, N., Hitoshi, Y., and Takatsu, K.,** Effect of anti-IL-5 monoclonal antibody on allergic bronchial eosinophilia and airway hyperresponsiveness in mice, *Life Sci.,* 53, 243, 1993.
158. **Nakajima, H., Iwamoto, I., Tomoe, S., Matsumura, R., Tamioka, H., Takatsu, K., and Yoshida, S.,** CD_4^+ T-lymphocytes and interleukin-5 mediate antigen-induced eosinophil infiltration into the mouse trachea, *Am. Rev. Respir. Dis.,* 146, 374, 1992.
159. **Murali, D. S., Kumar, A., Choi, H., Bansal, N. K., Fink, J. N., and Kurup, V. P.,** *Aspergillus fumigatus* antigen induced eosinophilia in mice is abrogated by anti-IL-5 antibody, *J. Leukemia Biol.,* 53, 264, 1993.
160. **Dent, L. A., Strath, M., Mellor, A. L., and Sanderson, C. J.,** Eosinophilia in transgenic mice expressing interleukin-5, *J. Exp. Med.,* 172, 1425, 1990.
161. **Hitoshi, Y., Yamaguchi, N., Korenaga, M., Mita, S., Tominaga, A., and Takatsu, K.,** *In vivo* administration of antibody to murine IL-5 receptor inhibits eosinophilia of IL-5 transgenic mice, *Int. Immunol.,*3, 135, 1991.
162. **Iwama, T., Nagai, H., Suda, H., Tsuruoka, N., and Koda, A.,** Effect of murine recombinant interleukin-5 on the cell population in guinea pig airways, *Br. J. Pharmacol.,* 105, 19, 1992.
163. **Iwama, T., Nagai, H., and Koda, A.,** Effects of NZ-107 on airway inflammation and cell activation in guinea pigs, *J. Pharm. Pharmacol.,* 45, 286, 1993.
164. **Iwama, T., Nagai, H.,Tsuruoka, N., and Koda, A.,** Effect of murine recombinant interleukin-5 on bronchial reactivity in guinea pigs, *Clin. Exp. Allergy,* 23, 32, 1993.

165. **Nagai, H., Yamaguchi, S., Kitagaki, N., Tsuruoka, N., and Koda, A.,** Priming with murine recombinant interleukin-5 resulted in the augmentation of PAF-induced airway hyperresponsiveness to histamine in guinea pigs, *Life Sci.,* 52, 147, 1993.
166. **Van Oosterhout, A. J. M., Van Ark, I., Hofman, G., Savelkoul, M. F. J., and Nijkamp, F. P.,** Recombinant interleukin-5 induces *in vivo* airway hyperresponsiveness to histamine in guinea pigs, *Eur. J. Pharmacol.,* 236, 379, 1993.
167. **Lilly, C. M., Hall, A. E., Kobzik, L., and Drazen, J. M.,** Interleukin-5 tracheal injection results in airway eosinophilia, *Am. J. Respir. Crit. Care Med.,* 149, A768, 1994.
168. **Pretolani, M., Lefort, J., Leduc, D., and Vargaftig, B. B.,** Human recombinant interleukin-5 induces *in vitro* lung hyperresponsiveness to platelet-activating factor in the guinea pig, *Br. J. Pharmacol.,* 106, 677, 1992.
169. **Pretolani, M., Lefort, J., and Vargaftig, B. B.,** Inhibition by nedocromil sodium of recombinant human interleukin-5-induced lung hyperresponsiveness to platelet-activating factor in actively sensitized guinea pigs, *J. Allergy Clin. Immunol.,* 91, 809, 1993.
170. **Chand, N., Harrison, J. E., Rooney, S., Pillar, J., Jakubicki, R., Nolan, K., Diamantis, W., and Sofia, R. D.,** Anti-IL-5 monoclonal antibody inhibits allergic late phase bronchial eosinophilia in guinea pigs: a therapeutic approach, *Eur. J. Pharmacol.,* 211, 121, 1992.
171. **Van Oosterhout, A. J. M., Ladenius, A. R. C., Savelkoul, H. F. J., Van Ark, I., Delsman, K. C., and Nijkamap, F. P.,** Effect of anti-IL-5 and IL-5 on airway hyperreactivity and eosinophils in guinea pigs, *Am. Rev. Respir. Dis.,* 147, 548, 1993.
172. **Van Oosterhout, A. J. M., Van Ark, I., Folkerts, G., Van Der Linden, H. J., Savelkoul, H. F. J., Verheyen, A. K. C. P., and Nijkamp, F. P.,** Antibody to interleukin-5 inhibits virus-induced airway hyperresponsiveness to histamine in guinea pigs, *Am. J. Respir. Crit. Care Med.,* 151, 177, 1995.
173. **Jacoby, D. B., Gleich, G. J., and Fryer, A. D.,** Human eosinophil major basic protein is an endogenous allosteric antagonist at the inhibitory muscarinic M_2 receptor, *J. Clin. Invest.,* 91, 1314, 1993.
174. **Elbon, C. L., Jacoby, D. B., and Fryer, A. D.,** Pretreatment with an antibody to interleukin-5 prevents loss of pulmonary M_2 muscarinic receptor function in antigen-challenged guinea pigs, *Am. J. Respir. Cell Mol. Biol.,* 12, 320, 1995.
175. **Sher, A., Coffman, R. L., Hieny, S., and Cheever, A. W.,** Ablation of eosinophil and IgE responses with anti-IL-5 or anti-IL-4 antibodies fails to affect immunity against *Schistosoma mansoni* in the mouse, *J. Immunol.,* 145, 3911, 1990.
176. **Sher, A., Coffman, R. L., Hieny, S., Scott, P., and Cheever, A. W.,** Interleukin-5 is required for the blood and tissue eosinophilia but not granuloma formation induced by infection with *Schistosoma mansoni, Proc. Natl. Acad. Sci. U.S.A.,* 87, 61, 1990.
177. **Korenaga, M., Hitoshi, Y., Yamaguchi, N., Soto, Y., Takatsu, K., and Tada, I.,** The role of interleukin-5 in protective immunity to *Strongyloides venezuelensis* infection in mice, *Immunology,* 72, 502, 1991.
178. **Herndon, F. J. and Kayes, S. G.,** Depletion of eosinophils by anti-IL-5 monoclonal antibody treatment of mice infected with *Trichinella spiralis* does not alter parasite burden or immunologic resistance to reinfection, *J. Immunol.,* 149, 3642, 1992.
179. **Sasaki, O., Sugaya, H., Ishida, K., and Yoshimura, K.,** Ablation of eosinophils with anti-IL-5 antibody enhances the survival of intracranial worms of *Angiostrongylus cantonensis* in the mouse, *Parasitol. Immunol.,* 15, 349, 1993.
180. **Bass, D. A. and Szejda, P.,** Eosinophils versus neutrophils in host defense. Killing of newborn larvae of *Trichinella spiralis* by human granulocytes *in vitro, J. Clin. Invest.,* 64, 1415, 1979.
181. **Haman, K. H., Barker, R. L., Loegering, D. A., and Gleich, G. J.,** Comparative toxicity of purified human eosinophil granule proteins for newborn larvae of *Trichinella spiralis, J. Parasitol.,* 73, 523, 1987.
182. **Okudaira, H., Nogami, M., Matsuzaki, G., Dohi, M., Suko, M., Kasuya, S., and Takatsu, K.,** T cell dependent accumulation of eosinophils in the lung and its inhibition by monoclonal anti-interleukin-5, *Int. Arch. Allergy Appl. Immunol.,* 94, 171, 1991.
183. **Charcot, J. M. and Robin, C.,** Observation de leocythemie, *C. R. Mem. Soc. Biol.,* 5, 44, 1853.
184. **Leyden, E.,** Zur Kenntnis des bronchial-asthma, *Arch. Pathol. Anat.,* 54, 324, 1872.

185. **Sanerkin, N. G. and Evans, D. M. D.,** The sputum in bronchial asthma: pathogonomic patterns, *J. Pathol. Bact.,* 89, 535, 1965.
186. **Dunnill, M. S.,** The pathology of asthma, with special reference to changes in the bronchial mucosa, *J. Clin. Pathol.,* 13, 27, 1960.
187. **Cupps, T. R. and Fauci, A. S.,** Corticosteroid-mediated immunoregulation in man, *Immunol. Rev.,* 65, 133, 1982.
188. **Barach, A. Z., Bickerman, H. A., and Beck, G. J.,** Clinical and physiological studies on the use of metacortandracin in respiratory disease. I. Bronchial asthma, *Dis. Chest,* 27, 575, 1955.
189. **Gynn, A. A. and Michaels, L.,** Bronchial biopsy in chronic bronchitis and asthma, *Thorax,* 15, 142, 1960.
190. **Rolfe, F. G., Hughes, J. M., Armour, C. L., and Sewell, W. A.,** Inhibition of interleukin-5 gene expression by dexamethasone, *Immunology,* 77, 494, 1992.
191. **Alexander, A. G., Barnes, N. C., and Kay, A. B.,** Trial of cyclosporin in corticosteroid-dependent chronic severe asthma, *Lancet,* 339, 324, 1992.
192. **Adair, J. R.,** Engineering antibodies for therapy, *Immunol. Rev.,* 130, 5, 1992.
193. **LoBuglio, A. F., Mansoor, N. S., Lee, J., Khazaeli, M. B., Carrano, R., Holden, H., and Wheeler, R. H.,** Phase I trial of multiple large doses of murine monoclonal antibody CO17–1A, *J. Natl. Cancer Inst.,* 80, 932, 1988.
194. **Carrasquillo, J. A., Sugarbaker, P., Colcher, D., Reynolds, J. C., Esteban, J., Bryant, G., Keenan, A. M., Perentesis, P., Yokoyama, K., Simpson, D. E., Ferroni, P., Farkas, R., Schlom, J., and Larson, S. M.,** Radioimmunoscintigraphy of colon cancer with iodine-131-labeled B72.3 monoclonal antibody, *J. Nucl. Med.,* 29, 1022, 1988.
195. **Goldberg, D., Morel, P., Chatenoud, L., Boitard, C., Menkes, C. J., Bertoye, P. H., Revillard, J. P., and Bach, J. F.,** Immunological effects of high dose administration of anti-CD antibody in rheumatoid arthritis patients, *J. Autoimmunol.,* 4, 617, 1991.
196. **Hafler, D. A., Ritz, J., Schlossman, F., and Weiner, H. L.,** Anti-CD4 and anti-CD2 monoclonal antibody infusions in subjects with multiple sclerosis, *J. Immunol.,* 141, 131, 1988.
197. **Horneff, G., Winkler, T., Kalden, J. R., Emmrich, F., and Burmester, G. R.,** Human anti-mouse antibody response induced by anti-CD4 monoclonal antibody therapy in patients with rheumatoid arthritis, *Clin. Immunol. Immunopathol.,* 59, 89, 1991.
198. **Ledermann, J. A., Begent, R. H. J., Massof, C., Kelly, A. M. B., Adam, T., and Bagshawe, K. D.,** A phase-I study of repeated therapy with radiolabelled antibody to carcinoembryonic antigen using intermittent or continuous administration of cyclosporin A to suppress the immune response, *Int. J. Cancer,* 47, 659, 1991.
199. **Saleh, M. N., Khazaeli, M. B., Wheeler, R. H., Dropcho, E., Liu, T., Urist, M., Miller, D. M., Lawson, S., Dixon, P., Russel, C. H., and LoBuglio, A. F.,** Phase I trial of the murine monoclonal anti-GD2 antibody 14G2A in metastatic melanoma, *Cancer Res.,* 52, 4342, 1992.
200. **Scheinberg, D. A., Lovett, D., Chaitanya, R. D., Graham, M. C., Berman, E., Pentlow, K., Feirt, N., Finn, R. D., Clarkson, B. D., Gee, T. S., Larson, S. M., Oettgen, H. F., and Old, L. J.,** A phase I trial of monoclonal antibody M195 in acute myelogenous leukemia: specific bone marrow targeting and internalization of radionuclide, *J. Clin. Oncol.,* 9, 478, 1991.
201. **Elliott, M. J., Maini, R. N., Feldmann, M., Fox, A. L., Charles, P., Bijl, H., and Woody, J. N.,** Repeated therapy with monoclonal antibody to tumour necrosis factor A (CA2) in patients with rheumatoid arthritis, *Lancet,* 344, 1125, 1994.
202. **Isaacs, J. D., Watts, R. A., Hazelman, B. L., Hale, G., Keogan, M. T., Cobbold, S. P., and Waldmann, H.,** Humanized monoclonal antibody therapy for rheumatoid arthritis, *Lancet,* 340, 748, 1992.
203. **Khazaeli, M. B., Saleh, M. N., Liu, T. P., Meredith, R. G., Wheeler, R. H., Baker, T. S., King, D., Secher, D., Allen, L., Rogers, K., Colcher, D., Schlom, J., Shochat, D., and LoBuglio, A. F.,** Pharmacokinetics and immune response of I-chimeric mouse/human B72.3 (human Y4) monoclonal antibody in humans, *Cancer Res.,* 51, 5461, 1991.
204. **LoBuglio, A. F., Wheeler, R. H., Trang, J., Haynes, A., Rogers, K., Harvey, E. B., Sun, L., Ghrayeb, J., and Khazaeli, M. B.,** Mouse/human chimeric monoclonal antibody in man. Kinetics and immune response, *Immunology,* 86, 4220, 1989.

205. **Moreland, L. W., Pratt, P. W., Sanders, M. E., and Koopman, W. J.,** Experience with a chimeric monoclonal anti-CD4 antibody in the treatment of refractory rheumatoid arthritis, *Clin. Exp. Rheumatol.,* Suppl. 8, 153, 1993.
206. **Saleh, M. N., Khazaeli, M. B., Wheeler, R. H., Allen, L., Tilden, A. B., Grizzle, W., Reisfeld, R. A., Yu, A. L., Gillies, S. D., and LoBuglio, A. F.,** Phase I trial of the chimeric anti-GD2 monoclonal antibody CH14.18 in patients with malignant melanoma, *Hum. Antibod. Hybridomas,* 3, 19, 1992.
207. **Brown, P. S., Jr., Parenteau, G. L., Frederick, M. D., Garsia, R. J., Goldman, C. K., Bukowski, M. A., Junghans, R. P., Queen, C., Hakimi, J., Benjamin, W. R., Clark, R. E., and Waldmann, T. A.,** Anti-Tac H, a humanized antibody to the interleukin-2 receptor, prolongs primate cardiac allograft survival, *Immunology,* 88, 2663, 1991.
208. **Hakimi, J., Chizzonite, R., Luke, D. R., Familletti, P. C., Bailon, P., Kondas, J. A., Pilson, R. S., Lin, P., Weber, D. V., Spence, C., Mondini, L. J., Tsien, W. H., Levin, J. L., Gallati, V. H., Korn, L., Waldmann, T. A., Queen, C., and Benjamin, W. R.,** Reduced immunogenicity and improved pharmacokinetics of humanized anti-Tac in cynomolgus monkeys, *J. Immunol.,* 147, 1352, 1991.
209. **Tinubu, S. A., Hakimi, J., Kondas, J. A. et al.,** Humanized antibody directed to the IL-2 receptor β-chain prolongs primate cardiac allograft survival, *J. Immunol.,* 153, 4330, 1994.
210. **Hale, G., Clark, M. R., Marcus, R., Winter, G., Dyer, M. J. S., Phillips, J. M., Riechmann, L., and Waldmann, H.,** Remission induction in non-hodgkin lymphoma with reshaped human monoclonal antibody campath-1H, *Lancet,* 2, 1394, 1988.
211. **Caron, P. C., Jurcic, J. G., Scott, A. M., Finn, R. D., Chaitanya, R. D., Martin, C., Graham, M. C., Jureidini, I. M., Sgouros, G., Tyson, D., Old, L. J., Larson, S. M., and Scheinberg, D. A.,** A phase 1B trial of humanized monoclonal antibody M195 (anti-CD33) myeloid leukemia: specific targeting without immunogenicity, *Blood,* 83, 1760, 1994.
212. **Elliott, M. J., Maini, R. N., Feldmann, M., Kalden, J. R., Antoni, C., Smolen, J. S., Leeb, B., Breedveld, F. C., Macfarlane, J. D., Bijl, H., and Woody, J. N.,** Randomized double-blind comparison of chimeric monoclonal antibody to tumour necrosis factor A (CA2) versus placebo in rheumatoid arthritis, *Lancet,* 344, 1105, 1994.
213. **Kabat, E. A., Wu, T. T., Perry, H. M., Gottesman, K. S., and Foeller, C.,** Tabulation and analysis of amino acid and nucleic acid sequences of precursors, V-regions, C-regions, J-chain, T-cell receptors for antigen, T-cell surface antigens, β_2-microglobulins major histocompatibility antigens, Thy-1, complement, C-reactive protein, thymopoietin, integrins, post-gamma globulin, α_2-macroglobulins, and other related proteins, in *Sequences of Proteins of Immunological Interest,* 5th ed., Vol. II, NIH Publication No. 91–3242, U.S. Department of Health and Human Services, Washington, D.C., 1991.
214. **Banchereau, J. and Rousset, F.,** Growing human B lymphocytes in the CD40 system, *Nature,* 353, 678, 1991.
215. **Yamamoto, K. R.,** Steroid receptor regulated transcription of specific genes and gene networks, *Annu. Rev. Genet.,* 19, 209, 1985.
216. **Ho, S. N., Boyer, S. H., Schreiber, S. L., Danishefsky, S. J., and Crabtree, G. R.,** Specific inhibition of formation of transcription complexes by a calicheamicin oligosaccharide. A paradigm for the development of transcriptional antagonists, *Proc. Natl. Acad. Sci. U.S.A.,* 91, 9203, 1994.
217. **Dinarello, C. A. and Thompson, R. C.,** Blocking IL-1: interleukin-1 receptor antagonist *in vivo* and *in vitro, Immunol. Today,* 12, 404, 1991.
218. **Aversa, G., Punnonen, J., Cocks, B. G., de Waal Malefyt, R., Vega, F., Zurawski, S. M., Zurawski, G., and de Vries, J.,** An interleukin-4 (IL-4) mutant protein inhibits both IL-4 or IL-13-induced human immunoglobulin G4 (IgG4) and IgE synthesis and B cell proliferation. Support for a common component shared by IL-4 and IL-13 receptors, *J. Exp. Med.,* 178, 2213, 1993.
219. **Welham, M. J., Dechert, U., Leslie, K. B., Jirik, F., and Schrader, J. W.,** Interleukin (IL)-3 and granulocyte/macrophage colony-stimulating factor, but not IL-4, induce tyrosine phosphorylation, activation, and association of SHPTP2 with Grb2 and phosphatidylinositol 3′-kinase, *J. Biol. Chem.,* 269, 23764, 1994.
220. **Anasetti, C., Hansen, J. A., Waldmann, T., Binger, M. H., Mounld, D., Satoh, H., Appelbaum, F. R., Deeg, H. G., Doney, K., Martin, P. J., Nash, R., Sullivan, K. M., Witherspoon, R. P., and Storb, R.,** Treatment of acute graft versus host disease with a humanized monoclonal antibody specific for the IL-2 receptor, *Blood,* 73, 2, 1994.

Modulation of Cytokine Signal Transduction

Chapter 12

SIGNAL TRANSDUCTION MECHANISMS OF INTERLEUKIN-1 AND TUMOR NECROSIS FACTOR

Jeremy Saklatvala

CONTENTS

0-8493-8381-1/96/$0.00+$.50

I. INTRODUCTION

Interleukin-1 (IL-1) and tumor necrosis factor (TNF) are major inflammatory cytokines. They were first isolated over a decade ago and there is copious literature on their biological actions (also reviewed in Chapters 1–3).[1,2] Their discovery was a landmark in understanding the inflammatory process because it revealed that substances produced by cells activated at sites of injury or infection not only had important local actions on vascular endothelium and connective tissue but also caused systemic features of inflammation. They induced the liver to make acute phase proteins, they acted on the brain to cause fever, they stimulated the hypothalamopituitary-adrenal axis, and they also regulated myeloproliferation and release of leukocytes from the bone marrow. We now know that some of these physiological actions are partly indirect: the acute phase response in the liver may be due more to circulating IL-6 than IL-1 or TNF, and changes in the bone marrow may depend more upon colony stimulating factors. However, IL-1 and TNF are major stimulants of production of these other cytokines by connective tissue cells. Thus, IL-1 and TNF can be thought of as coordinating and galvanizing the inflammatory process.

There is good reason to believe that interfering with the action of these cytokines may have anti-inflammatory effects. Attempts to blockade their receptors with small molecules have been unsuccessful and current approaches involve neutralizing them with antibodies, or soluble receptors (as reviewed in Chapters 9 and 10). In theory, it should also be possible to interfere with the action of the cytokines at the level of their intracellular signaling systems. The attractions of this approach are (1) there may be enzymes involved which are susceptible to inhibition by easily administered small molecules, (2) since IL-1 and TNF share actions and signaling mechanisms both could be inhibited simultaneously, and (3) if there are multiple pathways it may be possible to selectively block undesirable effects. Not enough is yet known about the signaling mechanisms to identify therapeutic targets, but the rate at which knowledge of cell signaling is advancing makes it likely that a rational scientific approach to interfering with cytokine action may not be far away. The purpose of this review is to collect and assess the significance of current ideas regarding how IL-1 and TNF signal.

Although structurally quite different, IL-1 and TNF elicit very similar responses whose nature varies with cell type. Many cells respond to them as a sole stimulus, for example, fibroblasts, vascular endothelial cells, chondrocytes, osteoblasts, and hepatocytes. Lymphocytes, on the other hand, only respond to them in association with additional stimuli such as antigens, or other cytokines. IL-1 and TNF are often called 'costimulants' of lymphocytes. There are subtle differences between the biological effects of IL-1 and TNF. IL-1 lacks TNF's marked cytotoxic effects.[3] IL-1 strongly stimulates prostaglandin synthesis in fibroblasts, while TNF, in comparison, has a relatively weak effect.[4] IL-1 stimulates resorption of cartilage and bone; TNF has similar effects, but on cartilage, for example, its effects are slower and weaker.[5] In both cartilage and bone,[6] a ten times higher dose of TNF is required to match IL-1.

Underlying many of the biological effects of the cytokines are changes in gene expression. There is a voluminous literature on both cytokines increasing synthesis of many proteins. Some of these are summarized in Table 1. The earliest increases in mRNA occur around 15 to 30 min after stimulation. Such 'early' genes include DNA-binding proteins such as c-Jun, NFIL-6, and NAK1, together with cytokines such as IL-6, IL-8, and gro. Other genes are more slowly induced; for example, mRNAs of collagenase, stromelysin, and cyclooxygenase increase after some hours of stimulation. Such delayed responses are usually prevented by inhibitors of protein synthesis, so it is thought that

TABLE 1
Proteins Induced by IL-1 and TNF

Transcription Factors	Secreted Proteins
c-Jun[7,8]	IL-2[21]
c-Fos[9]	IL-6[24]
c-Myc[9]	IL-8[25,26]
Interferon response factor-1[10]	Interferon β[40]
NAK1[11]	G-CSF, GM-CSF[27]
NFIL-6[12]	gro[28,29]
	NGF[30]
	PDGF A chain[31]
	Procollagenase[8,32]
	Prostromelysin[33]
	Plasminogen activator[34,35]
Intracellular Metabolism	**Other**
Glucose transporter[13]	A20[36]
Phospholipase A2 (group II)[14]	B94[37]
Cytosolic phospholipase A2[15]	
Cyclooxgenases I[16] and II[17,18]	
Inducible NO synthase[19]	
Hsp 70[20]	
Surface Membrane Proteins	
IL-2 receptor[21]	
E-selectin (ELAM)[22]	
I-CAM-1[23]	

induction of these late genes depends upon changes in expression of early genes coding for transcription factors such as c-Jun.

Although much is known about the physiological effects of the cytokines, and at the cellular level there is a large catalogue of changes in gene expression, much less is known about the signaling mechanisms that trigger cell responses.

II. INTERLEUKIN-1 RECEPTORS

Two IL-1 receptors have been identified (Figure 1). The type I is widely distributed. It has been characterized on lymphocytes, connective tissue cells, hepatocytes, and epidermal cells, and may be ubiquitous.[38–42] Some IL-1-sensitive lines express only a few hundred of these receptors per cell; fibroblasts may carry up to 5000 or more. Occupancy of relatively few receptors is sufficient for signaling. The receptor[43,44] is a member of the immunoglobulin superfamily and is unrelated to receptors for other cytokines or growth factors. The human protein comprises 552 amino acids.[44] It has a single membrane-spanning domain. The extracellular portion contains 302 amino acids organized in three immunoglobulin domains. This is glycosylated so that the protein has a molecular mass of about 80 kDa. The intracellular portion consists of 213 amino acids whose sequence gives no clue to its function, although it is essential for transducing a signal.[45] (The IL-1 receptor is also discussed in Chapter 9.)

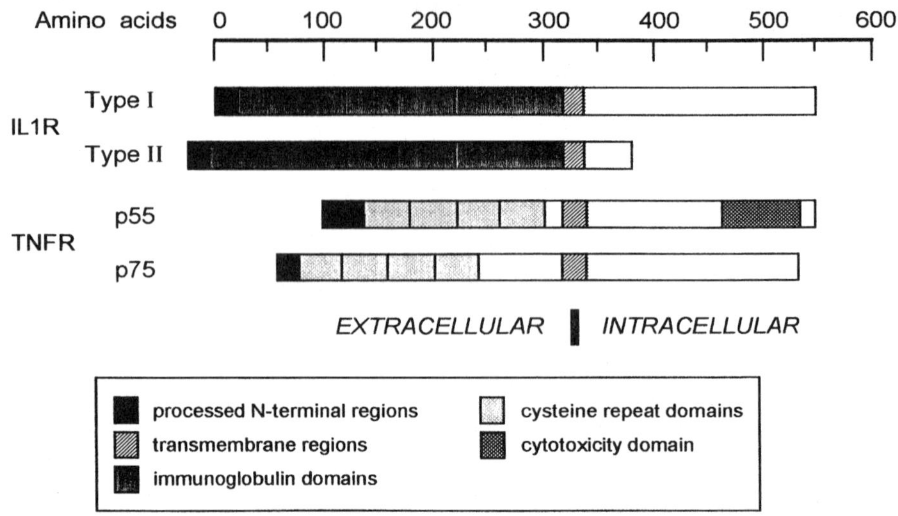

FIGURE 1. IL-1 and TNF receptors.

Monocytes, B lymphocytes, and neutrophils express a smaller receptor of about 60 kDa, called type II.[46] Its extracellular portion also consists of three immunoglobulin domains and is about 30% homologous to that of the type I receptor. It has a single membrane-spanning region but its cytoplasmic portion has only 29 amino acids. So far, no convincing intracellular signaling function has been found for this molecule. All the known effects of IL-1 that can be detected in cell culture appear to be mediated by the type I receptor.[47,48] A chimeric receptor made by joining cDNA encoding the extracellular and transmembrane parts of type II to that encoding the type I cytoplasmic portion had signaling functions similar to type I.[49] The type II receptor is shed by cells, particularly when they are activated. Its expression on polymorphonuclear leukocytes is increased by IL-4.[50] The receptor might act as a decoy for IL-1 and thus could underlie the ability of IL-4 to inhibit IL-1. The murine IL-1 receptors show about 70% homology in amino acid sequence to their human counterparts.[43]

Both receptor types bind the IL-1α or β forms or the IL-1 receptor antagonist protein with similarly high affinity ($K_{diss} \sim 10^{-10}$ to 10^{-11} M) and are probably univalent. The ligated type I receptor on porcine fibroblasts internalizes with $t^1/_2 \sim 10$ min,[51] but slower internalization was found in human fibroblasts.[52] Some IL-1 ligand maybe transported to the nucleus, possibly in association with receptor,[53,54] but the physiological significance of this is not understood. Neither receptor has been definitely shown to associate with any protein other than its cytokine, although porcine IL-1α could be chemically crosslinked to an unknown 100-kDa protein when liganded to the type I receptor on murine cells.[55]

III. TNF RECEPTORS

Two types of TNF receptor have been characterized and cloned (Figure 1). The extracellular parts of both were originally purified as TNF-binding proteins from urine.[56] A binding protein was also purified from the serum of cancer patients and an intact receptor was isolated from placenta.[58] Sequencing of these enabled the cloning of the cDNA for the smaller p55 receptor. The human receptor is translated as a protein of 455 amino acids and the N-terminal 40 amino acids are removed to form the mature protein. There is a single transmembrane domain and a cytoplasmic portion of 221 amino acids. A second TNF receptor, p75, was cloned at about the same time as the p55.[59] Its cDNA codes for a protein of 461 amino acids that is only about 20% homologous to the p55 in its extracellular region and shows no homology in its intracellular domain of 174 amino acids. The equivalent murine TNF receptors are 64 and 62%, respectively, homologous to their human counterparts.[60]

The TNF receptors belong to a superfamily of cellular and viral proteins characterized by six cysteine repeat motifs.[61] Two members are soluble pox viral products that bind TNF and contribute to virulence by neutralizing the cytokine. The others are cellular membrane proteins. They have single membrane-spanning portions and include CD30, CD40, CD27, 4–1BB, Fas, OX40, and the low-affinity nerve growth factor receptor. The ligands for these (apart from TNF and lymphotoxin) are also transmembrane proteins, but with their C-termini oriented extracellularly and their N-termini intracellularly. Some of these ligands are only expressed on activated lymphocytes and the subsequent ligand/receptor interactions provide costimulatory function.[61]

Both types of TNF receptor bind either TNFα or β (lymphotoxin) with similar affinity ($K_{diss} \sim 10^{-10}$ M). They are generally coexpressed and are widely distributed. Early studies with agonistic antibodies suggested that the p55 receptor mediated inflammatory effects and cytotoxity.[62] The intracellular part of the p55 receptor has a region of about 80 amino acids close to the C-terminus that is necessary for TNF cytotoxicity.[63] This region is only 28% homologous with the corresponding region of Fas. More recent work with an agonistic antiserum to p75 suggests that this larger receptor mediates the costimulatory proliferation of T cells treated with a mitogen such as Con A.[64] TNF mutants have been made showing high specificity for either receptor:[65,66] one p55-specific protein induced umbilical vein endothelial cells to produce IL-6 and GM-CSF. It also activated the transcription factor nuclear factor (NF) κB and was cytotoxic to a transformed cell line.[65] Another p55-specific mutant was found to have a similar set of effects but had less than 1% of the ability of the wild type to activate polymorphonuclear leukocytes.[66] Mutants specific for p75 did not have such effects, and did not modify the responses induced by the p55-specific ligands.[65,66] These findings all suggest that the receptors have different signaling functions. TNF exists as a trimer and can crosslink receptors. The fact that antibodies to the receptors can act as agonists suggests that receptor aggregation may provide the signal. It has also been suggested that p75 may pass TNF to p55 and so recruit the cytokine for signaling.[67]

IV. DO IL-1 AND TNF SIGNAL VIA KNOWN SECOND MESSENGERS?

There is a large volume of literature on the effects of IL-1 and TNF on known second messenger molecules. Many of these observations are unconfirmed and have been made in a variety of cell types. In some cases there are also contradictory reports, so the significance of the changes is hard to assess.

V. cAMP

Both IL-1[68,69] and TNF[70] have been found to cause rapid elevation of cAMP in fibroblasts and lymphocytic lines. There are also reports that IL-1 has no effect on cAMP levels.[71,72] One laboratory which had observed increases in cAMP in response to IL-1 also found that the cytokine activated a pertussis-sensitive GTP-binding protein in fibroblasts and a murine B cell line (70Z3) and some of its effects were inhibited by the toxin.[73] There are other reports that IL-1[74] and TNF[75] increase activity of GTP-binding proteins in membranes of EL4 (a murine thymoma) and HL60 (human monocytoid) cells, respectively, but not all have found such changes.[72] In the light of the structure of IL-1 and TNF receptors, they are unlikely to be coupled to proteins of the Gi or Gs types. It is possible they are affecting nucleotide binding of smaller G proteins, but as yet there is no evidence for or against the involvement of such molecules in IL-1 or TNF signaling. The ability of pertussis toxin to inhibit some effects of IL-1 resides in the binding oligomer of the toxin and is not dependent on the presence of the ADP-ribosyl transferase subunit.[76]

VI. PHOSPHOLIPID BREAKDOWN

Some of the actions of IL-1 and TNF resemble those of phorbol ester, which works by activating protein kinase (PK) C in cells. Physiologically, PKC is activated by its association with diacylglycerol, particularly when there is a rise in cytosolic free Ca^{++}. This occurs when ligands activate phosphoinositide-specific phospholipase C causing formation of inositol phosphates and diacylglycerol. Inositol trisphosphates trigger increases in intracellular free Ca^{++}. Generally, IL-1 and TNF have been found not to affect intracellular free Ca^{++},[77,78] although there have been reports that both increase phosphoinositide hydrolysis.[79–81]

Some forms of ligand-activated phospholipase C hydrolyze phosphatidyl choline, providing a mechanism for production of diacylglycerol without affecting intracellular Ca^{++}. IL-1 was found to cause hydrolysis of phosphatidyl choline in Jurkat cells, with accompanying elevation of diacylglycerol.[82] TNF may cause a similar effect.[83] IL-1 may also activate a nuclear phospholipase C.[84] Whether the cytokines cause activation of PKC is unclear. Translocation of PKC to the cell membrane has been observed in cells activated by IL-1[85] and by TNF,[86] but generally, the results of cell biological experiments in which PKC is inhibited or downregulated have strongly suggested that the cytokines do not signal via PKC.[87–89]

The positive reports of the breakdown of phosphoinositides and phosphatidyl choline by IL-1 are balanced by a negative one.[77]

VII. SPHINGOMYELIN HYDROLYSIS

The phospholipid sphingomyelin is a component of cellular membranes whose breakdown has been implicated in cell signaling. Originally, Hannun and colleagues detected hydrolysis of sphingomyelin to phosphoryl choline and ceramide in HL60 cells induced to differentiate with vitamin D_3.[90] The differentiating effect of the vitamin could be reproduced by treating the cells with sphingomyelinase (sphingomyelin is located predominantly in the outer face of the cell membrane), which suggested that sphingomyelin breakdown might serve as a signal.[91] Subsequently, TNFα and interferon-γ, whose effects on HL60 cells are similar to those of vitamin D_3, were also shown to cause sphingomyelin hydrolysis.[92] Ceramide superficially resembles diacylglycerol and might conceivably activate a protein kinase in a manner analogous to the way diacylglycerol activates protein kinase C. Kolesnick and colleagues have found such a ceramide-activatable kinase in a cell membrane fraction whose activity was increased in TNF-treated cells.[93,94] The enzyme had a MAP-kinase-like specificity in that it phosphorylated a synthetic peptide of the EGF receptor containing threonine T669. IL-1 was also found to increase sphingomyelin breakdown and the activity of the ceramide-activatable kinase, suggesting that it too might be using this signaling mechanism.[95] The kinase has been characterized as a 97-kDa protein and was able to phosphorylate another MAP kinase substrate, myelin basic protein.[96] Other observations suggesting possible signaling functions for sphingomyelin breakdown are that both sphingomyelinase and a membrane-permeating ceramide activate MAP kinase (see below) in HL60 cells,[97] sphingosine directly modulates kinase activities in cell extracts,[98] and ceramide activates heterotrimeric protein phosphatase 2A and an unidentified protein phosphatase.[99]

As a result of these findings, it has been proposed that IL-1 and TNF activate sphingomyelinase in the cell membrane and the ceramide generated serves as a second messenger to activate a protein kinase.[100,101] This hypothesis is supported by observations that sphingomyelinase treatment of cells mimics the effect of the cytokines: it augments IL-2 production,[96] induces prostaglandin synthesis,[102] increases cholesteryl ester synthesis,[103] and activates the latent transcription factor NF-κB.[104] Introduction of ceramide into permeabilized cells also activates NF-κB.[104] Treatment of cells with membrane-permeating C_2 ceramide reproduced the growth inhibitory effect of TNF, but did not of itself activate NF-κB, although it did enhance the activation by TNF.[105] Other investigators have not been able to induce NF-κB with ceramide, and have found that TNF induces NFκB in Jurkat cells without detectably altering the level of ceramide.[106]

In view of the conflicting nature of some of these reports, the significance of sphingomyelin hydrolysis for cytokine signaling is uncertain. A number of agents induce it that do not have the range of effects of IL-1 and TNF, but accumulating evidence points to it being involved in the response to cytokines.

VIII. IL-1 AND TNF ALTER PROTEIN KINASE ACTIVITY IN CELLS

The major general mechanism by which extracellular signals control intracellular processes is through protein phosphorylation. Surface receptors may themselves be protein kinases or they may associate with kinases, or they may generate second messengers which in turn are responsible for activating protein kinases.

One approach to find out how IL-1 and TNF signal has been to identify proteins rapidly phosphorylated in cells upon stimulation. The first protein that was shown to undergo an increase in phosphorylation upon IL-1 or TNF stimulation was the EGF receptor.[87] The phosphorylation involved serine and threonine, not tyrosine, and was associated with a loss of affinity of the receptor for EGF. Activation of PKC has a functionally similar effect on the EGF receptor, but the action of IL-1 was clearly not mediated by that enzyme. The effect of IL-1 was not sensitive to the PKC inhibitor, staurosporine, and it did not involve phosphorylation of T654, a site for PKC.[88]

A second protein that was identified as undergoing a strong increase in phosphorylation upon IL-1 and TNF stimulation was the small heat shock protein, hsp27.[107–110] Like the EGF receptor phosphorylation, the effect of the cytokines was not mediated by PKC.[89]

IX. IL-1 AND TNF ACTIVATE p42 MAP KINASE

One protein kinase implicated in phosphorylation of the EGF receptor is the p42 mitogen activated protein (MAP) kinase. This enzyme phosphorylates serine and threonine residues in the context of proline, a typical motif being PLTP, which is found around T669, one of the phosphorylation sites of the EGF receptor.[111] Bird and colleagues[112]

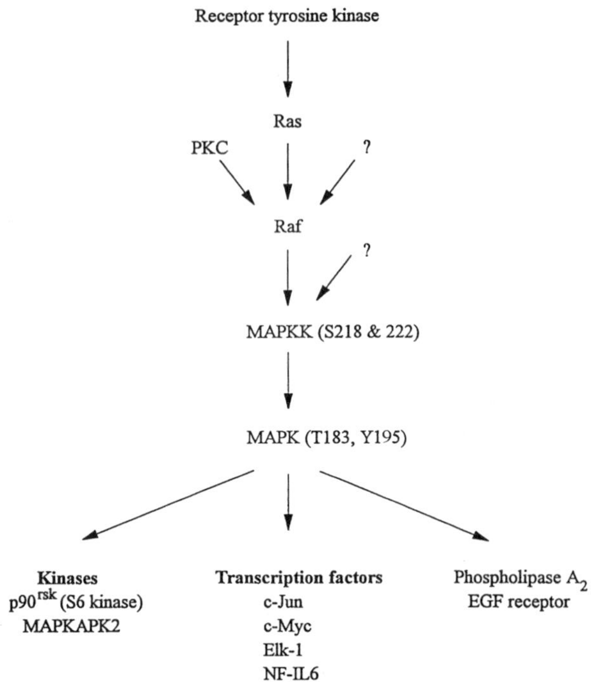

FIGURE 2. The p42 MAP kinase cascade.

found that extracts of IL-1-stimulated κB cells (and other cell types) contained increased amounts of kinase activity towards a synthetic peptide based on the sequence around T669. The enzyme was identified as p42/p44 MAP kinase by use of antibodies and chromatography. p42/p44 MAP kinase activation was also noted in fibroblasts stimulated with IL-1 or TNF in other laboratories.[113–115]

There are two ubiquitous MAP kinases of 42 and 44 kDa.[111] In human cells the p42 is the more abundant enzyme. They have very similar properties and are commonly assayed on myelin basic protein (MBP) or high molecular weight microtubule-associated protein 2. Some of their potential physiological substrates have been identified (Figure 2): these include the EGF receptor, at least two other protein kinases, several transcription factors (c-Jun, Elk-1, c-myc, NF-IL-6, and ATF-2), cytosolic phospholipase A_2, and the high molecular weight microtubule-associated protein 2. The phosphorylation of c-Jun occurs on Ser63 and 73 in the transactivation domain and increases its transcriptional activity.[116] The phosphorylations of Elk-1[117] and NF-IL-6[118] are also activating. The phosphorylation of cytosolic phospholipase A_2 increases its activity. The p42/p44 MAP kinases are activated by a very wide range of stimuli:[111] virtually all growth factors, insulin, phorbol ester, and many cytokines[120,121] including IL-1 and TNF.[115] Activation (Figure 2) occurs upon phosphorylation of threonine and tyrosine residues that are separated by a single glutamic acid residue. This is brought about by a dual-specificity 44-kDa enzyme, MAP kinase kinase,[122,123] which is activated by IL-1 or TNF in fibroblasts.[124] MAP kinase kinase is itself activated by phosphorylation upon serine residues.[125,126] This can be carried out by the kinase Raf-1,[127] which may be responsible for activation of the MAP kinase cascade by growth factors via the Ras pathway.[128] Protein kinase C can directly phosphorylate (and activate) Raf-1,[129] so Raf-1 represents a point on which the two systems converge to activate MAP kinase (Figure 2). How TNF and IL-1 activate MAP kinase kinase in fibroblasts remains to be established. It is possible that there are other regulated activators of MAP kinase kinase besides Raf-1.[130]

X. OTHER MAP KINASE-LIKE ENZYMES ACTIVATED BY IL-1 AND TNF: T669 KINASE AND p54 MAP KINASE

The p42/p44 MAP kinases are activated by IL-1 and TNF in fibroblasts.[115] On investigating KB cells, a transformed epidermoid cell line sensitive to the cytokines, we were unable to detect activation of these enzymes by the cytokines. However, another enzyme that phosphorylated the EGF receptor peptide T669 was induced. This was extensively purified and shown, like MAP kinase, to depend on tyrosine and serine/threonine phosphorylation for its activity. It did not react with p42/p44 MAP kinase antibodies, had very little activity on MBP, and could not be reactivated with MAP kinase kinase. The pure enzyme was identified on SDS gel electrophoresis as a phosphotyrosine-containing protein of 45 kDa.[131] This enzyme is distinct from p42/p44 MAP kinases, but it may be related to the p54 MAP kinases which are distant relatives of p42/p44.[132–135] p54 MAP kinase was discovered as an enzyme that phosphorylated microtubule-associated protein 2 in livers of rats injected with cycloheximide. There are three forms (α, β, γ) having 90% homology; they may exist in truncated forms and are expressed at low levels in many tissues.[134] They have similar substrate specificity to p42/p44 MAP kinase,[136] but strongly phosphorylate c-Jun[134,135] while only poorly phosphorylating MBP.[132] They are not activated by MAPKK, and in the activation site the intervening amino acid residue is proline, not glutamate.[134,135] Interestingly, p54 MAP kinase is induced by stressful stimuli such as heat, UV irradiation, or protein synthesis inhibitors, and by TNF, but only weakly or not at all by phorbol ester or mitogens.[134,135]

XI. hsp27 KINASE CASCADE REVEALS ANOTHER KINASE ACTIVATED BY DUAL PHOSPHORYLATION: p40

Hsp27 is a major cytosolic protein whose synthesis is strongly induced upon heat shock or other stressful stimuli.[137] It exists as large aggregates of 20 or more molecules, but its function is still not clear. It is a molecular chaperon[138] and an inhibitor of actin polymerization.[139] It protects actin from depolymerization by cytochalasin D, so it may be involved in regulation of the cytoskeleton.[140] When expressed at a high level it increases the ability of cells to withstand stressful stimuli.[141] Phosphorylation of human hsp27 occurs on serine (S)82, and to a lesser extent on S78 and S15.[142,143] This reduces its aggregation[144] and might promote its interaction with other proteins. Phosphorylation increases most markedly in response to IL-1 or TNF,[89] but also to some extent in response to many stimuli, including growth factors, phorbol ester,[145] and heat shock itself.[143]

Following the observation that IL-1 and TNF increased hsp27 phosphorylation in cells, it was shown that cytokine-stimulated cell extracts contained increased activity of an hsp27 kinase.[145] Purification of this enzyme from IL-1-stimulated KB cells revealed that it lies at the end of yet another kinase cascade (Figure 3).[147] The 50-kDa hsp27 kinase was activated by a 40-kDa protein by phosphorylation of threonine and serine residues. The 40-kDa protein was a kinase phosphorylated upon tyrosine and threonine residues; its activity was dependent upon both types of phosphorylation, suggesting a relationship with MAP kinases. It had little or no activity towards MBP, T669 peptide, or c-Jun and did not react with antibodies to other MAP kinases. The IL-1-stimulated κB cells also

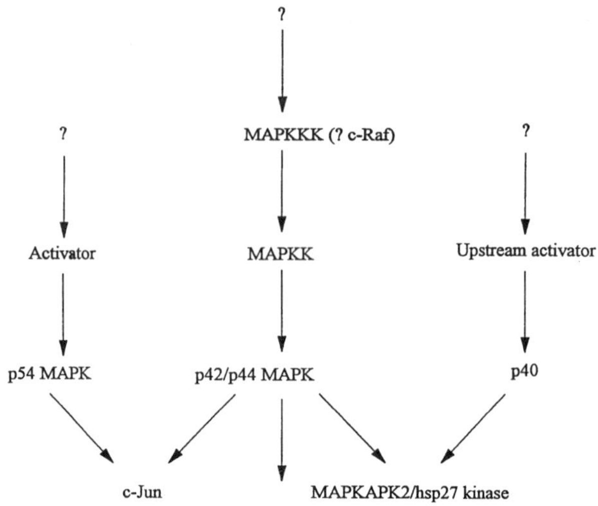

FIGURE 3. MAP kinase cascades activated by IL1 and TNF.

contained an activator of p40 which was a putative dual-specificity kinase distinct from MAP kinase kinase. It had a molecular mass of about 40 kDa and could be inactivated by protein phosphatase 2A. This hsp27 kinase cascade is similar to the MAP kinase cascade (Figure 3). However, p40 could not be activated by recombinant MAP kinase kinase, and the activator of p40 could not activate p42 MAP kinase. The hsp27 kinase itself, however, could be activated by MAP kinase as well as p40. It may be identical, or closely similar, to the MAP kinase activated protein kinase 2,[148–151] which is of similar size and readily phosphorylates hsp27. If so, then MAPKAPK 2 can be activated by two distinct cascades.

The relationship of p40 to the MAP kinases is not established. Whether the hsp phosphorylation is regulated by the MAP kinase or p40 cascades in fibroblasts is also unknown.

Thus depending upon cell type, IL-1 (and TNF) have been found to activate several enzymes regulated by tyrosine and threonine phosphorylation: p42/p44 MAP kinases, the p40 that phosphorylates hsp27 kinase, and p54 MAP kinase which is probably similar to the T669 kinase. These cascades are summarized in Figure 3.

XII. β CASEIN KINASE

The protein kinases activated by IL-1 and TNF appear mostly to be members of cascades of the MAP kinase type: all are regulated by phosphorylation. However, one kinase has been identified that is not, itself, regulated directly by phosphorylation, and so is obviously not part of a phosphorylation cascade. This enzyme phosphorylates β casein upon serine, and was first found by chance in fibroblasts stimulated with IL-1 or TNF,[115] but a similar enzyme was activated by the cytokines in vascular endothelial cells, chondrocytes, and Hela cells.[152] Among a very wide range of cell stimulants including serum, phorbol ester, heat stress, and sphingomyelinase, only IL-1 or TNF induced this kinase.[152] Such selective induction compared strikingly with the nonspecific activation of p42/p44 MAP kinase by many ligands, and suggests that there may be pathways for IL-1 and TNF that are relatively specific and which could be therapeutic targets. Its activity was unaffected by treatment with either protein phosphatase 2A or phosphotyrosine phosphatase.[152] A third interesting feature of the enzyme is that its specificity is different from those of the major protein kinase groups (see Reference 153 for review). Many kinases phosphorylate serine or threonine residues in motifs containing basic residues, usually arginine; examples of this are PKA and PKC. The MAP kinases and the cyclin-dependent kinases phosphorylate serine or threonine residues in the context of proline, while casein kinases I and II phosphorylate motifs characterized by the presence of glutamate or aspartate acidic residues. The β casein kinase does not phosphorylate the sites used by casein kinase I and II, and a peptide containing the site does not contain arginine or proline motifs.[152]

XIII. ACTIVATION OF TRANSCRIPTION FACTORS BY IL-1 AND TNF

Many of the genes listed in Table 1 are regulated by protein transcription factors binding to AP1 (phorbol ester response element), nuclear factor (NF)IL-6, and κB sites in their promoters. Some tentative connections can be made between these transcription factors and protein kinase activation. The principal known transcriptional activators binding to AP1 sites are homodimers of c-Jun, or its heterodimer with c-Fos. The cytokines strongly activate transcription of *c-jun,*[7,8] while their effects on *c-fos* are variable. IL-1 induced *c-fos* in one murine T cell line[7] but not in another.[154] Transcription of *c-jun* is controlled through a divergent AP1 site that is recognized by c-Jun homodimers[155] or heterodimers of c-Jun/ATF-2.[156] The activity of both complexes is stimulated by phosphorylation of c-Jun at residues S63 and S73, which can be carried out by p42/p44 or p54 MAP kinases.[116] Such phosphorylation of a preexisting pool of c-Jun leading to further transcription and translation of c-Jun might be the mechanism for cytokine activation of genes regulated by AP1 sites. However, analysis of AP-1 binding proteins activated in HepG2 cells by stimulation with IL-1 for 15 min suggested that they were immunologically distinct from c-Jun and c-Fos proteins.[157]

Similar mechanisms may be involved in regulation of NF-IL-6.[12,158] Synthesis of this transcription factor is induced by IL-1 and many other stimuli including IL-6 itself, and its ability to promote transcription is strongly increased by phosphorylation of T235 by MAP kinase.[158] Transcriptional efficacy of NF-IL-6 is also increased by phosphorylation of S105 by protein kinase C.[159] However, it has yet to be established that any preexisting pool of NF-IL-6 is activated upon cell stimulation.

One transcription factor that is preexistent is NF-κB. It is strongly activated by IL-1 and TNF,[160] as well as by many other stimuli including bacterial lipopolysaccharides, UV irradiation, oxygen radicals, viruses, and phorbol ester.[161] It comprises heterodimers of 50- and 65-kDa proteins which are bound to an inhibitor protein, IκB, in the cytoplasm. Upon cell activation IκB is rapidly degraded and the nuclear factor moves to the nucleus and binds to DNA κB sites.[162–166] Phosphorylation of the IκB has been implicated in the response,[165,166] possibly by inactivating the inhibitor and dissociating it from the nuclear factor, or targeting it for degradation. It is not yet possible to link NF-κB activation to a specific phosphorylation by cytokine-activated kinases.

XIV. CONCLUSIONS

IL-1 and TNF activate multiple protein kinases. Presumably, this rapidly diversifies the signal to produce pleiotropic responses. Three enzymes regulated by dual phosphorylation of tyrosine and threonine have been found. These MAP kinase-like enzymes are p42 MAP kinase itself, p54 MAP kinase, and p40 which phosphorylates hsp27 kinase. Another enzyme, T669 kinase, is probably related to p54 MAP kinase. These enzymes seem to have different activation pathways and it is likely that the pattern of kinases activated will vary with the cell type. One enzyme, β casein kinase, whose activation appears relatively specific for IL-1 and TNF, has yet to be shown to be part of these phosphorylation cascades.

The links between receptors and kinases remain to be established, but there is evidence sphingomyelin breakdown may be involved in MAP kinase activation.

REFERENCES

1. **Dinarello, C. A.,** Interleukin-1 and interleukin-1 antagonism, *Blood,* 77, 1627, 1991.
2. **Beutler, B.,** Tumour necrosis factors, in *The Molecules and Their Emerging Roles in Medicine,* Raven Press, New York, 1992.
3. **Onozaki, K., Matsushima, K., Aggarwal, B. B., and Oppenheim, J. J.,** Human interleukin-1 is a cytocidal factor for several tumor lines, *J. Immunol.,* 135, 3962, 1985.
4. **O'Neill, L. A. J. and Lewis, G. P.,** Interleukin-1 potentiates bradykinin and TNFα-induced PGE_2 release, *Eur. J. Pharmacol.,* 166, 131, 1989.
5. **Saklatvala, J.,** Tumour necrosis factor α stimulates resorption and inhibits synthesis of proteoglycan in cartilage, *Nature,* 322, 547, 1986.
6. **Bertolini, D. R., Nedwin, G. E., Bringman, T. S., Smith, D. D., and Mundy, G. R.,** Stimulation of bone resorption and inhibition of bone formation *in vitro* by human tumour necrosis factors, *Nature,* 319, 516, 1986.
7. **Muegge, K., Williams, J. M., Kant, J., Karin, M., Chiu, R., Schmidt, A., Siebenlist, U., Young, H. A., and Durum, S.,** Interleukin-1 costimulatory activity on the interleukin-2 promoter via AP-1, *Science,* 246, 249, 1989.
8. **Brenner, O. A., O'Hara, M., Angel, P., Chojkier, M., and Karin, M.,** Prolonged activation of *jun* and collagenase genes by tumour necrosis factor α, *Nature,* 337, 661, 1989.
9. **Kovacs, E. J., Oppenheim, J. J., and Young, H. A.,** Induction of c-fos and c-myc expression in T lymphocytes after treatment with recombinant interleukin-1-α, *J. Immunol.,* 137, 3649, 1986.
10. **Fujita, T., Reis, L. F. R., Watanabe, N., Kimura, Y., Taniguchi, T., and Vilček, J.,** Induction of the transcription factor IRF-1 and interferon-β mRNAs by cytokines and activators of second-messenger pathways, *Proc. Natl. Acad. Sci. U.S.A.,* 86, 9936, 1989.
11. **Rangnekar, V. V., Waheed, S., and Rangnekar, V. M.,** Interleukin-1 inducible tumor growth arrest is characterized by activation of cell type-specific "early" gene expression programmes, *J. Biol. Chem.,* 267, 6240, 1992.
12. **Akira, S., Isshiki, H., Sugita, T., Tanabe, O., Kinoshita, S., Nishio, Y., Nakajuna, T., Hirano, T., and Kishimoto, T.,** A nuclear factor for IL-6 expression (NF-IL-6) is a member of the C/EBP family, *EMBO J.,* 9, 1897, 1990.
13. **Bird, T. A., Davies, A., Baldwin, S. A., and Saklatvala, J.,** Interleukin-1 stimulates hexose transport in fibroblasts by increasing the expression of glucose transporters, *J. Biol. Chem.,* 265, 13578, 1990.
14. **Kerr, J. S., Stevens, T. M., Davis, G. L., McLaughlin, J. A., and Harris, R. R.,** Effects of recombinant interleukin-1 β on phospholipase A_2 activity, phospholipase A_2 mRNA and eicosanoid formation in rabbit chondrocytes, *Biochem. Biophys. Res. Commun.,* 165, 1079, 1989.
15. **Liu, L.-L., Liu, A. Y., and DeWitt, D. L.,** Interleukin-1 α induces the accumulation of cytosolic phospholipase A_2 and the release of prostaglandin E_2 in human fibroblasts, *J. Biol. Chem.,* 267, 23451, 1992.
16. **Maier, J. A. M., Hla, T., and Maciag, T.,** Cyclo-oxygenase is an immediate early gene induced by interleukin-1 in human endothelial cells, *J. Biol. Chem.,* 265, 10805, 1990.

17. **Ristimäki, A., Garfinkel, S., Wessendorf, J., Maciag, T., and Hla, T.,** Induction of cyclooxygenase-2 by interleukin-1α, *J. Biol. Chem.,* 269, 11769, 1994.
18. **Jones, D. A., Carlton, D. P., McIntyre, T. M., Zimmerman, G. A., and Prescott, S. M.,** Molecular cloning of human prostaglandin endoperoxide synthase type II and demonstration of expression in response to cytokines, *J. Biol. Chem.,* 268, 9049, 1993.
19. **Geller, D. A., Nussler, A. K., Di Silvio, M., Lowenstein, C. J., Shapiro, R. A., Wang, S. C., Simmons, R. L., and Billiar, T. R.,** Cytokines, endotoxin and glucocorticoids regulate the expression of inducible nitric oxide synthase in hepatocytes, *Proc. Natl. Acad. Sci. U.S.A.,* 90, 522, 1993.
20. **Cruz, T. F., Kandel, R. A., and Brown, I. A.,** Interleukin-1 induces the expression of a heat-shock gene in chondrocytes, *Biochem. J.,* 277, 327, 1991.
21. **Muegge, K. and Durum, S. K.,** Cytokines and transcription factors, *Cytokine,* 2, 1, 1990.
22. **Pober, J. S., Bevilacqua, M. P., Mendricke, D. L., Lapierre, L. A., Fiers, W., and Gimbrone, M. A., Jr.,** Two distinct monokines, interleukin-1 and tumour necrosis factor each independently induce biosynthesis and transient expression of the same antigen on the surface of cultured human vascular endothelial cells, *J. Immunol.,* 136, 1680, 1986.
23. **Dustin, M. L., Rothlein, R., Bhan, A. K., Dinarello, C. A., and Springer, T. A.,** Induction by IL-1 and interferon-γ: tissue distribution, biochemistry and function of a natural adherence molecule (ICAM-1), *J. Immunol.,* 137, 245, 1986.
24. **Van Damme, J., Opdenakker, G., Simpson, R. J., Rubira, M. R., Cayphas, S., Vink, A., Billiau, A., and Van Snick, J.,** Identification of the human 26 kDa protein, interferon β_2, as a B cell hybridoma/plasmacytoma growth factor induced by interleukin-1 and tumour necrosis factor, *J. Exp. Med.,* 165, 914, 1987.
25. **Matsushima, K., Morishita, K., Yoshimura, T., Lavu, S., Kobayishi, Y., Lew, W., Appella, E., Kung, H. F., Leonard, E. J., and Oppenheim, J. J.,** Molecular cloning of human monocyte-derived neutrophil chemotactic factor (MDNCF) and the induction of MDNCF mRNA by interleukin-1 and tumour necrosis factor, *J. Exp. Med.,* 167, 1883, 1988.
26. **Sica, A., Matsushima, K., Van Damme, J., Wang, J. M., Polentarutti, N., Dejana, E., Colotta, F., and Mantovani, A.,** IL-1 transcriptionally activates the neutrophil chemotactic factor/IL-8 gene in endothelial cells, *Immunology,* 69, 548, 1990.
27. **Seelentag, W. K., Mermod, J.-J., Montesano, R., and Vassali, P.,** Additive effects of interleukin-1 and tumour necrosis factor-α on the accumulation of the three granulocyte and macrophage colony-stimulating factor mRNAs in human endothelial cells, *EMBO J.,* 6, 2261, 1987.
28. **Anisowicz, A., Zajschowski, D., Stenman, G., and Sager, R.,** Functional diversity of *gro* gene expression in human fibroblasts and mammary epithelial cells, *Proc. Natl. Acad. Sci. U.S.A.,* 85, 9645, 1988.
29. **Wen, D., Rowland, D. A., and Derynck, R.,** Expression and secretion of gro/MGSA by stimulated human endothelial cells, *EMBO J.,* 8, 1761, 1989.
30. **Lindholm, D., Heumann, R., Hengerer, B., and Thoenen, A.,** Interleukin-1 increases stability and transcription of mRNA encoding nerve growth factor in cultured rat fibroblasts, *J. Biol. Chem.,* 263, 16348, 1988.
31. **Raines, E. W., Dower, S. K., and Ross, R.,** Interleukin-1 mitogenic activity for fibroblasts and smooth muscle cells is due to PDGF-AA, *Science,* 243, 393, 1989.
32. **Stephenson, M. L., Goldring, M. B., Birkhead, J. R., Krane, S. M., Rahmsdorf, H. J., and Angel, P.,** Stimulation of procollagenase synthesis parallels increases in cellular procollagenase mRNA in human articular chondrocytes exposed to recombinant interleukin-1 β or phorbol ester, *Biochem. Biophys. Res. Commun.,* 144, 583, 1987.
33. **Saus, J., Quinones, S., Otani, Y., Nagase, H., Harris, E. D., Jr., and Kurkinen, M.,** The complete primary structure of human matrix metalloproteinase 3 (MMP-3), identity with stromelysin, *J. Biol. Chem.,* 263, 6742, 1988.
34. **Leizer, R., Clarris, B. J., Ash, P. E., Van Damme, J., Saklatvala, J., and Hamilton, J. A.,** Interleukin-1β and interleukin-1α stimulate the plasminogen activator activity and prostaglandin E_2 levels of human synovial cells, *Arthritis Rheum.,* 30, 562, 1987.
35. **Bunning, R. A. D., Crawford, A., Richardson, H. J., Opdenakker, G., Van Damme, J., and Russell, R. G. G.,** Interleukin-1 preferentially stimulates the production of tissue type plasminogen activator by human articular chondrocytes, *Biochim. Biophys. Acta,* 924, 473, 1987.

36. **Opipari, A. W. J., Hu, H. M., Yabkowitz, R., and Dixit, V. M.,** The A20 zinc finger protein protects cells from tumor necrosis factor cytotoxicity, *J. Biol. Chem.,* 267, 12424, 1992.
37. **Wolf, F. W., Sarma, V., Seldin, M., Drake, S., Suchard, S. J., Shao, H., O'Shea, K. S., and Dixit, V. M.,** B94, a primary response gene inducible by tumor necrosis factor-α is expressed in hematopoietic tissues and the sperm acrosome, *J. Biol. Chem.,* 269, 3633, 1994.
38. **Dower, S. K., Kronheim, S. R., March, C. J., Conlon, P. J., Hopp, T. P., Gillis, S., and Urdal, D.,** Detection and characterization of high affinity plasma membrane receptors for human interleukin-1, *J. Exp. Med.,* 162, 505, 1985.
39. **Bird, T. A. and Saklatvala, J.,** Identification of a common class of high affinity receptors for both types of porcine interleukin-1 on connective tissue cells, *Nature,* 324, 263, 1986.
40. **Dower, S. K., Kronheim, S., Hopp, T. P., Cantrell, M., Deeley, M., Henney, C. S., Gillis, S., and Urdal, D. L.,** The cell surface receptors for interleukin-1α and interleukin-1β are identical, *Nature,* 324, 266, 1986.
41. **Dower, S. K., Call, S. M., Gillis, S., and Urdal, D. L.,** Similarity between the interleukin-1 receptors on a murine T-lymphoma cell line and a murine fibroblast line, *Proc. Natl. Acad. Sci. U.S.A.,* 83, 1060, 1986.
42. **Bird, T. A., Gearing, A. J. H., and Saklatvala, J.,** Murine interleukin-1 receptor: direct identification by ligand blotting and purification to homogeneity of an interleukin-1-binding glycoprotein, *J. Biol. Chem.,* 263, 12063, 1988.
43. **Sims, J. C., March, C. J., Cosman, D., Widmer, M. B., MacDonald, H. R., McMahan, C. J., Grubin, C. G., Wignall, J. M., Jackson, J. T., Call, S. M., Friend, D., Alpert, A. R., Gillis, S., Urdal, D. L., and Dower, S. K.,** cDNA expression cloning of the IL-1 receptor, a member of the immunoglobulin superfamily, *Science,* 241, 585, 1988.
44. **Sims, J. E., Acres, R. B., Grubin, C. E., McMahan, C. J., Wignall, J. M., March, C. J., and Dower, S. K.,** Cloning the interleukin-1 receptor from human T cells, *Proc. Natl. Acad. Sci. U.S.A.,* 86, 8946, 1989.
45. **Curtis, B. M., Gallis, B., Overall, R. W., McMahan, C. J., DeRoos, P., Ireland, R., Eisenman, J., Dower, S. K., and Sims, J. E.,** T-cell interleukin-1 receptor cDNA expressed in chinese hamster ovary cells regulates functional responses to interleukin-1, *Proc. Natl. Acad. Sci. U.S.A.,* 86, 3045, 1989.
46. **McMahan, C. J., Slack, J. L., Mosley, B., et al.,** A novel IL-1 receptor cloned from B cells by mammalian expression is expressed in many cell types, *EMBO J.,* 10, 2821, 1991.
47. **Stylianou, E., O'Neill, L. A. J., Rawlinson, L., Edbrooke, M. R., Woo, P., and Saklatvala, J.,** Interleukin-1 induces NF-KB through its type I but not its type II receptor in lymphocytes, *J. Biol. Chem.,* 267, 15836, 1992.
48. **Sims, J. E., Gayle, M. A., Slack, J. L., Alderson, M. R., Bird, T. A., Giri, J. G., Colotta, F., Re, F., Mantovani, A., Shanebeck, K., Grabstein, K. H., and Dower, S. K.,** Interleukin-1 signalling occurs exclusively via the type I receptor, *Proc. Natl. Acad. Sci. U.S.A.,* 90, 6155, 1993.
49. **Heguy, A., Balbari, C. T., Censini, S., Ghiara, P., and Telford, J. L.,** A chimeric type II/type I interleukin-1 receptor can mediate interleukin-1 induction of gene expression in T cells, *J. Biol. Chem.,* 268, 10490, 1993.
50. **Colotta, F., Re, F., Muzio, M., Bertini, R., Polentarutti, N., Sironi, M., Giri, J. G., Dower, S. K., Sims, J. E., and Mantovani, A.,** Interleukin-1 type II receptor: a decoy target for IL-1 that is regulated by IL-4, *Science,* 261, 472, 1993.
51. **Bird, T. A. and Saklatvala, J.,** Studies on the fate of receptor-bound ^{125}I-IL-1 in porcine synovial fibroblasts, *J. Immunol.,* 139, 92, 1987.
52. **Quarnstrom, E. E., Page, R. C., Gillis, S., and Dower, S. K.,** Binding, internalisation and intracellular localization of interleukin-1β in human diploid fibroblasts, *J. Biol. Chem.,* 263, 8261, 1988.
53. **Mizel, S. B., Kilian, P. L., Lewis, J. C., Paganelli, K. A., and Chizzonite, R. A.,** The interleukin-1 receptor. Dynamics of interleukin-1 binding and internalization in T cells and fibroblasts, *J. Immunol.,* 138, 2906, 1987.
54. **Curtis, B. M., Widmer, M. B., Deroos, P., and Quarnstrom, E. E.,** IL-1 and its receptor are translocated to the nucleus, *J. Immunol.,* 144, 1295, 1990.

55. **Bird, T. A., Gearing, A. J., and Saklatvala, J.,** Murine interleukin-1 receptor: differences in binding properties between fibroblastic and thymoma cells and evidence for a two chain receptor model, *FEBS Lett.,* 225, 121, 1987.
56. **Engelmann, H., Novick, D., and Wallach, D.,** Two tumor necrosis factor-binding proteins purified from human urine, *J. Biol. Chem.,* 265, 1531, 1990.
57. **Schall, T. J., Lewis, M., Koller, K. J., Lee, A., Rice, G. C., Wong, G. H. W., Gatanaga, T., Granger, G. A., Lentz, R., Raab, A., Kohr, W. J., and Goeddel, D. V.,** Molecular cloning and expression of a receptor for human tumor necrosis factor, *Cell,* 61, 361, 1990.
58. **Loetscher, H., Pan, Y.-C. E., Lahm, H.-W., Gentz, R., Brockhaus, M., Tabuchi, H., and Lesslauer, W.,** Molecular cloning and expression of the human 55 kDa tumor necrosis factor receptor, *Cell,* 61, 351, 1990.
59. **Smith, C. A., Davis, T., Anderson, D., Solam, L., Beckmann, M. P., Jerzy, R., Dower, S. K., Cosman, D., and Goodwin, R. G.,** A receptor for tumor necrosis factor defines an unusual family of cellular and viral proteins, *Science,* 248, 1019, 1990.
60. **Lewis, M., Tartaglia, L. A., Lee, A., Bennett, G. L., Rice, G. C., Wong, G. H. W., Chen, E. Y., and Goeddel, D. V.,** Cloning and expression of cDNAs for two distinct murine tumour necrosis factor receptors demonstrate one receptor is species specific, *Proc. Natl. Acad. Sci. U.S.A.,* 88, 2836, 1991.
61. **Smith, C. A., Farrah, T., and Goodwin, R. G.,** The TNF receptor superfamily of cellular and viral proteins: activation, costimulation and death, *Science,* 248, 1019, 1990.
62. **Engelmann, H., Holtmann, H., Brakebusch, C., Avni, S. Y., Sarov, I., Nophar, Y., Hadas, E., Leitner, O., and Wallach, D.,** Antibodies to a soluble form of a tumor necrosis factor receptor have TNF-like activity, *J. Biol. Chem.,* 265, 14497, 1990.
63. **Tartaglia, L. A., Ayres, T. M., Wong, G. H. W., and Goeddel, D. V.,** A novel domain within the 55 kDa TNF receptor signals cell death, *Cell,* 74, 845, 1993.
64. **Tartaglia, L. A., Goeddel, D. V., Reynolds, C., Figari, I. S., Welser, R. F., Fendly, B. M., and Palladino, M. A., Jr.,** Stimulation of human T-cell proliferation by specific activation of the 75 kDa tumor necrosis factor receptor, *J. Immunol.,* 151, 4637, 1993.
65. **Loetscher, H., Stueber, D., Banner, D., Mackay, F., and Lesslauer, W.,** Human tumor necrosis factor α mutants with exclusive specificity for the 55 kDa or 75 kDa receptors, *J. Biol. Chem.,* 268, 26350, 1993.
66. **Barbara, J. A. J., Smith, W. B., Gamble, J. R., Van Ostade, X., Vandenabeele, P., Tavernier, J., Fiers, W., Vada, M. A., and Lopez, A. F.,** Dissociation of TNF-α cytotoxic and proinflammatory activities by p55 receptor and p75 receptor selective mutants, *EMBO J.,* 13, 843, 1994.
67. **Tartaglia, L. A., Pennica, D., and Goeddel, D. V.,** Ligand passing: the 75-kDa tumor necrosis factor (TNF) receptor recruits TNF for signalling by the 55-kDa receptor, *J. Biol. Chem.,* 268, 18542, 1993.
68. **Shirakawa, F., Yamashita, U., Chedid, M., and Mizel, S. B.,** Cyclic AMP-an intracellular second messenger for interleukin-1, *Proc. Natl. Acad. Sci. U.S.A.,* 85, 8201, 1988.
69. **Munoz, E., Beutner, C. L., Zubiaga, A., and Huber, B. T.,** IL-1 activates two separate signal transduction pathways in T helper type II cells, *J. Immunol.,* 144, 964, 1990.
70. **Zhang, Y., Lin, J.-X., and Vilček, J.,** Synthesis of interleukin-6 in human fibroblasts is triggered by an increase in intracellular cyclic AMP, *J. Biol. Chem.,* 263, 6177, 1988.
71. **Shiroo, M. and Matsushima, K.,** Enhanced phosphorylation of 65 and 75 kDa proteins by tumor necrosis factor and interleukin-1 in human peripheral blood mononuclear cells, *Cytokine,* 2, 13, 1990.
72. **Ray, K., Thompson, N., Kennard, N., Rollins, P., Grenfell, S., Witham, S., Smithers, N., and Solari, R.,** Investigation of guanine-nucleotide-binding protein involvement and regulation of cyclic AMP metabolism in interleukin-1 signal transduction, *Biochem. J.,* 282, 59, 1992.
73. **Chedid, M., Shirakawa, F., Naylor, O., and Mizel, S.,** Signal transduction pathway for IL-2. Involvement of a pertussis-toxin-sensitive GTP-binding protein in the activation of adenylate cyclase, *J. Immunol.,* 142, 4301, 1989.
74. **O'Neill, L. A. J., Bird, T. A., Gearing, A. J. H., and Saklatvala, J.,** Interleukin-1 signal transduction. Increased GTP binding and hydrolysis in membranes of a murine thymoma line EL4, *J. Biol. Chem.,* 265, 3146, 1992.

75. **Imamura, K., Sherman, M. L., Spriggs, D., and Kufe, D.,** Effect of tumor necrosis factor on GTP binding and GTP-ase activity in HL60 and L929 cells, *J. Biol. Chem.,* 263, 10247, 1988.
76. **O'Neill, L. A. J., Ikebe, T., Sarsfield, S., and Saklatvala, J.,** The binding sub-unit of pertussis toxin inhibits IL-1 induction of IL-2 and prostaglandin production, *J. Immunol.,* 148, 474, 1992.
77. **Abraham, R. T., Ho, S. N., Barna, T. J., and Mckean, D. J.,** Transmembrane signalling during interleukin-1-dependent T cell activation, *J. Biol. Chem.,* 262, 2719, 1987.
78. **Pacifici, R., Civitelli, R., Rifas, L., Halstead, L., and Avioli, L. V.,** Does interleukin-1 affect intracellular calcium in osteoblast-like cells (UMR-106)?, *J. Bone Miner. Res.,* 3, 107, 1988.
79. **Wijelath, E. S., Kardasz, A. M., Drummond, R., and Watson, J.,** Interleukin-1 induced phospholipid breakdown in murine macrophages: a possible mechanism of receptor activation, *Biochem. Biophys. Res. Commun.,* 1552, 92, 1988.
80. **Rapuano, B. E. and Bockman, R. S.,** Tumor necrosis factor α stimulates phosphatidylinositol breakdown by phospholipase C to coordinately increase the levels of diacylglycerol, free arachidonic acid and prostaglandins in an osteoblast line (MC 3T3-E1), *Biochim. Biophys. Acta,* 1091, 374, 1991.
81. **Rotondo, D., Earl, C. R. A., McIntosh, G., McIntosh, F. S., Hepburn, A., Milton, A. S., and Davidson, J.,** The role of inositol lipids in the activation of monocytes by interleukin-1 and bacterial endotoxin, *FEMS Microbiol. Immunol.,* 105, 249, 1992.
82. **Rosoff, P. M., Savage, N., and Dinarello, C. A.,** Interleukin-1 stimulates diacylglycerol production in T lymphocytes by a novel mechanism, *Cell,* 54, 73, 1988.
83. **Schütze, S., Berkovic, D., Tomsung, O., Unger, C., and Kronke, M.,** Tumor necrosis factor induces rapid production of 1′,2′-diacylglycerol by a phosphatidylcholine-specific phosphidipase C, *J. Exp. Med.,* 174, 975, 1991.
84. **Marmiroli, S., Ognibene, A., Barelloni, A., Cinti, C., Cocco, L., and Maraldi, N. M.,** Interleukin-1α stimulates nuclear phospholipase C in human osteosarcoma Sa OS-2 cells, *J. Biol. Chem.,* 269, 13, 1994.
85. **Ostrowski, J., Meier, K. E., Stanton, T. H., Smith, L. L., and Bomsztyk, K.,** Interferon γ and interleukin-1α induce transient translocation of protein kinase C activity to membranes in a B lymphoid line, *J. Biol. Chem.,* 263, 13786, 1988.
86. **Schutze, S., Nottrolt, S., Pfizenmaier, K., and Kronke, M.,** Tumor necrosis factor signal transduction. Cell-type-specific activation and translocation of protein kinase C, *J. Immunol.,* 144, 2604, 1990.
87. **Bird, T. A. and Saklatvala, J.,** IL-1 and TNF transmodulate epidermal growth factor receptors by a protein kinase C-independent mechanism, *J. Immunol.,* 142, 126, 1989.
88. **Bird, T. A. and Saklatvala, J.,** Down modulation of epidermal growth factor receptor affinity in fibroblasts treated with interleukin-1 or tumor necrosis factor is associated with phosphorylation at a site other than threonine 654, *J. Biol. Chem.,* 265, 235, 1990.
89. **Saklatvala, J., Kaur, P., and Guesdon, F.,** Phosphorylation of the small heat shock protein is regulated by interleukin-1, tumour necrosis factor, growth factors, bradykinin and ATP, *Biochem. J.,* 277, 634, 1991.
90. **Okazaki, T., Bielawska, A., Bell, R. M., and Hannun, Y. A.,** Role of ceramide as a lipid mediator of 1α 25-dihydroxy vitamin D_3-induced HL-60 cell differentiation, *J. Biol. Chem.,* 265, 15823, 1990.
91. **Okazaki, T., Bell, R. M., and Hannun, Y. A.,** Sphingomyelin turnover induced by vitamin D_3 in HL-60 cells. Role in cell differentiation, *J. Biol. Chem.,* 264, 19076, 1991.
92. **Kim, M.-Y., Linardic, C., Obeid, L., and Hannun, Y.,** Identification of sphingomyelin turnover as an effector mechanism for the action of tumor necrosis factor α and γ-interferon, *J. Biol. Chem.,* 266, 484, 1991.
93. **Mathias, S., Dressler, K. A., and Kolesnick, R. N.,** Characterization of a ceramide-activated protein kinase: stimulation by tumour necrosis factor α, *Proc. Natl. Acad. Sci. U.S.A.,* 88, 10009, 1991.
94. **Dressler, K. A., Mathias, S., and Kolesnick, R. N.,** Tumor necrosis factor α activates the sphingomyelin signal transduction pathway in a cell-free system, *Science,* 255, 1715, 1992.
95. **Mathias, S., Younes, A., Kan, C.-C., Orlow, I., Joseph, C., and Kolesnick, R. N.,** Activation of the sphingomyelin signalling pathway in intact EL4 cells in a cell-free system by IL-1β, *Science,* 259, 519, 1993.

96. **Liu, J., Mathias, T., Yang, Z., and Kolesnick, R. N.,** Renaturation and tumor necrosis factor α stimulation of a 97 kDa ceramide-activated protein kinase, *J. Biol. Chem.,* 269, 3047, 1994.
97. **Rames, M. A., Kolesnick, R. N., and Golde, D. W.,** Sphingomyelinase and ceramide activate mitogen-activated protein kinase in myeloid HL60 cells, *J. Biol. Chem.,* 268, 14572, 1993.
98. **Puskareva, M. Y., Khan, W. A., Alessenko, A. V., Sahyoun, N., and Hannun, Y.,** Sphingosine activation of protein kinases in Jurkat T cells, *J. Biol. Chem.,* 267, 15246, 1992.
99. **Dobrowsky, R. T., Kamibayashi, C., Mumby, M. C., and Hannun, Y.,** Ceramide activates heterotrimeric protein phosphatase 2A, *J. Biol. Chem.,* 268, 15523, 1993.
100. **Kolesnick, R. and Golde, D. W.,** The sphingomyelin pathway in tumor necrosis factor and interleukin-1 signalling, *Cell,* 77, 325, 1994.
101. **Hannun, Y. A.,** The sphingomyelin cycle and the second messenger function of ceramide, *J. Biol. Chem.,* 269, 3125, 1994.
102. **Ballou, L. R., Chao, C. P., Holness, M. A., Barker, S. C., and Raghow, R.,** Interleukin-1-mediated PGE_2 production and sphingomyelin metabolism, *J. Biol. Chem.,* 267, 20044, 1992.
103. **Chatterjee, S.,** Neutral sphingomyelinase action stimulates signal transduction of tumor necrosis factor α in the synthesis of cholesteryl esters in human fibroblasts, *J. Biol. Chem.,* 269, 879, 1994.
104. **Schütze, S., Potthoff, K., Machleidt, T., Berkovic, D., Weigmann, K., and Kronke, M.,** TNF activates NF-KB by phosphatidylcholine-specific phospholipase C-induced 'acidic' sphingomyelin breakdown, *Cell,* 71, 765, 1992.
105. **Dhaito, G., Obeid, L. M., and Hannun, Y. A.,** Tumor necrosis factor-α signal transduction through ceramide. Dissociation of growth inhibitory effects of TNFα from activation of nuclear factor KB, *J. Biol. Chem.,* 268, 17762, 1993.
106. **Betts, J. C., Agranoff, A. B., Nabel, G. J., and Shayman, J. A.,** Dissociation of endogenous cellular ceramide from NF-KB activation, *J. Biol. Chem.,* 269, 8455, 1994.
107. **Kaur, P. and Saklatvala, J.,** Interleukin-1 and tumor necrosis factor increase phosphorylation of fibroblast proteins, *FEBS Lett.,* 241, 6, 1988.
108. **Kaur, P., Welch, W., and Saklatvala, J.,** Interleukin-1 and tumour necrosis factor increase phosphorylation of the small heat shock protein, *FEBS Lett.,* 258, 269, 1989.
109. **Schütze, S., Scheurich, P., Pfizenmaier, K., and Kronke, M.,** Tumor necrosis factor signal transduction. Tissue specific serine phosphorylation of a 26 kDa cytosolic protein, *J. Biol. Chem.,* 264, 3562, 1989.
110. **Robaye, B., Hepburn, A., Lecocq, R., Fiers, W., Boeynaems, J. M., and Dumont, J. E.,** Tumor necrosis factor α induces the phosphorylation of 28 kDa stress proteins in endothelial cells: possible role in protection against cytotoxicity?, *Biochem. Biophys. Res. Commun.,* 163, 301, 1989.
111. **Davis, R. J.,** The mitogen-activated protein kinase signal transduction pathway, *J. Biol. Chem.,* 268, 14553, 1993.
112. **Bird, T. A., Sleath, P. R., DeRoos, P. C., Dower, S. K., and Virca, G. D.,** Interleukin-1 represents a new modality for the activation of extracellular signal-regulated kinases/microtubule-associated protein-2 kinases, *J. Biol. Chem.,* 266, 22661, 1991.
113. **Guy, G. R., Chua, S. P., Wong, N. S., Ng, S. B., and Tan, Y. H.,** Interleukin-1 and tumor necrosis factor activate common multiple protein kinases in human fibroblasts, *J. Biol. Chem.,* 266, 14,343, 1991.
114. **Van Lint, J., Agostinis, P., Vandevoorde, V., Haegeman, G., Fiers, W., Merlevede, W., and Vandenheede, J. R.,** Tumour necrosis factor stimulates multiple serine/threonine protein kinases in Swiss 3T3 and L929 cells, *J. Biol. Chem.,* 267, 25916, 1992.
115. **Guesdon, F., Freshney, N., Waller, R. J., Rawlinson, L., and Saklatvala, J.,** Interleukin-1 and tumour necrosis factor stimulate two novel protein kinases that phosphorylate the heat shock protein hsp27 and β casein, *J. Biol. Chem.,* 68, 4236, 1993.
116. **Pulverer, B. J., Kyriakis, J. M., Avruch, J., Nikolakaki, E., and Woodgett, J. R.,** Phosphorylation of c-jun mediated by MAP kinases, *Nature,* 353, 670, 1991.
117. **Janknecht, R., Zinck, R., Ernst, W. H., and Nordheim, A.,** Functional dissection of the transcription factor Elk-1, *Oncogene,* 9, 1273, 1994.
118. **Nakajima, J., Kinoshita, S., Sasagawa, T., Sasaki, K., Naruto, M., Kishimoto, T., and Akira, S.,** Phosphorylation of threonine-235 by a ras-dependent mitogen-activated protein kinase cascade is essential for transcription factor NF-IL-6, *Proc. Natl. Acad. Sci. U.S.A.,* 90, 2207, 1993.

119. **Lin, L. L., Wartmann, M., Lin, A. Y., Knopf, J. L., Seth, A., and Davis, R. J.,** $cPLA_2$ is phosphorylated and activated by MAP kinase, *Cell,* 72, 269, 1993.
120. **Yin, T. and Yang, Y.-C.,** Mitogen-activated protein kinases and ribosomal S6 protein kinases are involved in signalling pathways shared by interleukin-11, interleukin-6, leukemia inhibitory factor, and oncostatin M in mouse 3T3-L1 cells, *J. Biol. Chem.,* 269, 3731, 1994.
121. **Thompson, H. L., Marshall, C. J., and Saklatvala, J.,** Characterization of two different phosphoforms of MAP kinase kinase induced in polymorphonuclear leukocytes following stimulation by Fmet-leu-phe or GM-CSF, *J. Biol. Chem.,* 269, 9486, 1994.
122. **Nakielny, S., Campbell, D. G., and Cohen, P.,** MAP kinase kinase from rabbit skeletal muscle, *FEBS Lett.,* 308, 183, 1992.
123. **Crews, C. M., Alessandrini, A., and Erikson, R. L.,** The primary structure of MEK, a protein kinase that phosphorylates the ERK gene product, *Science,* 258, 478, 1992.
124. **Saklatvala, J., Rawlinson, L. M., Marshall, C. J., and Kracht, M.,** Interleukin-1 and tumour necrosis factor activate the mitogen-activated protein (MAP) kinase kinase in cultured cells, *FEBS Lett.,* 334, 189, 1993.
125. **Alessi, D. R., Saito, Y., Campbell, D. G., Cohen, P., Sithanandam, G., Rapp, U., Ashworth, A., Marshall, C. J., and Cowley, S.,** Identification of the sites in MAP kinase kinase-1 phosphorylated by $p74^{raf-1}$, *EMBO J.,* 13, 1610, 1994.
126. **Zheng, C.-F. and Guan, K.-L.,** Activation of MEK family kinases requires phosphorylation of two conserved Ser/Thr residues, *EMBO J.,* 13, 1123, 1994.
127. **Kyriakis, J. M., App, H., Xhang, X.-F., Banerjee, P., Brautigan, D. L., Rapp, U. R., and Avruch, J.,** Raf-1 activates MAP kinase kinase, *Nature,* 358, 417, 1992.
128. **Howe, L. R., Leevers, S. J., Gomez, N., Nakielny, S., Cohen, P., and Marshall, C. J.,** Activation of the MAP kinase pathway by the protein kinase Raf, *Cell,* 71, 335, 1992.
129. **Kolch, W., Heidecker, G., Kochs, G., Hummel, R., Vahidi, H., Mlschak, H., Finkenzeller, G., Marme, D., and Rapp, U. R.,** Protein kinase Cα activates RAF-1 by direct phosphorylation, *Nature,* 364, 249, 1993.
130. **Gardner, A. M., Vaillancourt, R. R., Lange-Carter, C. A., and Johnson, G. L.,** MEK-1 phosphorylation by MEK kinase. Raf and mitogen-activated protein kinase: analysis of phosphopeptides and regulation of activity, *Mol. Biol. Cell,* 5, 193, 1994.
131. **Kracht, M., Shiroo, M., Marshall, C. J., Hsuan, J. J., and Saklatvala, J.,** Interleukin-1 activates a novel protein kinase that phosphorylates the EGF receptor peptide T669, *Biochem. J.,* 1994, 302, 897, 1994.
132. **Kyriakis, J. M., and Avruch, J.,** pp54 Microtubule-associated protein 2 kinase. A novel serine/threonine protein kinase regulated by phosphorylation and stimulated by poly-L-lysine, *J. Biol. Chem.,* 265, 17355, 1990.
133. **Kyriakis, J. M., Brautigan, D. L., Ingebritsen, T. S., and Avruch, J.,** pp54 Microtubule-associated protein-2 kinase requires both tyrosine and serine/threonine phosphorylation for activity, *J. Biol. Chem.,* 266, 10043, 1991.
134. **Kyriakis, J. M., Banerjee, P., Nikolakaki, E., Dai, T., Rubie, E. A., Ahmad, M. F., Avruch, J., and Woodgett, J. R.,** The stress-activated protein kinase subfamily of c-Jun kinases, *Nature,* 369, 156, 1994.
135. **Dérijard, B., Hibi, M., Wu, I.-H., Barrett, T., Su, B., Deng, T., Karin, M., and Davis, R. J.,** JNK1: a protein kinase stimulated by UV light and Ha-Ras that binds and phosphorylates the c-Jun activation domain, *Cell,* 76, 1025, 1994.
136. **Mukhopadhyay, N. K., Price, D. J., Kyriakis, J. M., Pelech, S., Sanghera, J., and Avruch, J.,** An array of insulin-activated, proline-directed serine/threonine protein kinases phosphorylate the p70 S6 kinase, *J. Biol. Chem.,* 267, 3325, 1992.
137. **Arrigo, P. and Welch, W. J.,** Characterization and purification of the small 28,000 dalton mammalian heat shock proteins, *J. Biol. Chem.,* 262, 15359, 1987.
138. **Jakob, U., Gaestel, M., Engel, K., and Buchner, J.,** Small heat shock proteins are molecular chaperones, *J. Biol. Chem.,* 268, 1517, 1993.
139. **Miron, T., Vancompernolle, K., Vandekerckhove, J., Wilchek, M., and Geiger, B.,** A 25 kDa inhibitor of actin polymerization is a low molecular mass heat shock protein, *J. Cell Biol.,* 114, 255, 1991.

140. **Lavoie, J. N., Hickey, E., Weber, L. A., and Landry, J.,** Modulation of actin microfilament dynamics and fluid phase pinocytosis by phosphorylation of heat shock protein 27, *J. Biol. Chem.,* 268, 24210, 1993.
141. **Lavoie, J. N., Gingras-Breton, G., Tanguay, R. M., and Landry, J.,** Induction of Chinese hamster hsp27 gene expression in mouse cells confers resistance to heat shock, *J. Biol. Chem.,* 268, 3420, 1993.
142. **Gaestel, M., Schroder, W., Benndorf, R., Lippmann, C., Buchner, K., Hucho, F., Erdmann, V. A., and Bielka, H.,** Identification of the phosphorylation sites of the murine small heat shock protein hsp25, *J. Biol. Chem.,* 266, 14721, 1991.
143. **Landry, J., Lambert, H., Zhou, M., Lavoie, J. N., Hickey, E., Weber, L. A., and Anderson, C. W.,** Human hsp27 is phosphorylated at serines 78 and 82 by heat shock and mitogen activated kinases that recognise the same amino acid motif as S6 kinase 11, *J. Biol. Chem.,* 267, 794, 1992.
144. **Kato, K., Hasegawa, K., Goto, S., and Inaguma, Y.,** Dissociation as a result of phosphorylation of an aggregated form of the small stress protein hsp27, *J. Biol. Chem.,* 269, 11274, 1994.
145. **Welch, W. J.,** Phorbol ester, calcium ionophore or serum added to quiescent rat embryo fibroblast cells all result in the elevated phosphorylation of two 28,000-dalton mammalian stress proteins, *J. Biol. Chem.,* 260, 3058, 1985.
146. **Guesdon, F. and Saklatvala, J.,** Identification of a cytoplasmic protein kinase regulated by IL-1 that phosphorylates the small heat shock protein, hsp27, *J. Immunol.,* 147, 3402, 1991.
147. **Freshney, N. W., Rawlinson, L., Guesdon, F., Jones, E., Cowley, S., Hsuan, J., and Saklatvala, J.,** Interleukin-1 activates a novel protein kinase cascade that results in the phosphorylation of hsp27, *Cell,* 78, 1039, 1994.
148. **Stokoe, D., Campbell, D. G., Nakielny, S., Hidaka, H., Leevers, S. J., Marshall, C. J., and Cohen, P.,** MAPKAP kinase-2: a novel protein kinase activated by mitogen-activated protein kinase, *EMBO J.,* 11, 3985, 1992.
149. **Stokoe, D., Engel, K., Campbell, D. G., Cohen, P., and Gaestel, M.,** Identification of MAPKAP kinase-2 as a major enzyme responsible for the phosphorylation of the small mammalian heat shock protein, *FEBS Lett.,* 313, 307, 1992.
150. **Stokoe, D., Caudwell, B., Cohen, P. T. W., and Cohen, P.,** The substrate specificity and structure of MAP kinase-activated protein kinase-2, *Biochem. J.,* 296, 843, 1993.
151. **Bird, T. A., Schule, H. D., Delaney, P., de Roos, P., Sleath, P., Dower, S. K., and Virca, G. D.,** The interleukin-1-stimulated protein kinase that phosphorylates the small heat shock protein hsp27 is activated by MAP kinase, *FEBS Lett.,* 338, 31, 1994.
152. **Guesdon, F., Waller, R. J., and Saklatvala, J.,** Specific activation of β casein kinase by the inflammatory cytokines interleukin-1 and tumour necrosis factor, *Biochem. J.,* 304, 761, 1994.
153. **Kennelly, P. J., and Krebs, E. G.,** Consensus sequences as substrate specificity determinants for protein kinases and phosphatases, *J. Biol. Chem.,* 266, 15555, 1991.
154. **Munoz, E., Zubiaga, A. M., and Huber, B. T.,** Interleukin-1 induces c-fos and c-jun gene expression in T helper type II cells through different signal transmission pathways, *Eur. J. Immunol.,* 22, 2101, 1992.
155. **Angel, P., Hattori, K., Smeal, T., and Karin, K.,** The *c-jun* oncogene is positively autoregulated by its product, Jun/AP-1, *Cell,* 55, 875, 1988.
156. **van Dam, H., Duyndam, M., Rottier, R., Bosch, A., de Vries-Smits, L., Herrlich, P., Zantema, A., Angel, P., and van der Eb, A. J.,** Heterodimer formation of c-Jun and ATF-2 is responsible for induction of c-Jun by the 243 amino acid adenovirus E1A protein, *EMBO J.,* 12, 479, 1993.
157. **Muegge, K., Vila, M., Gusella, G. L., Musso, T., Herrlich, P., Stein, B., and Durum, S. K.,** Interleukin-1 induction of the c-jun promoter, *Proc. Natl. Acad. Sci. U.S.A.,* 90, 7054, 1993.
158. **Kishimoto, T., Taya, T., and Akira, S.,** Cytokine signal transduction, *Cell,* 76, 253, 1994.
159. **Trautwein, C., Caelles, C., van der Geer, P., Hunter, T., Karin, M., and Chojkier, M.,** Transactivation by NF-IL-6/LAP is enhanced by phosphorylation of its activation domain, *Nature,* 364, 544, 1993.
160. **Osborn, L., Kunkel, S., and Nabel, G. J.,** Tumour necrosis factor α and interleukin-1 stimulate the human immunodeficiency virus enhancer by activation of the nuclear factor κB, *Proc. Natl. Acad. Sci. U.S.A.,* 86, 2336, 1989.

161. **Grimm, S. and Baeuerle, P. A.,** The inducible transcription factor NFκB: structure function relationship of its protein subunits, *Biochem. J.,* 290, 297, 1993.
162. **Sun, S.-C., Ganchi, A., Ballard, D. W., and Greene, W. C.,** NF-κB controls expression of inhibitor IκBα: evidence for an inducible auto-regulatory pathway, *Science,* 259, 1912, 1993.
163. **Brown, K., Park, S., Kanno, T., Franzoso, G., and Siebenlist, U.,** Mutual regulation of the transcriptional activator NF-κB and its inhibitor IκB-α, *Proc. Natl. Acad. Sci. U.S.A.,* 90, 2532, 1993.
164. **Henkel, T., Machleidt, T., Alkalny, I., Krönke, M., Ben-Neriah, Y., and Baeuerle, P. A.,** Rapid proteolysis of IKBα is necessary for activation of transcription factor NF-κB, *Nature,* 365, 182, 1993.
165. **Beg, A. A., Finco, T. S., Nantermet, P. V., and Baldwin, A. S., Jr.,** Tumor necrosis factor and interleukin-1 lead to phosphorylation and loss of IκBα: a mechanism for NF-κB activation, *Mol. Cell. Biol.,* 13, 3301, 1993.
166. **Sun, S.-C., Ganchi, P. A., Berand, C., Ballard, D. W., and Greene, W. C.,** Autoregulation of the NF-κB transactivator RelA (p65) by multiple cytoplasmic inhibitors containing ankyrin motifs, *Proc. Natl. Acad. Sci. U.S.A.,* 91, 1346, 1994.

Modulatory Cytokines

Chapter 13

TRANSFORMING GROWTH FACTOR-βs: ENDOGENOUS IMMUNOSUPPRESSIVE AND ANTI-INFLAMMATORY MOLECULES

Gerald M. Hochwald, Michael A. Palladino, and G. Jeanette Thorbecke

CONTENTS

0-8493-8381-1/96/$0.00+$.50

I. INTRODUCTION

Transforming growth factor-beta 1 (TGF-β_1) was originally named for its ability to induce a reversible phenotypic transformation of normal rat fibroblasts in soft agar.[1] Compelling data have recently been published to show that the TGF-β proteins provide regulatory signals for many cellular processes.[2] The conservation of the genes encoding the various TGF-β molecules, the sequence homologies among the TGF-β isotypes, and the observation that a wide range of cell types express TGF-β receptors support the fundamental roles of the TGF-β molecules in regulating biologic activities. Effects of TGF-β on gene expression are exerted at both transcriptional and postranscriptional levels and affect a great variety of cell types.[2] Whether repression or activation of a specific gene is induced often is dependent on the cell type affected — possibly related to an effect on different members of the c-jun gene family. Several reviews on TGF-β have appeared recently.[3–6] The present review is primarily directed at the immunologically relevant effects of TGF-β. Although we are in the early stages of understanding the potent immunoregulatory activities of TGF-β *in vivo,* the available data strongly suggest that these cytokines constitute a class of immunosuppressive molecules with the potential for broad clinical utility (see also Chapter 1).

II. THE TGF-β FAMILY

A. STRUCTURE

There are at present five identified TGF-β molecules that are encoded by different genes. The cDNA for TGF-β_1 was described originally by Derynck et al.[7] Seyedin et al.[8] isolated a protein from bone that stimulated type II collagen production. The cDNA for this protein (later termed TGF-β_2) was cloned by De Martin et al.[9] and Madisen et al.[10] The cDNA for TGF-β_3 was cloned by Ten Dijke et al.[11] TGF-β_1, TGF-β_2, and TGF-β_3 are synthesized as secretory precursor glycoproteins of 390, 414, and 412 amino acids, respectively, which require processing to a homodimeric molecule of 25 kDa for bioactivity.[12] Mammals express TGF-β_1, TGF-β_2, and TGF-β_3, and in view of the close homology it is understandable that the three isoforms also share functional activities. However, more recently, divergent functions and relative differences in potencies have been reported depending on the target cells and assay utilized. For example, TGF-β_1 has a 60- to 100-fold greater effect than TGF-β_2 in the inhibition of growth of bovine endothelial and of IL-3-stimulated murine hematopoietic cells.[13] Also, only TGF-β_2 has been shown to be active in mesoderm induction.[14]

A differential expression of both mRNA and protein of the three above TGF-β isoforms has been observed in different tissues as well as a dynamic expression of these proteins throughout embryogenesis and wound repair.[15,16] A fourth TGF-β is detectable only in chicken chrondrocytes and a fifth TGF-β in the frog *Xenopus laevis.* The five-TGF-βs are members of a larger gene family that also includes activins, inhibins, Mullerian inhibitory substance, bone morphogenic proteins, decapentaplegic gene complex of *Drosophila* embryo, Vg1, and Vgr 1.[17] The TGF-βs are functionally related in their ability to regulate cellular growth and differentiation, and are structurally similar, particularly in the C-terminal portion of the protein where seven out of nine cysteines are conserved in all the family members. These are important in the tertiary structure, as recently elucidated.[18,19] On the basis of their crystal structure, the TGF-βs have been placed, with NGF and PDGF, in a superfamily of growth factors called the "cystine knot" family.[20]

In spite of their sequence homology, members of the TGF-β gene family are not linked with respect to their chromosomal localization in human cells: TGF-β_1 on chromosome 19,[21] inhibin on chromosome 7,[22] and TGF-β_3 on chromosome 14.[23] This suggests that following amplification these genes have dispersed through the genome and diverged in sequence during evolution.

B. RECEPTORS

Cross-linking studies have revealed that most cells express three types of TGF-β receptors on their surface. The type I and type II receptors have recently been cloned by Ebner et al.[24] and Lin et al.,[25] respectively. These receptors share a conserved extracellular domain and, similar to the activin type II receptor, are serine-threonine kinases that form heteromeric kinase complexes.[26] It is interesting to note that most growth factor receptors are transmembrane tyrosine kinases or are associated with cytoplasmic tyrosine kinases, while the receptors for activin and TGF-β belong to the serine-threonine kinase family. Type I receptors need type II components for TGF-β or activin binding, whereas the type II receptors need type I components for signal transduction.[27] The type III receptor family, consisting of endoglins and betaglycan, are glycoproteins, usually with short cytoplasmic domains, that do not seem to mediate any of the known biological activities of TGF-β.[28] They may function as regulators of TGF-β activity by providing access to the signaling heteromeric kinase receptor complex formed by TGF-β receptor components I and II. Endoglins bind TGF-β_1 and -β_3, but not -β_2[29] while betaglycan binds all three TGF-β isoforms.[30] Signaling via the type I receptor seems to mediate the induction of several genes involved in cell-matrix interactions, while signaling via the type II receptors may be required for the antiproliferative effects of TGF-β and the hypophosphorylation of the retinoblastoma gene product pRB.[31]

C. FROM LATENCY TO BIOACTIVITY

Due to the widespread expression of the various TGF-β receptors and the fact that most cell types produce at least one type of TGF-β, the activation of this family of molecules must be a tightly controlled biological process. TGF-β is secreted in an inactive form that is unable to bind to its receptors because of its interaction with a masking protein.[32] A proposed scheme for the processing of TGF-β_1 is illustrated in Figure 1. However, the order of activation has not been completely delineated, and for simplicity these steps have been outlined in succession. A signal peptidase initiates processing by removing the hydrophobic N-terminal leader sequence from the TGF-β precursors (pre-pro-TGF-β) to generate pro-TGF-β. Carbohydrates are added, presumably at the three predicted N-linked glycosylation sites. Pro-TGF-β dimerizes by forming disulfide bonds linking two precursor remnants to each other and, in platelets, to one molecule of the 135 kDa TGF-β. It is unclear whether this 135-kDa molecule is also a component of the latent complex secreted by other TGF-β-producing cells. Proteolytic cleavage at dibasic residues Arg-Arg-Ala in the C-terminus of pro-TGF-β (position 278–280) occurs, but the mature TGF-β dimer remains noncovalently associated with the remaining complex.[5,33]

While linked with the precursor remnant or latency-associated peptide (LAP), TGF-β is inactive or latent, characterized by the inability to bind to its receptors. It is most likely of great importance for the effectiveness of TGF-β that it can be transported in latent form to distant sites and bound via the associated proteins in the latent complex to extracellular matrix and/or cell surfaces. It is then available for local activation when needed. A second latent form of TGF-β exists in serum and consists of active TGF-β complexed with α_2-macroglobulin. The extracellular matrix contains a variety of proteins

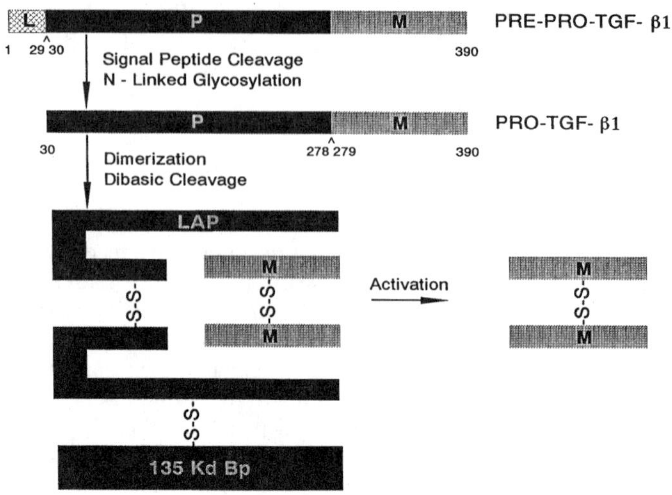

FIGURE 1. Schematic representation of the synthesis and structure of platelet-derived TGF-β_1. L — leader sequence; P — pro-region; M — mature; LAP — latency associated protein; BP — (platelet) binding protein.

able to bind TGF-β, including type IV collagen, osteonectin, fibronectin, and decorin.[12,34] *In vitro* activation of TGF-β with release of a mature TGF-β dimer from the latent complex can occur by treatment with extremes of pH, heat, chaotropic agents, and proteases such as plasmin, cathepsin D, and/or glycosidases.[35] The mechanisms that activate TGF-β *in vivo* are unknown, but presumably are enzyme mediated. Endothelial cells can activate the latent complex, but only when in contact with vascular pericytes.[36] The interaction appears to be mediated through mannose-6-phosphate receptors to which the pro-region of TGF-β can bind.[37] Once released from the latent complex, active TGF-β can bind to various extracellular molecules. Clearance of ^{125}I-labeled recombinant TGF-β after intravenous injection is extremely rapid, with a half-life of 2.2 min.[38] In addition, binding to α_2-macroglobulin also occurs rapidly, causing the availability of the active molecule to be restricted to the local environment where activation occurs.

The mechanisms by which latent TGF-β becomes active during immune responses are unknown. Earlier studies have demonstrated that macrophages have the ability to secrete both latent and active TGF-β.[39,40] The fact that macrophages as well as osteoclasts have acidic microenvironments and possess sialidase activity also suggests that the macrophage or macrophage-like cell types could play a role in the activation of TGF-β. Further support for this hypothesis is provided by the recent finding that IFN-γ-activated human macrophages have the ability to activate recombinant latent TGF-β.[41] Moreover, studies by Wallick et al.[42] have demonstrated that cell lines engineered to secrete latent TGF-β are immunosuppressive *in vivo,* again supporting an *in vivo* mechanism(s) to activate latent TGF-β.

D. ASSAYS

Both bioassays and immunoassays for TGF-β are complicated by the difficulties in differentiating between latent and active TGF-β, and the normal presence of (latent) TGF-β in serum and plasma proteins contaminated with α_2-macroglobulin. Bioassays should be performed in the absence of serum, since its latent TGF-β might get activated in the presence of certain cells — macrophages in particular. For example, the addition of neutralizing anti-TGF-β to cultures of lymphoid cells usually enhances proliferative responses and antibody formation *in vitro*. This can happen even if the proliferative responses are performed in the absence of serum, suggesting that the cells produce TGF-β with autocrine inhibitory effects.[43]

Bioassays consist of (1) growth inhibitory effects on mink lung epithelial cells (CCL64) either by ^{3}H-thymidine incorporation[44] or methylene blue staining and ELISA,[45] or (2) by inhibition of the binding of ^{125}I-TGF-β to its receptors on rat kidney cells NRK-49F.[46] Usually, fluids with unknown quantities of TGF-β are tested with and without acidification with 1 *N* HCl to activate latent TGF-β. However, little is known about the most effective method of converting latent to active and stable TGF-β in such fluids.

Immunoassays employ inhibition of precipitation of ^{125}I-TGF-β by antibodies to TGF-β[47] or ELISA.[43] It is relatively difficult to obtain neutralizing (and other) monoclonal antibodies to TGF-β, presumably because of the highly conserved structure of these proteins, but two groups of investigators have succeeded.[43,48] Antibodies that can be used for staining TGF-β in fixed and paraffin-embedded tissues may be obtained by immunization with TGF-β peptides bound to a carrier protein, immunogen. Judicious selection of the peptides can lead to antibodies specific for each of the TGF-β isotypes.[49,50]

III. IMMUNOREGULATORY EFFECTS OF TGF-β *IN VITRO*

A. EFFECTS ON T CELLS

1. Proliferation of Thymocytes and Peripheral CD4+ T Cells

TGF-β_1 and TGF-β_2 inhibit the proliferation of hematopoietic cells[51] and thymocytes.[52] A comparative analysis of growth-factor-driven mouse thymocyte proliferation showed that TGF-β_1 inhibits the costimulation of PHA with various cytokines.[53,54] Greater than 50% inhibition is obtained at low TGF-β concentrations (0.1 ng/ml) with IL-1, IL-6, and TNF-α. Concentrations of at least 1 ng/ml are needed for inhibition of IL-2, IL-4, or IL-7 costimulated thymocytes. TGF-β_1 and TGF-β_2 give identical dose response curves in these assays.

The precise mechanisms of how the various cytokines interact with TGF-β during lymphocyte activation are unknown. However, in many instances, TGF-β has biological properties that are opposite to those of IL-2 and TNF-α. Although TGF-βs also inhibit responses of human or mouse peripheral T cells to mitogens[55] and antigens,[50,56] IL-2 production is not inhibited. In contrast, responses induced by IL-2 are inhibited,[56] including the early increases in c-myc mRNA accumulation induced by IL-2 and IL-4[57] and the tyrosine phosphorylation of several proteins induced by mitogen + IL-2.[58] Of interest is the observation that TGF-β_1 also inhibits the IL-2-dependent phosphorylation of the retinoblastoma susceptibility gene product, which plays an important role in cell cycle progression.[58] It should be noted that the mechanism by which TGF-β causes cell cycle arrest in G1 in many cell types is thought to be due to the inhibition of a cyclin E-dependent kinase.[59]

TGF-β production has been shown by both CD4 and CD8 human T cells upon stimulation with con A *in vitro*.[60] Such peripheral blood-derived T cells express TGF-β mRNA shortly (1 to 2 h) after mitogen stimulation.[60,61] Secreted TGF-β, however, is not detectable in the culture supernatant until two to three days after mitogen stimulation, suggesting that the secreted TGF-β has a late effect on T cell proliferation and may function in an autocrine or paracrine manner to limit the extent of T cell clonal expansion and cytokine release. TGF-β also activates its own promoter, thereby positively regulating its production.[62] This could explain the enhancing effect of neutralizing mAbs to

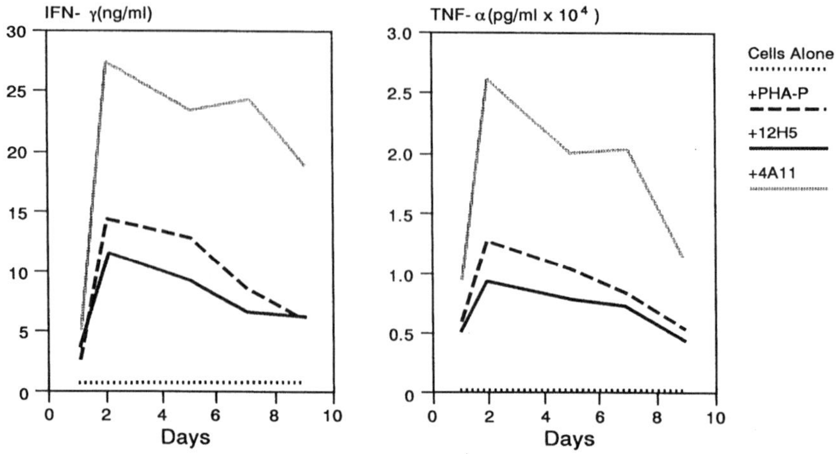

FIGURE 2. Effects of antibodies to TGF-β on cytokine production. Human peripheral blood mononuclear cells (PBMC) were isolated from buffy coats (Peninsula Memorial Blood Bank, Burlingame, CA) by Ficoll-Hypaque density sedimentation. The purified PBMC were washed three times in phosphate-buffered saline and adjusted to 4×10^6 cells/ml in AIM V medium (GIBCO, Grand Island, NY). Cultures were incubated at 37°C for 9 days with 10 μg/ml PHA-P (Difco Laboratory, Detroit, MI). Supernatants were removed on days 1, 3, 5, 7, and 9 for assessment of TNF-α and IFN-γ levels by ELISA. Monoclonal antibodies 12H5 (nonneutralizing) and 4A11 (neutralizing) were added to cultures at a concentration of 12.5 μg/ml, as previously described.[43]

TGF-β on T cell proliferation and on the production of TNF-α and IFN-γ (see Figure 2).[43] Recent studies also suggest that cyclosporin A promotes this autoinducing effect of TGF-β.[61]

IL-1α induces active TGF-β protein production in nonproliferating T cells, whereas stimulation by Con A + IL-1α does not.[63] The mechanism of this stimulation is apparently posttranscriptional, as IL-1α does not cause an increase in mRNA for TGF-β_1 in these cells, and the effect is inhibited by cycloheximide. It has been postulated on the basis of these findings that active TGF-β production by resting T cells may be a mechanism of controlling the growth response of T cells and may constitute a (partial) explanation for the need of a costimulus in the induction of T cell growth. IL-6 does not upregulate TGF-β production, but it has been shown to increase TGF-β receptor expression.[64] The nonspecific “natural suppressor” activity for cytotoxic T cell activation found in the supernatants of IL-2-activated lymphocytes also appears due to TGF-β secreted by the cells.[65]

2. Cytotoxic T and NK Cells

In regard to cytotoxic T-lymphocyte (CTL) development, TGF-β has been shown to significantly inhibit the generation, but not the cytotoxic activity, of CTL *in vitro* and *in vivo*.[66–68] In this regard, TGF-β and TNF-α exhibit opposing activities, suggesting that a balance between the negative and positive signals provided by growth factors and cytokines to lymphocyte populations is critical for the ultimate responses that are generated. Some of the inhibitory effects of TGF-β on activated T cells can be overcome by increasing the concentration of certain endogenously produced cytokines such as IL-2 and TNF-α.[66,69] Similar findings have been reported for the proliferation and IL-2R expression of natural killer (NK) and lymphokine activated killer (LAK) cells.[69–72] In addition, however, the cytotoxic functions of LAK and NK cells are also inhibited by TGF-β.[72–75]

Expression of CD8 is somehow related to suppressor functions of T cells.[76] It is of interest in this regard that incubation with TGF-β causes an enhancement in CD8 expression by both mature T cells[77] and thymocytes.[77,78] It has, in fact, been suggested that TGF-βs mediate some of the effects of thymic epithelial cells on thymocyte maturation from $CD4^-CD8^{lo}$ to $CD4^+CD8^+$ cells.[79] In addition, TGF-β costimulates the proliferation of CD8 cells in response to anti-CD3.[80] TGF-β has been shown to maintain and/or activate suppressor activity of $CD8^+$ T cells.[81,82] This effect may be mediated by enhancement of CD8 expression and/or increased TGF-β production by the T cells, as it is known that TGF-β enhances its own production.[62]

3. T Cell Subset Differentiation

TGF-β also appears to have, with both murine and human T cells *in vitro,* a different effect on Th1 as compared to Th2 CD4 T cell differentiation. Exogenously added as well as endogenously produced TGF-β was shown to have a preferential inhibitory effect on Th2 cells, while actually promoting Th1 memory cell differentiation.[83–85] This effect is partly mediated through inhibition of IL-4 production, causing a significant increase in the ratio of IFN-γ/IL-4 production by the $CD4^+$ T cells responding either to *S. enterotoxin* superantigen or to anti-CD3.[84,85] However, the effect of exogenously added TGF-β is not completely overcome by addition of IL-4. In addition, TGF-β promotes the proliferation and mature $CD45RB^{lo}CD44^{hi}$ (memory) phenotype development of isolated anti-CD3 stimulated $CD8^+$ T cells *in vitro.*[80] More recent studies suggest that the effect of TGF-β on T cell subsets is more complicated than was first realized, as IL-12-dependent IFN-γ production, and therefore also Th1 cell development from naive $CD4^+$ T cells, is strongly inhibited by TGF-β.[86]

Another interesting effect of TGF-β on T lymphocytes is that during the activation of T cells *in vitro* it causes an increase in the expression of L-selectin and cutaneous lymphocyte-associated antigen, the receptors that mediate homing into lymph nodes and skin, respectively.[87,88]

B. B CELLS

TGF-β inhibits the pre-B cell[89] as well as normal mature B cell proliferation induced by both thymus-dependent and thymus-independent stimuli.[90,91] The cells are arrested in G1,[92,93] suggesting that the mechanism of inhibition may resemble the one recently elucidated for lung epithelial cells, where the G1 block was in part due to down-modulation of cyclin-dependent kinase 4.[94] Of the protooncogenes/transcription factors activated during B cell stimulation, TGF-β inhibits B-myb induction, but not c-myc and c-myb,[95] and it blocks the B cell activation in a step following PKC activation.[96] TGF-β counteracts the effect of the 12-kDa B cell growth factor on normal human B cells,[97,98] but EBV-transformed B cells are completely resistant to TGF-β, probably due to a loss of TGF-β receptors.[99,100] Various human and murine lymphoma and plasmacytoma cell lines, as well as normal B cells, produce TGF-β[90,101,102] spontaneously and on stimulation. Under these conditions, TGF-β can have autocrine negative effects on growth[101,102] as well as a suppressive effect on the immune response of the host.[103]

In addition, TGF-β influences the differentiation of B cells. It inhibits T-B cell interaction[104] and promotes the induction of growth arrest by anti-Ig in immature B lymphoma cells by preventing the reversal of the anti-Ig effect by T cell factors.[105] In synergy with IL-2 and IL-5, TGF-β also greatly promotes IgA production by activated B cells.[106–108] Analysis of the isotype switch mechanism in such TGF-β-treated B cells suggests that the cytokine promotes switching to IgA and enhances the germline transcripts of Cα.[109–112] In LPS-activated B cells from some mouse strains, TGF-β has also been shown to promote IgG2b production.[113] Moreover, the naturally high IgG2b production in LPS-stimulated B cells from DBA/2 mice is inhibited by addition of neutralizing anti-TGF-β.[114] Inasmuch as isotype switching in B cells occurs predominantly within germinal centers in lymphoid tissue,[115,116] it is perhaps relevant that a high concentration of intracellular TGF-β_1 is found in germinal center B cells (L. Santambrogio, L. Gold, and G.J. Thorbecke, unpublished observations).

C. MACROPHAGES AND OTHER MYELOID CELLS

TGF-β_1, but not TGF-β_2, is very inhibitory to the growth of multipotential hematopoietic progenitor cells *in vitro.*[13,117] Megakaryopoiesis is also inhibited.[118] On more mature myeloid cells, TGF-β_1 synergizes with GM-CSF to stimulate granulopoiesis.[119] The effect of TGF-β_1 on myeloid leukemic cells is somewhat variable, but ranges from induction of apoptosis in synergy with GM-CSF[120] and interference with proliferation through downregulation of the receptors for human stem cell factor[121] to induction of terminal differentiation in synergy with vitamin D_3.[122] Induction of apoptosis by TGF-β is also observed with eosinophils.[123] Similar to its effect on various epithelial cells,[124–126] the apoptosis induced by TGF-β appears secondary to an interference with production of and/or response to growth factors.

TGF-β is highly chemotactic for monocytes,[127] neutrophils,[128] and fibroblasts.[129] The concentrations required for these activities of TGF-β are 10^3 to 10^5 times lower than those needed for the inhibition of T and myeloid cell proliferation *in vitro.* At similar concentrations, TGF-β augments 5-lipoxygenase synthesis in HL60 cells that are differentiating into neutrophils,[130] and of fibronectin synthesis in macrophages.[131] Fibronectin is thought to play an important role in the chemotactic response of neutrophils to TGF-β.[132]

Higher concentrations of TGF-β (≥1 ng/ml) affect monocyte functions, resulting in the induction of IL-1β gene expression[127] and in a reduction in the intracellular killing of bacteria by lymphokine-pretreated macrophages.[133] TGF-β is much more effective in preventing the activation of macrophages by IFN-γ than in reversing the cytotoxic effector functions of previously activated cells.[133] Similarly, TGF-β_1, -β_2, and -β_3 are capable of inhibiting the induction of reactive nitrogen intermediates such as NO_2^- in macrophages by IFN-γ and by (IFN-γ + TNF-α), but do not affect the induction of nitric oxide by IFN-γ and LPS.[134] IL-10 acts synergistically with TGF-β to inhibit both reactive nitric oxide production and killing of parasites such as *T. cruzi.*[135,136] The mechanism by which TGF-β affects the synthesis of nitric oxide appears to be by regulating the production of IFN-γ-inducible nitric oxide synthase through a destabilizing effect on the mRNA for this enzyme and on the protein itself.[137] Interestingly, certain parasites, such as *Leishmania,* that need to multiply in the macrophages of the host, induce the production of active TGF-β in the host cells. Treatment of genetically resistant mice with TGF-β causes the mice to become susceptible to *L. braziliensis,* while disease progression in susceptible mice can be halted by treatment with neutralizing anti-TGF-β. Thus, TGF-β may be regarded as a virulence-determining factor in the *Leishmania*-host interaction.[138,139]

TGF-β has been shown to decrease the replication of HIV virus in monocytic cell lines or primary macrophage/monocyte cultures.[140,141] It is of interest that the upregulation of HIV-associated reverse transcriptase synthesis induced by PMA or IL-6 is counteracted by TGF-β_1, -β_2, and -β_3, but the more potent induction by TNF-α is not inhibited. Moreover, TGF-β appears to have a dichotomous effect on HIV expression in primary macrophage/monocyte cultures, as pretreatment of the cells with TGF-β increases the subsequent multiplication of HIV virus in the cells.[140]

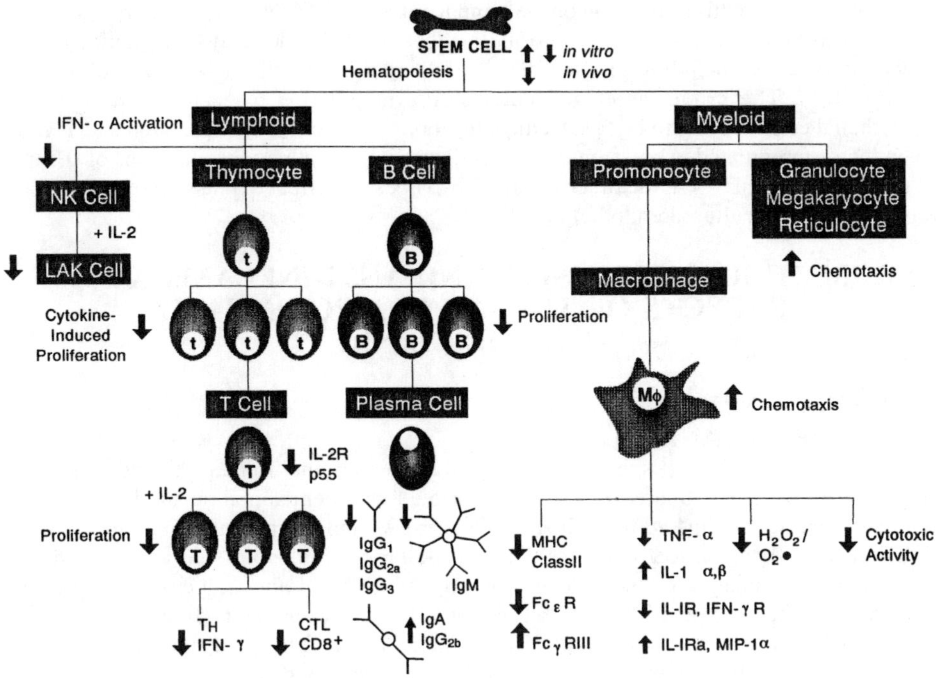

FIGURE 3. Overview of the multiple immunoregulating effects of TGF-β_1.

Some reports emphasize that pretreatment of macrophages with TGF-β lowers their sensitivity to IFN-γ. Indeed, TGF-β probably downregulates the expression of receptors for IFN-γ on macrophages.[142] The expression of Fc receptors on monocytes is also affected by TGF-β: the FcεRII (CD23) is downregulated[143] and the FcγRIII (CD16) is upregulated,[144,145] both in antagonism to the effects of IL-4. The upregulation of class II MHC expression on Langerhans cells (the dendritic cells of the epidermis) by cytokines such as IL-1, IFN-γ, and TNF-α, is abolished by the simultaneous presence of TGF-β ($\geq$10 ng/ml), although the effects of IL-6 and IL-2 in this respect are not counteracted by TGF-β.[146] Background levels of class II MHC expression on these cells are not affected by TGF-β, even at levels of 100 ng/ml.

It has also been observed that preincubation of peritoneal cells with TGF-β and antigen prior to injection into mice causes tolerance rather than sensitization to that antigen, a phenomenon similar to that observed with macrophages harvested from TGF-β-rich sites in the body (such as the anterior chamber of the eye).[147] Some or all of these effects of TGF-β on macrophages and dendritic cells may be involved when TGF-β causes functional abnormalities in the presentation of antigens to syngeneic cells and in the stimulation of allogeneic T cells.[148] An overview of the multiple effects of TGF-β on the individual components of the immune system is presented in Figure 3.

D. EFFECTS ON THE EXPRESSION OF ADHESION MOLECULES

The expression of β1 integrins is enhanced by TGF-β on both fibroblasts and monocytes,[149,150] causing increased adherence of the cells to extracellular matrices. In contrast, however, TGF-β inhibits the adhesion of lymphocytes[151–154] and neutrophils[155] to endothelial cells *in vitro*. In this respect, TGF-β_1 antagonizes the effects of TNF-α and IFN-γ on the endothelial cells rather than on the lymphocytes.[151,152,154,156] Several reports indicate that TNF-α enhances the expression of both adhesion molecules and MHC antigens on a variety of cells, including glial cells[157,158] and endothelial cells (ICAM and MHC[159]; VCAM-1[160]). The prime target molecule for downregulation by TGF-β_1 and -β_2 on endothelial cells appears to be E-selectin, since both basal levels of expression and TNF-α-induced upregulated expression of E-selectin are inhibited by concentrations of 0.2 to 2 ng TGF-β per ml.[161] On the other hand, TGF-β does not affect expression of VCAM-1 and ICAM-1 on cultured endothelial cells.[151]

IV. IMMUNOSUPPRESSIVE AND ANTI-INFLAMMATORY EFFECTS OF TGF *IN VIVO:* POTENTIAL FOR THE CLINIC

A. IMPORTANCE OF ENDOGENOUSLY PRODUCED TGF-β IN NORMAL ANIMALS

With the use of TGF-β-specific monoclonal antibodies and immunoperoxidase staining, TGF-β_1, -β_2, and -β_3 are found to be associated with hematopoietic and lymphopoietic cell populations in the fetal liver, bone marrow, and peripheral lymphoid tissues. TGF-β_1 and -β_3 are also found in thymic medullary reticuloepithelial cells[162] and in activated macrophages in the spleen after injection of certain antigens.[163] Although specific for the isotype of TGF-β,[49] staining for the various TGF-βs usually cannot distinguish between active and latent TGF-β. Therefore, the functional significance of the presence of TGF-βs in the lymphoid tissue is not always evident.

Several forms of immunosuppression *in vivo* have been attributed, at least in part, to production of TGF-β by the cells responsible for the suppression, such as the "natural suppressor cell" in bone marrow,[164] lymphokine activated killer cells,[65] alveolar macrophages,[165] $CD4^+$ T cells in antigenic competition[166] and in resistance to autoimmune disease,[167] and $CD8^+$ T cells in orally induced tolerance.[168] The TGF-β produced by activated T cells is thought to contribute to the immune deficiency accompanying certain viral infections, including HIV infection. Peripheral blood-derived lymphocytes from HIV-infected patients show increased levels of TGF-β, mRNA, and TGF-β production, and proliferative responses of these lymphocytes to recall antigens can be significantly restored *in vitro* by neutralizing polyclonal antibodies to TGF-β.[169] A high IL-6 concentration in the circulation, such as is seen in various infections, autoimmune diseases, and after traumatic injury, may play a role, because in the presence of macrophages, IL-6 induces an inhibition of T cell activation that is prevented by anti-TGF-β.[170]

An important role for TGF-β_1 in homeostasis in mice is indicated by the observation that animals homozygous for a mutated TGF-β_1 allele succumb at three weeks of age to a wasting syndrome accompanied by mixed lymphoid and inflammatory cell infiltrates and by tissue necrosis in multiple organs[171,172] (see Chapter 1 for further discussion of cytokine transgenics). TGF-β_1 appears, therefore, to possess potent systemic anti-inflammatory effects. It is of great interest that despite the normal widespread distribution of three TGF-β molecules with very similar properties, knockout of TGF-β_1 alone causes profound disturbances, suggesting an important role for TGF-β_1 in the normal homeostasis of the immune system that is not performed by TGF-β_2 or -β_3. It has been suggested[173] that upregulation of class I and II MHC molecules on a variety of cells is of importance in causing the generalized inflammatory lesions occurring in TGF-$\beta_1^{-/-}$ mice. However, neither the lymphoid infiltrates nor the enhanced MHC class II expression are observed in TGF-$\beta^{-/-}$ SCID mice (T. Doetschman, personal communication). This observation strongly suggests that lymphocytes are essential for the initiation of the inflammatory lesions in TGF-$\beta^{-/-}$ mice. Therefore, the lymphoid infiltrates may be the result of an "autoimmune" response, due to defective negative selection of T cells in the thymus of mice lacking TGF-β. It is of interest in this regard that TGF-β has been shown to influence the differentiation of $CD4^-8^+$ T cells to $CD4^+8^+$ cells in the thymus.[79] Alternately, the lymphoid infiltration into various organs could be caused by an enhanced adhesiveness of endothelial cells for lymphocytes in the absence of TGF-β. Further experiments are needed to differentiate between these possibilities.

Recent studies have also demonstrated that treatment with exogenous TGF-β can significantly decrease virally and bacterially induced CNS disease as a result of its nonspecific anti-inflammatory effects.[174,175] It is therefore not clear whether the beneficial effects of TGF-β in autoimmune diseases (described below) are due to a specific modification of autoimmune processes or primarily a useful manifestation of its anti-inflammatory effects in general.

B. GRAFT REJECTION

Exposure of macrophage-like cells to active TGF-β in "immune privileged" sites, such as the eye, has been suggested to play a role in the phenomenon known as immune deviation, resulting in partial tolerance.[147] The ability of TGF-β to counteract the effects of IL-1α, IFN-γ, and TNF-α in the upregulation of MHC-class II expression on Langerhans cells[146] and other cells[176] may be relevant to the observation that, together with anti-IFN-γ, TGF-β promotes acceptance of pancreatic islet xenographs.[177]

Although the *in vitro* data on the immunosuppressive activities of TGF-β warrant further *in vivo* testing, a major question to be addressed is whether a molecule with such pleotropic activities can be administered chronically at a dose that will not evoke unwanted biological effects. Initial *in vivo* studies were focused on determining whether TGF-β_1 could suppress the rejection of a heart allograft in mice.[42] The animal model measured the prolonged survival of neonatal BALB/c hearts implanted subcutaneously in the ears of adult C3H/KM mice. This model is useful in characterizing the immunosuppressive activities of newly identified biological modifiers.[178] Normal C3H/KM mice were injected intraperitoneally with various doses of recombinant human TGF-β_1. Spleen cells removed at various times after the TGF-β_1 treatment showed significantly reduced NK activity against YAC-1 target cells and a reduced ability to proliferate to con A. No significant side effects were noted.[42] In subsequent studies, C3H/KM recipients of BALB/c neonatal hearts were treated intraperitoneally with 1 μg of TGF-β_1 (40 μg/kg) on days 1 through 13. Heart transplant survival was significantly prolonged in the TGF-β_1-treated mice as compared to the saline-treated controls.[42] A similar level of graft prolongation after treatment with cyclosporin A required a dose of 6 mg/kg. Although TGF-β_1 demonstrated a statistically significant effect, the absence of continued allograft survival did not support the use of TGF-β_1 in this area.

C. AUTOIMMUNITY

The treatment of autoimmune diseases such as rheumatoid arthritis and multiple sclerosis still presents an enormous clinical challenge. Immunosuppressive and anti-inflammatory agents are of limited value and many have serious side effects that limit their effectiveness (see Chapters 5, 6, 9, and 10 for further discussion of these points). In animal models of collagen-induced arthritis (CIA) and relapsing experimental autoimmune encephalomyelitis (EAE), immunosuppressive drugs such as cyclosporin A and cyclophosphamide suppress only when administered during the induction of the disease. The development of biological agents that can suppress the inflammatory effects of IL-1 and TNF-α are important recent approaches to the treatment of autoimmune diseases. The use of such an approach is suggested by observations in patients with rheumatoid arthritis that disease severity is linked to enhanced production of multiple cytokines in their synovial fluid, including IL-1, GM-CSF, IFN-γ, and TNF-α. Although TGF-β did not exhibit profound suppression in the CTL-mediated allograft model, it was considered worthwhile to expand the potential role for TGF-β into other more chronic T cell-mediated models. Recent studies have clearly demonstrated the protective effects of systemically administered TGF-β in T cell-mediated models of autoimmunity in animals.

EAE is a chronic autoimmune disease characterized by inflammation and demyelination in the central nervous system. Studies by Kuruvilla et al.,[179] Johns et al.,[180] and Racke et al.[181] have demonstrated that TGF-β can significantly reduce the induction and severity of EAE in mice, either induced by active immunization with myelin basic protein (MBP) in complete Freunds adjuvant (CFA) or by passive immunization through the transfer of MBP-specific T cell lines. Recent studies have also demonstrated that systemically administered TGF-β can reduce the incidence of CIA in mice[179,182] and streptococcal-induced arthritis in rats.[183] Rheumatoid synovial fluids contain significant amounts of TGF-β.[184] However, therapeutic efficacy has been reported only after systemic administration of TGF-β, whereas local injection into the joint actually induces an acute inflammatory response.[185] Whether the TGF-β produced in the joint contributes to joint destruction and synovial fibroblast hyperplasia, or functions to suppress further joint destruction by the immune system needs further study. In contrast to the protective effects of systemic TGF-β, systemically administered TNF-α and IL-1 have been shown to enhance the severity of type II collagen-induced arthritis.[182,186]

Although both TGF-β_1 and TGF-β_2 inhibit the proliferative responses of T cells to the sensitizing antigen or to mitogens *in vitro* (see above), including responses to MBP and PLP,[50] both TGF-β isoforms, given on days 5 through 9 after immunization with spinal cord in CFA, protect against the development of EAE in SJL mice without a detectable inhibitory effect on the generation of MBP- and PLP-sensitized T cells in the lymph nodes or peripheral blood of the mice.[50] In spite of the presence of sensitized cells freely circulating in both treated and control mice, lymphoid cell infiltration in CNS is reduced or absent in TGF-β_1 protected mice. This suggests that in spite of their presence in the circulation, the sensitized cells fail to enter and/or to proliferate in the CNS. These results are in agreement with those of others[180,181] who have shown that TGF-β treatment interferes with the passive transfer of EAE and prevents the development of perivascular mononuclear cell infiltrates and demyelination in CNS. Thus, under circumstances where TGF-β cannot have interfered with the production of sensitized cells, infiltration of the CNS is still prevented. In addition, preincubation of a MBP-specific cell line with TGF-β was shown to render the cells more, rather than less, effective in the transfer of EAE.[187] It is of interest to note in this regard that TGF-β_1 also inhibits expression of contact sensitivity in sensitized mice when it is injected at the time of challenge with antigen,[146,188] whereas TGF-β_1 does not inhibit induction of contact sensitivity when it is injected during sensitization.[146] Thus, both in EAE and CIA,[182] and in contact sensitivity, TGF-β protects against the late induction/effector phase of the immune response rather than against the early T cell induction phase.

Lymphoid cell infiltration in CNS in EAE is thought to depend on the adherence of peripheral blood cells to the brain capillary endothelia,[189–191] with resulting damage to the endothelial cells[192,193] and migration of lymphoid cells into the perivascular spaces.[194,195] The adhesion proteins implicated in these lymphocyte-endothelial cell interactions are VLA-4 and VCAM-1.[190,196] Anti-VLA-4 prevents transfer of EAE by T cell clones in rats[190] and mice.[197] The encephalitogenicity of the T cell clones correlates with their ability to express VLA-4,[197] as well as with their capacity to produce TNF-α.[198] In view of the inhibitory effect of TGF-β on the expression of E-selectin, discussed above,[161] it is likely that TGF-β treatment of the recipient protects against the development of EAE by inhibition of the entry of lymphoid cells into the CNS through downregulation of the expression of this and/or other adhesion molecules on the endothelial cells. In addition, TGF-β also inhibits the upregulation of class II MHC on CNS endothelial cells during EAE.[181]

In addition to the protective effect of exogenously administered TGF-β_1, recent results from our laboratory indicate that endogenous production of TGF-β also plays a major role in limiting the severity of experimental autoimmune diseases, both in CIA and in acute and relapsing EAE. There appears to be a relatively precise time window after the initial immunization used to induce CIA or EAE, during which both the exogenously administered TGF-β and neutralization of endogenously produced TGF-β have the greatest effects. This is during the third week after the immunization with collagen type II and CFA in CIA induction, and during the second week after immunization with spinal cord and CFA in acute EAE induction. In both cases this period is shortly before the appearance of symptoms. Earlier administration is much less effective and continued administration throughout the period of immunization is not more effective. Later administration of the neutralizing mAb to TGF-β has no detectable effect on EAE. Thus, while recovery from the acute EAE episode does not appear to be solely dependent on endogenous TGF-β production, during the second week after immunization there is an important protective effect of endogenously produced TGF-β. Similarly, treatment with TGF-β protects against EAE relapses while anti-TGF-β accelerates the appearance and increases the severity of relapsing EAE in SJL mice.[50,179,182] An augmenting effect of anti-TGF-β on EAE induced by transferred T cells has been reported by Racke et al.[199]

It is not clear which cells produce the TGF-β that is providing the protective effect. In rats recovered from acute EAE and resistant to reinduction of the disease, $CD4^+$ T cells are detected in the lymphoid tissue that produce TGF-β on stimulation, inhibiting the production of IFN-γ by other $CD4^+$ T cells.[167] In orally tolerized rats, $CD8^+$ cells are detected that transfer suppression of EAE and are inhibited from doing so by anti-TGF-β treatment of recipients.[168] In addition, cytokines such as IL-1 and TNF-α can induce TGF-β production in endothelial cells,[200] while M-CSF and GM-CSF induce TGF-β production in macrophages.[201] Thus, during inflammatory responses such as those which occur in EAE, a variety of cell types may be upregulated with respect to TGF-β production. Indeed, perivascular mononuclear cells, endothelial cells, and astrocytes contain TGF-β in EAE lesions, mostly TGF-β_1 and -β_3, both at the peak of disease and after the symptoms have subsided.[50,180,199]

The antagonism between TNF-α and TGF-β in various immunological reactions may relate to their respective roles in autoimmune diseases. Not only does TGF-β inhibit the production of TNF-α,[202–205] but it also frequently has opposite effects from TNF-α on target cells.[53,54,66,146,203,204] As shown by various investigators,[50,182,206] the effect of anti-TNF-α on acute and relapsing EAE and on CIA is protective, in contrast to that of anti-TGF-β, suggesting that the endogenous production of TNF contributes to the severity of these diseases. This could be on the basis of upregulation of adhesion molecules or MHC class II antigen, but other mechanisms by which TNF could aggrevate these autoimmunc diseases should also be considered. TNF with or without IFN-γ activates macrophages to produce toxic mediators such as H_2O_2 and NO_2.[207] This effect on macrophages is inhibited by TGF-β, as discussed above.[134,135] Normal rat microglia have been reported to exhibit natural cytotoxicity towards oligodendrocytes *in vitro* which is upregulated by IFN-γ and inhibited by TGF-β.[208] Direct cytotoxicity of TNF-α and TNF-β on oligodendrocytes has also been shown.[209] In view of the finding that IFN-β has an ameliorating effect on multiple sclerosis,[210] it is of interest that, like TGF-β, IFN-α and IFN-β suppress TNF-α and TNF-β production by mitogen-activated human peripheral blood cells.[211] A recent report suggests that IFN-β increases TGF-β production by, and suppressor activity of, peripheral blood mononuclear cells from multiple sclerosis patients *in vitro*.[212] It is therefore possible that the effect of IFN-β on these patients is mediated via induction of TGF-β formation *in vivo*.

We propose that TGF-β may protect against the induction of acute EAE and against relapses in chronic EAE by antagonizing both the production of TNF-α and its effects, thereby interfering (1) with the entry of lymphoid cells into CNS through inhibition of the upregulation of adhesion molecule expression by TNF on CNS endothelial cells, and (2) with subsequent inflammatory processes inside the CNS.

D. HOST-TUMOR RELATIONSHIPS

Loss of autocrine inhibition through either decreased production or a loss of responsiveness to TGF-β may be important during carcinogenesis. In support of this are recent studies directed at determining the role for TGF-β in hyperproliferative responses in the skin. Glick et al.[213] have shown that TGF-β expression and response to it are compartmentalized and that TGF-β_1 and TGF-β_2 may have distinct functions *in vivo*. Normal epidermis, tumor promoter-treated epidermis, and skin papillomas at low risk for malignant conversion express TGF-β_1 in the basal cell compartment and TGF-β_2 in the suprabasal layer. In contrast, the majority of high-risk papillomas and squamous cell carcinomas are devoid of both TGF-β_1 and TGF-β_2. Whether TGF-β can be utilized clinically to treat the human equivalent of these diseases awaits further studies.

The escape from negative regulation by TGF-β may also play a role in the growth of neoplastic lymphoid cells. The effects of TGF-β on lymphoid leukemic cells has therefore been examined. In contrast to normal B and T cells, a number of lymphoid tumor cells lines were found to be insensitive to the antiproliferative effects of TGF-β_1 and TGF-β_2.[101] Binding and cross-linking with radioiodinated TGF-β_1 demonstrated either low or absent expression of all three TGF-β receptor species on three B cell tumor lines, but T cell and non-T, non-B cell tumors expressed large numbers of receptors. Treatment of the B cell lines with PMA induced the expression of TGF-β receptors and inhibited proliferation in all three B cell lines in a dose- and time-dependent manner. The cell lines constitutively produced TGF-β mRNA and released small amounts of latent TGF-β; however, PMA induced increased expression of TGF-β mRNA and, more importantly, a release of active TGF-β. A neutralizing antibody to TGF-β was able to reverse the PMA-induced growth inhibition, and addition of exogenous TGF-β reversed the effects of the neutralizing antibody.[214] Thus, some lymphoma cells appear to have escaped from TGF-β's negative regulation by failing to express functional TGF-β receptors and/or failing to secrete active TGF-β. In these experiments, PMA inhibits lymphoma cell growth by inducing the expression of TGF-β receptors and the secretion of active TGF-β, thereby reestablishing an autocrine growth inhibitory loop.

However, recent studies have shown that Epstein-Barr virus-transformed B cells and Burkitt's lymphoma cells can be either TGF-β sensitive or insensitive, regardless of the presence or absence of TGF-β receptors on the cell surface.[215] Also, all HTLV-1 transformed T cells tested could bind TGF-β but could not be inhibited by TGF-β, suggesting that there are several ways to escape regulation by TGF-β.

Several human lymphomas, including CLL[216] and sclerosing Hodgkin's disease,[217,218] produce TGF-β and have receptors for TGF-β. It is not immediately clear in these cases in how far there is an autocrine effect slowing tumor growth *in vivo*.

In addition to the ability of tumor cells to escape negative growth regulation by TGF-β, there is mounting evidence that tumor-induced immunosuppression is mediated through TGF-β and that cancer immunotherapy can be adversely affected by TGF-β. Patients whose glioblastomas secrete active TGF-β_2 have a generalized immunosuppression.[214] Several studies have previously shown that immunogenic tumors which secrete TGF-β escape immune surveillance, and grow progressively in the host.[219] It is clear that in tumor-bearing animals, there is marked immunosuppression. Tada et al.[220] have shown that tumors can make active TGF-β detectable in a local environment, and the suppression of $CD4^+$ helper activity can be completely overcome by the addition of neutralizing antibodies to TGF-β. This provides a mechanism for the previously observed selective inhibition of $CD4^+$ T cells in tumor-bearing animals. Despite the ability of LAK cells to infiltrate the tumors of many patients,[220] they frequently do not display much cytotoxic activity *in vivo* to tumor cells which they can kill *in vitro.* In addition, many tumor lines produce more active TGF-β than do their normal cellular counterparts and have lost their receptors for TGF-β and thus their sensitivity to the antiproliferative effect of TGF-β. The release of active TGF-β into the vicinity of such tumor cells has been proposed to result in a more hospitable environment for tumor growth, thus explaining the increased tumorigenesis and metastases associated with increased expression of TGF-β.[221,222]

Cellular and humoral immune responses including helper and suppressor T-cell functions, NK activity, and Ig synthesis are markedly suppressed in acute T cell leukemia patients. The etiologic agent for acute T cell leukemia as well as tropical spastic paraparesis is HTLV-1. The HTLV-tax gene product can stimulate the TGF-β_1 promoter through AP-1 binding sites. It should be noted that freshly isolated leukemia cells from patients with acute T cell leukemia constitutively release high levels of TGF-β_1, suggesting that the immunosuppression in these patients may in part be TGF-β mediated and that inhibiting TGF-β activity may be a useful adjunct therapy for this disease.

V. SUMMARY

Transforming growth factor-β_1 (TGF-β_1) belongs to a family of highly pleotrophic proteins that can affect the growth and differentiation of primitive, bone marrow-derived stem cells as well as the functions of fully differentiated immune effector cells. Although early studies identified primarily the immunosuppressive activities of TGF-β_1, recent evidence supports the concept that TGF-β_1 can both inhibit and stimulate the immune system. In addition, the autocrine and paracrine actions of TGF-β_1 on immune functions indicate that immune cells produce and activate the latent secreted forms of TGF-β, and that endogenously produced TGF-β is important in regulating normal immune activity. In this chapter, we reviewed recent data concerning the effects of TGF-β on individual components of the immune system *in vitro* and on the immune response *in vivo* under normal and pathological conditions. Its role in immunosuppression associated with various pathologies such as cancer and AIDS, as well as in the control of autoimmune diseases received particular attention.

ACKNOWLEDGMENTS

This work was supported in part by a grant from the National Multiple Sclerosis Society.

We thank Dr. T. Doetschman for permission to quote his as of then unpublished work.

REFERENCES

1. **Seyedin, S. M., Segarini, P. R., Rosen, D. M., Thompson, A. Y., Bentz, H., and Graycar, J.,** Cartilage-inducing factor-B is a unique protein structurally and functionally related to transforming growth factor-beta, *J. Biol. Chem.,* 262, 1946, 1987.
2. **Edwards, D. R. and Heath, J. K.,** Regulation of transcription by transforming growth factor-β, in *The Hormonal Control Regulation of Gene Transcription,* Cohen, P. and Foulkes, J. G., Eds., Elsevier, Amsterdam, 1991, 333.
3. **Roberts, A. B. and Sporn, M. B.,** The transforming growth factor βs, in *Peptide Growth Factors and Their Receptors,* Sporn, M. B. and Roberts, A. B., Eds., Springer-Verlag, Heidelberg, 419, 1990.
4. **Massague, J., Cheifetz, S., Laiho, M., Ralph, D. A., Weis, F. M. B., and Zentella, A.,** Transforming growth factor-β, *Cancer Surv.,* 12, 81, 1992.
5. **Miyazono, K., Hellman, U., Wernstedt, C., and Heldin, C.-H.,** Latent high molecular weight complex of transforming growth factor β1: purification from human platelets and structural characterization, *J. Biol. Chem.,* 263, 6407, 1988.
6. **McCartney-Francis, N. L. and Wahl, S. M.,** Transforming growth factor β: a matter of life and death, *J. Leuk. Biol.,* 55, 401, 1994.
7. **Derynck, R., Jarrett, J. A., Chen, E. Y., Eaton, D. H., Bell, J. R., Assoian, R. K., Roberts, A. B., Sporn, M. B., and Goeddel, D. V.,** Human transforming growth factor-β complementary DNA sequence and expression in normal and transformed cells, *Nature (London),* 316, 701, 1985.
8. **Seyedin, S. M., Thompson, A. Y., Bentz, H., Rosen, D. M., McPherson, J. M., Conti, A., Siegel, N. R., Galluppi, G. R., and Piez, K. A.,** Cartilage-inducing Factor-A. Apparent identity to transforming growth factor-β, *J. Biol. Chem.,* 261, 5693, 1986.
9. **De Martin, R., Haendler, B., Hofer-Warbinek, R., Gaugitsch, H., Wrann, M., Schlusener, H., Seifert, J. M., Bodmer, S., Fontana, A., and Hofer, E.,** Complimentary DNA for human glioblastoma-derived T cell suppressor factor, a novel member of the transforming growth factor-β gene family, *EMBO J.,* 6, 3673, 1987.
10. **Madisen, L., Webb, N. R., Rose, T. M., Marquardt, H., Ikeda, T., Twardzik, D., Seyedin, S., and Purchio, A. F.,** Transforming growth factor-beta-2: cDNA cloning and sequence analysis, *DNA,* 7, 1, 1988.
11. **Ten Dijke, P., Geurts van Kessel, A. H. M., Foulkes, J. G., and Le Beau, M. M.,** Transforming growth factor β3 maps to human chromosome 14, region q23-q24, *Oncogene,* 3, 721, 1988.
12. **Miyazono, K., Ten Dijke, P., Ichijo, H., and Heldin, C.-H.,** Receptors for transforming growth factor-β, *Adv. Immunol.,* 55, 181, 1993.
13. **Ohta, M., Greenberger, J. S., Anklesaria, P., Bassols, A., and Massague, J.,** Two forms of transforming growth factor-β distinguished by multipotential haematopoietic progenitor cells, *Nature,* 329, 539, 1987.
14. **Roberts, A. B., Kondaiah, P., Rosa, F., Watanabe, S., Good, P., Danielpour, D., Roche, N. S., Rebert, M. L., Dawid, I. A., and Sporn, M. B.,** Mesoderm induction in *Xenopus laevis* distinguishes between the various TGF-β3 TGF-β isoforms, *Growth Factors,* 3, 227, 1990.
15. **Thompson, N. L., Flanders, K. C., Smith, J. M., Ellingsworth, L. R., Roberts, A. B., and Sporn, M. B.,** Expression of transforming growth factor-β1 in specific cells and tissues of adult and neonatal mice, *J. Cell Biol.,* 108, 661, 1989.
16. **Levine, J. H., Moses, H. L., Gold, L. I., and Nanney, L. B.,** Spatial and temporal patterns of immunoreactive TGF-β1, TGF-β2, and TGF-β3 during excisional wound repair, *Am. J. Pathol.,* 143, 268, 1992.
17. **Robertson, D. M.,** Transforming growth factor β/inhibin family, *Brailliere's Clin. Endocrinol. Metab.,* 5, 615, 1991.
18. **Daopin, S., Piez, K. A., Ogawa, Y., and Davies, D. R.,** Crystal structure of transforming growth factor-β2: an unusual fold for the superfamily, *Science,* 257, 369, 1992.
19. **Schlunegger, M. P. and Grutter, M. G.,** An unusual feature revealed by the crystal structure at 2.2A resolution of human transforming growth factor-β2, *Nature,* 358, 430, 1992.
20. **McDonald, N. Q. and Hendrickson, W. A.,** A structural superfamily of growth factors containing a cystine knot motif, *Cell,* 73, 421, 1992.

21. **Fujii, D., Brissenden, J. E., Derynck, R., and Franke, U.,** Transforming growth factor β gene maps to human chromosome 19 long arm and to mouse chromosome 7, *Som. Cell Mol. Genet.,* 12, 281, 1986.
22. **Barton, D. E., Yang-Feng, T. L., Mason, A. J., Seeburg, P. H., and Francke, U.,** INHA, INHBA, and INHBB, the loci for the three subunits of inhibin, mapped in mouse and man, *Cytogenet. Cell Genet.,* 46, 578, 1987.
23. **Ten Dijke, P., Hansen, P., Iwata, K. K., Pieler, C., and Foulkes, J. G.,** Identification of another member of the transforming growth factor type β gene family, *Proc. Natl. Acad. Sci. U.S.A.,* 85, 4715, 1988.
24. **Ebner, R., Chen, R.-H., Shum, L., Lawler, S., Zioncheck, T. F., Lee, A., Lopez, A. R., and Derynck, R.,** Cloning of a type I TGF-β receptor and its effect on TGF-beta binding to the type II receptor, *Science,* 260, 1344, 1993.
25. **Lin, H. Y., Wang, X.-F., Ng-Eaton, E., Weinberg, R. A., and Lodish, H. F.,** Expression cloning of the TGF-β type II receptor, a functional transmembrane serine/threonine kinase, *Cell,* 68, 775, 1992.
26. **Attisano, L., Carcamo, J., Ventura, F., Weis, F. M. B., Massague, J., and Wrana, J. L.,** Identification of human activin and TGF-β type I receptors that form heteromeric kinase complexes with type II receptors, *Cell,* 75, 671, 1993.
27. **Wrana, J. L., Attisano, L., Carcamo, J., Zentella, A., Doody, J., Laiho, M., Wang, X.-F., and Massague, J.,** TGF-β signals through a heteromeric protein kinase receptor complex, *Cell,* 71, 1003, 1992.
28. **Lopez-Casillas, F., Wrana, J. L., and Massague, J.,** Betaglycan presents ligand to the TGFβ signaling receptor, *Cell,* 73, 1435, 1993.
29. **Cheifetz, S., Bellon, T., Cales, C., Vera, S., Bernabeu, C., Massague, J., and Letarte, M.,** Endoglin is a component of the TGF-β receptor system in human endothelial cells, *J. Biol. Chem.,* 267, 19027, 1992.
30. **Cheifetz, S., Hernandez, H., Laiho, M., Ten Dijke, P., Iwata, K. K., and Massague, J.,** Distinct transforming growth factor-β receptor subsets as determinants of cellular responsiveness to three TGF-β isoforms, *J. Biol. Chem.,* 265, 20533, 1990.
31. **Chen, R.-H., Ebner, R., and Derynck, R.,** Inactivation of the type II receptor reveals two receptor pathways for the diverse TGF-β activities, *Science,* 260, 1335, 1993.
32. **Tsuji, T., Okada, F., Yamaguchi, K., and Nakamura, T.,** Molecular cloning of the large subunit of transforming growth factor type β masking protein and expression of the mRNA in various rat tissues, *Proc. Natl. Acad. Sci. U.S.A.,* 87, 8835, 1990.
33. **Wakefield, L. M., Smith, D. M., Flanders, K. C., and Sporn, M. B.,** Latent transforming growth factor-β from human platelets, *J. Biol. Chem.,* 263, 7646, 1988.
34. **Yamaguchi, Y., Mann, D. M., and Ruoslahti, E.,** Negative regulation of transforming growth factor-β by the proteoglycan decorin, *Nature,* 346, 281, 1990.
35. **Flaumenhaft, R., Kojima, S., Abe, M., and Rifkin, D. B.,** Activation of latent transforming growth factor β, *Adv. Pharmacol.,* 24, 51, 1993.
36. **Antonelli-Orlidge, A., Saunders, K. B., Smith, S. R., and D'Amore, P. A.,** An activated form of transforming growth factor β is produced by cocultures of endothelial cells and pericytes, *Proc. Natl. Acad. Sci. U.S.A.,* 86, 4544, 1989.
37. **Rifkin, D. B., Kojima, S., Abe, M., and Harpel, J. G.,** TGF-beta: structure, function and formation, *Thromb. Haem.,* 70, 177, 1993.
38. **Coffey, R. J., Jr., Kost, L. J., Lyons, R. M., Moses, H. L., and LaRusso, N. F.,** Hepatic processing of transforming growth factor β in the rat uptake, metabolism and biliary excretion, *J. Clin. Invest.,* 80, 750, 1987.
39. **Assoian, R. K., Fleurdelys, B. E., Stevenson, H. C., Miller, P. J., Madtes, D. K., Raines, E. W., Ross, R., and Sporn, M. B.,** Expression and secretion of type β transforming growth factor by activated human macrophages, *Proc. Natl. Acad. Sci. U.S.A.,* 84, 6020, 1987.
40. **Nelson, B. J., Danielpour, D., Rossio, J. L., Turpin, J., and Nacy, C. A.,** Interleukin-2 suppresses activated macrophage intracellular killing activity by inducing macrophages to secrete TGF-β, *J. Leuk. Biol.,* 55, 81, 1994.

41. **Twardzik, E., Mikovits, J., Ranchalis, J., Purchio, A., Ellingsworth, L., and Ruscetti, F.,** Gamma interferon induced activation of latent transforming growth factor beta by human monocytes, *Ann. N. Y. Acad. Sci.,* 593, 237, 1990.
42. **Wallick, S. C., Figari, I. S., Morris, R. E., Levinson, A. D., and Palladino, M. A.,** Immunoregulatory role of transforming growth factor β (TGF-β) in development of killer cells: comparison of active and latent TGF-β_1, *J. Exp. Med.,* 172, 1777, 1990.
43. **Lucas, C., Bald, L. N., Fendly, B. M., Mora-Worms, M., Figari, I. S., Patzer, E. J., and Palladino, M. A.,** The autocrine production of transforming growth factor-β_1 during lymphocyte activation, *J. Immunol.,* 145, 1415, 1990.
44. **Danielpour, D., Dart, L. L., Flanders, K. C., Roberts, A. B., and Sporn, M. B.,** Immunodetection and quantitation of the two forms of transforming growth factor-β (TGF-β 1 and TGF-β 2) secreted by cells in culture, *J. Cell. Physiol.,* 138, 79, 1989.
45. **Absher, M., Baldor, L., and Kelly, J.,** A rapid colorimetric bioassay for transforming growth factor β (TGF-beta) using a microwell plate reader, *J. Immunol. Meth.,* 138, 301, 1991.
46. **Wakefield, L. M., Smith, D. M., Masui, T., Harris, C. C., and Sporn, M. B.,** Distribution and modulation of the cellular receptor for transforming growth factor-β, *J. Cell Biol.,* 105, 965, 1987.
47. **Keski-Oja, J., Lyons, R. M., and Moses, H. L.,** Immunodetection and modulation of cellular growth with antibodies against native transforming growth factor-β, *Cancer Res.,* 47, 6451, 1987.
48. **Dasch, J. R., Pace, D. R., Waegell, W., Inenaga, D., and Ellingsworth, L.,** Monoclonal antibodies recognizing transforming growth factor-β. Bioactivity neutralization and transforming growth factor beta2 affinity purification, *J. Immunol.,* 142, 1536, 1989.
49. **Pelton, R. W., Saxena, B., Jones, M., Moses, H. L., and Gold, L. I.,** Immunohistochemical localization of TGFβ1, TGFβ2, and TGFβ3 in the mouse embryo. Expression patterns suggest multiple roles during embryonic development, *J. Cell Biol.,* 115, 1091, 1991.
50. **Santambrogio, L., Hochwald, G. M., Saxena, B., Leu, C. H., Martz, J. E., Carlino, J. A., Ruddle, N. H., Palladino, M. A., Gold, L. I., and Thorbecke, G. J.,** Studies on the mechanisms by which TGF-β protects against allergic encephalomyelitis: antagonism between endogenously produced TGF-β and TNF, *J. Immunol.,* 151, 1116, 1993.
51. **Ottmann, O. G. and Pelus, L. M.,** Differential proliferative effects of transforming growth factor-β on human hematopoietic progenitor cells, *J. Immunol.,* 140, 2661, 1988.
52. **Ellingsworth, L. R., Nakayama, D., Segarini, P., Dasch, J., Carrillo, P., and Waegell, W.,** Transforming growth factor-βs are equipotent growth inhibitors of interleukin-1-induced thymocyte proliferation, *Cell. Immunol.,* 114, 41, 1988.
53. **Chantry, D., Turner, M., and Feldmann, M.,** Interleukin-7 (murine pre-B cell growth factor/lymphopoietin 1) stimulates thymocyte growth: regulation by transforming growth factor β, *Eur. J. Immunol.,* 19, 783, 1989.
54. **Kim, K. J., Abrams, J., Alphonso, M., Pearce, M., Thorbecke, G. J., and Palladino, M. A.,** Role of endogenously produced interleukin-6 as a second signal in murine thymocyte proliferation induced by multiple cytokines: regulatory effects of transforming growth factor-β_1, *Cell. Immunol.,* 131, 261, 1990.
55. **Wahl, S. M., Hunt, D. A., Bansal, G., McCartney-Francis, N., Ellingsworth, L., and Allen, J. B.,** Bacterial cell wall-induced immunosuppression. Role of transforming growth factor β, *J. Exp. Med.,* 168, 1403, 1988.
56. **Stoeck, M., Ruegg, C., Miescher, S., Carrel, S., Cox, D., Von Fliedner, V., and Alkan, S.,** Comparison of the immunosuppressive properties of milk growth factor and transforming growth factors β1 and β2, *J. Immunol.,* 143, 3258, 1989.
57. **Ruegemer, J. J., Ho, S. N., Augustine, J. A., Schlager, J. W., Bell, M. P., McKean, D. J., and Abraham, R. T.,** Regulatory effects of transforming growth factor-β on IL-2- and IL-4-dependent T cell-cycle progression, *J. Immunol.,* 144, 1767, 1990.
58. **Ahuja, S. S., Paliogianni, F., Yamada, H., Balow, J. E., and Boumpas, D. T.,** Effect of transforming growth factor-β on early and late activation events in human T cells, *J. Immunol.,* 150, 3109, 1993.
59. **Koff, A., Ohtsuki, M., Polyak, K., Roberts, J. M., and Massague, J.,** Negative regulation of G1 in mammalian cells: inhibition of cyclin E-dependent kinase by TGF-β, *Science,* 260, 536, 1993.

60. **Kehrl, J. H., Wakefield, L. M., Roberts, A. B., Jakowlew, S., Alvarez-Mon, M., Derynck, R., Sporn, M. B., and Fauci, A. S.,** Production of transforming growth factor β by human T lymphocytes and its potential role in the regulation of T cell growth, *J. Exp. Med.,* 163, 1037, 1986.
61. **Li, B., Sehajpal, P. K., Khanna, A., Vlassara, H., Cerami, A., Stenzel, K. H., and Suthanthiran, M.,** Differential regulation of transforming growth factor β and interleukin-2 genes in human T cells: demonstration by usage of novel competitor DNA constructs in the quantitative polymerase chain reaction, *J. Exp. Med.,* 174, 1259, 1991.
62. **Kim, S.-J., Jeang, K.-T., Glick, A. B., Sporn, M. B., and Roberts, A. B.,** Promoter sequences of the human transforming growth factor-β_1 gene responsive to transforming growth factor-β_1 autoinduction, *J. Biol. Chem.,* 264, 7041, 1989.
63. **Bristol, L. A., Ruscetti, F. W., Brody, D. T., and Durum, S. K.,** IL-1-alpha induces expression of active transforming growth factor-β in nonproliferating T cells via a post-transcriptional mechanism, *J. Immunol.,* 145, 4108, 1990.
64. **Zhou, D., Chrest, F. J., Adler, W., Munster, A., and Winchurch, R. A.,** Age-related changes in the expression of the TGF-β receptor on $CD4^+$ and $CD8^+$ subsets of T cells, *AGING: Immunol. Infect. Dis.,* 3, 217, 1992.
65. **Yamamoto, H., Hirayama, M., Genyea, C., and Kaplan, J.,** TGF-β mediates natural suppressor activity of IL-2-activated lymphocytes, *J. Immunol.,* 152, 3842, 1994.
66. **Ranges, G. E., Figari, I., Espevik, S. T., and Palladino, M. A., Jr.,** Inhibition of cytotoxic T cell development by transforming growth factor β and reversal by recombinant tumor necrosis factor alpha, *J. Exp. Med.,* 166, 991, 1987.
67. **Fontana, A., Frei, K., Bodmer, S., Hofer, E., Schreier, M. H., Palladino, M. A., Jr., and Zinkernagel, R. M.,** Transforming growth factor-β inhibits the generation of cytotoxic T cells in virus-infected mice, *J. Immunol.,* 143, 3230, 1989.
68. **Inge, T. H., Hoover, S. K., Susskind, B. M., Barrett, S. K., and Bear, H. D.,** Inhibition of tumor-specific cytotoxic T-lymphocyte responses by transforming growth factor β_1, *Cancer Res.,* 52, 1386, 1992.
69. **Kasid, A., Bell, G. I., and Director, E. P.,** Effects of transforming growth factor-β on human lymphokine-activated killer cell precursors. Autocrine inhibition of cellular proliferation and differentiation to immune killer cells, *J. Immunol.,* 141, 690, 1988.
70. **Jin, B., Scott, J. J. L., Vadas, M. A., and Burns, G. F.,** TGFbeta downregulates TLiSA1 expression and inhibits the differentiation of precursor lymphocytes into CTL and LAK cells, *Immunology,* 66, 570, 1989.
71. **Su, H. C., Leite-Morris, K. A., Braun, L., and Biron, C. A.,** A role for transforming growth factor-β-1 in regulating natural killer cell and T lymphocyte proliferative responses during acute infection with lymphocytic choriomeningitis virus, *J. Immunol.,* 147, 2717, 1991.
72. **Rook, A. H., Kehrl, J. H., Wakefield, L. M., Roberts, A. B., Sporn, M. B., Burlington, D. B., Lane, H. C., and Fauci, A. S.,** Effects of transforming growth factor β on the functions of natural killer cells. Depressed cytolytic activity and blunting of interferon responsiveness, *J. Immunol.,* 136, 3916, 1986.
73. **Ortaldo, J. R., Mason, A. T., O'Shea, J. J., Smyth, M. J., Falk, L. A., Kennedy, I. C. S., Longo, D. L., and Ruscetti, F. W.,** Mechanistic studies of transforming growth factor-β inhibition of IL-2 dependent activation of $CD3^-$ large granular lymphocyte functions. Regulation of IL-2Rbeta (p75) signal transduction, *J. Immunol.,* 146, 3791, 1991.
74. **Smyth, M. J., Strobl, S. L., Young, H. A., Ortaldo, J. R., and Ochoa, A. C.,** Regulation of lymphokine-activated killer activity and pore-forming protein gene expression in human peripheral blood $CD8^+$ T lymphocytes. Inhibition by transforming growth factor-β, *J. Immunol.,* 146, 3289, 1991.
75. **Geller, R. L., Smyth, M. J., Strobl, S. L., Bach, F. H., Ruscetti, F. W., Longo, D. L., and Ochoa, A. C.,** Generation of lymphokine-activated killer activity in T cells. Possible regulatory circuits, *J. Immunol.,* 146, 3280, 1991.
76. **Hambor, J. E., Kaplan, D. R., and Tykocinski, M. L.,** CD8 functions as an inhibitory ligand in mediating the immunoregulatory activity of $CD8^+$ cells, *J. Immunol.,* 145, 1646, 1990.
77. **Inge, T. H., McCoy, K. M., Susskind, B. M., Barrett, S. K., Zhao, G., and Bear, H. D.,** Immunomodulatory effects of transforming growth factor-β on T lymphocytes, *J. Immunol.,* 148, 3847, 1992.

78. **Suda, T. and Zlotnik, A.,** *In vitro* induction of CD8 expression on thymic pre-T cells. I. Transforming growth factor-β and tumor necrosis factor-alpha induce CD8 expression on $CD8^-$ thymic subsets including the $CD25^+CD3^-CD4^-CD8^-$ pre-T cell subset, *J. Immunol.,* 148, 1737, 1992.
79. **Takahama, Y., Letterio, J. J., Suzuki, H., Farr, A. G., and Singer, A.,** Early progression of thymocytes along the CD4/CD8 developmental pathway is regulated by a subset of thymic epithelial cells expressing transforming growth factor β, *J. Exp. Med.,* 179, 1495, 1994.
80. **Lee, H.-M. and Rich, S.,** Differential activation of $CD8^+$ T cells by transforming growth factor-β1, *J. Immunol.,* 151, 668, 1993.
81. **Quere, P. and Thorbecke, G. J.,** Multiple suppressive effects of transforming growth factor β1 on the immune response in chickens, *Cell. Immunol.,* 129, 468, 1990.
82. **Gray, J. D. and Horwitz, D. A.,** A role for transforming growth factor beta, facilitated by NK cells, in activating $CD8^+$ T lymphocytes to suppress antibody production, *FASEB J.,* 8, A479, 1994.
83. **Swain, S. L., Huston, G., Tonkonogy, S., and Weinberg, A.,** Transforming growth factor-β and IL-4 cause helper T cell precursors to develop into distinct effector helper cells that differ in lymphokine secretion pattern and cell surface phenotype, *J. Immunol.,* 147, 2991, 1991.
84. **Fargeas, C., Wu, C. Y., Nakajima, T., Cox, D., Nutman, T., and Delespesse, G.,** Differential effect of transforming growth factor β on the synthesis of T_h1- and T_h2-like lymphokines by human T lymphocytes, *Eur. J. Immunol.,* 22, 2173, 1992.
85. **Nagelkerken, L., Gollob, K. J., Tielemans, M., and Coffman, R. L.,** Role of transforming growth factor-β in the preferential induction of T helper cells of type 1 by staphylococcal enterotoxin B, *Eur. J. Immunol.,* 23, 2306, 1993.
86. **Schmitt, E., Hoehn, P., Huels, C., Goedert, S., Palm, N., Rude, E., and Germann, T.,** T helper type 1 development of naive $CD4^+$ T cells requires the coordinate action of interleukin-12 and interferon-gamma and is inhibited by transforming growth factor-β, *Eur. J. Immunol.,* 24, 793, 1994.
87. **Picker, L. J., Treer, J. R., Ferguson-Darnell, B., Collins, P. A., Buck, D., and Terstappen, L. W. M. M.,** Control of lymphocyte recirculation in man. I. Differential regulation of the peripheral lymph node homing receptor L-selectin on T cells during the virgin to memory cell transition, *J. Immunol.,* 150, 1105, 1993.
88. **Picker, L. J., Treer, J. R., Ferguson-Darnell, B., Collins, P. A., Bergstresser, P. R., and Terstappen, L. W. M. M.,** Control of lymphocyte recirculation in man. II. Differential regulation of the cutaneous lymphocyte-associated antigen, a tissue-selective homing receptor for skin-homing T cells, *J. Immunol.,* 150, 1122, 1993.
89. **Lee, G., Namen, A. E., Gillis, S., Ellingsworth, L. R., and Kincade, P. W.,** Normal B cell precursors responsive to recombinant murine IL-7 and inhibition of IL-7 activity by transforming growth factor-β, *J. Immunol.,* 142, 3875, 1989.
90. **Kehrl, J. H., Roberts, A. B., Wakefield, L. M., Jakowlew, S., Sporn, M. B., and Fauci, A. S.,** Transforming growth factor β is an important immunomodulatory protein for human B lymphocytes, *J. Immunol.,* 137, 3855, 1986.
91. **Kehrl, J. H., Taylor, A. S., Delsing, G. A., Roberts, A. B., Sporn, M. B., and Fauci, A. S.,** Further studies of the role of transforming growth factor-β in human B cell function, *J. Immunol.,* 143, 1868, 1989.
92. **Smeland, E. B., Blomhoff, H. K., Holte, H., Ruud, E., Beiske, K., Funderud, S., Godal, T., and Ohlsson, R.,** Transforming growth factor type beta (TGF β) inhibits G_1 to S transition, but not activation of human B lymphocytes, *Exp. Cell Res.,* 171, 213, 1987.
93. **Cross, D. and Cambier, J. C.,** Transforming growth factor beta 1 has differential effects on B cell proliferation and activation antigen expression, *J. Immunol.,* 144, 432, 1990.
94. **Ewen, M. E., Sluss, H. K., Whitehouse, L. L., and Livingston, D. M.,** TGF-β inhibition of Cdk4 synthesis is linked to cell cycle arrest, *Cell,* 74, 1009, 1993.
95. **Golay, J., Cusmano, G., and Introna, M.,** Independent regulation of c-myc, B-myb, and c-myb gene expression by inducers and inhibitors of proliferation in human B lymphocytes, *J. Immunol.,* 149, 300, 1992.

96. **Barrett, T. B., Shu, G. L., Draves, K. E., Pezzutto, A., and Clark, E. A.,** Signaling through CD19, Fc receptors or transforming growth factor-β: each inhibits the activation of resting human B cells differently, *Eur. J. Immunol.,* 20, 1053, 1990.
97. **Flescher, E., Fossum, D., Ballester, A., Maizel, A., Sharma, S., and Talal, N.,** Characterization of B cell growth in systemic lupus erythematosus. Effects of recombinant 12-kDa B cell growth factor, interleukin-4 and transforming growth factor-β, *Eur. J. Immunol.,* 20, 2425, 1990.
98. **Petit-Koskas, E., Genot, E., Lawrence, D., and Kolb, J.-P.,** Inhibition of the proliferative response of human B lymphocytes to B cell growth factor by transforming growth factor-β, *Eur. J. Immunol.,* 18, 111, 1988.
99. **Kumar, A., Rogers, T., Maizel, A., and Sharma, S.,** Loss of transforming growth factor β receptors and its effects on the growth of EBV-transformed human B cells, *J. Immunol.,* 147, 998, 1991.
100. **Janssen, O., Gillis, S., and Kabelitz, D.,** *In vitro* transformation by Epstein-Barr virus induces a switch in growth factor and anti-IgM responsiveness in a human leukemic B cell clone, *Eur. J. Immunol.,* 20, 7, 1990.
101. **Sing, G. K., Ruscetti, F. W., Beckwith, M., Keller, J. R., Ellingsworth, L., Urba, W. J., and Longo, D. L.,** Growth inhibition of a human lymphoma cell line: induction of a transforming growth factor-β-mediated autocrine negative loop by phorbol myristate acetate, *Cell Growth Different.,* 1, 549, 1990.
102. **Newcom, S. R., Tagra, K. K., and Kadin, M. E.,** Neutralizing antibodies against transforming growth factor β potentiate the proliferation of Ki-1 positive lymphoma cells, *Am. J. Pathol.,* 140, 709, 1992.
103. **Berg, D. J. and Lynch, R. G.,** Immune dysfunction in mice with plasmacytomas. I. Evidence that transforming growth factor-β contributes to the altered expression of activation receptors on host B lymphocytes, *J. Immunol.,* 146, 2865, 1991.
104. **Bartlett, W. C., Purchio, A., Fell, H. P., and Noelle, R. J.,** Cognate interactions between helper T cells and B cells. VI. TGF-β inhibits B cell activation and antigen-specific, physical interactions between T_h and B cells, *Lymphokine Cytokine Res.,* 10, 177, 1991.
105. **Gold, M. R., Gajewski, T. F., and DeFranco, A. L.,** Regulation of anti-immunoglobulin-induced B lymphoma growth arrest by transforming growth factor β 1 and dexamethasone, *Int. Immunol.,* 3, 1091, 1991.
106. **Coffman, R. L., Lebman, D. A., and Shrader, B.,** Transforming growth factor β specifically enhances IgA production by lipopolysaccharide-stimulated murine B lymphocytes, *J. Exp. Med.,* 170, 1039, 1989.
107. **Kim, P.-H. and Kagnoff, M. F.,** Transforming growth factor β1 increases IgA isotype switching at the clonal level, *J. Immunol.,* 145, 3773, 1990.
108. **Sonoda, E., Matsumoto, R., Hitoshi, Y., Ishii, T., Sugimoto, M., Araki, S., Tominaga, A., Yamaguchi, N., and Takatsu, K.,** Transforming growth factor β induces IgA production and acts additively with interleukin-5 for IgA production, *J. Exp. Med.,* 170, 1415, 1989.
109. **Matsuoka, M., Yoshida, K., Maeda, T., Usuda, S., and Sakano, H.,** Switch circular DNA formed in cytokine-treated mouse splenocytes: evidence for intramolecular DNA deletion in immunoglobulin class switching, *Cell,* 62, 135, 1990.
110. **Lebman, D. A., Nomura, D. Y., Coffman, R. L., and Lee, F. D.,** Molecular characterization of germ-line immunoglobulin A transcripts produced during transforming growth factor type β-induced isotype switching, *Proc. Natl. Acad. Sci. U.S.A.,* 87, 3962, 1990.
111. **Islam, K. B., Nilsson, L., Sideras, P., Hammarstrom, L., and Smith, C. I. E.,** TGF-β 1 induces germ-line transcripts of both IgA subclasses in human B lymphocytes, *Int. Immunol.,* 3, 1099, 1991.
112. **Shockett, P. and Stavnezer, J.,** Effect of cytokines on switching to IgA and alpha germline transcripts in the B lymphoma 1.29 mu. Transforming growth factor-β activates transcription of the unrearranged C-alpha gene, *J. Immunol.,* 147, 4374, 1991.
113. **McIntyre, T. M., Klinman, D. R., Rothman, P., Lugo, M., Dasch, J. R., Mond, J. J., and Snapper, C. M.,** Transforming growth factor β_1 selectively stimulates immunoglobulin G2b secretion by lipopolysaccharide-activated murine B cells, *J. Exp. Med.,* 177, 1031, 1993.

114. **Snapper, C. M., Waegell, W., Beernink, H., and Dasch, J. R.,** Transforming growth factor-β1 is required for secretion of IgG of all subclasses by LPS-activated murine B cells *in vitro*, *J. Immunol.,* 151, 4625, 1993.
115. **Butcher, E. C., Rouse, R. V., Coffman, R. L., Nottenburg, C. N., Hardy, R. R., and Weissman, R. L.,** Surface phenotype of Peyer's patch germinal center cells: implications for the role of germinal centers in B cell differentiation, *J. Immunol.,* 129, 2698, 1982.
116. **Lebman, D. A., Griffin, P. M., and Cebra, J. J.,** Relationship between expression of IgA by Peyer's patch cells and functional IgA memory cells, *J. Exp. Med.,* 166, 1405, 1987.
117. **Keller, J. R., Mcniece, I. K., Sill, K. T., Ellingsworth, L. R., Quesenberry, P. J., Sing, G. K., and Ruscetti, F. W.,** Transforming growth factor β directly regulates primitive murine hematopoietic cell proliferation, *Blood,* 75, 596, 1990.
118. **Ishibashi, T., Miller, S. L., and Burstein, S. A.,** Type β transforming growth factor is a potent inhibitor of murine megakaryocytopoiesis *in vitro*, *Blood,* 69, 1737, 1987.
119. **Keller, J. R., Jacobsen, S. E. W., Sill, K. T., Ellingsworth, L. R., and Ruscetti, F. W.,** Stimulation of granulopoiesis by transforming growth factor β: synergy with granulocyte/macrophage-colony-stimulating factor, *Proc. Natl. Acad. Sci. U.S.A.,* 88, 7190, 1991.
120. **Taetle, R., Payne, C., Dos Santos, B., Russell, M., and Segarini, P.,** Effects of transforming growth factor β1 on growth and apoptosis of human acute myelogenous leukemia cells, *Cancer Res.,* 53, 3386, 1993.
121. **Vos, De S., Brach, M. A., Asano, Y., Ludwig, W.-D., Bettelheim, P., Gruss, H.-J., and Herrmann, F.,** Transforming growth factor-β 1 interferes with the proliferation-inducing activity of stem cell factor in myelogenous leukemia blasts through functional down-regulation of the c-kit proto-oncogene product, *Cancer Res.,* 53, 3638, 1993.
122. **Testa, U., Masciulli, R., Tritarelli, E., Pustorino, R., Mariani, G., Martucci, R., Barberi, T., Camagna, A., Valtieri, M., and Peschle, C.,** Transforming growth factor-β potentiates vitamin D3-induced terminal monocytic differentiation of human leukemic cell lines, *J. Immunol.,* 150, 2418, 1993.
123. **Alam, R., Forsythe, P., Stafford, S., and Fukuda, Y.,** Transforming growth factor β abrogates the effects of hematopoietins on eosinophils and induces their apoptosis, *J. Exp. Med.,* 179, 1041, 1994.
124. **Yanagihara, K. and Tsumuraya, M.,** Transforming growth factor β_1 induces apoptotic cell death in cultured human gastric carcinoma cells, *Cancer Res.,* 52, 4042, 1992.
125. **Rotello, R. J., Lieberman, R. C., Purchio, A. F., and Gerschenson, L. E.,** Coordinated regulation of apoptosis and cell proliferation by transforming growth factor β1 in cultured uterine epithelial cells, *Proc. Natl. Acad. Sci. U.S.A.,* 88, 3412, 1991.
126. **Lin, J.-K. and Chou, C.-K.,** *In vitro* apoptosis in the human hepatoma cell line induced by transforming growth factor β1, *Cancer Res.,* 2, 385, 1992.
127. **Wahl, S. M., Hunt, D. A., Wakefield, L. M., McCartney-Francis, N., Wahl, L. M., Roberts, A. B., and Sporn, M. B.,** Transforming growth factor type β induces monocyte chemotaxis and growth factor production, *Proc. Natl. Acad. Sci. U.S.A.,* 84, 5788, 1987.
128. **Reibman, J., Meixler, S., Lee, T. C., Gold, L. I., Cronstein, B. N., Haines, K. A., Kolasinski, S. L., and Weissmann, G.,** Transforming growth factor β1, a potent chemoattractant for human neutrophils, bypasses classic signal-transduction pathways, *Proc. Natl. Acad. Sci. U.S.A.,* 88, 6805, 1991.
129. **Postlethwaite, A. E., Keski-Oja, J., Moses, H. L., and Kang, A. H.,** Stimulation of the chemotactic migration of human fibroblasts by transforming growth factor β, *J. Exp. Med.,* 165, 251, 1987.
130. **Steinhilber, D., Radmark, O., and Samuelsson, B.,** Transforming growth factor β upregulates 5-lipoxygenase activity during myeloid cell maturation, *Proc. Natl. Acad. Sci. U.S.A.,* 90, 5984, 1993.
131. **Strassmann, G., Cone, J. L., Arthur, P. M., Guertin, M., and Herrfeldt, J.,** Effect of platelet-derived transforming growth factor (TGF) type β1 on murine inflammatory mononuclear phagocytes: increased fibronectin production, *Cell. Immunol.,* 121, 306, 1989.
132. **Parekh, T., Saxena, B., Reibman, J., Cronstein, B. N., and Gold, L. I.,** Neutrophil chemotaxis in response to TGF-β isoforms (TGF-β1, TGF-β2, TGF-β3) is mediated by fibronectin, *J. Immunol.,* 152, 2456, 1994.

133. **Nelson, B. J., Ralph, P., Green, S. J., and Nacy, C. A.,** Differential susceptibility of activated macrophage cytotoxic effector reactions to the suppressive effects of transforming growth factor-β1, *J. Immunol.,* 146, 1849, 1991.
134. **Ding, A., Nathan, C. F., Graycar, J., Derynck, R., Stuehr, D. J., and Srimal, S.,** Macrophage deactivating factor and transforming growth factors -β1, -β2, and -β3 inhibit induction of macrophage nitrogen oxide synthesis by IFN-gamma, *J. Immunol.,* 145, 940, 1990.
135. **Oswald, I. P., Gazzinelli, R. T., Sher, A., and James, S. L.,** IL-10 synergizes with IL-4 and transforming growth factor-β to inhibit macrophage cytotoxic activity, *J. Immunol.,* 148, 3578, 1992.
136. **Gazzinelli, R. T., Oswald, I. P., Hieny, S., James, S. L., and Sher, A.,** The microbicidal activity of interferon-gamma-treated macrophages against *Trypanosoma cruzi* involves an L-arginine-dependent, nitrogen oxide-mediated mechanism inhibitable by interleukin-10 and transforming growth factor-β, *Eur. J. Immunol.,* 22, 2501, 1992.
137. **Vodovotz, Y., Bogdan, C., Paik, J., Xie, Q.-W., and Nathan, C.,** Mechanisms of suppression of macrophage nitric oxide release by transforming growth factor β, *J. Exp. Med.,* 178, 605, 1993.
138. **Barral, A., Barral-Netto, M., Yong, E. C., Brownell, C. E., Twardzik, D. R., and Reed, S. G.,** Transforming growth factor β as a virulence mechanism for *Leishmania braziliensis, Proc. Natl. Acad. Sci. U.S.A.,* 90, 3442, 1993.
139. **Barral-Netto, M., Barral, A., Brownell, C. E., Skeiky, Y. A. W., Ellingsworth, L. R., Twardzik, D. R., and Reed, S. G.,** Transforming growth factor-β in *Leishmanial* infection: a parasite escape mechanism, *Science,* 257, 545, 1992.
140. **Poli, G., Kinter, A. L., Justement, J. S., Bressler, P., Kehrl, J. H., and Fauci, A. S.,** Retinoic acid mimics transforming growth factor β in the regulation of human immunodeficiency virus expression in monocytic cells, *Proc. Natl. Acad. Sci. U.S.A.,* 89, 2689, 1992.
141. **Poli, G., Kinter, A. L., Justement, J. S., Bressler, P., Kehrl, J. H., and Fauci, A. S.,** Transforming growth factor β suppresses human immunodeficiency virus expression and replication in infected cells of the monocyte/macrophage lineage, *J. Exp. Med.,* 173, 589, 1991.
142. **Pinson, D. M., LeClaire, R. D., Lorsbach, R. B., Parmely, M. J., and Russell, S. W.,** Regulation by transforming growth factor-β1 of expression and function of the receptor for IFN-gamma on mouse macrophages, *J. Immunol.,* 149, 2028, 1992.
143. **Tanaka, M., Lee, K., Yodoi, J., Saito, H., Iwai, Y., Kim, K.-M., Morita, M., Mayumi, M., and Mikawa, H.,** Regulation of Fc_e receptor 2 (CD23) expression on a human eosinophilic cell line EoL3 and a human monocytic cell line U937 by transforming growth factor β, *Cell. Immunol.,* 122, 96, 1989.
144. **Wong, H. L., Welch, G. R., Brandes, M. E., and Wahl, S. M.,** IL-4 antagonizes induction of FcgammaRIII (CD16) expression by transforming growth factor-β on human monocytes, *J. Immunol.,* 147, 1843, 1991.
145. **Phillips, J. H., Chang, C., and Lanier, L. L.,** Platelet-induced expression of FcgammaRIII (CD16) on human monocytes, *Eur. J. Immunol.,* 21, 895, 1991.
146. **Epstein, S. P., Baer, R. L., Thorbecke, G. J., and Belsito, D. V.,** Immunosuppressive effects of transforming growth factor β: inhibition of the induction of Ia antigen on Langerhans cells by cytokines and of the contact hypersensitivity response, *J. Invest. Dermatol.,* 96, 832, 1991.
147. **Wilbanks, G. A., Mammolenti, M., and Streilein, J. W.,** Studies on the induction of anterior chamber-associated immune deviation (ACAID). III. Induction of ACAID depends upon intraocular transforming growth factor-β, *Eur. J. Immunol.,* 22, 165, 1992.
148. **Demidem, A., Taylor, J. R., Grammer, S. F., and Streilein, J. W.,** Comparison of effects of transforming growth factor-β and cyclosporin A on antigen-presenting cells of blood and epidermis, *J. Invest. Dermatol.,* 96, 401, 1991.
149. **Heino, J., Ignotz, R. A., Hemler, M. E., Crouse, C., and Massague, J.,** Regulation of cell adhesion receptors by transforming growth factor-β, *J. Biol. Chem.,* 264, 380, 1989.
150. **Wahl, S. M., Allen, J. B., Weeks, B. S., Wong, H. L., and Klotman, P. E.,** Transforming growth factor β enhances integrin expression and type IV collagenase secretion in human monocytes, *Proc. Natl. Acad. Sci. U.S.A.,* 90, 4577, 1993.
151. **Gamble, J. R. and Vadas, M. A.,** Endothelial cell adhesiveness for human T lymphocytes is inhibited by transforming growth factor-β_1, *J. Immunol.,* 146, 1149, 1991.

152. **Cai, J.-P., Falanga, V., and Chin, Y.-H.,** Transforming growth factor-β regulates the adhesive interactions between mononuclear cells and microvascular endothelium, *J. Invest. Dermatol.,* 97, 169, 1991.
153. **Chin, Y.-H., Cai, J.-P., and Xu, X.-M.,** Transforming growth factor-β 1 and IL-4 regulate the adhesiveness of Peyer's patch high endothelial venule cells for lymphocytes, *J. Immunol.,* 148, 1106, 1992.
154. **Bereta, J., Bereta, M., Coffman, F. D., Cohen, S., and Cohen, M. C.,** Inhibition of basal and tumor necrosis factor-enhanced binding of murine tumor cells to murine endothelium by transforming growth factor-β, *J. Immunol.,* 148, 2932, 1992.
155. **Gamble, J. R. and Vadas, M. A.,** Endothelial adhesiveness for blood neutrophils is inhibited by transforming growth factor-β, *Science,* 242, 97, 1988.
156. **Chin, Y. H. and Xu, X. M.,** IL-4 regulates transforming growth factor-beta receptor type 1 and type II expression on lymph node high endothelial venule cells, *FASEB J.,* 8, A972, 1994.
157. **Satoh, J., Kim, S. U., Kastrukoff, L. F., and Takei, F.,** Expression and induction of intercellular adhesion molecules and major histocompatibility complex antigens on cultured murine oligodendrocytes and astrocytes, *J. Neurosci. Res.,* 29, 1, 1991.
158. **Kraus, E., Schneider-Schaulies, S., Miyasaka, M., Tamatani, T., and Sedgwick, J.,** Augmentation of major histocompatibility complex class I and ICAM-1 expression on glial cells following measles virus infection: evidence for the role of type-1 interferon, *Eur. J. Immunol.,* 22, 175, 1992.
159. **Ritchie, A. J., Johnson, D. R., Ewenstein, B. M., and Pober, J. S.,** Tumor necrosis factor induction of endothelial cell surface antigens is independent of protein kinase C activation or inactivation, *J. Immunol.,* 146, 3056, 1991.
160. **Taichman, D. B., Cybulsky, M. I., Djaffer, I., Longenecker, B. M., Toxido, J., Rice, G. E., Arufo, A., and Bevilacqua, M. P.,** Tumor cell surface alpha 4 beta 1 integrin mediates adhesion to vascular endothelium: demonstration of an interaction with the N-terminal domains of INCAM-110/VCAM-1, *Cell Regul.,* 2, 347, 1991.
161. **Gamble, J. R., Khew-Goodall, Y., and Vadas, M. A.,** Transforming growth factor-beta inhibits E-selectin expression on human endothelial cells, *J. Immunol.,* 150, 4494, 1993.
162. **Schluns, K. S., Grutkoski, P. S., Engelmann, G. L., and Le, P. T.,** Expression of the TGF-β family and the TGF-β receptor in the human thymic epithelial cells (TEC), *FASEB J.,* 8, A479, 1994.
163. **Wahl, S. M., Hunt, D. A., Wong, H. L., Dougherty, S., McCartney-Francis, N., Wahl, L. M., Ellingsworth, L., Schmidt, J. A., Hall, G., Roberts, A. B., and Sporn, M. B.,** Transforming growth factor-β is a potent immunosuppressive agent that inhibits IL-1-dependent lymphocyte proliferation, *J. Immunol.,* 140, 3026, 1988.
164. **Moore, S. C., Shaw, M. A., and Soderberg, L. S. F.,** Transforming growth factor-β is the major mediator of natural suppressor cells derived from normal bone marrow, *J. Leuk. Biol.,* 52, 596, 1992.
165. **Roth, M. D. and Golub, S. H.,** Human pulmonary macrophages utilize prostaglandins and transforming growth factor β1 to suppress lymphocyte activation, *J. Leuk. Biol.,* 53, 366, 1993.
166. **Stach, R. M. and Rowley, D. A.,** A first or dominant immunization. II. Induced immunoglobulin carries transforming growth factor β and suppresses cytolytic T cell responses to unrelated alloantigens, *J. Exp. Med.,* 178, 841, 1993.
167. **Karpus, W. J. and Swanborg, R. H.,** CD4+ suppressor cells inhibit the function of effector cells of experimental autoimmune encephalomyelitis through a mechanism involving transforming growth factor-β 1, *J. Immunol.,* 146, 1163, 1991.
168. **Miller, A., Lider, O., Roberts, A. B., Sporn, M. B., and Weiner, H. W.,** Suppressor T cells generated by oral tolerization to myelin basic protein suppress both *in vitro* and *in vivo* immune responses by the release of transforming growth factor β after antigen-specific triggering, *Proc. Natl. Acad. Sci. U.S.A.,* 89, 421, 1992.
169. **Kekow, J., Wachsman, W., McCutchan, J. A., Cronin, M., Carson, D. A., and Lotz, M.,** Transforming growth factor β and noncytopathic mechanisms of immunodeficiency in human immunodeficiency virus infection, *Proc. Natl. Acad. Sci. U.S.A.,* 87, 8321, 1990.
170. **Zhou, D., Munster, A., and Winchurch, R. A.,** Pathologic concentrations of interleukin-6 inhibit T cell responses via induction of activation of TGF-beta, *FASEB J.,* 5, 2582, 1991.

171. **Shull, M. M., Ormsby, I., Kier, A. B., Pawlowski, S., Diebold, R., Yin, J. M., Allen, R., Sidman, C., Proetzel, G., Calvin, D., Annunziata, N., and Doetschman, T.,** Targeted disruption of the mouse transforming growth factor-β1 gene results in multifocal inflammatory disease, *Nature,* 359, 693, 1992.
172. **Kulkarni, A. B., Huh, C.-G., Becker, D., Geiser, A., Lyght, M., Flanders, K. C., Roberts, A. B., Sporn, M. B., Ward, J. M., and Karlsson, S.,** Transforming growth factor β_1 null mutation in mice causes excessive inflammatory response and early death, *Proc. Natl. Acad. Sci. U.S.A.,* 90, 770, 1993.
173. **Geiser, A. G., Letterio, J. J., Kulkarni, A. B., Karlsson, S., Roberts, A. B., and Sporn, M. B.,** Transforming growth factor β 1 (TGF-β_1) controls expression of major histocompatibility genes in the postnatal mouse. Aberrant histocompatibility antigen expression in the pathogenesis of the TGF-beta_1 null mouse phenotype, *Proc. Natl. Acad. Sci. U.S.A.,* 90, 9944, 1993.
174. **Stitz, L., Planz, O., Bilzer, T., Frei, K., and Fontana, A.,** Transforming growth factor-β modulates T cell-mediated encephalitis caused by borna disease virus, *J. Immunol.,* 147, 3581, 1991.
175. **Pfister, H.-W., Frei, K., Ottnad, B., Koedel, U., Tomasz, A., and Fontana, A.,** Transforming growth factor β_2 inhibits cerebrovascular changes and brain edema formation in the tumor necrosis factor alpha-independent early phase of experimental pneumococcal meningitis, *J. Exp. Med.,* 176, 265, 1992.
176. **Czarniecki, C. W., Chiu, H. H., Wong, G. H. W., McCabe, S. M., and Palladino, M. A.,** Transforming growth factor-β 1 modulates the expression of class II histocompatibility antigens on human cells, *J. Immunol.,* 140, 4217, 1988.
177. **Carel, J.-C., Schreiber, R. D., Falqui, L., and Lacy, P. E.,** Transforming growth factor β decreases the immunogenicity of rat islet xenografts (rat to mouse) and prevents rejection in association with treatment of the recipients with a monoclonal antibody to interferon-gamma, *Proc. Natl. Acad. Sci. U.S.A.,* 87, 1591, 1990.
178. **Babany, G., Morris, R. E., Babany, I., and Kates, R. E.,** Evaluation of the *in vivo* dose-response relationship of immunosuppressive drugs using a mouse heart transplant model: application of cyclosporin, *J. Pharmacol. Exp. Ther.,* 244, 259, 1988.
179. **Kuruvilla, A. P., Shah, R., Hochwald, G. M., Liggitt, H. D., Palladino, M. A., and Thorbecke, G. J.,** Protective effect of transforming growth factor beta_1 on experimental autoimmune diseases in mice, *Proc. Natl. Acad. Sci. U.S.A.,* 88, 2918, 1991.
180. **Johns, L. D., Flanders, K. C., Ranges, G. E., and Sriram, S.,** Successful treatment of experimental allergic encephalomyelitis with transforming growth factor, *J. Immunol.,* 147, 1792, 1991.
181. **Racke, M. K., Dhib-Jalbut, S., Cannella, B., Albert, P. S., Raine, C. S., and McFarlin, D. E.,** Prevention and treatment of chronic relapsing experimental allergic encephalomyelitis by transforming growth factor-β_1, *J. Immunol.,* 146, 3012, 1991.
182. **Thorbecke, G. J., Shah, R., Leu, C. H., Kuruvilla, A. P., Hardison, A. M., and Palladino, M. A.,** Involvement of endogenous TNF-alpha and TGF-β produced during the induction of collagen type II arthritis in mice, *Proc. Natl. Acad. Sci. U.S.A.,* 89, 7375, 1992.
183. **Brandes, M. E., Allen, J. B., Ogawa, Y., and Wahl, S. M.,** TGF-β-1 suppresses leukocyte recruitment and synovial inflammation in experimental arthritis, *J. Clin. Invest.,* 87, 1108, 1991.
184. **Wahl, S. M., Allen, J. B., Wong, H. L., Dougherty, S. F., and Ellingsworth, L. R.,** Antagonistic and agonistic effects of transforming growth factor-β and IL-1 in rheumatoid synovium, *J. Immunol.,* 145, 2514, 1990.
185. **Allen, J. B., Manthey, C. L., Hand, A. R., Ohura, K., Ellingsworth, L., and Wahl, S. M.,** Rapid onset synovial inflammation and hyperplasia induced by transforming growth factor β, *J. Exp. Med.,* 171, 231, 1990.
186. **Hom, J. T., Butler, L. D., Riedl, P. E., and Bendele, A. M.,** The progression of the inflammation in established collagen-induced arthritis can be altered by treatments with immunological or pharmacological agents which inhibit T cell activities, *Eur. J. Immunol.,* 18, 881, 1988.
187. **Weinberg, A. D., Whitman, R., Swain, S. L., Morrison, W. J., Wyrick, G., Hoy, C., Vandenbark, A. A., and Offner, H.,** Transforming growth factor-β enhances the *in vivo* effector function and memory phenotype of antigen-specific T helper cells in experimental autoimmune encephalomyelitis, *J. Immunol.,* 148, 2109, 1992.

188. **Meade, R., Askenase, P. W., Geba, G. P., Neddermann, K., Jacoby, R. O., and Pasternak, R. D.,** Transforming growth factor-β_1 inhibits murine immediate and delayed type hypersensitivity, *J. Immunol.,* 149, 521, 1992.
189. **Male, D., Pryce, G., Hughes, C., and Lantos, P.,** Lymphocyte migration into brain modelled *in vitro:* control by lymphocyte activation, cytokines, and antigen, *Cell. Immunol.,* 127, 1, 1990.
190. **Yednock, T. A., Cannon, C., Fritz, L. C., Sanchez-Madrid, F., Steinman, L., and Karin, N.,** Prevention of experimental autoimmune encephalomyelitis by antibodies against alpha4β1 integrin, *Nature,* 356, 63, 1992.
191. **Cannella, B., Cross, A. H., and Raine, C. S.,** Upregulation and coexpression of adhesion molecules correlate with relapsing autoimmune demyelination in the central nervous system, *J. Exp. Med.,* 172, 1521, 1990.
192. **Sedgwick, J. D., Hughes, C. C., Male, D. K., MacPhee, I. A. M., and ter Meulen, V.,** Antigen-specific damage to brain vascular endothelial cells mediated by encephalitogenic and nonencephalitogenic $CD4^+$ T cell lines *in vitro, J. Immunol.,* 145, 2474, 1990.
193. **Claudio, L., Kress, Y., Factor, J., and Brosnan, C. F.,** Mechanisms of edema formation in experimental autoimmune encephalomyelitis. The contribution of inflammatory cells, *Am. J. Pathol.,* 137, 1, 1990.
194. **Cross, A. H., Cannella, B., Brosnan, C. F., and Raine, C. S.,** Homing to central nervous system vasculature by antigen-specific lymphocytes. I. Localization of C-labeled cells during acute, chronic, and relapsing experimental allergic encephalomyelitis, *Lab. Invest.,* 63, 162, 1990.
195. **Merrill, J. E., Kono, D. H., Clayton, J., Ando, D. G., Hinton, D. R., and Hofman, F. M.,** Inflammatory leukocytes and cytokines in the peptide-induced disease of experimental allergic encephalomyelitis in SJL and B10.PL mice, *Proc. Natl. Acad. Sci. U.S.A.,* 89, 574, 1992.
196. **May, M. J. and Ager, A.,** ICAM-1-independent lymphocyte transmigration across high endothelium: differential up-regulation by interferon-gamma, tumor necrosis factor-alpha and interleukin-1β, *Eur. J. Immunol.,* 22, 219, 1992.
197. **Baron, J. L., Madri, J. A., Ruddle, N. H., Hashim, G. C., and Janeway, A., Jr.,** Surface expression of alpha 4 integrin by CD4 cells is required for their entry into brain parenchyma, *J. Exp. Med.,* 177, 57, 1992.
198. **Powell, M. B., Mitchell, D., Lederman, J., Buckmeier, J., Zamvil, S. S., Graham, M., Ruddle, N. H., and Steinman, L.,** Lymphotoxin and tumor necrosis factor-alpha production by myelin basic protein-specific T cell clones correlates with encephalitogenicity, *Int. Immunol.,* 2, 8178, 1990.
199. **Racke, M. K., Cannella, B., Albert, P., Sporn, M., Raine, C. S., and McFarlin, D. E.,** Evidence of endogenous regulatory function of transforming growth factor-β in experimental allergic encephalomyelitis, *Int. Immunol.,* 4, 615, 1992.
200. **Phan, S. H., Gharaee-Kermani, M., McGarry, B., Kunkel, S. L., and Wolber, F. W.,** Regulation of rat pulmonary artery endothelial cell transforming growth factor-β production by IL-1β and tumor necrosis factor-alpha, *J. Immunol.,* 149, 103, 1992.
201. **Celada, A. and R. A. Maki,** Transforming growth factor-β enhances the M-CSF and GM-CSF-stimulated proliferation of macrophages, *J. Immunol.,* 148, 1102, 1992.
202. **Chantry, D., Turner, M., Abney, E., and Feldmann, M.,** Modulation of cytokine production by transforming growth factor-beta 1, *J. Immunol.,* 142, 4295, 1989.
203. **Dunham, D. M., Arkins, S., Edwards, C. K., III, Dantzer, R., and Kelley, K. W.,** Role of interferon-gamma in counteracting the suppressive effects of transforming growth factor-β2 and glucocorticoids on the production of tumor necrosis factor-alpha, *J. Leuk. Biol.,* 48, 473, 1990.
204. **Espevik, T., Figari, I. S., Ranges, G. E., and Palladino, M. A., Jr.,** Transforming growth factor-β_1 (TGF-β_1) and recombinant human tumor necrosis factor-alpha reciprocally regulate the generation of lymphokine-activated killer cell activity. Comparison between natural porcine platelet-derived TGF-β_1 and TGF-β_2, and recombinant human TGF-$beta_1$, *J. Immunol.,* 140, 2312, 1988.
205. **Tsunawaki, S., Sporn, M., Ding, A., and Nathan, C.,** Deactivation of macrophages by transforming growth factor-β, *Nature,* 334, 260, 1988.
206. **Ruddle, N. H., Bergman, C. M., McGrath, K. M., Lingenheld, E. G., Grunnet, M. L., Padula, S. J., and Clark, R. B.,** An antibody to lymphotoxin and tumor necrosis factor prevents transfer of experimental allergic encephalomyelitis, *J. Exp. Med.,* 172, 1193, 1990.

207. **MacMicking, J. D., Willenborg, D. O., Weidemann, M. J., Rockett, K. A., and Cowden, W. B.,** Elevated secretion of reactive nitrogen and oxygen intermediates by inflammatory leukocytes in hyperacute experimental autoimmune encephalomyelitis: enhancement by the soluble products of encephalitogenic T cells, *J. Exp. Med.,* 176, 303, 1992.
208. **Merrill, J. E. and Zimmerman, R. P.,** Natural and induced cytotoxicity of oligodendrocytes by microglia is inhibited by TGF-beta, *Glia,* 4, 327, 1991.
209. **Selmaj, K. W. and Raine, C. S.,** Tumor necrosis factor mediates myelin and oligodendrocyte damage *in vitro, Ann. Neurol.,* 23, 339, 1988.
210. **Jacobs, L., Salazar, A. M., Herndon, R. et al.,** Intrathecally administered natural human fibroblast interferon reduces exacerbations of multiple sclerosis. Results of a multicenter, double-blind study, *Arch. Neurol.,* 44, 589, 1987.
211. **Abu-khabar, K. S., Armstrong, J. A., and Ho, M.,** Type I interferons (IFN-alpha and -beta) suppress cytotoxin (tumor necrosis factor-alpha and lymphotoxin) production by mitogen-stimulated human peripheral blood mononuclear cells, *J. Leuk. Biol.,* 52, 165, 1992.
212. **Noronha, A., Jensen, M., and Toscas, A.,** TGF-beta activity in MS: effect of IFN-beta, *Neurology,* 43, A355, 1993.
213. **Glick, A. B., Kulkarni, A. B., Tennenbaum, T., Hennings, H., Flanders, K. C., O'Reilly, M., Sporn, M. B., Karlsson, S., and Yuspa, S. H.,** Loss of expression of transforming growth factor β in skin and skin tumors is associated with hyperproliferation and a high risk for malignant conversion, *Proc. Natl. Acad. Sci. U.S.A.,* 90, 6076, 1993.
214. **Miescher, S., Whiteside, T. L., Carrel, S., and von Fliedner, V.,** Functional properties of tumor infiltrating and blood lymphocytes in patients with solid tumors: effects of tumor cells and their supernatants on proliferative responses of lymphocytes, *J. Immunol.,* 136, 1899, 1986.
215. **Altiok, A., Bejarano, M. T., Ruscetti, F., Altiok, E., Klein, G., and Klein, E.,** Effect of transforming growth factor-beta 1 and -beta 2 on the proliferation of Burkitt lymphoma and lymphoblastoid cell lines, *Growth Factors,* 4, 117, 1991.
216. **Lotz, M., Ranheim, E., and Kipps, T. J.,** Transforming growth factor β as endogenous growth inhibitor of chronic lymphocytic leukemia B cells, *J. Exp. Med.,* 179, 999, 1994.
217. **Kadin, M. E., Agnarsson, B. A., Ellingsworth, L. R., and Newcom, S. R.,** Immunohistochemical evidence of a role for transforming growth factor beta in the pathogenesis of nodular sclerosing Hodgkin's disease, *Am. J. Pathol.,* 136, 1209, 1990.
218. **Millan, F. A., Denhez, F., Kondaiah, P., and Akhurst, R. J.,** Embryonic gene expression patterns of TGF β1, β2 and β3 suggest different developmental functions *in vivo, Development,* 111, 131, 1991.
219. **Torre-Amione, G., Beauchamp, R. D., Koeppen, H., Park, B. H., Schreiber, H., Moses, H. L., and Rowley, D. A.,** A highly immunogenic tumor transfected with a murine transforming growth factor type β cDNA escapes immune surveillance, *Proc. Natl. Acad. Sci. U.S.A.,* 87, 1486, 1990.
220. **Tada, T., Ohzeki, S., Utsumi, K., Takiuchi, H., Muramatsu, M., Li, X.-F., Shimizu, J., Fujiwara, H., and Hamaoka, T.,** Transforming growth factor-β-induced inhibition of T ceil function, *J. Immunol.,* 146, 1077, 1991.
221. **Arrick, B. A., Lopez, A. R., Elfman, F., Ebner, R., Damsky, C. H., and Derynck, R.,** Altered metabolic and adhesive properties and increased tumorigenesis associated with increased expression of transforming growth factor β-1, *J. Cell Biol.,* 118, 715, 1992.
222. **Schwarz, L. C., Wright, J. A., Gingras, M.-C., Kondaiah, P., Danielpour, D., Pimentel, M., Sporn, M. B., and Greenberg, A. H.,** Aberrant TGF-β production and regulation in metastatic malignancy, *Growth Factors,* 3, 115, 1990.

Chapter 14

INTERLEUKIN-10

Jacques Banchereau, Francine Brière, Christophe Caux, Anne-Catherine Fluckiger, Pierre Merville, and Françoise Rousset

CONTENTS

0-8493-8381-1/96/$0.00+$.50

I. INTRODUCTION

Interleukin-10 (IL-10) was identified in 1989 as an activity produced by T helper cell subset 2 (TH_2) inhibiting the synthesis of cytokines, such as interferon γ by TH_1 cells, hence its early name CSIF for Cytokine Synthesis Inhibitory Factor.[1] Based on this activity, cDNA clones encoding mouse IL-10 were isolated in 1990[2] which allowed the isolation by cross-hybridization of cDNA clones encoding human IL-10.[3] It should also be noted that IL-10 was identified early and independently as a product of murine $CD5^+$ B cell lymphomas able to support the survival of mast cell clones[4] and to enhance the proliferation of thymocytes,[5] and later as a product of human T cell lines able to inhibit the production of IL-1β and TNF by stimulated mononuclear cells.[6] IL-10 has now been identified as a pleiotropic factor acting on T cells, B cells, NK cells, mast cells, macrophage/monocyte/dendritic cells, and polymorphonuclear leukocytes. A receptor for IL-10 was identified in 1993 for both mouse and man.[7,8] The present review will primarily focus on human IL-10 (hIL-10) and the most recent results obtained *in vivo* with murine IL-10 (mIL-10). Readers are strongly encouraged to consult the excellent recent reviews by Moore et al.[9] and Rennick et al.[10] The relationships between IL-10 and other interleukins is surveyed in Chapter 2.

II. IL-10 MOLECULE

A. IL-10 PROTEIN AND GENE STRUCTURE

The human and murine IL-10 cloned cDNAs encode quite similar open reading frames (73% amino acid homology; >80% DNA sequences homology). The mIL-10 molecule has 157 AA with two potential N-glycosylation sites which generate polypeptides of 17, 19, and 21 kDa. Only one of the two N-linked glycosylation sites is used in mIL-10 (the N-terminus proximal site). If this site is mutated away, only one polypeptide is seen, as for h, vIL-10 (R. Kastelein and K. Moore, personal communication). The hIL-10 protein consists of 160 AA with one potential N-glycosylation site which is not required for biological activity, and has a molecular weight of approximately 18 kDa. Both IL-10s contain two disulfide bonds linking the first cysteine residue in the sequence with the third and the second with the fourth.[11] IL-10 forms a 37-kDa noncovalently linked dimer.[12] Far-UV circular dichroism analysis indicates that hIL-10 and mIL-10 are composed of ≈60% α-helix, most likely divided into four α-helical motifs as observed with other cytokines such as human growth hormone[13] or α/γ interferon.[14,15] The IL-10 genes are located on chromosome 1 in both mouse and man. The murine IL-10 gene is composed of five exons spanning ≈5.1 kb of genomic DNA. The promoter contains motifs associated with regulation of gene transcription such as glucocorticoid-responsive elements, cAMP-responsive elements, and candidate AP-1 and NF-kB binding sites.[16] This promoter displays several characteristics of the IL-6 gene and IL-10 expression shares features of IL-6 expression. Transcription of the IL-10 gene generates a single species mRNA of 1.4 kb (mIL-10) and 2.0 kb (hIL-10).

Both mIL-10 and hIL-10 exhibit marked homology to an open reading frame in the EBV genome, BCRF-1.[2,3,17] This relationship is most pronounced in the mature protein-coding sequences, where hIL-10 and BCRF-1 are 84% identical, and does not appear in the signal sequence or flanking untranslated regions. Of the mouse, human, and viral sequences, hIL-10 and BCRF-1 are the most closely related at the protein level, while mIL-10 and hIL-10 are significantly more homologous in their DNA sequences (≈80% throughout both the coding and untranslated region). These relationships suggest that the mIL-10 and hIL-10 genes evolved from a common ancestor. The BCRF-1 amino acid sequence is clearly constrained to resemble that of the human cytokine, and the ancestor of BCRF-1 may have been a captured, processed host IL-10 gene. The equine herpesvirus type 2 (EHV2) also encodes a viral homolog of IL-10 with differences that lie in the N-terminal region as for the other IL-10s.[18]

B. IL-10-PRODUCING CELLS

mIL-10 is secreted by TH_2 clones, B lymphomas, activated mast cell lines, T cells, $CD5^+$ B cells, and activated macrophages.[9] Murine keratinocytes have been shown to release significant amounts of IL-10, in particular following UV irradiation.[19,20] hIL-10 is produced by naive ($CD45RA^+$) and memory ($CD45RO^+$) $CD4^+$ T cells and $CD8^+$ T cells, as well as TH_0, TH_1, TH_2, $CD4^+$ T cell clones and $CD8^+$ T cell clones.[21,22] B lymphocytes also release IL-10 following activation, either through their antigen receptor or their CD40 antigen.[23] B lymphocytes from patients with systemic lupus erythematosus, rheumatoid arthritis, and Sjögren syndrome spontaneously secrete IL-10.[24] EBV-transformed B cell lines also appear to produce large amounts of human IL-10.[23,25,26] Viral IL-10 is secreted in minor amounts by some B cell lines, but vIL-10 seems to play a crucial role early during the transformation process of B cells by EBV. IL-10 is also produced by monocytes[27,28] whose IL-10 production is in fact inhibited by IFNγ and IFNα. Many carcinomas and melanoma cell lines also produce IL-10.[29]

C. IL-10 IN BIOLOGICAL FLUIDS

IL-10 has been detected in human biological fluids under various pathological circumstances. In particular, IL-10 was found in the serum of 47/101 patients with active non-Hodgkin's lymphomas while it was present only in 3 of 52 patients with partial or complete response.[30] While serum IL-10 was identified in all subtypes of non-Hodgkin's lymphomas and in all clinical stages, its presence in the serum of patients with intermediate or high-grade lymphomas was associated with a poor survival. Recent studies have indicated that IL-10 is produced by the leukemic cells (Blay, J.-Y., personal communication), as observed earlier with AIDS lymphomas.[31] IL-10 has also been detected in serum of patients suffering from early stage myelomas, particularly those with solitary myelomas, but not in the serum of patients with advanced myelomas.[32] Finally, the ascites of patients with ovarian and other intra-abdominal cancers also contain IL-10.[33]

IL-10 is also present in the biological fluids of patients suffering from infections. Very high levels of IL-10 have been measured in the serum of patients suffering from malaria, most particularly during the cerebral and severe cases.[34] The cerebrospinal fluid of the majority of patients suffering from *Neisseiria meningitidis* infections also contains IL-10, and those patients with no detectable IL-10 in their CSF had a poorer prognosis than those with IL-10 (unpublished data).

III. IL-10 RECEPTORS

Human IL-10 acts on both human and mouse cells, whereas mIL-10 is species specific. This is due to the specificity of binding to respective receptors.[7,8,12] hIL-10 binds to a single class of high-affinity receptors (50 to 200 p*M*) which are expressed at low levels (100 to 1000 per cell). A cDNA coding for a human IL-10 receptor was isolated by expression cloning using a protein composed of an 8-amino acid FLAG sequence fused to the N-terminus of hIL-10 and a library made from the Burkitt lymphoma cell line, BJAB. The hIL-10 receptor cDNA contains an open reading frame encoding a protein of 578 AA including a signal peptide of 21 AA, an extracellular domain of 215 AA, a transmembrane segment of 25 AA, and a cytoplasmic domain of 317 AA. This corresponds to a protein backbone of 61 kDa contrasting with the size of 90 to 110 kDa observed by crosslinking of labeled IL-10. It is likely that the hIL-10R is glycosylated at several of the six potential N-glycosylation sites of the extracellular domain. The hIL-10R cDNA is 70% homologous to the mIL-10R cDNA in the untranslated regions and the open reading frame. The predicted amino acid sequences of hIL-10R and mIL-10R are 60% identical and 73% similar. The IL-10R is a member of the cytokine receptor family[35] and is most similar to the smaller group of class II receptors which includes interferon receptors, a viral IFNR homolog, and tissue factor. The extracellular portion of hIL-10R may be considered as two homologous segments of 110 AA. As are other cytokine receptors, the hIL-10R is certainly composed of several components. In particular, binding studies show that vIL-10 at a 100-fold molar excess does not detectably compete with hIL-10 for binding to cells. Furthermore, labeled vIL-10 does not bind to cells transfected with the hIL-10R cDNA.[8] Cross-linking studies indicate that iodinated IL-10 binds to components of 50, 190, and 210 kDa. Whether these elements represent degradation or polymerization products of the 90-kDa component, respectively, remains to be established. The hIL-10R mRNA of 3.6 kb is expressed at relatively high levels in monocytes, B cells, and NK cells. It is expressed at low levels in many nonhematopoietic tissues, but may then represent a contamination from blood cells. The hIL-10R gene was mapped on chromosome 11.

IV. IL-10 INHIBITS ANTIGEN-DEPENDENT T LYMPHOCYTE ACTIVATION

A. *IN VITRO* STUDIES

1. Inhibition of Antigen-Dependent T Cell Proliferation

The early study describing the cytokine synthesis inhibitory activity of IL-10 indicated that murine IL-10 was not directly acting on T cells but rather on antigen-presenting cells.[1] Likewise human IL-10 was found to inhibit the proliferation of human $CD4^+$ and $CD8^+$ T lymphocytes induced by autologous antigen pulsed monocytes[27,36] or allogeneic monocytes.[37] IL-10 decreases the production of IL-2 and the expression of IL-2 receptors by T cells cultured in the presence of monocytes. Accordingly, addition of exogenous IL-2 partially reverses the IL-10-dependent inhibition of T cell proliferation. IL-10 also inhibits the generation of cytotoxic $CD8^+$ cells induced during mixed lymphocyte cultures. Inasmuch as dendritic cells[38] are professional antigen-presenting cells characterized by their ability to stimulate primary antigen-specific responses *in vitro* and *in vivo,* the effects of IL-10 on dendritic cell APC function have been examined recently. Experiments with murine cells have demonstrated an IL-10-dependent inhibition of cytokine (IFNγ) production during dendritic cell/T cell interactions. One study using spleen dendritic cells failed to detect altered T cell proliferation,[39] while another study, using enriched epidermal Langerhans cells in suspension, indicated an IL-10-dependent inhibition of TH_1 cell proliferation leading to tolerance.[40] Our studies with human dendritic cells have demonstrated the profound inhibitory effect of IL-10 on dendritic cell-dependent T cell proliferation. This includes the proliferative response of blood T lymphocytes to allogeneic skin Langerhans cells,[41] or allogeneic tonsil interdigitating dendritic cells, or dendritic cells generated *in vitro* by culturing $CD34^+$ hematopoietic progenitors in the presence of GM-CSF and TNFα.[42,43] IL-10 also inhibits the primary alloreactive proliferation of highly purified $CD4^+$ and $CD8^+$ T cells as well as the secondary proliferative response of alloreactive T cell clones elicited by target dendritic cells, or tetanus toxoid-specific T cell clones to antigen pulsed dendritic cells. The primary cytotoxic response elicited by monocytes[39] was also inhibited by IL-10. These studies have concluded that T lymphocyte proliferation elicited by immobilized anti-CD3 or EBV-transformed B cell lines was not affected by IL-10, thus further confirming the role of IL-10 in the human antigen-presenting cells. IL-10 is likely to act both on the secretion of cytokines[44] and on the expression of surface antigens, as discussed later. IL-10 also inhibits monocyte/macrophage-dependent cytokine synthesis (IFNγ and TNFα) by human NK cells stimulated by IL-2, although it has no effect on the small IFNγ production of NK cells stimulated by IL-2 without accessory cells.[45]

2. Effects of IL-10 on Antigen-Presenting Cells

IL-10 profoundly affects both murine and human monocytes/macrophages but its effects on dendritic cells are not yet as well documented. IL-10 downregulates the spontaneous and cytokine-induced expression of HLA class II antigens on human monocytes.[27] It also partially blocks the upregulation of HLA class II antigens on skin Langerhans cells, which spontaneously occurs in culture.[41] Probably more important to the inhibition of antigen presentation is the IL-10-induced blocking of B7–1 expression on activated macrophages.[46] However, it is not known whether IL-10 affects either B7–2 or B7–3.[47,48] IL-10 does not alter the expression of integrins ($CD11_{abc}$/CD18), CD54 (ICAM-1), or CD58 (LFA-3). However, it stimulates the expression on monocytes of the high affinity $Fc\gamma R_I$/CD64.[49] Thus, IL-10 shares this property with IFNγ while these two cytokines demonstrate contrasting properties on the regulation of MHC class II antigen expression. The increased expression of $Fc\gamma R_I$/CD64 by IL-10-treated monocytes correlates with an enhanced capacity to lyse anti-D-coated human rhesus-positive erythrocytes. While IL-4 and IL-10 share the property of decreasing the secretion of pro-inflammatory cytokines by monocytes/macrophages, as discussed below, IL-10 is able to prevent downregulation of all three $Fc\gamma R_s$ on monocytes that are induced by IL-4.

The production of cytokines (IL-1, IL-6, IL-8, TNFα, GM-CSF, IL-10) by activated human monocytes or murine macrophages is strongly inhibited by IL-10.[27,50] Interestingly, activated monocytes produce IL-10 in a relatively delayed fashion when compared to the other monokines, and its neutralization by blocking antibodies results in an increased production of the other monokines.[44] In addition to blocking the production of pro-inflammatory cytokines, IL-10, as well as IL-4, increases the production of the anti-inflammatory cytokine IL-1ra.

IL-10 also inhibits the production of reactive oxygen (H_2O_2) and nitrogen (NO) intermediates following activation of monocytes/macrophages.[51–54] The blocking of NO production may be indirect and dependent on the inhibition of TNFα production. Whatever the mechanism, this results in decreased killing of parasites by IFNγ-activated monocytes/macrophages.[55,56] Indeed, parasite development appears to be associated with increased *in vivo* expression of IL-10,[55,57–59] which may block the development of TH_1 responses that contribute to parasite clearance. IL-10 also inhibits the intracellular killing of *Leishmania* by human macrophages activated with recombinant migration inhibitory factor (MIF).[61]

3. Effects of IL-10 on T Lymphocytes

IL-10 was identified early on as a B cell product able to induce the proliferation of murine thymocytes and mature T lymphocytes in the presence of IL-2 and IL-4.[5,61] IL-10 has also been found to enhance the development of cytotoxic T cells.[62] Studies on human T lymphocytes indicate that IL-10 can directly inhibit T cell proliferation, as shown with anti-CD3-activated blood T cells[36] or T cell clones.[63] The inhibition appears to be mostly due to a direct blocking of IL-2 secretion, and addition of exogenous IL-2 or IL-4 reversed the inhibition of T cell proliferation. However, other studies showed that proliferation of T lymphocytes in response to allogeneic EBV-transformed B cells[37,42] or anti-CD3[1,42] is not inhibited by IL-10. Differences in culture conditions and considered T cell populations may explain these apparent discrepancies. Indeed, a good example of the differential effect of IL-10 on T cell subpopulations is its chemotactic activity towards $CD8^+$ T cells, which contrasts with its lack of chemotactic activity towards $CD4^+$ T cells, monocytes, or neutrophil granulocytes.[64] Interestingly, IL-10 inhibits the chemotactic response of $CD4^+$ T cells towards IL-8. Furthermore, the rare $CD4^+$ $CD7^-$ circulating T cell subset was recently found to proliferate directly in response to IL-10 (P. Debré, personal communication). Finally, IL-10 promotes the survival of IL-2-dependent T cells otherwise destined to die by apoptosis.[65] A direct effect of IL-10 on cytokine synthesis by T cells is also shown by the IL-10-induced inhibition of migration inhibitory factor secretion by a human T cell hybridoma line.[60]

B. *IN VIVO* STUDIES

1. Administration of Exogenous IL-10

A single injection of recombinant murine IL-10 is able to protect mice from bacterial lipopolysaccharide-induced shock[66,67] and from staphylococcal enterotoxin B-induced shock which involves polyclonal T cell activation.[68] The protective effect of IL-10 correlates with a substantial decrease in endotoxin-induced TNFα release. The proinflammatory cytokines are reviewed in Chapter 3.

mIL-10 administered *in vivo* also displays a protective effect on IgG or IgA immune complex-induced alveolitis in rats which respectively involve macrophages and neutrophils or macrophages only.[69] *In vivo* administration of IL-10 in the form of UV-irradiated keratinocyte supernatants inhibits the sensitization of DTH response.[20] In combination with IL-4, IL-10 also inhibits the *Leishmania*-induced footpad swelling of mice immune to this parasite.[70] Thus, administration of IL-10 *in vivo* can inhibit both the initiation and effector phases of DTH responses.

Continuous administration of IL-10 to NOD mice prevents the spontaneous onset of diabetes,[71] though mice display a clear cellular infiltration of Langerhans islets but without the classic signs of destruction found in diabetes. Interestingly, transgenic mice carrying the IL-10 gene under the control of an insulin promoter also display insulitis without signs of diabetes, indicating that IL-10 may promote accumulation of inflammatory cells either because of its chemotactic activity or because of endothelial cell activation resulting in expression of new adhesion molecules allowing inflammatory cell extravasation.[72] A marked increase in IL-10 mRNA is also observed in spinal tissue during the recovery phase of induced autoimmune encephalomyelitis in mice.[73] Taken together, these various models indicate that IL-10 may participate *in vivo* in the control of T cell-dependent inflammation.

IL-10 may also be involved in the peripheral tolerance generated experimentally at the time of heart or spleen cells transplants.[74,75] The high levels of IL-10 secreted by blood cells from human SCID patients successfully reconstituted with allogeneic hematopoietic progenitors may also explain the successful chimeric state of these individuals.[76]

2. Blocking of Endogenous IL-10

Mice treated with neutralizing anti-IL-10 antibody display elevated levels of circulating TNFα and, in many cases, IL-6. Such animals are particularly susceptible to death by endotoxin-induced shock, thus confirming the natural anti-inflammatory role of IL-10 discussed above. IL-10-deficient mice, generated by gene targeting, are growth retarded and anemic and suffer from chronic enterocolitis.[77] Alterations in the intestine include extensive mucosal hyperplasia, inflammatory reactions, and aberrant expression of major histocompatibility complex class II molecules on epithelia. In contrast, mutants kept under specific pathogen-free conditions develop only a local inflammation limited to the proximal colon. These results indicate that the bowel inflammation in the mutants originates from uncontrolled immune responses stimulated by enteric antigens, and that IL-10 is an essential immunoregulator in the intestinal tract.

V. IL-10 INDUCES B CELL DIFFERENTIATION

A. IL-10 AND B CELL SURVIVAL

mIL-10 augments expression of class II MHC on small resting mouse splenic B cells and is a viability maintenance factor for these cells in culture.[78] Human IL-10 was also found to enhance the survival of splenic B lymphocytes during a 15-day culture period, an effect associated with enhanced expression of the bcl-2 protein.[79] In our studies, hIL-10 did not modify expression of HLA class II antigens on human resting tonsillar B cells, nor did it permit their increased short-term survival.

When chronic lymphocytic leukemia (CLL) B cells were considered, IL-10 was found to inhibit the spontaneous DNA synthesis observed in 3 of 12 samples and to decrease the viable cell recovery of all B-CLL samples. After 1 week some B-CLL populations cultured with IL-10 were lost, while those cultured without IL-10 survived. Flow cytometric analysis, DNA gel electrophoresis, and Giemsa staining all revealed that IL-10 induced B-CLL cells to die from apoptosis. Addition of IL-2, IL-4, IFNγ, and anti-CD40 mAb prevented the IL-10-mediated apoptosis of B-CLL cells. None of the malignant B cell populations obtained from eight non-Hodgkin's lymphomas and three hairy cell leukemias underwent apoptosis after IL-10 treatment. It will be important to determine whether a subpopulation of normal B lymphocytes is sensitive to the pro-apoptotic effects of IL-10.

B. IL-10 AND B CELL GROWTH

hIL-10 enhanced DNA replication of B lymphocytes stimulated through their antigen receptors (surface Ig) either by insolubilized anti-IgM antibody or *Staphylococcus aureus* strain Cowan (SAC) particles.[80] The costimulation was of lower magnitude than that observed with either IL-2 or IL-4. In contrast, when B cells were stimulated by cross-linking of their CD40 antigen with anti-CD40 antibody and mouse L cells expressing human FcγRII (CD32),[81,82] hIL-10 markedly enhanced B cell growth and was as effective as IL-4 over a 10-day period. hIL-10 induced a 10- to 25-fold multiplication of B cells within 2 weeks. IL-4 and hIL-10 were additive in this system, their combination resulting in a 60- to 100-fold expansion of the number of viable B cells. IL-10 also costimulates with IL-2 for the proliferation of B cells cultured in the CD40 system, an effect which is particularly striking as CD40-activated B cells respond poorly to IL-2. This synergy is likely to be due to the IL-10-induced increase of high-affinity IL-2 receptors on CD40-activated B lymphocytes which is associated to an increased Tac/CD25 expression.[83]

hIL-10 also enhances EBV-induced B cell proliferation, whereas IL-2 and IL-4 do not. Interestingly, B cells are induced to produce significant amounts of hIL-10 after EBV transformation, and anti-hIL-10 neutralizing antibodies reduce considerably EBV-induced B cell proliferation although they do not prevent the establishment of EBV-transformed cell lines.[23] B lymphocytes infected with EBV display vIL-10 transcripts a few hours after infection,[26] while earlier studies had indicated vIL-10 to be a "late" viral gene expressed only during the lytic phase of virus replication.[84] The early expression of vIL-10 appears critical to the transformation process, as antisense oligonucleotides for vIL-10 mRNA added at the time of viral infection prevented subsequent B cell transformation.[26]

C. IL-10 AND B CELL DIFFERENTIATION

IL-10 induces B cells activated with particles of *Staphylococcus aureus* strain Cowan (SAC) to produce IgG, IgA, and IgM at levels above those obtained with either IL-2 or IL-4.[80] The differentiation effect of IL-10 is particularly striking when B cells are stimulated in the CD40 system. Unseparated B cells then produce large amounts of IgM, IgA, IgG_1, IgG_2, and IgG_3, but virtually never IgG_4 or IgE. The IL-10-induced B cell differentiation did not require the presence of L cells, as soluble anti-CD40 and IL-10 were sufficient to induce Ig secretion. Morphologic studies have shown that cells differentiate into plasma blasts full of intracytoplasmic immunoglobulins. Eventually, CD40-activated B cells stop proliferating after several days in the presence of IL-10, possibly as a consequence of the full differentiation. Combination of IL-10 with IL-2 or IL-3 induces, respectively, SAC-preactivated B cells or T cell-preactivated B cells to terminally differentiate into plasma cells when secondarily cultured onto peculiar microenvironments, such as rheumatoid arthritis synoviocytes or bone marrow stroma. Naive $sIgD^+$ B cells isolated from tonsils, spleen, cord blood, or from patients with hyper-IgM syndrome produce large amounts of IgM, as well as IgA,[86] IgG_1, and IgG_3.[87] This observation suggests that IL-10 (together with CD40 signals?) may act as a switch factor for these Ig classes, most particularly since costimulation of these naive B cells with SAC and the CD40 system results in the production of large amounts of IgM without the other isotypes, thus excluding the possible contribution of contaminating post-switch B cells. Combination of IL-10 and TGFβ were found to induce naive CD40-activated B cells to produce large amounts of IgA.[86] Similar patterns of Ig secretion have been obtained using CD40 ligand transfected L cells.

Recent studies in IgA-deficient individuals have shown that their B lymphocytes can be induced to secrete IgA following CD40 activation in the presence of IL-10.[87] This demonstrates the critical role of this cytokine in the regulation of this isotype. Blood mononuclear cells from these individuals can, however, produce IL-10 although at significantly reduced levels.

D. IL-10 AND B CELLS *IN VIVO*

Studies in mice have also indicated a possible role of IL-10 in B cell growth and differentiation. In particular, $CD5^+$ B cells have been shown to preferentially produce IL-10.[4,89]

Mice treated with anti-IL-10 have decreased serum IgM and IgA[90,91] and increased IgG_{2a} and IgG_{2b} levels, a pattern which is compatible with increased IFNγ levels detected in these animals. These animals display a depletion of $CD5^+$ B cells which is reversed following discontinuation of antibody treatment. These animals cannot produce antibodies to phosphorylcholine and α1,3-dextran but develop normal antibody responses to TNF-Ficoll. The depletion of $CD5^+$ B cells in these animals appears to depend on increased levels of IFNγ.

Neutralizing IL-10 may be beneficial in B-cell-mediated autoimmune disease such as systemic lupus erythematosus. NZB/W F1 mice, which spontaneously develop an SLE-like syndrome with lethal immune complex-dependent glomerulonephritis, display a delayed onset of autoimmunity when treated with anti-IL-10 and conversely. Injection of IL-10 accelerates the disease process.[92] Thus, the high levels of IL-10 spontaneously produced by B cells from SLE patients may represent a key element of this disease.

Recent experiments with mice rendered IL-10 deficient by gene targeting appear to invalidate the above results.[77] IL-10-deficient mice thus do not display alteration in T and B lymphocyte development and display normal levels of the various serum isotypes. Furthermore, when these animals develop chronic enterocolitis, they display increased serum IgA levels as a consequence of numerous IgA plasma cells infiltrating the inflamed mucosa. The reasons for these discrepancies are not clear and may involve the well-recognized redundancy that exists within the cytokine system.[93] It is possible that a developing mouse embryo will compensate for the loss of a critical cytokine gene by amplifying expression of a gene that codes for a protein of comparable function.

VI. IL-10 AND OTHER CELL TYPES

The modulatory effects of IL-10 on hematopoiesis have been studied in the mouse but not in humans. While IL-10 does not display direct growth-promoting activity, it functions as a costimulant for the development of cells of several lineages. In particular, IL-10 enhances the proliferation of murine stem cells induced by the stem cell factor used alone or with the costimulants IL-1, IL-6, and G-CSF.[94] IL-10 neither enhances nor inhibits the generation of neutrophils, eosinophils, or macrophages in the presence of various growth factors, but synergizes with IL-3 for the generation of megakaryocytes. Furthermore, like IL-4, IL-10 enhances the growth of mast cells in response to IL-3.[95] IL-10 induces mouse mast cells to express the mouse mast cell protease 2 (MMCP2) which is typical of mast cells of the mucosal phenotype by contrast to those of connective tissue.[96]

IL-10 decreases B lymphopoiesis in Whitlock-Witte cultures and permits the emergence of mast cells, as well as very large, highly refractile semiadherent cells which may be histiocytes that participate in regenerative processes in bone marrow (Rennick, D., personal communication).

IL-10 enhances the proliferation of fetal human progenitor B lymphocytes cultured in a CD40 system together with IL-3 and IL-7, thus further exemplifying IL-10 costimulatory properties.[97] Daily injections of IL-10 to mice treated with 5-fluorouracil, which is toxic to actively proliferating (bone marrow) cells but not to quiescent stem cells, accelerates the regeneration of early hematopoiesis (increased colony forming units) but has no effect on the release of mature elements (Rennick, D., personal communication). It should be stressed that IL-10-deficient mice have normal development of T and B lymphocytes but severe anemias, which may be indirect as a consequence of excess circulating IL-1 and TNF.[77] Polymorphonuclear leukocytes represent another target of IL-10 as it blocks their production of pro-inflammatory cytokines (TNF, IL-1, IL-8) following LPS stimulation.[98] This comes in addition to the other anti-inflammatory properties of IL-10.

VII. CONCLUDING REMARKS

The numerous studies performed in the 5 years (1989 to 1993) that followed IL-10 discovery as a TH_2 cell product inhibiting TH_1 cells, confirmed the concept that presided over its discovery in the sense that IL-10 has proved to play a major role in suppressing acute inflammation and cell-mediated immunity characteristic of TH_1 responses, and stimulating humoral immunity and mast cell development characteristic of TH_2 responses (Figure 1). The suppression of inflammatory responses may prove useful in the management of acute inflammatory responses as in septic shock or in chronic inflammation such as inflammatory bowel diseases or rheumatoid arthritis. In contrast, it may be beneficial to antagonize an excessive IL-10 response, which may induce a block of cellular immunity

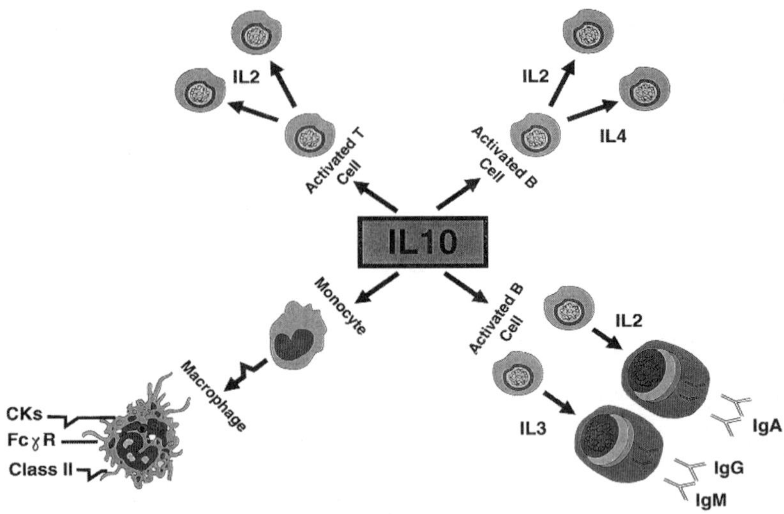

FIGURE 1. Multiple biological effects of human IL-10; —^→ : inhibitory effect; → stimulatory effect.

as observed in certain cancers or which may be associated to stimulated humoral responses as observed in HIV-associated immunodeficiency[99] and patients suffering from systemic lupus erythematosus. It is interesting to consider interferons or IL-12 as IL-10 antagonists as these agents have been involved in blocking IL-10 secretion and/or stimulating TH_1 cell development.

Furthermore, studies aimed at (1) unraveling the structure of the IL-10 receptor, (2) identifying the intracellular pathways transducing IL-10 signals, and (3) understanding the mechanisms controlling IL-10 secretion will hopefully permit chemists to design chemical entities for altering the IL-10 balance.

REFERENCES

1. **Fiorentino, D. F., Bond, M. W., and Mosmann, T. R.,** Two types of mouse T helper cell. IV. Th2 clones secrete a factor that inhibits cytokine production by Th1 clones, *J. Exp. Med.,* 170, 2081, 1989.
2. **Moore, K. W., Vieira, P., Fiorentino, D. F., Trounstine, M. L., Khan, T. A., and Mosmann, T. R.,** Homology of the cytokine synthesis inhibitory factor (IL-10) to the Epstein-Barr virus gene BCRFI, *Science,* 248, 1230, 1990.
3. **Vieira, P., de Waal-Malefyt, R., Dang, M. N., Johnson, K. E., Kastelein, R., Fiorentino, D. F., de Vries, J. E., Roncarolo, M. G., Mosmann, T. R., and Moore, K. W.,** Isolation and expression of human cytokine synthesis inhibitory factor (CSIF/IL-10) cDNA clones: homology to Epstein-Barr virus open reading frame BCRF1, *Proc. Natl. Acad. Sci. U.S.A.,* 88, 1172, 1991.
4. **O'Garra, A., Stapleton, G., Dhar, V., Pearce, M., Schumacher, J., Rugo, H., Barbis, D., Stall, A., Cupp, J., Moore, K., Vieira, P., Mosmann, T., Whitmore, A., Arnold, L., Haughton, G., and Howard, M.,** Production of cytokines by mouse B cells: B lymphomas and normal B cells produce interleukin-10, *Int. Immunol.,* 2, 821, 1990.

5. **Suda, T., O'Garra, A., MacNeil, I., Fischer, M., Bond, M., and Zlotnik, A.,** Identification of a novel thymocyte growth promoting factor derived from B cell lymphomas, *Cell. Immunol.,* 129, 228, 1990.
6. **Ralph, P., Nakoinz, I., Sampson-Johannes, A., Fong, S., Lowe, D., Min, H.-Y., and Lin, L.,** IL-10, T lymphocyte inhibitor of human blood cell production of IL-1 and tumor necrosis factor, *J. Immunol.,* 148, 808, 1992.
7. **Ho, A. S.-Y., Liu, Y., Khan, T. A., Hsu, D.-H., Bazan, J. F., and Moore, K. W.,** A receptor for interleukin-10 is related to interferon receptors, *Proc. Natl. Acad. Sci. U.S.A.,* 90, 11267, 1993.
8. **Liu, Y., Wei, S. H.-Y., Ho, A. D.-Y., de Waal Malefyt, R., and Moore, K. W.,** Expression cloning and characterization of a human interleukin-10 receptor, *J. Immunol.,* 152, 1821, 1994.
9. **Moore, K. W., O'Garra, A., de Waal Malefyt, R., Vieira, P., and Mosmann, T. R.,** Interleukin-10, *Annu. Rev. Immunol.,* 11, 165, 1993.
10. **Rennick, D., Berg, D., and Holland, G.,** Interleukin-10: an overview, *Prog. Growth Factor Res.,* 4, 207, 1993.
11. **Windsor, W. T., Syto, R., Tsarbopoulos, A., Zhang, R., Durkin, J., Baldwin, S., Paliwal, S., Mui, P. W., Pramanik, B., Protta, P. P., and Tindall, S. H.,** Disulfide bond assignments and secondary structure analysis of human and murine interleukin-10, *Biochemistry,* 32, 8807, 1993.
12. **Tan, J. C., Indelicato, S. R., Narula, S. K., Zavodny, P. J., and Chou, C.-C.,** Characterization of interleukin-10 receptors on human and mouse cells, *J. Biol. Chem.,* 268, 21053, 1993.
13. **de Vos, A. M., Ultsch, M., and Kossiakoff, A. A.,** Human growth hormone and extracellular domain of its receptor: crystal structure of the complex, *Science,* 255, 306, 1992.
14. **Senda, T., Shimazu, T., Matsuda, S., Kawano, G., Shimizu, H., Nakamura, K. T., and Mitsui, Y.,** Three-dimensional crystal structure of recombinant murine interferon-β, *EMBO J.,* 11, 3193, 1992.
15. **Ealick, S. E., Cook, W. J., Vijay-Kumar, S., Carson, M., Nagabhushan, T. L., Trotta, P. P., and Bugg, C. E.,** Three dimensional structure of recombinant human interferon-γ, *Science,* 252, 698, 1991.
16. **Kim, J. M., Brannan, C. I., Copeland, N. G., Jenkins, N. A., Khan, T. A., and Moore, K. W.,** Structure of the mouse IL-10 gene and chromosomal localization of the mouse and human genes, *J. Immunol.,* 148, 3618, 1992.
17. **Baer, R., Bankier, A. T., Biggin, M. D., Deininger, P. L., Farrell, P. J., Bigson, T. J., Hatfull, G., Hudson, G. S., Satchwell, S. C., Séguin, C., Tuffnell, P. S., and Barrell, B. G.,** DNA sequence and expression of the B95–8 Epstein-Barr virus genome, *Nature,* 310, 207, 1984.
18. **Rode, H. J., Janssen, W., Rosen-Wolff, A., Bugert, J. J., Thein, P., Becker, Y., and Darai, G.,** The genome of equine herpesvirus type 2 harbors an interleukin-10 (IL-10)-like gene, *Virus Genes,* 7, 111, 1993.
19. **Enk, A. H. and Katz, S. I.,** Identification and induction of keratinocyte-derived IL-10, *J. Immunol.,* 149, 92, 1992.
20. **Rivas, J. M. and Ullrich, S. E.,** Systemic suppression of delayed-type hypersensitivity by supernatants from UV-irradiated keratinocytes, *J. Immunol.,* 149, 3865, 1992.
21. **Yssel, H., de Waal Malefyt, R., Roncarolo, M.-G., Abrams, J. S., Lahesmaa, R., Spits, H., and de Vries, J. E.,** IL-10 is produced by subsets of human CD4+ T cell clones and peripheral blood T cells, *J. Immunol.,* 149, 2378, 1992.
22. **Del Prete, G., de Carli, M., Almerigogna, F., Giudizi, M. G., Biagiotti, R., and Romagnani, S.,** Human IL-10 is produced by both type 1 helper (Th1) and type 2 (Th2) T cell clones and inhibits their antigen-specific proliferation and cytokine production, *J. Immunol.,* 150, 353, 1993.
23. **Burdin, B., Péronne, C., Banchereau, J., and Rousset, F.,** Epstein-Barr virus-transformation induces B lymphocytes to produce human interleukin-10, *J. Exp. Med.,* 177, 295, 1993.
24. **Llorente, L., Richaud-Patin, Y., Wijdenes, J., Alcocer-Varela, J., Maillot, M. C., Durand-Gasselin, I., Morel-Fourrier, B., Galanaud, P., and Emilie, D.,** Spontaneous production of interleukin-10 by B lymphocytes and monocytes in systemic lupus erythematosus, *Eur. Cytokine Network,* 4, 1993.
25. **Benjamin, D., Knobloch, T. J., and Dayton, M. A.,** Human B-cell interleukin-10: B cell lines derived from patients with acquired immunodeficiency syndrome and Burkkit's lymphoma constitutively secrete large quantities of interleukin-10, *Blood,* 80, 1289, 1992.

26. **Miyazaki, I., Cheung, R. K., and Dosch, H.-M.,** Viral interleukin-10 is critical for the induction of B cell growth transformation by Epstein-Barr virus, *J. Exp. Med.,* 178, 439, 1993.
27. **de Waal Malefyt, R., Haanen, J., Spits, H., Roncarolo, M. G., te Velde, A., Figdor, C., Johnson, K., Kastelein, R., Yssel, H., and de Vries, J. E.,** Interleukin-10 (IL-10) and viral IL-10 strongly reduce antigen-specific human T cell proliferation by diminishing the antigen-presenting capacity of monocytes via downregulation of class II major histocompatibility complex expression, *J. Exp. Med.,* 174, 915, 1991.
28. **Chomarat, P., Rissoan, M.-C., Banchereau, J., and Miossec, P.,** Interferon γ inhibits interleukin-10 production by monocytes, *J. Exp. Med.,* 177, 523, 1993.
29. **Gastl, G. A., Abrams, J. S., Nanus, D. M., Oosterkamp, R., Silver, J., Liu, F., Chen, M., Albino, A. P., and Bander, N. H.,** Interleukin-10 production by human carcinoma cell lines and its relationship to interleukin-6 expression, *Int. J. Cancer,* 55, 96, 1993.
30. **Blay, J.-Y., Burdin, N., Rousset, F., Lenoir, G., Biron, P., Philip, T., Banchereau, J., and Favrot, M.,** Serum interleukin-10 in non-Hodgkin's lymphoma. A prognostic factor, *Blood,* 82, 2169, 1993.
31. **Emilie, D., Touitou, R., Raphael, M., Peuchmaur, M., Devergnee, O., Rea, D., Coumbraras, J., Crevon, M.-C., Edelman, L., Joab, I., and Galanaud, P.,** *In vivo* production of interleukin-10 by malignant cells in AIDS lymphomas, *Eur. J. Immunol.,* 22, 2937, 1992.
32. **Merville, P., Rousset, F., Banchereau, J., Klein, B., and Bataille, R.,** Serum interleukin-10 in early stage multiple myeloma, *Lancet,* 340, 1544, 1992.
33. **Gotlieb, W. H., Abrams, J. S., Watson, J. M., Velu, T. J., Berek, J. S., and Martinez-Maza, O.,** Presence of interleukin-10 (IL-10) in the ascites of patients with ovarian and other intra-abdominal cancers, *Cytokine,* 4, 385, 1992.
34. **Peyron, F., Burdin, N., Ringwald, P., Vuillez, J.-P., Rousset, F., and Banchereau, J.,** High levels of circulating IL-10 in human malaria, *Clin. Exp. Immunol.,* 95, 300, 1994.
35. **Miyajima, A., Kitamura, T., Harada, N., Yokota, T., and Arai, K.,** Cytokine receptors and signal transduction, *Annu. Rev. Immunol.,* 10, 295, 1992.
36. **Taga, K. and Tosato, G.,** IL-10 inhibits human T cell proliferation and IL-2 production, *J. Immunol.,* 148, 1143, 1992.
37. **Bejarano, M.-T., de Waal Malefyt, R., Abrams, J. S., Bigler, M., Bacchetta, R., de Vries, J. E., and Roncarolo, M.-G.,** Interleukin-10 inhibits allogeneic proliferative and cytotoxic T cell responses generated in primary mixed lymphocyte cultures, *Int. Immunol.,* 4, 1389, 1992.
38. **Steinman, R. M.,** The dendritic cell system and its role in immunogenicity, *Annu. Rev. Immunol.,* 9, 271, 1991.
39. **Macatonia, S. E., Doherty, T. M., Knight, S. C., and O'Garra, A.,** Differential effect of interleukin-10 on dendritic cell-induced T cell proliferation and interferon-γ production, *J. Immunol.,* 150, 3755, 1993.
40. **Enk, A. H., Angeloni, V. L., Udey, M. C., and Katz, S. I.,** Inhibition of Langerhans cell antigen-presenting function by IL-10, *J. Immunol.,* 151, 2390, 1993.
41. **Péguet-Navarro, J., Moulon, C., Caux, C., Dalbiez-Gauthier, C., Banchereau, J., and Schmitt, D.,** Interleukin-10 inhibits the primary allogeneic T cell response to human epidermal Langerhans cells, *Eur. J. Immunol.,* in press, 1994.
42. **Caux, C., Massacrier, C., Vanbervliet, B., Barthelemy, C., Liu, Y.-J., and Banchereau, J.,** Interleukin-10 inhibits T cell alloreaction induced by human dendritic cells, *Int. Immunol.,* 6, 1177, 1994.
43. **Caux, C., Dezutter-Dambuyant, C., Schmitt, D., and Banchereau, J.,** GM-CSF and TNF-α cooperate in the generation of dendritic Langerhans cells, *Nature,* 360, 258, 1992.
44. **de Waal Malefyt, R., Abrams, J., Bennett, B., Figdor, C. G., and de Vries, J. E.,** Interleukin-10 (IL-10) inhibits cytokine synthesis by human monocytes: an autoregulatory role of IL-10 produced by monocytes, *J. Exp. Med.,* 174, 1209, 1991.
45. **Hsu, D. H., Moore, K. W., and Spits, H.,** Differential effects of IL-4 and IL-10 on IL-2-induced IFN-γ synthesis and lymphokine-activated killer activity, *Int. Immunol.,* 4, 563, 1992.
46. **Ding, L., Linsley, P. S., Huang, L.-H., Germain, R. N., and Shevach, E. M.,** IL-10 inhibits macrophage costimulatory activity by selectively inhibiting the upregulation of B7 expression, *J. Immunol.,* 151, 1224, 1993.

47. **Freeman, G. F., Bribben, J. G., Boussiotis, V. A., Ng, J. W., Restivo, V. A., Jr., Lombard, L. A., Gray, G. S., and Nadler, L. M.,** Cloning of B7–2: a CTLA-4 counterreceptor that costimulates human T cell proliferation, *Science,* 262, 909, 1993.
48. **Boussiotis, V. A., Freeman, G. J., Gribben, J. G., Daley, J., Gray, G., and Nadler, L. M.,** Activated human B lymphocytes express three CTLA-4 counterreceptors that costimulate T-cell activation, *Proc. Natl. Acad. Sci. U.S.A.,* 90, 11059, 1993.
49. **Te Velde, A. A., de Waal Malefijt, R., Huijbens, R. J. F., de Vries, J. E., and Figdor, C. G.,** IL-10 stimulates monocyte FcγR surface expression and cytotoxic activity. Distinct regulation of antibody-dependent cellular cytotoxicity by IFN-γ, IL-4 and IL-10, *J. Immunol.,* 149, 4048, 1992.
50. **D'Andrea, A., Aste-Amezaga, M., Valiante, N. M., Ma, X., Kubin, M., and Trinchieri, G.,** Interleukin-10 (IL-10) inhibits human lymphocyte interferon γ production by suppressing natural killer cell stimulatory factor/IL-12 synthesis in accessory cells, *J. Exp. Med.,* 178, 1041, 1993.
51. **Gazzinelli, R. T., Makino, M., Chattopadhyay, S. K., Snaper, C. M., Sher, A., Hügin, A. W., and Morse, H. C.,** $CD4^+$ subset regulation in viral infection. Preferential activation of Th2 cells during progression of retrovirus-induced immunodeficiency in mice, *J. Immunol.,* 148, 182, 1992.
52. **Cunha, F. Q., Moncada, S., and Liew, F. Y.,** Interleukin-10 (IL-10) inhibits the induction of nitric oxide synthiase by interferon-γ in murine macrophages, *B.B.R.C.,* 182, 1155, 1992.
53. **Bogdan, C., Vodovotz, Y., and Nathan, C.,** Macrophage deactivation by interleukin-10, *J. Exp. Med.,* 174, 1549, 1991.
54. **Oswald, I. P., Wynn, T. A., Sher, A., and James, S. L.,** Interleukin-10 inhibits macrophage microbicidal activity by blocking the endogenous production of tumor necrosis factor α required as a costimulatory factor for interferon γ-induced activation, *Proc. Natl. Acad. Sci. U.S.A.,* 89, 8676, 1992.
55. **Silva, J. S., Morrissey, P. J., Grabstein, K. H., Mohler, K. M., Anderson, D., and Reed, S. G.,** Interleukin-10 and interferon γ regulation of experimental *Trypanosoma cruzi* infection, *J. Exp. Med.,* 175, 169, 1992.
56. **Gazzinelli, R. T., Oswald, I. P., James, S. L., and Sher, A.,** IL-10 inhibits parasite killing and nitrogen oxide production by IFN-γ-activated macrophages, *J. Immunol.,* 148, 1792, 1992.
57. **Ghalib, H. W., Piuvezam, M. R., Skeiky, Y. A. W., Siddig, M., Hashim, F. A., El-Hassam, A. M., Russo, D. M., and Reed, S. G.,** Interleukin-10 production correlates with pathology in human *Leishmania donovani* infections, *J. Clin. Invest.,* 92, 324, 1993.
58. **Karp, C. L., El-Safi, S. H., Wynn, T. A., Satti, M. M. H., Kordofani, A. M., Hashim, F. A., Hag-Ali, M., Neva, F. A., Nutman, T. B., and Sacks, D. L.,** *In vivo* cytokine profiles in patients with Kala-azar, *J. Clin. Invest.,* 91, 1644, 1993.
59. **Sher, A., Gazzinelli, R. T., Oswald, I. P., Clerici, M., Kullberg, M., Pearce, E. J., Berzofsky, J. A., Mosmann, T. R., James, S. L., Morse, H. C., III, and Shearer, G. M.,** Role of T-cell derived cytokines in the down-regulation of immune responses in parasitic and retroviral infection, *Immunol. Rev.,* 127, 183, 1992.
60. **Wu, J., Cunha, F. Q., Liew, F. Y., and Weiser, W. Y.,** IL-10 inhibits the synthesis of migration inhibitory factor and migration inhibitory factor-mediated macrophage activation, *J. Immunol.,* 151, 4325, 1993.
61. **MacNeil, I., Suda, T., Moore, K. W., Mosmann, T. R., and Zlotnik, A.,** IL-10: a novel cytokine growth cofactor for mature and immature T cells, *J. Immunol.,* 145, 4167, 1990.
62. **Chen, W.-F. and Zlotnik, A.,** IL-10: a novel cytotoxic T cell differentiation factor, *J. Immunol.,* 147, 528, 1991.
63. **de Waal Malefyt, R., Yssel, H., and de Vries, J. E.,** Direct effects of IL-10 on subsets of human CD4+ T cell clones and resting T cells, *J. Immunol.,* 150, 4754, 1993.
64. **Jinquan, T., Larsen, C. G., Gesser, B., Matsushima, K., and Thestrup-Pedersen, K.,** Human IL-10 is a chemoattractant for $CD8^+$ T lymphocytes and an inhibitor of IL-8-induced $CD4^+$ T lymphocyte migration, *J. Immunol.,* 151, 4545, 1993.
65. **Taga, K., Cherney, B., and Tosato, G.,** IL-10 inhibits apoptotic cell death in human T cells starved of IL-2, *Int. Immunol.,* 5, 1599, 1993.
66. **Howard, M., Muchamuel, T., Andrade, S., and Menon, S.,** Interleukin-10 protects mice from lethal endotoxemia, *J. Exp. Med.,* 177, 1205, 1993.

67. **Gérard, C., Bruyns, C., Marchant, A., Abramowicz, D., Vandenabeele, P., Delvaux, A., Fiers, W., Goldman, M., and Velu, T.,** Interleukin-10 reduces the release of tumor necrosis factor and prevents lethality in experimental endotoxemia, *J. Exp. Med.,* 177, 547, 1993.
68. **Bean, A. G. D., Freiberg, R. A., Andrade, S., Menon, S., and Zlotnik, A.,** Interleukin-10 protects mice against staphylococcal enterotoxin B-induced lethal shock, *Infect. Immunol.,* 61, 4937, 1993.
69. **Mullingan, M. S., Jones, M. L., Vaporciyan, A. A., Howard, M. C., and Ward, P. A.,** Protective effects of IL-4 and IL-10 against immune complex-induced lung injury, *J. Immunol.,* 151, 5666, 1993.
70. **Powrie, F., Menon, S., and Coffman, R. L.,** Interleukin-4 and interleukin-10 synergize to inhibit cell-mediated immunity *in vivo*, *Eur. J. Immunol.,* 23, 2223, 1993.
71. **Pennline, K. J., Roque-Gaffney, E., and Monahan, M.,** Recombinant human IL-10 (rHuIL-10) prevents the onset of diabetes in the nonobese diabetic (NOD) mouse, *Clin. Immunol. Immunopathol.,* 71, 169, 1994.
72. **Wogensen, L., Huang, X., and Sarvetnick, N.,** Leukocyte extravasation into the pancreatic tissue in transgenic mice expressing interleukin-10 in the islets of Langerhans, *J. Exp. Med.,* 178, 175, 1993.
73. **Kennedy, M. K., Torrance, D. S., Picha, K. S., and Mohler, K. M.,** Analysis of cytokine mRNA expression in the central nervous system of mice with experimental autoimmune encephalomyelitis reveals that the IL-10 mRNA expression correlates with recovery, *J. Immunol.,* 149, 2496, 1992.
74. **Takeuchi, T., Lowry, R. P., and Konieczny, B.,** Heart allografts in murine systems. II. Differential activation of Th2-like effector cells in peripheral tolerance, *Transplantation,* 53, 1281, 1992.
75. **Abramowicz, D. D. P., Gerard, C., Donckier, V., Amraoui, Z., Velu, T., and Goldman, M.,** Neonatal induction of transplantation tolerance in mice is associated with in vivo expression of IL-4 and IL-10 mRNAs, *Transplant. Proc.,* 25, 312, 1993.
76. **Baccelta, R., Bigler, M., Touraine, J.-L., Parkman, R., Tovo, P.-A., Abrams, J., de Waal Malefyt, R., de Vries, J. E., and Roncarolo, M.-G.,** High levels of IL-10 production *in vivo* are associated with tolerance in SCID patients transplanted with HLA mismatched hematopoietic stem cells, *J. Exp. Med.,* 179, 493, 1994.
77. **Kühn, R., Löhler, J., Rennick, D., Rajewsky, K., and Müller, W.,** Interleukin-10-deficient mice develop chronic enterocolitis, *Cell,* 75, 263, 1993.
78. **Go, N. F., Castle, B., Barrett, R., Kastelein, R., Dang, W., Mosmann, T. R., Moore, K. W., and Howard, M.,** Interleukin-10, a novel B cell stimulatory factor. Unresponsiveness of X chromosome-linked immunodeficiency B cells, *J. Exp. Med.,* 172, 1625, 1990.
79. **Levy, Y., and Brouet, J.-C.,** Interleukin-10 prevents spontaneous death of germinal center B cells by induction of the bcl-2 protein, *J. Clin. Invest.,* 93, 424, 1994.
80. **Rousset, F., Garcia, E., Defrance, T., Péronne, C., Vezzio, N., Hsu, D. H., Kastelein, R., Moore, K. W., and Banchereau, J.,** Interleukin-10 is a potent growth and differentiation factor for activated human B lymphocytes, *Proc. Natl. Acad. Sci. U.S.A.,* 89, 1890, 1992.
81. **Banchereau, J., de Paoli, P., Vallé, A., Garcia, E., and Rousset, F.,** Long term human B cell lines dependent on interleukin-4 and anti-CD40, *Science,* 251, 70, 1991.
82. **Banchereau, J. and Rousset, F.,** Growing human B lymphocytes in the CD40 system, *Nature,* 353, 678, 1991.
83. **Fluckiger, A.-C., Garrone, P., Durand, I., Galizzi, J. P., and Banchereau, J.,** IL-10 upregulates functional high affinity IL-2 receptors on normal and leukemic B lymphocytes, *J. Exp. Med.,* 178, 1473, 1993.
84. **Hudson, G. S., Bankier, A. T., Satchwell, S. C., and Barrell, B. G.,** The short unique region of the B95–8 Epstein-Barr virus genome, *Virology,* 147, 81, 1985.
85. **Garrone, P., Galibert, L., Rousset, F., Fu, S. M., and Banchereau, J.,** Regulatory effects of prostaglandin E2 on the growth and differentiation of human B lymphocytes activated through their CD40 antigen, *J. Immunol.,* 152, 4282, 1994.
86. **Defrance, T., Vanbervliet, B., Brière, F., Durand, I., Rousset, F., and Banchereau, J.,** Interleukin-10 and Transforming Growth Factor β cooperate to induce anti-CD40-activated naive human B cells to secrete Immunoglobulin A, *J. Exp. Med.,* 175, 671, 1992.

87. **Brière, F., Bridan, J.-M., Chevet, D., Souillet, G., Bienvenue, F., and Banchereau, J.,** Interleukin 10 induces B lymphocytes from IgA deficient patients to secrete IgA, *J. Clin. Invest.,* 94, 97, 1994.
88. **Brière, F., Servet-Delprat, C., Bridan, J.-M., Saint-Rémy, J.-M., and Banchereau, J.,** Human interleukin-10 induces naive $sIgD^+$ B cells to secrete IgG_1 and IgG_3, *J. Exp. Med.,* 179, in press.
89. **O'Garra, A., Chang, R., Go, N., Hastings, R., Haughton, G., and Howard, M.,** Ly-1 B (B-1) cells are the main source of B cell-derived interleukin-10, *Eur. J. Immunol.,* 22, 711, 1992.
90. **Ishida, H., Hastings, R., Kearney, J., and Howard, M.,** Continuous anti-IL-10 antibody administration depletes mice of Ly-1 B cells but not conventional B cells, *J. Exp. Med.,* 175, 1213, 1992.
91. **Ishida, H., Hastings, R., Thompson-Snipes, L., and Howard, M.,** Modified immunological status of anti-IL-10 treated mice, *Cell. Immunol.,* 148, 371, 1993.
92. **Ishida, H., Muchamuel, T., Sakaguchi, S., Andrade, S., Menon, S., and Howard, M.,** Continuous administration of anti-IL-10 antibodies delays onset of autoimmunity in NZB/W F1 mice, *J. Exp. Med.,* in press.
93. **Paul, W. E.,** Pleiotropy and redundancy: T cell-derived lymphokines in the immune response, *Cell,* 57, 521, 1989.
94. **Rennick, D., Hunte, B., Dang, W., Thompson-Snipes, L., and Hudak, S.,** IL-10 promotes the growth of megakaryocyte, mast cell and multi-lineage colonies; analysis with committed progenitors and Thillo Scal* stem cells, *Exp. Hematol.,* 22, 136, 1994.
95. **Thompson-Snipes, L., Dhar, V., Bond, M. W., Mosmann, T. R., Moore, K. W., and Rennick, D.,** Interleukin-10: a novel stimulatory factor for mast cells and their progenitors, *J. Exp. Med.,* 173, 507, 1991.
96. **Ghildyal, N., Friend, D. S., Nicodemus, C. F., Austen, K. F., and Stevens, R. L.,** Reversible expression of mouse mast cell protease 2 mRNA and protein in cultured mast cells exposed to IL-10, *J. Immunol.,* 141, 3206, 1993.
97. **Saeland, S., Duvert, V., Moreau, I., and Banchereau, J.,** Human B cell precursors proliferate and express CD23 after CD40 ligation, *J. Exp. Med.,* 178, 113, 1993.
98. **Cassatella, M. A., Meda, L., Bonora, S., Ceska, M., and Constantin, G.,** Interleukin-10 (IL-10) inhibits the release of proinflammatory cytokines from human polymorphonuclear leukocytes. Evidence for an autocrine role of tumor necrosis factor and IL-1β in mediating the production of IL-8 triggered by lipopolysaccharide, *J. Exp. Med.,* 178, 2207, 1993.
99. **Clerici, M. and Shearer, G. M.,** A T_H1/T_H2 switch is a critical step in the etiology of HIV infection, *Immunol. Today,* 14, 107, 1993.

Index

INDEX

C

D

E

F

G

H

I

N

O

P

V

W

X